East-Slavic Religions and Religiosity: Mythologies, Literature and Folklore: A Reassessment

East-Slavic Religions and Religiosity: Mythologies, Literature and Folklore: A Reassessment

Editor

Dennis Ioffe

MDPI • Basel • Beijing • Wuhan • Barcelona • Belgrade • Manchester • Tokyo • Cluj • Tianjin

Editor
Dennis Ioffe
Université libre de Bruxelles
Belgium

Editorial Office
MDPI
St. Alban-Anlage 66
4052 Basel, Switzerland

This is a reprint of articles from the Special Issue published online in the open access journal *Religions* (ISSN 2077-1444) (available at: https://www.mdpi.com/journal/religions/special_issues/East-Slavic).

For citation purposes, cite each article independently as indicated on the article page online and as indicated below:

LastName, A.A.; LastName, B.B.; LastName, C.C. Article Title. *Journal Name* **Year**, *Volume Number*, Page Range.

ISBN 978-3-0365-2025-4 (Hbk)
ISBN 978-3-0365-2026-1 (PDF)

Contents

About the Editor

Dennis Ioffe (PhD University of Amsterdam, 2009) is an Associate Professor of Russian Studies (Titulaire de la Chaire de langue et littérature Russe), Université libre de Bruxelles. Before coming to ULB Dr Ioffe served as a post-doctoral Research Fellow and Assistant Professor/'Doctor-Assistent' at the Department of Languages and Cultures (Slavic & East-European), The Faculty of Arts, Ghent University, Belgium. He also served as a Research Fellow at the University of Amsterdam's Department of Slavic Studies. Since January 2016 Dr Ioffe is Co-Editor-in-Chief, of "Russian Literature", Elsevier Science BV.

Preface to "East-Slavic Religions and Religiosity: Mythologies, Literature and Folklore: A Reassessment"

The focus of this Special Issue is on Slavic religion and mythology, as reflected in theory, literature, and folklore. Our volume will address various forms of pre-Christian religious beliefs, myths, and ritual practices of the Slavs seeping through into "double beliefs" (as in alleged "dvoeverie" remnants). Additionally, we also address the uneasy ways Christian Orthodoxy handled the various challenges traditionally posed by popular beliefs and mythologies. Most existing studies focused on these questions are relatively dated, and there is a growing need for fresh scholarly approach and reassessment. The scope of our volume extends to various forms of Slavic religions, mythology, folklore, and their intersections and interaction with literature and other creative arts. This is examined in reference to historical as well as contemporary material. We seek to contribute to the scholarship in these areas with regard to both officially sanctioned and heterodox religious practices. By doing so, we bring together archaic forms of religious spirituality and modern literary worlds, embracing folklore analysis along with philosophical and theological ideas.

Dennis Ioffe
Editor

 religions

Editorial

East-European Critical Thought: Myth, Religion, and Magic versus Literature, Sign and Narrative

Dennis Ioffe

Département D'enseignement de Langues et Lettres, Faculté de Lettres, Traduction et Communication, Université Libre de Bruxelles, Campus du Solbosch, CP 175, Avenue F.D. Roosevelt, 50, 1050 Bruxelles, Belgium; Denis.Ioffe@ulb.be

Abstract: The Introductory article offers a general overview of the highly complicated topic of religious and mythological consciousness discussed in sub-species narrative critique and literary theory. It also provides a detailed context for the wide array of religious matters discussed in this special volume of Religions. Each of the nineteen papers is positioned within its own particular thematic discourse.

Keywords: Russian Paganism; mythology; theory; narrative; semiotics & semiosis

check for
updates

Citation: Ioffe, Dennis. 2021. East-European Critical Thought: Myth, Religion, and Magic versus Literature, Sign and Narrative. *Religions* 12: 717. https://doi.org/10.3390/rel12090717

Received: 23 August 2021
Accepted: 27 August 2021
Published: 2 September 2021

Publisher's Note: MDPI stays neutral with regard to jurisdictional claims in published maps and institutional affiliations.

Our special edition concentrates on Slavic religions and mythology as evidenced in theory, literature and folklore. The nineteen articles deal with various forms of pre-Christian religious beliefs, myths and ritual practices of the Slavs, which infiltrate into 'dual beliefs' (such as remnants of 'dvoeverie'). Also addressed are the uneasy ways in which Christian Orthodoxy has handled the various challenges posed by popular belief and mythology. Most existing studies are somewhat dated, prompting a growing need for fresh scholarly approaches and reassessments concerning both officially sanctioned and heterodox religious practices. A basic aim, therefore, is to bring together archaic forms of religious spirituality, creative literature, folklore, philosophy and theology, a synthesis which, we hope, will illumine the ways in which mythologies and religious traditions inform ideas and artistic practices, past and present.

One of the underlying and unifying elements here is myth. After all, *religion begins and ends with myth*, while the contemporary endurance of *myth* and the complex cultural theoretical suggestions associated with its cultural and literary usage remain entirely relevant and topical. In the prevailing concept of myth, one can discern a general totality of sacred truths, as it were, parallel to daily human routine. A pioneer of myth theory as applied to the social fabric, French classical scholar Vernant (1996) contrasted *myth* with the practical notion of *logos*, even though, initially, they coincided in the *etymona* of their original meaning. Subsequently, λóγος came to denote the ordered structural ability of thinking, signifying—eventually—even the (Kantian) isolated mind, while myth drifted away into the *Imaginarium* of sacredness and narrative fiction.

Myth(ological) and 'mythogenic' modes of consciousness have driven extensive research in the humanities, generating a variety of interpretations in dictionaries, encyclopedias, reference books, scientific tracts and popular books and articles, deriving its primary school of thought from Soviet Russia[1]. Formulating a parallel question in their primary investigations (rather [quint]essential for Russian-language research on 'mythologism') and providing a new definition of the ambiance existing between myth and literature, prominent Soviet theorist Yuri Lotman and his colleagues and co-authors Zara Mintz and Eleazar Meletinsky discussed the gradual diffusion and submergence of 'the field of mythological and historical and everyday narrative texts' into those general literary studies researching "sacral-magical function peculiar to myth" (Lotman et al. 1987)—while, on the

1

other hand, semiotically, we are treating of the practical tasks inherent in codified meanings of the secondary type.

In their provisional definitions of myth, mythological consciousness and language, various commentators have focused on the contingent semiotic signification of the codified message (Barthes 2004),[2] the cultural meaning and perception of the individual and the surrounding symbiotic environment, including political conflict or propaganda (as, for example, in Eliade 2000). Others have touched on the duality of the mythic as a special kind of narrative, including its connection with ritual—which represents two important aspects (mythological and iconic) of primitive culture (see Tokarev and Meletinsky 1987; Turner 1983). When using the term 'myth', we tend to refer to extant ancient narratives about the nature of the gods, the creation of the world, and the origin/genesis of all beginnings, where cosmogonic narratives or "myths about the origin of the world, the universe, and man" (Toporov 1987), describe the space–time parameters of the multifaceted universe. Creationist myths cover the very emergence of the universe, its minor planets, all physical and material beings, images and things, as well as the primary division of the main elements: water, fire, earth and air. Vladimir Toporov distinguished inter alia two working structural stages in the mythological histories of human civilizations (2010, p. 29): the first, as a rule, deals with what was *before* the ultimate act of creation, while the second deals with the logical division of the world and the creation of the primary elements of the physical universe. Myth, therefore, is primarily an attempt to seek and find a satisfactory answer to questions about *the origin of the nature of things* in the broadest sense of these words. Indeed, one of the most crucial ideas of mythology is to bring order to disorder or transform 'unsigned chaos into semiotic cosmos' (Toporov 2010, p. 22; see also Grygar 2007; Meletinsky 2000, p. 27).

One unifying and multifaceted *syncretism* characteristic of mythological consciousness is, in Eliade's terms, the 'aspects' of myth and *the mythical* (whether ritualistic, artistic or just narrative) and the original status of 'authorship' in the mythological system of coordinates. The 'absent' author in myth is, in fact, not only 'unconscious' but in a sense even '(self)-uncreated' in the emergence of the magical myth itself. It is no coincidence that one of the authors of the literary concept, autocide (see Ioffe 2008b)—Barthes (2004) was also a 'mythologist' of the new and special science of 'mythology[3]' (not to be confused with Claude Levi-Strauss' 'Mythologies'). The nameless collectivity of narrative creationism is an important hallmark of any classical *mythmaking*. Any descriptor or 'author of myth', whether Hesiod or (pseudo) Apollodorus, is only a *retranslator* of something 'alien' (in Bakhtinian terms), merely transmitting a tradition already in existence. At the same time, the main goal of myth remains unchanged, i.e., it is always the creation of a kind of *ordered cosmos of naming* out of the chaos of the nameless (and wordless) natural forces, although whether myth is yet another genre of literature or folklore, or whether, on the contrary, it is a special form of original philosophy or scientific ideology lies beyond the scope of this enquiry. After all, the traditional understandings of myth within the framework of the so-called 'anthropological school' (i.e., Taylor–Lang–Fraser)[4] or the 'mythological school' (i.e., Muller and the Grimm brothers) are well known, so there is little sense in describing them here.

The sacral function of myth emerges parallel to the various paradigmatic aspects of magic. In fact, to perceive magic as a set of ritual-symbolic relations, rules, practices, and related representations, actions, texts and objects aimed at manipulative contact with the supernatural (sacred, or *numinous*, in the terminology of Rudolf Otto; see Otto 2008) is entirely feasible—yet another mode of influencing the visible world by mobilizing the forces of the invisible. Magic differs both from religion and from scientific or rational knowledge, but shares characteristics with both (see Versnel 1991; Tambiah 1990; Wax and Wax 1963; Thorndike 1958). The pragmatic purpose of ritual sacral magic is to try to alter the material ('visible') world by manipulating individual forces and energies of the spiritual ('invisible') world by applying special registers and laws, as well as ad hoc principles and rules, whereby these forces and energies must be subservient. It is the *instrumental* and manipulative "techno"-aspect of magic which differs from religious and mystical–philosophical teachings and practices aimed at contemplative experience—a special kind of knowledge of the sacred and a unity therewith, but one devoid of utilitarian application.

The mythological essence of magic can be clarified in part by reference to the etymology of this term itself: the Greek μάγος (mágos, plural magoi) meant magician or sorcerer and was borrowed from the Iranian (compare Old Persian maguš, Avest. Moγu-), where it first meant a representative of one of the Median tribes and later a member of the estate or religious caste of healers and (Zoroastrian) priests (Buck 1949, pp. 1494–96; Frisk 1960, pp. 156–57). The term goes back to the root *magh-*, i.e., "to be able to articulate power". The priests, thus, were characterized as "powerful", "having (power)", and magical in their acquisition of (supernatural) power, allowing one to achieve goals, purposeful in their original, religious sense. Mágos is close to the Greek word *góēs*—the ancient designation of magician, as well as astrologer, and, in general, bearer of any esoteric and occult knowledge, including early alchemy. The earliest mention of the term mágos is found in fragments of the writings of Heraclitus, 6th century BC (Ioffe et al. 2017).

Mytho-magical practices have been known since ancient times. At the same time, there is no strong evidence to suggest that magic corresponds to any of the specific stages of the development of a society (no matter how we define these stages), i.e., magic might be present at a certain point, but recedes at another. Magical actions, texts and objects were part of various pagan cults, although it is virtually impossible to draw a clear line between magic and religion (especially pagan). Pagan religions create pantheons of characters, coupled with stable legends (myths) pertaining to their deeds, creating sacred precedents, and establishing codices of rituals and rules aimed at influencing these characters and interacting with them. In this context paganism intersects with magic, something which also determines the rules of this interaction and, more broadly, the ways in which it can influence the invisible world, and thereby, the visible one.

The mythogenic (myth- and narrative-generating) legacy of world literature—a basis for further neoreligious neomythologism—can be observed in the culture of European classicism: *Medea* and *Oedipus* by Corneille or *Thebes, Andromache, Iphigenia in Aulis, Phaedra* by Racine. In the more complex, pre-modern period of the Enlightenment, we find a quaint version of 'rational neomythology', for example, in some of Voltaire's texts such as *Merope, Mahomet, Oedipus*, or, conversely, in Friedrich Gottlieb Klopstock's *Messiad*, which, as we know, served as a magnificent epic of a new-age German myth, being crucial to the evolution of Germany as a political nation, the deities of the resurrected German-Scandinavian 'Olympus' inspiring new and heroic myths and deeds. (Vojvodic and Ioffe 2019). All this is to emphasize that the suggestive role of neomythological literature in 19th century Western Europe was truly immense, from Goethe's *Prometheus* and *Ganymede* to Schiller's *Triumph of Victors* and *Ceres' Complaint*.

Our special volume reinforces the study of ideas focused on "new religious ambiance" and "new myth", i.e., a modified mythological substance of *the New Age* (as opposed to classical antiquity and the *Middle Ages*), at least, in part. In a sense we are developing the mythocentric tradition initiated by F. Schelling and the Jena school of German mystical Romanticism, reflected in the critical (re)mythologizing of Wagner and the 'new philosophy' of Nietzsche, bastioned further by the new 'mythological criticism'—from J. Fraser to C.G. Jung (later followed by F. Raglan and R. Graves), and to the more recent Russian (originally Soviet) studies by Vladimir Toporov and his colleagues such as Zara Mints, Eleazar Meletinsky, Boris Uspensky and Sergei Nekliudov.[5]

The dominant point of this *mythocriticism* (whence neomythologism would evolve) was the dominant role of the myth-narrative in enriching the cultural and creative history of humanity. The influential monograph by Canadian literary scholar Northrop Frye (*Anatomy of Criticism*), published in the late fifties of the twentieth century, is also of crucial importance, here, inasmuch as he was one of those who brought the notion of a mythological substratum to modern academic scholarship on culture and literature. A modern Russian researcher of mythological discourse in folklore, Sergei Nekliudov, rightly observes that when speaking about textual realization of myth we should remember that it is 'multidimensional' and that "mythological messages are transmitted on different levels of tradition" (cf. Nekliudov 2000). Speaking of the semiotic component of mytho-research, Nekliudov explains how myth-encoding depends directly upon the variety of 'cultural texts' (referring to the Toporov's and Lotman's pioneering scholarship).

The narrative concept of 'magic' appears to parallel the concepts of 'myth' and 'religion', embracing both folk or peasant belief systems as well as Western European 'intellectual' mythology, artificial religion and magic, per se, in the complex esoteric tradition of both the Renaissance and the New and Modern eras. European culture has created an "intellectual" magical tradition which is rather mystical–occult, spiritualistic, hypnotic, remedial, artistic and even charlatan. It has also influenced Russian literature and writers, as well as the theurgy practiced by many Russians of the Symbolist period: a synthesis of music, poetry, color effects and ritual action capable, manifestly, of changing both time and the physical world, like a magical ritual (see a detailed collection of sources on theurgy in Lewy 2011; also, Petrov 2003). Separate avenues of enquiry are theosophy and anthroposophy, which also aim at a rational synthesis of the sciences so as to manage the explicit and hidden subjects of the universe (see on the topic of theosophy and various currents connected with it in Russia: Carlson 1993).

It is challenging enough to draw a clear historical and theoretical border between myth, magic and religion on the one hand, and esoteric science, philosophy, medicine, art on the other. An important topic is the literal and metaphorical attribution of magical functions to a person of the arts (in particular, literature), who could be perceived as a special kind of 'magician' (Kris and Kurz [1934] 1979), even if magic occupies a much larger place in literature (cf. the shocking 'bourgeois consciousness' narrative of Apuleius with his descriptions of transgressive (often erotic) magic in the *Golden Ass*[6]) than it does in our daily life. There are many different forms or 'faces' of magic in the canonical literature of Western Europe. Consider, for example, one of the main protagonists of Shakespeare's play *The Tempest*, wherein the art of Prospero is to achieve existential world supremacy through incantatory magic. In the celebrated work of John Keats, 'Ode to a Nightingale' written in 1817 we read the virtually modernist lines: "Charm'd magic casements, opening on the foam" (Ioffe et al. 2017). In addition to the metaphorical and religio/mythological use of the word *magic* and its derivatives, we are talking about a special kind of 'magic air' that controls the movement of material objects, in this case, the ship itself. In Keats' poetical universe, magic is always associated with the imagination of the author, for he uses words as hidden spells—indeed, one of Keats's sources may have been Edmund Spencer's famous poem *The Faerie Queene* (1590). One may also recall the special sylph-sylphs described by Paracelsus, the magical *spirits of the air*, mentioned along with the gnomes as the infernal spirits of the earth, as well as Alexander Pope's odd nymphs and salamanders in his mock-heroic and magical–ironic work, *The Rape of the Lock* (1712) (cf.: McIntos 1998, p. 108).

This kind of quasi-romantic use of magic in 'high' literature is characteristic of the Renaissance and modern times. A celebrated German occult polymath Heinrich Cornelius Agrippa von Nettesheim (1486–1535) turned magic into the worldview of an active and vital individual, giving rise to a special (neo) platonic synthetic motif. Subsequently, in 1818 the "Agrippian" tradition leads into the debut of Mary Shelley and the narrative images of Victor Frankenstein. In this context, we also recall Christopher Marlowe and his celebrated text "The Tragical History of the Life and Death of Doctor Faustus" (1593), based on the popular book about Doctor Faust (see Zhirmunsky 1978). The same archetypal plot leads to Goethe and his mytho-tragedy *Faust* published in 1829. Similarly, Agrippa is in the semantic kernel of the novel *The Fiery Angel*, written by Russian "modernist/symbolist magician" Valery Bryusov. We might also mention William Butler Yeats with his intellectual national-folklore magical hypostasis of occultism. Ultimately, European "fantastic realism" is also important (with authors such as Jan Pototsky, Gerard de Nerval, Théophile Gautier, Villiers de L'Isle-Adam) as well as the latest South American magic realism (Miguel Angel Asturias and Gabriel García Márquez, and, with somewhat less relevance, also Julio Cortazar and Jorge Luis Borges) (Vojvodic and Ioffe 2019).

For Ernst Cassirer (Savodnik 2003) and Levi-Strauss (1999) the esoterica of myth crystallizes as a kind of separate creation, parallel to the signs of verbal language, the enzymatic substance of a metaphysical 'story of a special kind'. For Claude Lévi-Strauss, as his numerous works demonstrate, it was not so much the semantic as the ideographic isolation of specific pre-elements that, in fact, fund the spectral existence of the UR-myth, per se. Lévi-Strauss often proposed to designate such elementary protoreligious units as special *myths*, or 'elementals', which, to a greater or lesser degree, are responsible for the visibility of *the* mytho-narrative *structure*. In turn, this intricate composition of mythological fabric proves to be similar to the musical (melodic and melismatic) orchestrated system of signs. The narrative of myth can thus recreate the complex sonoric *score* of an anonymous form of hidden authorship. Intricate relations between myth and music remain quite intriguing ones. (Levi-Strauss 1999).

The semiotic domain of signs and *signification* mentioned above would seem to be crucial for understanding the conceptualization of *neomythologism* in Slavic literatures and cultures over the last hundred and fifty years or so. Alexey Losev furnished a successful definition of myth as *an unfolded magic name*, serving further as a basis for his idea that 'myth is a word/name "[which] unfolded towards meaning and idea,—*the name* given as a contemplated, sculpted semantic picture of the essence and its destiny in the other-being". Losev continues this line of thought, linking primordial magic with the *mytho-word* as such: "Magic is *the name* moving in the direction of Sophia, *the name* given as the realized reality and life of otherness; *euchology* is the name moving in the direction of pure energy, the name meaningfully proceeding from essence to otherness, meaningfully and energetically, intelligently transforming the fragments of intelligence of this otherness and returning to essence together with an intelligently transformed otherness" (Losev 1992, p. 231). Losev also offers a rather typical example of a special magical myth-name, quoting an incantatory prayer in the rite of the expulsion of the demonic forces (from the *Trebnik of Peter of the Grave*): " . . . here I will have to finish the phenomenological–dialectical concept of myth. Myth as *an expanded magical name* can no longer be analyzed further . . . Here is the final, indivisible, and central *point* of the meaning *of myth* per se." (Losev 1992).

Scholarship and science cannot, as the 19th-century positivists hoped, simply 'displace mythology' because "science does not solve such general metaphysical problems as the meaning of life, the purpose of history, the mystery of death, etc., while mythology still claims to solve them" (Meletinsky 2000, p. 5). Myth intends to resolve what is beyond the possibilities of science, it subsumes the metaphysical problems of human life, questions of birth and death, or again, the problematic questions of the existence of the absolute beginning and the absolute end. As Toporov notes, society is interested in unification "precisely because it consciously or unconsciously distinguishes in it certain guarantees of its security, an opportunity to overcome existential crisis, to manage and control members of this collective" (2010, p. 12). Some features of mythological thinking are preserved in mass consciousness, together with elements of philosophical and scientific knowledge (Tokarev and Meletinsky 1987, p. 15), but the endeavor to revive myth as a functioning system in its archaic form in modern society is impossible, if not utopian. It can, however, be an issue of influential 'remnants of the primordial mentality' (Eliade 2000, p. 171), which still form an important component of human existence.

Given the meaning of archaic myth and its understanding in contemporary society and culture, the concept of neomythologism could be linked with the culture of the twentieth and twenty-first centuries. A 'neomythological consciousness' appears to be one of the main directions of cultural vogue in the entire twentieth century—during the deliberations about the omnipotence of *the* (divine) *Logos* versus the crisis of natural scientific knowledge, eliciting interest in the irrational and the unconscious. It is in the development of all kinds of simulation practices and the crisis of logocentrism (for example, with Lyotard) that the culture of the 20th and the 21st centuries directs attention to the basic principles of mythological thinking. Contemporary technology supports this, at least, in part.

The rudimentary traces of neomythological and semi-religious thinking inherent in modern times are initially revealed in the complex poetics of the Russian modernists. Zara Mints in her pioneering article "On 'Neo-Mythological' Texts in the Oeuvre of Russian Symbolists" (2004) discusses the poetics of Blok, Merezhkovsky, Bely and Sologub, emphasizing that in the 'neomythological' texts of Russian symbolism the 'plan of expression' is determined by valves of 'contemporary or historical life' or the pictorial 'history of the lyrical *self*', while the 'content plan forms the correlation of the depicted with myth'.[7] Accordingly, myth assumes the function *of a language,* or even of a *cipher–code,* revealing the implicit meaning of what is occurring within any given narrative (2004, p. 67). Mintz distinguishes folklore fiction during the period of Romanticism from the (neo)mythological world of the Symbolists, when the universe of the artistic text is endowed with 'ontological' existence and an 'ineffable truth', i.e., is equated with myth, affirming that the first mythic texts of the Symbolists were essentially narrative novels understood as 'myths about the world' and requiring a broad, discursive vision. Similarly, one of the pioneers of the post-war academic study of myth, Eliade, observes that the novel as a kind of institution in modern society has earned the place of a special *mythological narrative* (2000, p. 179)— reminiscent of the ideas which Meletinsky expressed in his influential collection *From Myth to Literature* (Meletinsky 2000, p. 130). Meletinsky considers Mann, Joyce, and Kafka to be the initiators of such a 'mythological' novel, because they use mythopoetical models and critical narrative elements; in fact, mythology becomes their 'instrument of narrative structuring' (Ibid.)

Our special volume opens with a detailed scholarly essay by academician Andrey Toporkov (Russian Academy of Sciences, Alexey M. Gorky Institute of World Literature) "The Carol About the Pagan Rite of Sacrifice of a Goat and Its Interpretation in Russian Scholarship of the 19th–20th Centuries". The essay discusses the publications of Russian folklore along with authentic texts and several literary stylizations based on folklore. The article traces the history of one such pseudo-folkloric text—a carol that was first published by Ivan Petrovich Sakharov in 1837 and which attracted particular interest on the part of researchers of Slavic mythology, because it described an old man sacrificing a goat. In this pseudo-folkloric text, several generations of historians saw a description of a ritual which pagan Slavs performed in ancient times. Considering the carol as a historical document, researchers of mythology elaborated their interpretations based on the supposed time of its appearance, the nature of its genre, plot and the individual details. In this way, Sakharov's pseudo-folkloric creation found an eager audience among scholars, stimulating their imagination in picturing the life of pagan Rus'.

Toporkov's paper is followed by academician Svetlana M. Tolstaia (Russian Academy of Sciences, Institute for Slavic Studies, InSlav): "Eastern Christianity and Popular Culture: Mechanisms of Interaction". According to Tolstaia, in traditional Slavic folk culture, Christianity is a foreign, borrowed cultural model, while the oral tradition is native and familiar. The different areas of folk culture were influenced in varying degrees by the Christian tradition. The most forward area of Slavic folk culture dependent upon Christianity was the calendar, the cardinal elements of Christian content of which, in many cases, it accepted, albeit superficially, reinterpreting them in concordance with traditional mythological conceptions and notions. The same can be observed about the folk cult of saints. The Christian saints replaced pagan gods which over time entered the system of folk ideas, beliefs and rituals. The mechanism for regulating the balance between man and the world is a system of prohibitions, the violation of which is recognized as sin and is punished by natural disasters, death, disease and human misfortunes. The Slavic folk tradition adapted not only the individual elements, structures and semantic categories of Christianity, but also the texts, plots, motifs, and themes elaborated in various genres of folklore. Therefore, the pre-Christian folk tradition of the Slavs assimilated many Christian concepts, symbols, and texts, translated them into its language and filled them with its own content.

What follows in our collection is a scholarly paper by Dieter Stern (Ghent University) "Ruthenian Sacred Geography of the Baroque Age—the Role of Devotional Songs in Shaping Landscapes of Popular Worship in the Polish-Lithuanian Commonwealth". The article explores how the newly founded and highly contested Christian confession of the Greek Catholics or Uniates employed strategies of mass mobilization so as to establish and maintain their position within a contested confessional terrain. The Greek Catholic clerics, above all, monks of the Basilian order, fostered an active policy of acquiring, founding and promoting Marian places of grace in order to create a sense of belonging among their flock. The author argues that folk ideological notions concerning the spatial and physical conditions for the working of miracles were seized upon by the Greek Catholic faithful to establish a mental map of the grace of their own. Especially, the Basilian order took care to organize collective events such as annual pilgrimages and crowning celebrations for miraculous images and promote Marian devotion via reports of miracles and songs about icons in an attempt to define what it means to be a Greek Catholic in terms of sacred territoriality. The paper is followed by Aleksey Yudin (also from Ghent University) with his essay "Christian Saints in Russian Incantations: Names, Images and Functions". This paper discusses the Christian saints most often mentioned in Russian incantations: such as George, Nicholas, Florus and Laurus, Kossma and Damian, Zosima and Savvaty of Solovki, as well as the semi-apocryphal saints Sisinius and Solomonia, the first six being among the most popular saints in Russian folk Orthodoxy. The article describes the names and attributes of the saints and their functions in Russian folk magic. Depending on their magical function, the protagonists of the incantations can act as helpers, protectors and healers, assisting in various practical areas of life and protecting against real and magical dangers and helping to heal diseases and wounds.

A related topic is explored by Aleksandra Ippolitova (Russian Academy of Sciences, Alexey M. Gorky Institute of World Literature) in "Sacred and Profane: Religious Taboo in Russian Incantations and Herbals". This paper explores linguistic taboos (euphemisms, omissions, and others)—an essential part of Slavic verbal and written culture—, analyzing cryptography as a taboo mechanism in the magical texts of the manuscript tradition of the 17th and 18th centuries such as handwritten incantations and herbals. In the latter, for example, cryptography taboos "sinful" or artful topics (love magic, magic used against courts and authorities, sorcery, jinx, and 'secret' knowledge), and in the texts which were wont to bear a sacral meaning (incantations and prayers).

With the detailed research paper by Nicoletta Misler (Istituto Orientale, Napoli, Italy) and John E Bowlt (University of Southern California, Los Angeles, CA, USA) "Painting is Believing: Pavel Filonov and His Icon of St. Catherine" we open a different corpus of essays. The authors discuss the Orthodox icon which Pavel Filonov (1883–1941) painted in 1908 or 1909 for his sister, Ekaterina, placing it within the broader context of his oeuvre, his family and his understanding of 'religiosity'. Referring to Filonov's system of Analytical Art and to what he called 'madness', the authors focus on the particular technical devices which he used in the icon and on the *podlinnik* (or primer) from which he copied the main elements. Reference is also made to other religious motifs in Filonov's art such as the Magi, Flight into Egypt and Crucifixion.

Dorota Walczak-Delanois (Université Libre de Bruxelles) presents a detailed account of the "Poems by Polish Poetesses and the Burning Issue of Religion" which helps uncover the presence of religion and the particular evolution of lyrical matrixes connected to religion in the Polish poems of female poets, a notable presence of women lying at the roots of the Polish literary and lyrical traditions—for centuries, the image of a woman with a pen in her hand was one of the most important imponderabilia, although, until the 19th century, Polish female poets continued to be rare. In any case, whenever women poets appear, they are linked to institutions such as convents, where female intellectuals were able to enjoy relative liberty and refuge, even if many of the poetic forms they used in the 16th, late 17th and 18th centuries were typically male in origin and followed established models. In the 19th century, the specific image of the mother as a link to the religious portrait of the Madonna and the Mother of God (the first Polish poem presents *Bogurodzica, the Virgin Mary, the Mother of Jesus*) reinforces the new female presence. Indeed, after Adam Mickiewicz's poem *Do matki Polki* ('To Polish Mother'), the term "Polish mother" becomes a separate literary, epistemological and sociological category. Throughout the 20th century (with some exceptions), the impact of Romanticism and its poetical and religious models remained alive, even if they underwent modifications. The period of communism, as during the Period of Partitions and the Second World War, privileged established models of lyric, where the image of women reproduced Romantic schema from 19th-century poetical canons, closely linked to religion.

The same territory is explored in a seminal paper by Petra James (Université Libre de Bruxelles), "Myth, Ethics and Sacrifice in Jan Patočka's Aesthetic Thinking". Intense and systematic scholarly interest in the relation of Patočka's phenomenology to religion and Christianity is recent, attracting particular attention over the last decade. Thus far, the topic has mainly been studied from philosophical and theological perspectives and the extensive body of Patočka's cultural writings has failed to attract the intellectual community at large. This article focuses on the analysis of cultural archetypes in Patočka's cultural writings related to the topic of religion and Christianity: the archetypes of the Faustian figures of Patočka's creative writings, whether Goethe's Faust, Goethe's Marguerite, or Mann's Adrian Leverkühn, are all Socratic–Christic avatars that personify Patočka's philosophical concept of "care for the soul" in the modern age. The legacy of Plato's Greek philosophy and that of Western Christianity, as presented by Patočka, insists on the universally shared existential experience of finitude that should be grasped as a positive challenge in the quest for meaning. Patočka's "titanism" and the archetypal titanic figures of his writings are manifestations of this universal effort, indicating that a culturological approach to Patočka's thinking on religion and Christianity might prove to be especially relevant.

Raymond Detrez (Ghent University) presents a paper "Tensions Among Religious Communities in Interwar Bulgaria. Observations by Flemish Missionaries in the Diocese of Nicopolis". Premodern Ottoman society consisted of four major religious communities, Muslims, Orthodox Christians, Armenian Christians and Jews; the Muslim and Christian communities also included various ethnic groups, as did the Muslims, Arabs and Turks, Orthodox Christian Bulgarians, Greeks and Serbs who identified, in the first place, with their religious communities and considered ethnic identity of secondary importance. Living together for centuries, albeit segregated within the borders of the Ottoman Empire, Bulgarians and Turks (to a considerable extent) shared the same worldview and moral value system and tended to react uniformly to various events. For example, the Bulgarian attitudes toward natural disasters, on which this contribution focuses, seem not to have differed essentially from those of their Turkish neighbors, as both proceed from the basic idea of God's providence being intrinsic to these disasters. Despite the (overwhelmingly Western) perception of Muslims as passive and fatalistic, the problem of whether or not one had the right to escape the wrath of God was also deliberated in similar terms. However, in addition to a comparable religious mental make-up, social circumstances and administrative measures, the conditions of both religious communities, rather than cross-cultural influences, seem to provide a more plausible explanation for these parallels.

With the paper by Alexander Zholkovsky (University of Southern California, Los Angeles), we turn to Russian classics. His paper "To Be or Not to be God: Debating the Authorial Power in Dostoevsky" is focused on the issue of God in the oeuvre of the legendary Russian author, discussing the now widely accepted concept of Dostoevsky's dialogism—which alleges the author's equal empowerment of all his characters. Using examples from *Crime and Punishment* and *The Brothers Karamazov*, Zholkovsky focuses on instances of scene-staging based on the 'scripts' devised and enacted by some characters, and 'read', with varying success, by their targets. He documents the resulting 'discursive combat' among the characters, paying special attention to those 'playing at god' and, thus, the more 'authorial' among them. In several cases, the would-be 'divine' manipulation is shown to be consistently subverted by the Dostoevskian narrative. However, in one instance, where Aliosha Karamazov charitably scripts Captain Snegirev's behavior (the ensuing discussion of this episode) in Aliosha's conversations with Lise Khokhlakova, upholds Aliosha's right to play God with the Other—"for the Other's good", of course (not unlike the Grand Inquisitor).

Another academic from California, Igor Pilshchikov (University of California, Los Angeles) offers a fundamental paper "Gogol's The Nose Between Linguistic Indecency and Religious Blasphemy". Focused on Nikolai Gogol's absurdist narrative tale, "The Nose" (1835), the article is an investigation into the concealed representation of suppressed and marginalized libertine and antireligious discourses in nineteenth-century Russian literature. The author identifies overlooked idiomatic phraseology, forgotten specificities of the Imperial hierarchy (the Table of Ranks) and allusions to religious customs and Christian rituals that would have been apparent to Gogol's readers, while demonstrating how some were camouflaged to escape censorship in successive drafts of the work. Pilshchikov's research builds on the approaches to Gogol's language, imagery and plot developed by the Russian Formalists, Tartu–Moscow semioticians, and other scholars, who revealed the latent obscenity of Gogol's "rhinology" and the sacrilegious meaning of the very specific chronotope within the tale. An integrated analysis of these aspects helps us understand what the details of the story reveal about Gogol's religious and psychological crisis of the mid-1830s, revealing, for example, how he aggregated indecent Shandyism, social satire, and religious blasphemy into a single quasi-oneiric narrative.

Monika Spivak (Russian Academy of Sciences, Alexey M. Gorky Institute of World Literature; Andrey Belyi Museum, State Alexander Pushkin Museum) offers a paper titled "The 'Christology' of Belyi the anthroposophist: Andrei Bely and the Apostle Paul". This text focuses on Rudolf Steiner's perception of the Gospels and its impact on Bely's works. The latter had long valued Steiner's lectures on Christ and the Fifth Gospel, the "Anthroposophic" (relating to the philosophy of human genesis, existence, and outcome) Gospel, the knowledge of which had been received in a visionary way. In addition, Bely was an esoteric follower of Steiner who often quoted from Apostle Paul's 2 Corinthians, "Ye are our epistle written in our hearts, known and read of all men". The citation occurs in Bely's philosophical works (The History of the Formation of the Self-Conscious Soul, "Crisis of Consciousness"), autobiographical prose (Reminiscences of Steiner), the essay "Why I Became a Symbolist . . . ", and letters to Ivanov-Razumnik and Fedor Gladkov. Bely's own anthroposophic and esoteric ideas relating to the gospel dicta are also examined. Resorting to a single quotation, Spivak shows the specifics of Bely's stance vis-à-vis the anthroposophist's perception of Christian texts. This furnishes a methodological meaning which helps us understand other Biblical quotations and images in the works of Bely, anthroposophical Christology also being the key to their deciphering

A scholar from St. Petersburg, Andrey A. Astvatsaturov (State University of St Petersburg SPBGU) presents "Franny's Jesus Prayer: Salinger and Russian Orthodox Spirituality". The text discusses *The Way of a Pilgrim* and *The Pilgrim Continues His Way*—a Russian Hesychast text that was first published in 1881 and translated into English in 1931. It has gained popularity in the English-speaking world thanks to J.D. Salinger, who mentions and re-narrates it in his stories 'Franny' and 'Zooey', and the reference has often been noted in both critical works on Salinger and studies dedicated to the book The Way of a Pilgrim. However, this is the first time that a scholar has attempted to provide a fundamental analysis of the textual interconnections between Salinger's stories and the radical Hesychast religious mind. In the article, the text of The Way of a Pilgrim is read within the framework of Salinger's stories and is interpreted as being significant to his later texts. From the Hesychast book, Salinger borrows several images and presents its philosophy as a spiritual ideal. At the same time, he approaches it with a certain irony, exposing several pitfalls and misconceptions in interpretations of the Jesus prayer, as illustrated by Franny, one of Salinger's characters. After illuminating Franny's mistakes and her peccant intention, Salinger reestablishes the Hesychast ideal and connects it with Søren Kierkegaard's principle of theistic existentialism

Vadim Polonsky (Russian Academy of Sciences, Alexey M. Gorky Institute of World Literature) offers "Merezhkovsky's NeoChristianity of the Third Testament: From the Symbolist Historiosophy to the Right-Wing Politics", an article which places Dmitry Merezhkovsky's Chiliastic concept of the Three Testaments within a unified structure. The author analyzes the writer's integral system of Christological, anthropological, *historio-sophic idiomyths* and meta-symbols, investigating the religious, philosophical and aesthetic geneses of the semantic transformation of traditional theological constructions and the doctrinal compilation of the dominant elements within Russian fin de siècle culture. We discover how modernist mythmaking alters political reality in Merezhkovsky's mind and draws him towards radical ideologies of the extreme left and right.

Henrietta Mondry (University of Canterbury, New Zealand) presents a paper on the related period "Synthesizing Religions: Vasily Rozanov's Phallic Christianity", Rozanov being one of the first Russian writers of the fin de siècle to create a nexus between the study of the history of world religions and the history of sexuality. Rozanov viewed Christian asceticism as a source of the disintegration of the contemporary family and Mondry examines his strategy to synthesize religions and to use pre-Christian religions of the Middle East as proof of common physical and metaphysical essence in celestial, human, animal and mythological human/animal/divine bodies. Mondry argues that while his rehabilitation of physical life by endowing it with religious value was socially positive, his self-proclaimed 'mission of sexuality', when politically motivated, was manipulative and incorporated the notion of the atavistic 'survivals'. In conclusion, the author explains that Rozanov's monistic search for the divine in the physical body, as well as his strategy for religious synthesis, were additionally driven by his doubts in the preeminence of Christian eschatology.

Irina Sakhno (Higher School of Economics, Moscow) presents another paper focused on the art of the avant-garde, "Apophatic Theology and Sacred Signs in the Russian Avant-Garde". The author discusses Kazimir Malevich's suprematist art in the context of negative (apophatic) theology and as a crucial tool in analyzing both the artist's theoretical conclusions and his new visual optics. Analysis indicates that the artist moved intuitively towards recognizing the ineffability of the multidimensional universe and perceiving God as the spiritual absolute. In his attempt to see the invisible in the formula of emptiness and nothingness, Malevich turned to the primary forms of geometric abstraction—the square, circle and cross—which he endowed with symbolic concepts and meanings, for the artist treated of his *Suprematism* as a method of perceiving the ineffability of the absolute. With the *Black Square* seen as a face of God, the patterns of negative theology rise to become a philosophical formula of primary importance. Malevich's series called Mystical Suprematism (1920–1922) confirms the presence of complex metaphysical reflection and apophatic thought, for not only does the series contain icon paraphrases and the Christian symbolism of the cross and mandorla, but it also advances the formula of the apophatic faith of the modern times, *Suprematism* presenting primary forms as universals of "the image of the future" and the energy of non-objective art.

Svetlana Efimova (Ludwig–Maximilians–Universität München) debates the question of God in Boris Parsternak in her paper entitled "The October Revolution as the Passion of Christ: Revisiting *Doctor Zhivago* and Its Contexts". The article offers a new interpretation of Doctor Zhivago in the cultural and historical context of the first half of the 20th century, emphasizing the interrelationship between religion and philosophy of history. Doctor Zhivago is analyzed as a condensed representation of a religious conception of Russian history between 1901 and 1953 and as a cyclical repetition of the Easter narrative. The bipartite narrative consists of the Passion and Resurrection of Christ as symbols of violence and renewal (liberation), while the novel as a whole cycles through this narrative several times, symbolically connecting the 'Easter' revolution (March 1917) and the Thaw (the spring of 1953). The sources of Pasternak's Easter narrative include the Gospels, Leo Tolstoy's philosophy of history, and pre-Christian mythology. The model of cyclical time in the novel brings together the sacred, natural and historical cycles. This concept of a cyclical renewal of life differs from the linear temporality of the apocalypse as an expectation of the end of history.

Our volume concludes with yet another paper on the relation between the Avant-garde and religious philosophy, Evgeny Pavlov (University of Canterbury, New Zealand) presenting a paper entitled "God is Everywhere, Possibly: Aleksandr Vvedensky's Tropes of Theology". Bringing vast textological evidence, this article focuses on the complicated questions of God, time and the new poetics of transcendence which find their intellectual roots in Russian Kantianism and epistemology of temporality.

Funding: This research received no external funding.

Conflicts of Interest: The author declares no conflict of interest.

Notes

[1] See, for example, Lotman et al. (1987); Kozlov (2001); Tolstoi (1995); Toporkov (1997); Toporov (1967); Toporov (1988); Uspenskii (2018); Vernant and Vidal-Naquet (1996)

[2] On the complex-dialectical development of this whole complex of Barthian myth-ideas, see in particular Ioffe (2008a, 2008c).

[3] Of the relatively recent interdisciplinary studies of the theory of myth and the mythical in culture, we note the following: Doty (2004); Adamenko (2007); Ellwood (2008); Bennett and McDougall (2013); Marderness (2009); Coupe (1997).

[4] For a further look at myth in an anthropological vein, see also Malinowski (2004) and Durkheim (2018) as well as Moss (1996). See many theoretical and applied variations on the definition of myth and the mythic in: Barthes (2004); Weiman (1975); Grigoriev (1987); Diakonov (1990); Dumezil (1986); Durkheim (2018); Ivanov (2005); Ioffe (2006a, 2007, 2009); Caillois (2003)); Kanevskaia (2000); Lifshitz (1980); Levi-Strauss (1985, 1999); Lord (1994); Losev (1982, 1992, 1994); Lotman and Mintz (1981); Lotman and Ouspensky (1973); Meletinsky (1995); Nekliudov (2000); Turner (1983); Tolstoy (1995); Toporov (1995); Hübner (1996); Buonanno (2018); Cassirer (1946); Cassirer (1955); Coupe (1997); Csapo (2005); Cunningham (1973); Thomas (1978); Day (1984); Ioffe (2006b); Kirk (1970); Liszka (1989); Segal (1998); Strenski (1987); Vickery (1966). See, in particular, Mintz (2004); Toporov (2010); Meletinsky (1995); Nekliudov (2000). See also Vojvodic and Ioffe (2019).

[5] See also the relevant provisions in the work of Lotman and Ouspensky (1973), which, in a certain sense, contrasts with and complements the mytho-philosophical views and the entire ideological position of Losev. See context and details in Losev (1992, pp. 231–32).

[6] The preferecne of the word-choice for *Ass* instead of say *Donkey* appears to be shocking enough on its own.

[7] See many corresponding later texts in our collection focused on modernism and the avant-garde: Ioffe and White (2012). See also Ioffe (2006a, 2017).

References

Adamenko, Victoria. 2007. *Neo-Mythologism in Music: From Scriabin and Schoenberg to Schnittke and Crumb*. Hillsdale: Pendragon Press.

Barthes, Roland. 2004. Mif segodnia. In *Mifologii*. Moskva: Izdatel'stvo Sabashnikovykh, pp. 265–323.

Bennett, Pete, and Julian McDougall, eds. 2013. *Barthes' Mythologies Today: Readings of Contemporary Culture*. London: Routledge.

Buonanno, Michael. 2018. *The Meaning of Myth in World Cultures*. Jefferson: McFarland.

Buck, C. D. 1949. *A Dictionary of Selected Synonyms in the Principal Indo-European Languages. A Contribution to the History of Ideas*. Chicago and London: University of Chicago Press.

Caillois, Roger. 2003. *Myth, Chelovek i sakralnoe*. Moscow: OGI.

Carlson, Maria. 1993. *No Religion Higher Than Truth: A History of the Theosophical Movement in Russia, 1875–1922*. Princeton: Princeton UP.

Cassirer, Ernst. 1946. *Language and Myth*. Translated by Susanne K. Langer. New York: Dover Publications.

Cassirer, Ernst. 1955. *The Philosophy of Symbolic Forms vol. 2 Mythical Thought*. New Haven: Yale UP.

Coupe, Laurence. 1997. *Myth*. London: Routledge.

Csapo, Eric. 2005. *Theories of Mythology*. Hoboken: Wiley-Blackwell.

Cunningham, Adrian, ed. 1973. *The Theory of Myth. Six Studies*. London: Sheed & Ward.

Day, Martin S. 1984. *The Many Meanings of Myth*. Lanham: University Press of America.

Diakonov, M. Igor. 1990. *Archaicheskie Mythy Vostoka i Zapada*. Moscow: Nauka.

Doty, William G. 2004. *Myth: A Handbook*. Westport: Greenwood Press.

Dumezil, Georges. 1986. *The Supreme Gods of the Indo-Europeans*. Moscow: Nauka.

Durkheim, Émile. 2018. *Elementarnye Formy Religioznoi Zhizni*. Moskva: Delo.

Eliade, Mircea. 2000. *Aspekty Mifa*. Moskva: Akademicheskii proekt.

Ellwood, Robert S. 2008. *Myth: Key Concepts in Religion*. London and New York: A&C Black.

Frisk, H. 1960. *Griechisches etymologisches Wörterbuch, Band I*. Heidelberg: Carl Winter.

Grigoriev, Nikolai V. 1987. Sootnosheniia mifa i iskusstva v kul'ture. In *Iskusstvo v sisteme kul'tury*. Leningrad: Nauka.

Grygar, Mojmír. 2007. *Znakotvorchestvo. Semiotics of the Russian Avant-garde*. St. Petersburg: Akademicheskii proekt.

Hübner, Kurt. 1996. *Istina Mytha*. Moscow: Respublika.

Ioffe, Dennis, and Frederick White, eds. 2012. *The Russian Avant-Garde and Radical Modernism*. Boston: Academic Studies Press.

Ioffe, Dennis, Andrey Toporkov, and Aleksey Yudin. 2017. Magic, Folklore, Literature: An Introduction. *Russian Literature* 93–94: 2–45.

Ioffe, Dennis. 2006a. Modernism in the Context of Russian Life-Creation: Lebenskunst and the Theory of Life ⇔ Text Sign Systems. *New Zealand Slavonic Journal* 40: 22–55.

Ioffe, Dennis. 2007. Russian Religious Criticism and the Problem of Imiaslavie (o. Pavel Florenskii, o. Sergii Bulgakov, A.F. Losev). *Kritika i Semiotika* 11: 109–73.

Ioffe, Dennis. 2008a. The Problem of Author as Scriptor and the Postmodern Theory of Literature. *Mirgorod: Journal, Posviashchennyi Voprosam Epistemologii Literaturovedenii (Journal of the Epistemology of Literary Research)* 1.1: 93–108.

Ioffe, Dennis. 2008b. Velimir Khlebnikov, Islamic Mysticism, and Oriental Discourse. In *Doski Sudby Velimira Khlebnikova: Tekst i Konteksty*. Moscow: Tri Kvadrata, pp. 547–637.

Ioffe, Dennis. 2008c. Gustave Chpet, la religion et le probleme du signe. In *Gustave Chpet et Son Heritage Aux Sources Russes Du Structuralisme Et De La Semiotique, Slavica Occitania*. Toulouse: Université Toulouse-Jean Jaurès, vol. 26.

Ioffe, Dennis. 2009. Sign Theory and the Role of Ancient Egypt in the Context of Khlebnikov. In *Russian Anthropological School. Russian State University for Humanities*. Moscow: RGGU, vol. 6, pp. 241–59.

Ioffe, Dennis. 2017. The Politics of Modernism in Russia: A Reassessment. *Russian Literature* 92: 3–17.

Ioffe, Jeremiah. 2006b. Symbolism' (mystical idealism); Constructivism; Expressionism and Surrealism. In *Selected: 1920s–30s. From Culture and Style, Synthetic Art Theory, Synthetic Art Studies and Sound Cinema*. Compiled and Edited by. M. S. Kagan. St. Petersburg: Petropolis, pp. 197–251.

Ivanov, Viach. 2005. Egypt of the Amarna Period in Kharms and Khlebnikov. In Paper presented at the Centenary of Daniil Kharms: Proceedings of the International Scientific Conference, St. Petersburg, Russia, August 24–26; pp. 80–90.

Kanevskaia, Marina. 2000. History and Myth in the Postmodern Russian Novel. *From the Proceedings of the Russian Academy of Sciences. Series of Literature and Language* 59: 37–47.

Kirk, Geoffrey S. 1970. *Myth: Its Meaning and Functions in Ancient and Other Cultures*. Cambridge: Cambridge UP.

Kozlov, Alexander S. 2001. Myth and Mythological Criticism. In *Literary Encyclopedia of Terms and Concepts*. Edited by Alexander N. Nikolyukin. Moscow: Intelvalk, pp. 559–62.

Kris, Ernst, and Otto Kurz. 1979. *Legend, Myth, and Magic in the Image of the Artist*. New Haven: Yale University Press. First published 1934.

Levi-Strauss, Claude. 1985. How Myths Die. In *Foreign Studies in the Semiotics of Folklore*. Moscow: Universitetskaia kniga.

Levi-Strauss, Claude. 1999. *Mythologiques*. 4 vols. Moscow: Universitetskaia kniga.

Lewy, H. 2011. *Chaldaean Oracles and Theurgy: Mysticism, Magic and Platonism in the Later Roman Empire*, 3rd ed. par Michel Tardieu, avec un supplément 'Les oracles chaldaïques'. Paris: Institut des Etudes Augustiniennes.

Lifshitz, Mikhail A. 1980. *Mythology Ancient and Modern*. Moscow: Iskusstvo.

Liszka, James Jakob. 1989. *The Semiotic of Myth: A Critical Study of the Symbol*. Bloomington: Indiana University Press.

Lord, Albert B. 1994. *The Singer of Tales*. Moscow: Nauka.

Losev, Aleksei F. 1982. *Sign—Symbol—Myth*. Moscow: MGU.

Losev, Aleksei F. 1992. Myth is an Expanded Magical Name. *Symbol* 28: 231–32.

Losev, Aleksei F. 1994. *Myth. Number. Essence*. Moscow: Mysl'.

Lotman, Yury M., and Boris A. Ouspensky. 1973. Myth—Name—Culture. *Trudy Po Znakovym Sistemam* VI: 282–303.

Lotman, Yury M., and Zara G. Mintz. 1981. Literature and Mythology. *Proceedings: Sign Systems* 13: 35–55.

Lotman, Yury M., Zara G. Mintz, and Eleazar M. Meletinsky. 1987. Literature and Myths. In *Encyclopedia: Mythy Narodov Mira*. Moscow: Sovetskaia Encyclopedia.

Malinowski, Bronisław. 2004. *The Argonauts of the Western Pacific*. London: Routledge and Sons.

Marderness, William. 2009. *How to Read a Myth*. New York: Humanity Books.

McIntos, Charles. 1998. *The Rosicrucians: The History, Mythology, and Rituals of an Esoteric Order*. Boston: Weiser Books.

Meletinsky, Eleazar M. 1995. *Poetics of Myth*. Moscow: Nauka.

Meletinsky, Eleazar M. 2000. *From Myth to Literature. Textbook on the Theory of Myth and Historical Poetics of Narrative Genres*. Moscow: Vostochnaia literatura.

Mintz, Zara. 2004. On Some Neomythological Texts in the Work of Russian Symbolists. In *Blok and Russian Symbolism: Selected Works in Three Volumes*. St. Petersburg: Iskusstvo, Available online: http://www.ruthenia.ru/mints/papers/neomifologich.html (accessed on 14 May 2017).

Moss, Marcel. 1996. *Society. Exchange. Personality. Works on Social Anthropology*. Moscow: Nauka.

Nekliudov, Sergei Yu. 2000. *The Structure and Function of Myth. Myths and Mythology in Modern Russia*. Edited by Karl Eimermacher, Folk Bomsdorff and Gennadii Bordugov. Moscow: AIRO-XX, pp. 17–38.

Otto, Rudolph. 2008. *Sviaschennoe. Das Heilige: Uber das Irrationale in der Idee des Gottlichen und sein Verhaltnis zum Rationalen*. Sankt Petersburg: SPBGU University Press.

Petrov, A.V. 2003. *Fenomen theurgii. Filosofia i magia v antichnosti*. Sankt Peterburg: RHGI.

Savodnik, Peter. 2003. Ernst Cassirer's Theory of Myth. A Critical Review. *A Journal of Politics and Society* 15: 447–59.

Segal, Robert A. 1998. *The Myth and Ritual Theory*. Hoboken: Wiley-Blackwell.

Strenski, Ivan. 1987. *Four Theories of Myth in Twentieth-Century History: Cassirer, Eliade, Levi Strauss and Malinowski*. Iowa City: University of Iowa Press.

Tambiah, Stanley J. 1990. *Magic, Science, Religion, and the Scope of Rationality*. Cambridge: University Press.

Thomas, Keith. 1978. *Religion and the Decline of Magic: Studies in Popular Beliefs in Sixteenth and Seventeenth-Century England*. London: Weidenfeld & Nicolson.

Thorndike, Lynn. 1958. *A History of Magic and Experimental Science*. 8 vols. New York: Macmillan.

Tokarev, Sergei A., and E. M. Meletinsky. 1987. Mythology. In *Myths of the Peoples of the World*. Moscow: Sovetskaia Encyclopedia, pp. 11–22.

Tolstoi, N. I. 1995. *Iazyk i narodnaia kul'tura. Ocherki po slavianskoi mifologii i etnolingvistike*. Moskva: Indrik.

Tolstoy, Nikita I. 1995. *Language and Popular Culture. Essays on Slavic Mythology and Ethnolinguistics*. Moscow: Indrik.

Toporkov, Andrey. 1997. *Teoriia Mifa V Russkoi Filologicheskoi Nauke XIX veka*. Moscow: Indrik.

Toporov, Vladimir N. 1967. Towards the Reconstruction of the Myth of the World Egg (on the Material of Russian Fairy Tales). In *Trydy Po Znakovym Systemam*. Tartu: University Press, vol. III.

Toporov, Vladimir N. 1987. Cosmogonic Myths. In *Myths of the Peoples of the World*. Moscow: Sovetskaia Encyclopedia.

Toporov, Vladimir N. 1988. On Ritual. Introduction to the Problematics. In *Archaic Ritual in Folklore*. Moscow: Nauko.

Toporov, Vladimir N. 1995. *Myth. Ritual. Symbol. Image: A Study in the Field of Mytho-Poetic*. Moscow: Progress.

Toporov, Vladimir N. 2010. *World Tree. Universal Sign Systems*. Moscow: RPDR, vols. 1 and 2.

Turner, William. 1983. *Symbol and Ritual*. Moscow: Nauka.

Uspenskii, Boris A. 2018. *Issledovaniia po russkoi literature, fol'kloru i mifologii*. Moskva: Common Place.

Vernant, Jean-Pierre, and Pierre Vidal-Naquet. 1996. *Myth and Tragedy in Ancient Greece*. New York: Zone Book.

Vernant, Jean-Pierre. 1996. *Myth and Society in Ancient Greece*. Translated by Janet Lloyd. New York: Zone Book.

Versnel, Henk. 1991. Some Reflections on the Relationship Magic-Religion. *Numen* 38: 177–97. [CrossRef]

Vickery, John B., ed. 1966. *Myth and Literature: Contemporary Theory and Practice*. Lincoln: University of Nebraska Press.

Vojvodic, Jasmina, and Dennis Ioffe. 2019. K voprosu o neomifologizme v literature. *Russian Literature* 107–108: 2–31.

Wax, Murray, and Rosalie Wax. 1963. The Notion of Magic. *Current Anthropology* 4: 495–512. [CrossRef]

Weiman, Robert. 1975. Myth and Ritual. In *Literary History and Mythology. Essays on the Methodology and History of Literature*. Moscow: Progress Publishers, pp. 279–85.

Zhirmunsky, Victor M. 1978. *Legenda o doktore Fauste*. Edited by Victor M. Zhirmunsky. Moscow and Leningrad: Nauka.

 *religions* 

Article

The Carol about the Pagan Rite of Sacrifice of a Goat and Its Interpretation in Russian Scholarship of the 19th to 20th Centuries

Andrey Toporkov

Gorky Institute of the World Literature, Russian Academy of Sciences, 121069 Moscow, Russia; atoporkov@mail.ru

Abstract: In publications of Russian folklore, along with authentic texts there are a number of literary stylizations based on folklore. The article traces the history of one such pseudo-folkloric text—a carol which was first published by Ivan Petrovich Sakharov (1807 to 1863) in 1837. It has been established that this carol is a montage of two texts: the first is a carol, printed in 1817 by I.E. Sreznevsky in the *Ukrainian Bulletin*, and the second is a song included in the *Tale of Brother Ivanushka and his Sister Alyonushka* (SUS 450). Such contamination is unique and is found only in this one text, which was later reprinted many times. Taking into account Sakharov's reputation as a falsifier of folklore, there is no reason to doubt that it was he who composed this carol; such contamination of works belonging to different folkloric genres is also characteristic of other of Sakharov's publications. The carol that Sakharov published attracted the particular interest of researchers of Slavic mythology due to the fact that it described how an old man was going to sacrifice a goat. Several generations of historians saw in this pseudo-folkloric text a description of a ritual that pagan Slavs performed in ancient times. Considering the carol as an historical document, researchers of mythology built their interpretations based on the supposed time of its appearance, the nature of its genre, plot, and individual details. Thus, Sakharov's pseudo-folkloric creation found an eager audience among scholars, and it stimulated their imagination in picturing the life of pagan Rus'.

Keywords: folklore; Ivan Petrovich Sakharov; literary pastiche; a hoax; a fake; Christmas carols; the sacrifice of a goat

Citation: Toporkov, Andrey. 2021. The Carol about the Pagan Rite of Sacrifice of a Goat and Its Interpretation in Russian Scholarship of the 19th to 20th Centuries. *Religions* 12: 366. https://doi.org/10.3390/rel12050366

Academic Editor: Dennis Ioffe

Received: 19 April 2021
Accepted: 17 May 2021
Published: 20 May 2021

Publisher's Note: MDPI stays neutral with regard to jurisdictional claims in published maps and institutional affiliations.

1. Introduction: I. P. Sakharov as Publisher of Folklore

Researchers of the pre-Christian culture of ancient Rus from the late 1830s to almost our time have regarded I. P. Sakharov's carol ("Za rekoiu, za bystroiu . . . " [Beyond the river, beyond the swift]) as evidence of the pagan ritual of animal sacrifice. Sakharov first published the carol in 1837, and it was reprinted the following year by I.M. Snegirev. It was then reproduced in many publications and was accepted by many scholars of Slavic mythology. However, the evidence indicates that this work was Sakharov's own creation, adapted from carols and other material from folklore. This article examines the origins of this falsification and its subsequent history.

Ivan Petrovich Sakharov (1807 to 1863), a local historian, paleographer, archaeologist, and publisher of historical and folkloric materials, occupies a special place in the history of Russian culture. Sakharov came from the clergy; his father, a priest, died when the boy was six years old. In 1815, Sakharov entered the Tula district religious school, and in 1822, the Tula seminary. After graduating in 1830, Sakharov went to Moscow, where he entered the medical faculty of Moscow University. Upon graduation, Sakharov first worked as a doctor in Moscow, and in February 1836 moved to St. Petersburg. For many years, Sakharov remained a practicing physician, devoting all his free time to studying folklore and the book tradition.

In 1835, Sakharov prepared the first part of his *Tales of the Russian People About the Family Life of Their Ancestors*, which was published in 1836, after he moved to St. Petersburg. A second edition came out in 1837, and in the same year, the second and third parts of *Tales of the Russian People* were published. After that, Sakharov began preparing a new expanded edition, but his plan was only partially implemented. Instead of the seven volumes in 30 parts that he had conceived, two volumes in eight parts were published. In 1838 to 1839, Sakharov also published a collection of *Songs of the Russian People* in five parts.

In the 1830s to 1840s, the distinction between publications of folklore and works of art on themes from folk life was not yet clearly understood. The lines between a story from folk life, a "physiological" sketch and an ethnographic account were blurred. Several Russian and Ukrainian writers of the time collected and published folklore and at the same time used it in their own artistic works. In journals intended for a wide audience, pictures of folk life were fictionalized and generously embellished with fantasy.

Sakharov had his own publishing house; it published numerous books and may be considered a successful commercial undertaking. Judging by contemporaries' admiring reviews, his publications in which folklore was stylized in a pseudo-folk spirit were popular with the reading public. Ideologically, Sakharov adhered to the doctrine of "Official Nationality" and his appeal to ancient legends, fairytales, songs, icon painting, and church singing was motivated not only by the desire to preserve them for posterity, but also to counteract the values of Western civilization.

Folklorists, ethnographers, and literary historians have confirmed that Sakharov composed fairytales which he passed off as authentic, using the plots of genuine *byliny* and referring to non-existent manuscripts. He also edited the texts of charms (*zagovory*) and riddles and made insertions in the texts of ancient Russian monuments. Although Sakharov claimed to publish mostly his own recordings of folklore and asserted that he reproduced them with great accuracy, doubts about the reliability of his materials were expressed, even during his lifetime.

Nevertheless, collections of folklore materials prepared by Sakharov continue to be republished. It should be borne in mind that not everything in them should be considered "fakelore". He took some texts from existing editions of folklore. Many collectors of the time also gave materials to Sakharov. The letters of the Tikhvin merchant G.I. Parikhin, who sent Sakharov the texts of riddles, songs and charms, have been preserved. On the basis of this correspondence, one can gain an idea of Sakharov's relationship with his local correspondents and of how he reworked the original materials he published (Toporkov 2014). Sakharov made extensive use of a kind of montage in which fragments of authentic texts were combined with sections that he himself composed. In Sakharov's editions, there are texts that are completely authentic as well as those that underwent only slight stylistic editing. At the same time, there are also texts entirely composed by him, albeit with subject matter taken from folklore. The pseudo-folkloric texts that Sakharov composed were subsequently reprinted in other publications; as is the case with "Za rekoiu, za bystroiu ...", scholars have used them to reconstruct pagan mythology, so poorly represented in authentic historical sources.

For evaluating Sakharov's publishing activity, the history of the perception of his texts in the subsequent tradition is thus no less important than the history of their first appearance. In this article, we will try to answer two main questions. Firstly, what is the origin of Sakharov's carol—what comes from folklore and what does not? Secondly, how was the carol perceived by readers and how did scholars of Slavic mythology regard it in the 19th and 20th centuries?

2. The Origins of Sakharov's Carol

Sakharov published the carol under consideration three times. The first was in the second part of the first edition of *Tales of the Russian People* in the section "Russian Family Songs" (Sakharov 1837, pp. 257–58, no. 63). The second was in the first part of *Songs of the Russian People*, in the section "Carols" (*Pesni koliadskie*) (Sakharov 1838, pp. 94–95, no. 2), and the third time in the third book of the first volume of *Tales of the Russian People*, also in the section "Carols" (Sakharov 1841, p. 16, no. 2; this edition is labeled "third" on the cover, although in fact it is not). The first publication of the carol was printed without comment, while the second and third publications gave the place of alleged recording and cited another carol, published in 1817 by I.E. Sreznevsky (the father of the famous Slavist), as a parallel.

Here is the text from the publication in the first part of *Songs of the Russian People* (Sakharov 1838):

За рекою, за быстрою,
Ой коліодка! ой коліодка!
Леса стоят дремучие,
Во тех лесахогни горят,
Огни горят великие,
5 Вокругогней скамьи стоят,
Скамьи стоят дубовыя,
На тех скамьях добры молодцы,
Добры молодцы, красны девицы,
Поют песни каліодушки.
Ой коліодка! ой коліодка!
10 В средине их старик сидит,
Он точит свой булатной нож.
Котел кипит горючий,
Возле котла козел стоит;
Хотят козла зарезати.
Ой коліодка! ой коліодка!
15 Ты, братец, Иванушко,
Ты выди, ты выпрыгни!
Я рад бы выпригнул(так!),
Горюч камень
К котлу тянет,
20 Желты пески
Сердце высосали.
Ой коліодка! Ой коліодка!

(Sakharov 1838, pp. 94–95, № 2)

[Across the fast river, / Oh, *koliodka*! Oh, *koliodka*! / Bright lights are burning, / Benches stand around the lights, / Oak benches stand, / On those benches there are good young people, / Young people [and] beautiful girls, / They sing songs to the *kaliodushka*. / Oh, *koliodka*! Oh, *koliodka*! / An old man sits among them, / He sharpens his steel knife, / The cauldron is boiling hot, / Near the cauldron a goat stands, / They want to slaughter the goat. / Oh, *koliodka*! Oh, *koliodka*! / You, brother Ivanushka, / Come out, jump out! / I'd love to jump out, / [However] the burning stone / Pulls [me down] into the cauldron, / Yellow sands sucked out [my] heart. / Oh, *koliodka*! Oh, *koliodka*!]

In numbering the lines, we did not count the chorus ("*Oi koliodka! Oi koliodka!*"). Judging by Sakharov's publication, the chorus was to be performed four times; choruses are often repeated after each verse when performing ritual songs. Such exclamatory refrains are found in Russian Christmas songs, which are usually called "carols" (*koliadki*). In these same songs "*koleda*" (or "*koliada*", "*koliadushka*", "*kaliodushka*", "*koliodka*", also meaning "carol") is personified and her actions are depicted. For example, a carol recorded in the village of Vasilevo, Dmitrovsky district, Moscow province, begins with the words:

Koljoda, Koljoda!
Koleda came

> To the master's yard
> On Christmas Eve ...

([Shejn 1898](), p. 306, no. 1035)

It is noted that "boys call to *koleda* on the evening before Christmas" ([Shejn 1898](), p. 306, no. 1035). Similarly, a carol from the village of Storozhi, Efremovsky district, Tula province, begins:

> *Koljodushka-koleda!*
> *Koleda* walked about
> On Holy Evenings.

([Shejn 1898](), p. 306, no. 1033)

Sakharov notes that the carol was recorded "from the words of Tula villagers" ([Sakharov 1838](), p. 163, note 18). In the section "Variants of Christmastide Songs", he claims that only one version of the song exists, the one printed by Sreznevsky in the *Ukranian Bulletin* in 1817:

> За рекою, за быстрою, ой каліодка!
> Леса стоят дремучие,
> Во тех лесахогни горят,
> Огни горят великие,
> 5 Вокругогней скамьи стоят,
> Скамьи стоят дубовыя;
> На тех скамьях добры молодцы,
> Добры молодцы, красны девицы
> Поют песни каліодушке;
> 10 В средине их старик сидит,
> Он точит свой булатной нож;
> Возле его козел стоит.

[Beyond the river, beyond the fast one, oh *kaliodka*! / The dense forests stand. / In those forests fires are burning, / Great fires are burning, / Benches stand around the fires, / Oak benches stand; / On those benches good fellows, pretty girls / Sing songs to *kaliodushka*; / In the middle of them an old man sits, / He sharpens his steel knife; / A goat stands beside him.] ([Sakharov 1838](), pp. 129–30)

Sakharov states that "Something is missing from this ritual song. The song in our collection fills in all the omissions ..."

In Sreznevsky's article "Slavic Mythology, or About Russian Pagan Worship" of 1817, Sakharov's carol is cited in the context of pagan myths and rituals:

> Koliada, according to some, was the Slavic god of peace and corresponded to the Roman Janus. His feast was celebrated on December 24, which was celebrated with celebrations and feasts. The temple of this god (or perhaps goddess) was in Kiev; we do not know anything about how it looked.
>
> In some places, superstitious people still celebrate this day with almost the same rituals as the ancients. As a child, I happened to hear one hymn to Koliada, of which I can only remember the beginning; here it is [...]. This beginning demonstrates the rite of ancient sacrifice itself.

([Sreznevskij 1817](), pp. 19–20)

Thus, Sakharov's assertion that "Something is missing in this ritual song" is based not only on acquaintance with the song itself, but also on Sreznevsky's direct testimony that he could only remember its beginning.

As a child, Ivan Evseevich Sreznevsky (1770 to 1819) lived in the village of Sreznevo, Spassky district, Ryazan province, and then studied at the Ryazan religious school. This means that, most likely, he heard the carol that he recited from memory in Ryazan or its environs. Sakharov's statements that the text was written "from the words of Tula villagers" and that "Something is missing in this ritual song. The song in our collection fills in all the omissions" are to some extent contradictory, and also at odds with the fact that the first part of Sakharov's carol is almost identical to that of Sreznevsky. It is difficult to imagine that the "Tula villagers" really knew a text that, on the one hand, almost literally repeats the carol published in 1817, and on the other, "fills in all the omissions" of the carol. If such a text were really performed orally, it would inevitably have contained some differences from the carol that Sreznevsky heard in childhood (probably in Ryazan province). It is much easier to imagine that Sakharov himself took the carol Sreznevsky published and added to it. He was familiar with the publication in the *Ukrainian Bulletin* and modified other texts of the oral tradition, such as those collected by Parikhin (Toporkov 2014).

The first 11 lines of the carols of Sakharov and Sreznevsky coincide. Then, Sakharov has as the 12th line, absent in Sreznevsky, "The cauldron is boiling hot". Verse 12 in Sreznevsky's text is: "A goat stands beside him", which in Sakharov corresponds to the 13th verse: "Near the cauldron the goat stands" with the replacement of the pronoun "him" by the noun "cauldron". Note that the word "cauldron", which appears twice in Sakharov's carol, does not occur at all in Sreznevsky's. The next eight verses are found only in Sakharov's text.

The question also inevitably arises whether we can trust Sreznevsky. Is his version not also a fake? To answer this question, editions of authentic folklore offer guidance. Three carols, two from the Orenburg province and one from the Omsk region, are quite close to Sreznevsky's. The first carol was recorded in the village Podgornaya Pokrovka in the Orenburg district (Celebrating Christmas carols 1888). The second was written down by A.P. Kuznetsov in the same Orenburg province (without specifying the place Shejn 1898, p. 309, No. 1046; reprint: Zemtsovskij 1970, p. 71, No. 20; see ibid. p. 550). The carol was also recorded by N. Kravets in 1975 in the village of Loginovka, Pavlogradsky District, Omsk Region (Bolonev et al. 1997, p. 51, No. 1). In all three carols, there are steep mountains, fast rivers, and dense forests; people are standing around burning fires, but there is no old man, no goat, no steel knife, and people are not going to kill anything but are just singing carols.

There are also parallels to the carols under consideration in Belarusian ritual poetry; however, there is no theme of sacrifice there either. For example, in a carol from the town of Molodusha Rechitskiy in Minsk province, fires and cauldrons are also described, although they are not located in the forest but in the courtyard of the owner of the house. Wine, beer and *"koledka"* are brewed in cauldrons (Shejn 1887, p. 76, no. 67; reprint: Mazhejka 1975, p. 220, no. 276). In ritual songs with a similar plot, images of grandfathers sometimes appear. For example, in the Belarusian Easter song (*volochebnaia pesnia*), old men melt wax for candles in cauldrons (Bartashevich and Salavej 1980, p. 209, no. 116). The motif of the impending sacrifice of a goat has parallels in the Ukrainian and Belarusian rituals of "leading a goat". These Christmas rituals stage the killing of an animal which then miraculously comes back to life.[1] In some of the songs accompanying the ceremony, the goat says that it is afraid of its old grandfather (Romanov 1912, p. 105).

Thus, motifs that are found in Sreznevsky's text have parallels in three Russian carols and in Belarusian carols and Easter songs. The motif of killing a goat, which is only suggested in Sreznevsky's text, is absent in the other carols; however, as noted, it does appear in the Belarusian and Ukrainian ritual of "leading a goat". Hence, we may state that Sreznevsky's carol fits quite organically into the Christmastide ritual poetry of the Eastern Slavs.

Despite the coincidence of Sakharov and Sreznevsky's texts, the former's edits are not a reworking of the original. Instead, Sakharov amends the texts with elements of a fairy tale. In one song from a fairytale, "a brother, walking with his sister on the road, drinks water from under a goat's hoof and turns into a kid; the sister is getting married, but the witch drowns her and replaces her (with herself or with her daughter), and they want to slaughter the kid; [but] everything is then revealed" (Barag et al. 1979, p. 135, no. 450).

The contamination of a carol with a song from a fairytale is unique and only occurs in Sakharov's text.

In the earliest known version of this fairytale, published in the collection *An Old Tune in a New Setting* (1795), a boy who has turned into a lamb addresses his sister with the words: "Sister! Alyonushka! Come here, say goodbye to me: they are sharpening steel knives, cast-iron cauldrons are boiling, they want to kill me, a lamb!" And his sister answers him from under the water: "Oh, my brother Ivanushka! I would be glad to come out to you, [but] spring water is washing into my eyes, a burning stone pulls me to the bottom!" (An Old Tune in a New Setting 2003, p. 130, no. 12). In some later published versions of the tale, the brother and sister's dialogue grows in length, and is graphically presented like poetry. Sometimes it is stipulated that the performers must sing these lines. One of the most extensive versions reads:

> Попросился козлёночек перед смертью к воде, напиться, пришел на бережок
> и плачет:
>
> Алёнушка,
> Сестрица моя!
> Выплынь, выплынь,
> На бéрежок,
> Ты выдь ко мне,
> Промолвь со мной:
> Костры кладут
> Высокие,
> Котлы висят
> Глубокие,
> Огни горят
> Горючие,
> Смолы кипят,
> Кипучия,
> Ножи точáт
> Булатные,
> Хотят меня
> Зарезати!
>
> А сестрица Алёнушка отвечает ему со дна:
>
> Ты братец мой,
> Иванушка,
> Иванушка,
> Козлёночек!
> Я рада бы
> Помóчь тебе,
> Тебе тошнó,
> А мне тошней:
> Тяжол камень
> Ко дну тянет;
> Шелковá трава
> На руках свилáсь,
> Ноги спутала;
> Желты пески
> На грудь легли;
> Лютá змея
> Сердце высосала;
> Белá рыба
> Глаза выела!

Погорюет, погорюет козлёночек, а как видит себе скорую смерть, до трех раз просился к воде и звал сестрицу. (Bessonov 1868, pp. 122–24)

[Before dying, the little goat asked for water to drink, came to the shore, and cries:

Alyonushka, / My sister! / Swim out, swim out, / Onto the shore. / Come out to me, / Say with me: / Bonfires are made, / Tall ones, / Cauldrons hang / Deep ones, / They burn / Hot, / The pitch boils / Boiling, / The knives are sharpened / Of steel, / They want / To slaughter me!

And his dear sister Alyonushka answers him from the deep:

You, my dear brother / Ivanushka, / Little goat! / I would be glad / To help you. / You feel sick, / But I am sicker: / A heavy stone / Is pulling me to the bottom;/ Silken grass encircled my arms, / It caught my legs, / Yellow sands / Lay down on my chest; / A fierce snake / Sucked out my heart, / A white fish / Ate out my eyes!

The little goat grieves and grieves, and as he sees death approaching, he asked to go up to the water up to three times and called his sister.]

The fourteenth line from Sakharov's carol ("They want to slaughter a goat") has a parallel in a fairytale from Saratov province published by A.N. Afanasyev: "The little goat walks to the riverbank and cries bitterly, saying: "Olenushka, my dear sister! Come out to me, take a look: I am your brother, Ivanushka, I have come to you with unhappy news, they want to kill me, [a goat], to slaughter me ..." Olenushka answers him from under the water: "Oh, my brother Ivanushka! I would be glad to look out at you, [but] a heavy stone is pulling me to the bottom ..." (Afanas'yev 1985, p. 255, no. 263).

In verses 15 to 21, Alyonushka's response to her brother is paraphrased, which is directly indicated in line 15: "You, brother, Ivanushko ... ". The words from the carol "You go out, you jump out!" in the fairytale quoted above from Afanasyev's collection correspond to: "You come out to me, you look out ..."—but this is Ivanushka's remark addressed to Alyonushka, and not vice versa, as in Sakharov's text.

In the fairytale, Ivanushka asks his sister to look out from under the water, and she explains why she cannot, while in Sakharov's carol, Alyonushka asks Ivanushka to come out from somewhere and to jump out. However, this is not motivated, since neither Ivanushka nor Alyonushka are located in the water. Sakharov's text is strange: "A burning stone // Pulls [me down] into the cauldron" instead of "to the bottom". By the way, V. Ia. Propp, quoting the carol, replaced "pulls into the boiler" by "pulls to the bottom" and noted: "[This is] corrected from an obvious error" (Propp 1995, p. 57, note 15). In this case, however, we are dealing not with an "obvious error" but with a deliberate change to the original text. Several semantic inconsistencies also indicate the contamination of texts of different genres.

If in Sreznevsky's carol there was only a hint of a possible future killing of a goat, in Sakharov's text this turns into a whole scene of preparation for the murder of a child who has been turned into a kid, complete with dramatic remarks from the proposed victim. In this context, the killing of a goat is on a par with human sacrifice, and the whole situation takes on a menacing character that is not at all characteristic of carols. The motif of killing a goat is not found in carols, but in songs accompanying the Christmas ritual of driving a goat to sacrifice.

Thus, Sakharov's carol is clearly contaminated and shows inconsistencies in the dialogue between Ivanushka and his sister. However, despite the eclecticism and imperfection of the text, it has repeatedly been reprinted and taken seriously in studies on Slavic mythology.

3. Subsequent Use of Sakharov's Carol as Historical Evidence

Ivan Mikhailovich Snegirev was the first to reprint Sakharov's carol in the section "Carols and Vinograd'e Songs" in the second part of his *Russian Folk Holidays and Rites* (Snegirev 1838, pp. 68–69, no. 4). In a note to the song Snegirev describes its contents as follows: "Here is depicted a sacrifice in which sand is poured, into which among Northern peoples the blood of the victim was spilled; this is expressed by the words the sands have sucked out the heart. See the song in the first part of this book, p. 103" (Snegirev 1838, p. 69, note 1). Snegirev thus refers the reader to the first issue of *Russian Folk Holidays and Rituals,* where first the sacrifice of a goat by Lithuanians is described and then Sreznevsky's carol is cited as evidence that such rituals were known in Russia: "Observation reveals that this celebration with more or less similar rituals and sacrifices was common not only throughout Lithuania, but also in Lithuanian Rus. According to the testimony of Mr. Sreznevsky, in southern Russia they sang the following song, which depicts a sacrificial ritual similar to the Lithuanian one" (Snegirev 1837, p. 103).

In his book *Slavic* Mythology (1847), Nikolai Ivanovich Kostomarov wrote with reference to Sakharov's material that during the winter holidays "sacred games were performed, as can be seen from winter round dances, and sacrifices were made. Thus, in one Russian Christmas song, an obviously pagan sacrifice is portrayed: fires are burning in the forest, benches are standing around them, young men and women are sitting on benches" (Kostomarov 1847, pp. 100–1).

Izmail Ivanovich Sreznevsky referred to our carol in his book *Studies on the Pagan Worship of the* Ancient Slavs (1848). He cited the text of the carol in a review of materials on pagan sacrifice: "There is a legend about the sacrifice of a goat preserved in a folk ritual song: 'Beyond the river [...] They want to slaughter a goat.' According to the testimony of Dlugosh, the Poles also made sacrifices: sheep and bulls were sacrificed during festivities at which people gathered. They also occurred among the Baltic Slavs" (Sreznevskij 1848, p. 73).

The historian Sergei Mikhailovich Soloviev, in his "Sketch of the Customs, Rites and Religion of the Slavs, Mainly the Eastern, in Pagan Times" (1850), noted that "among the carols the following is remarkable" and quoted Sakharov's text (Solov'ev 1850, p. 30). He further considered the ritual sacrifice of a goat in comparative context (Solov'ev 1850, p. 31).

Fyodor Ivanovich Buslaev brought up Sakharov's carol in one of his lectures on the history of Russian literature that were read to the heir, Tsarevich Nikolai Alexandrovich, in 1859 to 1860. Speaking about Christmastide rites, Buslaev noted: "It goes without saying that in carols the Nativity of Christ is recalled, but beyond the Christian element, the most ancient pagan material is also evident. Especially important in the latter respect is the carol about the sacrifice of a goat, which has survived to this day in Southern Russia ..." (Buslaev 1990, p. 434). Having quoted Sakharov's text in full, Buslaev continued: "We already know that the goat was dedicated to Toor-Perun, the deity of agriculture and family settlement. He rode in a cart pulled by two goats" (Buslaev 1990, p. 435).

In the article "Kupala and Kolyada in Relation to the Folk Life of the Russian Slavs", included in his book *Russian Nationality in its Beliefs*, Rituals and Fairytales (1862), Dmitry Ottovich Shepping quoted Sakharov's text, making some adjustments. Most importantly, he considered the text not a carol, but a *kupala* (summer solstice) song, and in this connection he replaced the line "Singing songs of *Kolyudushka*" with "Singing songs of *Kupalushka*" (Shepping 1862, p. 40). Shepping associated Ivanushka's name with the night of Ivan Kupala: "That this song does not belong to the carols, but on the contrary, to the celebrations of Ivanovka night, is clearly indicated by the name of the goat Ivanushka, which is why we consider ourselves entitled to replace the word *Kolyudushka* with the word *Kupalushka*" (Shepping 1862, p. 40, note 1). In fact, the name of Ivanushka in the carol is borrowed from a fairytale and has nothing to do with Ivan Kupala.

According to Shepping, Sakharov's song testified that in antiquity some living creatures were burned in the Kupala bonfire: "It is likely that during pagan times live sacrifices were also made to the Bathing Fire (*Kupal'nomu ogniu*), as we see from this ancient song . . ."

(Shepping 1862, p. 39). Shepping identified the old man who appears in the carol with a pagan priest: "The very production of living fire, as in the lighting of Ivanovo bonfires, is usually entrusted to old people, who, as can be seen from this rite—one of a purely communal life, and from the above mentioned song, performed the office of a priest in the public worship of pagan Russia" (Shepping 1862, pp. 40–41).

Alexander Nikolaevich Afanasyev, in his three-volume study *Poetic Views of the Slavs on Nature*, noted that: "A remembrance of the sacrificial slaughter of a goat is preserved in a carol, which is all the more curious because it also conveys the very setting of the ceremony . . . " (Afanas'yev 1868, pp. 257–58). Citing Sakharov's text, Afanasyev continues: "The sacrifice was accompanied by the singing of ritual songs; the goat was cut by the elder and its meat was cooked in a cauldron, just as was done among the Germanic tribes and the Scythians. Among the Lithuanians, the priest, stabbing a goat with a sharp knife, invoked God's blessing on those praying; blood was collected in a jug or bowl, and then people, cattle and even dwellings were sprinkled with it; the meat was eaten while singing songs and trumpet playing, and what remained uneaten was buried in the ground at the crossroads so that not a single animal could touch the sacred food" (Afanas'yev 1868, p. 258).

The 19th-century tradition of mythological research was revived in the 1960s to 1980s by the archaeologist and academician Boris Alexandrovich Rybakov. In his monograph *The Paganism of Ancient Rus* (1st ed.—1987), he cited the carol from Snegirev's collection and noted that this text "reveals the essence of the ritual ceremony—the sacrifice of a goat . . . " (Rybakov 1987, p. 85). Rybakov, following Shepping, wrote that the setting of the sacrifice is reminiscent of the Kupala holidays: "The name of brother Ivanushka may indicate a ceremony on the night of Ivan Kupala; then sister Alyonushka is Kupala herself, a victim doomed to become 'engulfed in water.' On the night of Kupala, 'great fires burn,' and rituals are performed near the water, imitating the drowning of the victim: bathing a girl dressed up as Kupala or immersing a stuffed doll depicting Kupala in the water" (Rybakov 1987, p. 95). It is hard to agree with this reasoning. As we have already noted, the name Ivanushka is borrowed from a fairytale and has nothing to do with the night of Ivan Kupala; Alyonushka is identified with the mythical Kupala only in Rybakov's imagination.

4. Conclusions

Thus, researchers have regarded Sakharov's carol "Za rekoiu, za bystroiu . . . " as authentic evidence of pagan rituals of pre-Christian Rus. In order for the carol to be seen as an historical document, they have interpreted its generic nature, time of appearance, plot, and individual details in a particular way. First, the song was viewed not as a traditional carol, performed to congratulate homeowners on the feast of the Nativity of Christ, but as a miraculously preserved description of a real ritual practiced in ancient Russia, as if it were a fragment of a chronicle or a work by an Arab or Byzantine author of the tenth century. Second, scholars have seen in the work a description of the sacrifice of a goat, although Sreznevsky's text speaks more about preparing a goat for slaughter, and Sakharov's carol— about the imminent killing of child who has been changed into a goat. The modern reader sees in the refrain "Oh, *koledka*!" an appeal to the holiday of *Kolyada*, as Christmas Eve was called, contrary to Sreznevsky's assertion that Kolyada "was the Slavic god of peace and corresponded to the Roman Janus [. . .] The temple of this god (or maybe goddess) was in Kiev . . . " (Sreznevskij 1817, pp. 19–20; italics in the original). Indeed, Sreznevsky published the carol as "a hymn to Kolyada". He did not directly assert that the goat was intended as a sacrifice to Kolyada, but implied it. Thus, the connection between the carol and pagan rites had already been suggested in Sreznevsky's publication; that is, Sakharov was following in a certain tradition which had formed even before he composed his carol. Finally, in the third place, scholars interpreted the fires as sacrificial fires, and the old man who "sharpens his steel knife" as a pagan priest.

Given Sakharov's record as a falsifier of folklore, there is no reason to doubt that it was he who created this carol. Oddly enough, there is not a single line in it that Sakharov composed by himself: they are taken either from the carols of I.E. Sreznevsky or from versions of the folktale about Alyonushka and her brother Ivanushka. Despite the pseudo-folkloric nature of Sakharov's carol, several generations of historians and philologists have seen it as a description of a ritual performed by pagan Slavs in ancient times. The material considered in this article leads to the conclusion that the use of folklore as historical data requires extreme care and a close examination of the sources involved.

Funding: This research was funded by the Russian Foundation for Basic Research (grant number 20-012-00117).

Institutional Review Board Statement: I choose to exclude this statement if the study did not involve humans or animals.

Informed Consent Statement: I choose to exclude this statement as the study did not involve humans.

Data Availability Statement: I choose to exclude this statement as the study did not report any data.

Acknowledgments: I am thanking Marcus Levitt for his translation work.

Conflicts of Interest: The author declare no conflict of interest.

Note

[1] In the ceremony of "leading a goat" (*vozhdenie kozy ili kozla*) a group of mummers (often teenagers) sings a song about a goat. One of the mummers acts as the goat and the others lead him from house to house. The song usually tells how when a goat is out walking with its young children it is attacked by a wolf. The goat seems to have been killed but then comes back to life, as in the song minor harmonies change into major, joyful ones.

References

Afanas'yev, Aleksandr Nikolaevich. 1868. Поэтические воззрения славян на природу: Опыт сравнительного изучения славянских преданий и верований в связи с мифическими сказаниями других родственных народов. Т. 2. [*Poetic Views of the Slavs on Nature: An Attempt at Comparative Study of Slavic Traditions and Beliefs in Connection with the Mythical Tales of Other Related Peoples. Vol. 2*]. Moscow: K. Soldatenkov.

Afanas'yev, Aleksandr Nikolaevich, ed. 1985. Народные русские сказкиА.Н. Афанасьева в трех томах, Издание подготовили Л.Г. Бараг, Н.В. Новиков. Т. 2. [*Russian Fairytales by A. N. Afanasyev in Three Volumes, Vol. 2*]. Edited by Lev Grigorievich Barag and Nikolay Vladimirovich Novikov. Moscow: Nauka.

An Old Tune in a New Setting. 2003. Старая погудка на новый лад: Русская сказка в изданиях конца XVIII века. 2003. [*An Old Tune in a New Setting: The Russian Fairytale in Publications of the Late Eighteenth Century*]. St. Petersburg: Tropa Troyanova, ISBN 978-5-9025993-95.

Barag, Lev G., Ivan P. Berezovskij, Konstantin P. Kabashnikov, and Nikolaj V. Novikov. 1979. Сравнительный УказательСюжетов: ВосточнославянскаяСказка. [*Comparative index of plots: The East Slavic Fairytale*]. Leningrad: Nauka, Leningradskoe otdelenie.

Bartashevich, Galina A., and Liya M. Salavej, eds. 1980. Валачобныя песні [*Valachobnyya Songs*]. Minsk: Navuka i tekhnika.

Bessonov, Petr A. 1868. Детские песни. [*Children's Songs*]. Moscow: Tipografiya P. Bakhmet'eva.

Bolonev, Firs Fedosovich, Mikhail Nikiforovich Mel'nikov, and Natal'ya Vladimirovna Leonova. 1997. Русский календарно-обрядовый фольклорСибири и Дальнего Востока: Песни. Заговоры. [*Russian Calendar and Ritual Folklore of Siberia and the Far East: Songs. Verbal Charms*]. Novosibirsk: Nauka, ISBN 5-02-030895-1.

Buslaev, Fedor. 1990. О литературе: Исследования; Статьи, Составление, вступительная статья, примечания Э. Афанасьева. [*About Literature: Research; Articles*]. Edited by Eduard Afanasyev. Moscow: Khudozhestvennaya Literature, ISBN 5-280-00724-2.

Celebrating Christmas carols. 1888. Празднование коляды в Оренбургском уезде[Celebrating Christmas Carols in the Orenburg District]. Orenburgskij Listok [Orenburg Sheet]. N. 52, p. 2.

Kostomarov, Nikolay Ivanovich. 1847. Славянская мифология. [*Slavic Mythology*]. Kiev: Tipografiya I. Val'nera.

Mazhejka, Zinaida Yakovlevna, ed. 1975. Зімовыя песні. Калядкі і шчадроўкі. [*Winter Songs. Carols and Wedrowki*]. Minsk: Navuka i tekhnika.

Propp, Vladimir Yakovlevich. 1995. Русскиеаграрные праздники: (Опыт историко-этнографического исследования). [*Russian Agrarian Holidays: (An Attempt at Historical and Ethnographic Research)*]. St. Petersburg: Terra–Azbuka, ISBN 5-300-00114-7.

Romanov, Evdokim Romanovich. 1912. Белорусский сборник. Вып. 8–9: Быт белоруса; Опыт словаря условных языков Белоруссии. [*Belarusian Collection. Issue 8-9: Life of the Belarusian; Attempt at a Dictionary of the Conditional Languages of Belarus*]. Vil'na: Tipografiya A. G. Syrkina.

Rybakov, Boris Alexandrovich. 1987. Язычество Древней Руси. [*The Paganism of Ancient Rus*]. Moscow: Nauka.

Sakharov, Ivan Petrovich, ed. 1837. Сказания русского народа о семейной жизни своих предков, собранные И.П. Сахаровым. Ч. 2. История русской народной литературы. [*Russian Folktales about the Family Life of Their Ancestors, Collected by I. P. Sakharov. Part 2. History of Russian Folk Literature*]. St. Petersburg: Guttenbergova tipografiya.

Sakharov, Ivan Petrovich, ed. 1838. Песни русского народа. Ч. 1. [*Songs of the Russian People. Part 1*]. St. Petersburg: Tipografiya Sakharova.

Sakharov, Ivan Petrovich, ed. 1841. Сказания русского народа о семейной жизни своих предков, собранные И.П. Сахаровым. Т. 1, Кн. 3. История русской народной литературы. [*Russian Folktales about the Family Life of Their Ancestors, Collected by I. P. Sakharov. Vol. 1, Book 3. History of Russian Folk Literature*]. St. Petersburg: Tipografiya Sakharova.

Shejn, Pavel Vasil'evich. 1887. Материалы для изучения быта и языка русского населенияСеверо-Западного края. Т. 1, ч. 1: Бытовая и семейная жизнь белоруса вобрядах и песнях. [*Materials for Studying the Life and Language of the Russian Population of the North-Western Region. Vol. 1, Part 1: Household and Family Life of the Belarusian in Rites and Songs*]. St. Petersburg: Tipografiya Imperatorskoj Akademii nauk.

Shejn, Pavel Vasil'evich. 1898. Великорусс в своих песнях, обрядах, обычаях, верованиях, сказках, легендах и т.п. Т. 1, вып. 1. [*The Great Russian in His Songs, Rites, Customs, Beliefs, Fairytales, Legends, etc. Vol. 1, Issue 1*]. St. Petersburg: Tipografiya Imperatorskoj Akademii nauk.

Shepping, Dmitrij Ottovich. 1862. Русская народность в ее поверьях, обрядах и сказках. [*The Russian Nationality in Its Beliefs, Rituals and Fairytales*]. Moscow: Tipografiya Bakhmet'eva.

Snegirev, Ivan Mikhailovich. 1837. Русские простонародные праздники и суеверныеобряды. Вып. 1. [*Russian Folk Festivals and Superstitious Rituals. Issue 1*]. Moscow: Universitetskaya tipografiya.

Snegirev, Ivan Mikhailovich. 1838. Русские простонародные праздники и суеверныеобряды. Вып. 2. [*Russian Folk Festivals and Superstitious Rituals. Issue 2*]. Moscow: Universitetskaya tipografiya.

Solov'ev, Sergej Mikhailovich. 1850. Очерк нравов, обычаев и религии славян, преимущественно восточных, во времена языческие. In Архив историко-юридических сведений, относящихся до России, издаваемый Н. Калачовым. [*"An Essay on the Mores, Customs and Religion of the Slavs, Mainly Eastern, in Pagan Times", in: An Archive of Historical and Legal Information Relating to Russia, Published by N. Kalachov*]. Moscow: Tipografiya Alexandra Semyona, vol. 1, pp. 1–54.

Sreznevskij, Ivan Evseevich. 1817. Славянская мифология, илио богослужении русском в язычестве[*Slavic Mythology, or About Russian Pagan Worship*]. *Ukrainskij vestnik [Ukrainian Bulletin], Khar'kov* 6: 3–24.

Sreznevskij, Izmail Ivanovich. 1848. Исследованияо языческом богослужении древних славян. [*Studies on the Pagan Worship of the Ancient Slavs*]. St. Petersburg: Tipografiya K. Zhernakova.

Toporkov, Andrey L'vovich. 2014. «В нашихСказаниях не все то помещено, что известно в селениях» (фольклорные записи изархивного собрания И.П. Сахарова) [«*In Our Tales, Not Everything is Included That Is Known in the Villages*» (*Records of Folklore from the Archive Collection of I. P. Sakharov*)]. *Traditsionnaya kul'tura [Traditional Culture]* 4: 141–54.

Zemtsovskij, Izalij Iosifovich, ed. 1970. Поэзия крестьянских праздников. [*The Poetry of Peasant Holidays*]. Leningrad: Sovetskij pisatel', Leningradskoe otdelenie.

Article

Christianity and Slavic Folk Culture: The Mechanisms of Their Interaction

Svetlana M. Tolstaya

The Institute of Slavic Studies of the Russian Academy of Sciences, 119334 Moscow, Russia; smtolstaya@yandex.ru

Abstract: In Slavic folk culture, Christianity is a foreign, borrowed cultural model, while the oral tradition is native and familiar. The different areas of folk culture were influenced to varying degrees by the Christian tradition. The most dependent area of Slavic folk culture on Christianity was the calendar. In many cases, it only superficially accepted the Christian content of calendar elements and reinterpreted it in accordance with the traditional mythological notions. The same can be said about the folk cult of saints. The Christian saints replaced pagan gods and over time were included in the system of folk ideas, beliefs and rituals. The mechanism for regulating the balance between man and the world is a system of prohibitions, the violation of which is recognized as sin and is punished by natural disasters, death, disease and human misfortunes. The Slavic folk tradition adapted not only the individual elements, structures and semantic categories of Christianity, but also the whole texts, plots, motifs, and themes developed in various folklore genres. Therefore, the pre-Christian folk tradition of the Slavs was able to assimilate many Christian concepts, symbols, and texts, translate them into its own language and fill them with its own content.

Keywords: Christianity; Slavic folk culture; mythology; beliefs; rites; folklore

Citation: Tolstaya, Svetlana M. 2021. Christianity and Slavic Folk Culture: The Mechanisms of Their Interaction. *Religions* 12: 459. https://doi.org/10.3390/rel12070459

Academic Editor: Dennis Ioffe

Received: 4 June 2021
Accepted: 13 June 2021
Published: 23 June 2021

Christianity and the oral folk tradition represent two different cultural models that have coexisted in the same ethno-cultural space since the adoption of Christianity by the Slavs. They differ in several important ways. The oral folk tradition has its roots in deep antiquity, and Christianity, as a cultural model, emerged in more recent historical times—although it, too, draws largely on folklore. In Slavic culture, Christianity is a foreign, borrowed cultural model, while the oral tradition is native and familiar. These two cultural models contrast in several different ways: written form—oral form; codified—non-codified; official—non-official; closed—open; reflection (the presence of theory and metalanguage)—non-reflection (the lack of theory and metalanguage); and consciousness (the conscious maintenance of tradition)—spontaneity (the unconscious adherence to tradition). The first characteristic in each pair relates to the Christian tradition, while the second relates to folk culture. These characteristics determine the relationships between the two cultural types. While the two cultural systems can be contrasted as religious and mythological (mythopoetic, cosmological), they are very similar in "genre" composition. Both combine a certain range of ideas about the world (the Christian cosmology and anthropology and the folkloric mythopoetic picture of the world, respectively,) with developed ritual forms (religious cult and folk rites); both are equipped with a large number of texts (verbal, musical, visual).

Historically, the direction of influence has been mostly one-sided from Christianity to the folk tradition: that is, from a stronger, more organized, strictly codified and ideologically shaped closed system to a less organized, amorphous, non-rigid and "open" system. The reverse effect was much weaker in both scope and significance and did not affect the structural and ideological foundations of Christianity; although Christian demonology was influenced by folk ideas and some church rituals adopted elements of folk customs. Therefore, first of all, we should talk about the mechanisms of adaptation of Christian elements in folk culture.

Every researcher of folk culture and folklore has to separate the pre-Christian, primordial and pagan elements from those imported from Christianity. However, the interpenetration and inevitable transformation of these elements over the centuries have created such complex forms of cultural syncretism that separating one from the other is a very difficult and sometimes impossible task.

Specialists in the field of Slavic folk culture are often accused of excessive fascination with paganism and attributing antiquity to the ritual and non-ritual folklore forms, texts and symbols recorded in records and observations of the 19th and 20th centuries. In some specific cases, this reproach may be justified, but, in general, it is likely to be a function of the ambiguity of the very concept of paganism. We must distinguish between paganism as a historical form of Slavic culture stemming from pre-Christian times and as a cultural model with its own specific "picture of the world", axiology, patterns of behaviour and symbols. Paganism as a cultural model adapts and assimilates material from many other traditions, including "foreign" belief systems such as Christianity. This assimilation sometimes happens to such an extent that, strictly speaking, it no longer makes sense to talk of paganism at all.

Therefore, the concept of dual or even triple-belief can be applied to traditional Slavic culture, primarily or almost exclusively in the genetic sense: it is, indeed, a special fusion of primordial, pre-Christian, pagan elements, Christian elements and elements of late antiquity and non-Slavic cultures that came with Christianity or from elsewhere. In part, the concept of dual faith may refer to the functional distribution of genetically pagan and genetically Christian (or non-Christian) forms in some specific situations (for example, in fortune telling on the one hand, and in church rituals on the other).

However, in terms of typology and "ideology", many areas of folk culture remained pagan in the nineteenth century and, where folk culture is preserved as an integral tradition, remain so even today. A powerful layer of Christian ritual forms, motifs, images, characters, symbols and concepts and, finally, texts assimilated by the folk tradition was in many cases subjected to mythological reinterpretation and adaptation in accordance with the traditional pre-Christian picture of the world. At the same time, different areas of folk culture were influenced to varying degrees by the Christian tradition.

1. Folk Calendar

The most dependent area of Slavic folk culture on Christianity was the calendar, which adopted the Christian system of holidays and weekdays, fasts and meat-eating as the structural basis of the entire ritual annual cycle. While folk culture borrowed from the Christian calendar the composition, order and names of festivals, in many cases, it only superficially accepted the Christian content of calendar elements and reinterpreted it in accordance with the traditional mythological notions of calendar time and in relation to the magical elements of agricultural practice.

Thanks to the practice of church worship, Christian influence on the national calendar is not erased, but remains constantly relevant. At the same time, the popular interpretation and extra-ecclesiastical ritual pragmatics of the calendar are mostly mythological in nature. Some characteristic features of the national calendar illustrate this point well.

The mythological model of time in the Slavic folk calendar is based on the opposition of pure, good and sacred time and impure, evil, unkind and dangerous time. Days and weeks can be defined by the epithets holy, Divine, bright, pure, great, joyful and good on the one hand, and evil, filthy, unclean, "unbaptized", harmful, terrible, dangerous, crooked, empty and vain on the other. Days (and holidays) can be male and female, young and old, short and lame, mad and stupid, rotten and deaf, dead and cheerful, strong and heavy, rich and poor, tough and lazy, fat and hungry, warm and cold, dry and wet, fire and water or hail and wind.

In Slavic folk chrononyms, different symbolic codes are used—colours, plants, animals and others. Days, weeks and holidays can be white, red, black, green or variegated (cf. White week, White Fryday, Green Yury, Green week, Red week, Black Wednesday,

Colorful Shrovetide and many others). Among holidays, we can see Apple Spas[1], Palm Week, Coloured (Flower) Sunday, Maple Saturday, Nut Spas, Grass Friday and similar names. In the folk calendar of the Bulgarians and the Serbs, there are autumn "wolf days", which are marked so that wolves do not attack livestock (SD 1995–2012, vol. 1, pp. 427–28), as well as "mouse days", which are marked so that mice do not eat grain in the fields and barns (SD 1995–2012, vol. 3: pp. 346–47). In the Polish and Kashubian calendar, one can find such chrononyms as *Matka Boska Niedźwiedzia* 'Bear Mother of God' (Candlemas), *Matka Boska Żabiczna* 'Frog Mother of God' (Annunciation) or *Matka Boska Zielna* 'Herb Mother of God' (Assumption of the Virgin). Calendar days and holidays may be related to each other: for example, as father, mother or nephew. In Belarus, St. Barbara's Day is the mother of St. Nicholas's Day and St. Sava's Day is the father of Nicholas's Day, while, in Russia, Shrovetide is Semik's[2] niece.

In the popular interpretation, a holiday is a stop or break in the chain of time, a moment when the border between two worlds opens. A holiday is a dangerous time fraught with threats to the inhabitants of our world and, as such, necessitates special protective measures. The meaning of individual holidays and segments of the annual cycle also often has a mythological and primitive magical basis. In the popular consciousness, even as big an annual holiday as the Annunciation often loses its original Christian meaning and "the good news" is understood as the news of the coming of spring. Similarly, Christmas is associated primarily with the advent of the new year—the magic of the beginning, the first time or the sacred proto-event.

The Russian word *prazdnik*, 'holiday', is of Church Slavonic origin. This name reflects one of the relevant features of the concept of holiday, namely the meaning 'idle, empty, free from labour'. The same word is used in all South Slavic languages. In addition, in the West Slavic and western part of the East Slavic languages, another Proto-Slavic word is used, namely **svęto,* that literally means 'holy' (Ukrainian and Belarusian *sviato, sviatok,* Polish *święto,* Czech *svátek*). This word actualizes another relevant feature of this concept—the sacredness of the holiday. Both ideas have a great weight in traditional folk culture. It is known that, in the popular consciousness, the main feature of the concept of holiday is the ban on work in general or on certain types of work and the violation of these prohibitions is considered a great sin. In turn, the sanctity of the holiday, which goes back to the pre-Christian era and is supported by the Christian canon, serves as the main reason for the honouring of holidays and justifies the prohibitions and regulations that are observed. Both names contrast the holidays as a sacred time dedicated to the divine powers with the profane time of everyday life.

In the folk tradition, it is not only etymologically that the concept of a holiday and the concept of holiness are connected. The Christian calendar, which forms the structural basis of the Slavic folk calendar, prescribes the veneration of most of the holidays as days of remembrance of a particular saint. Accordingly, the canonical names of holidays are a kind of calendar formula, including the name of the saint or the name of the venerated event: for example, "the day of the prophet Elijah", "the day of Apostle and Evangelist Mark" or "Entry into Jerusalem". In real church and folk practice, this formula was simplified and rolled up to just the name of the venerated saint (Mark, Peter and Paul, Elijah, etc.) or a two-word construction such as Peter's Day, Elijah's day, etc. Such constructions and even one-word anthroponomic nominations of holidays have been known since ancient times. In the Novgorod birch bark letters, starting from the XII century, we can see "on Peter's day" and even "before Proclus" (XV century) (Zaliznyak 1995, pp. 611, 568).

This method of denomination leads first to the identification of the name of the holiday and the name of the Saint, then to the identification of the holiday and the person of the Saint and, finally, to the complete personification of a holiday. Hence, Holiday can come to a house to punish for violation of the prohibition on work and other rules. The object of personification may even be a holiday, which in the Christian calendar is dedicated to some event, and in its name does not contain the name of the Saint (Tolstaya 2005, pp. 377–84; Tolstaya 2011).

We can see the identification of the saint and the day dedicated to him, for example, in the Belarusian Easter songs. They tell about the order and distribution of roles between the holidays-saints. For example: "Saint Boris carries sheaves and Saint Anna takes them home" (VP 1980, p. 116). In such texts, it is not the saints themselves who act, but the personified feasts. This is indicated by the transformation of female names into male names and vice versa (for example, Saint Eudokia is called in the masculine gender "Saint Eudoky", and Saint Nicholas is called by the feminine name Saint Nikola). The use of the names of saints in the plural form (for example, Barbaras, Savvas, Avdotyas, Androses) can also be seen, as well as the pairing (doubling) of the names of saints to whom the same day is dedicated (for example, Peter-Paul). But the most important thing is that among the saints-holidays there are such characters as the Saint Christmas, the Saint Annunciation, the Saint Forty Martyrs, the Saint Ascension, the Saint Intercession and the Saint Assumption. Such quasi-saints participate in the distribution of functions in the same way as the holidays dedicated to real saints.

Here are a few examples from Belarus: "The saint Eudoxias broke the winter; the saint Forty Martyrs do not walk on the river; the Annunciation begins to plough with black horses and a golden plough in its right hand"; "Oh, the first feast is Saint Yuri, the second feast is Saint Nikola (fem.), the third feast is Saint Peter, the fourth feast is Saint Elijah and the fifth feast is the Saint *Spas* (Transfiguration); Saint Yuri drives the cattle out to pasture; Saint Nicola sows the field; Saint Peter puts bees in the hive and Saint *Spas* (Transfiguration) sows wheat" (VP 1980, pp. 135, 186). In Serbian ritual songs, the quite human character *Božić* (Christmas) appears. He greets people "from behind the water (river)", shouts to them from a high mountain or from a spring or bridge and asks them to give him wine and rakiya (vodka), Christmas smoked bacon, figs and other dishes from the festive table. He wishes everyone fun and joy." (Moroz 1998).

The anthropomorphising and personification of holidays, however, is not only due to the linguistic form and pragmatics of calendar discourse. It is connected primarily with the general, mythological concept of time in folk culture. This concept gives units and time periods a positive or negative rating. A person is afraid of the elements of time and seeks to master time and subdue it by magical manipulations; for example, by compressing, stretching or deceiving time (Tolstaya 1997). The anthropomorphising of holidays is one way of mythologizing time. There is evidence of this in both language data and ritual practice.

It seems clear that the main feature of holidays is their danger to humans. According to popular beliefs, all holidays are dangerous, and the larger and holier the holiday, the more dangerous it is. In Polesie,[3] they say *varovity* day or holiday and in Serbia *varovni dan*. People are no less afraid of holidays than they are of evil spirits, and it is out of fear of them that they strictly observe prohibitions, restrictions and regulations and seek to protect themselves from them in the same ways and with the same care as they do from demons. The Serbs in Bolevac believe that: a child born on the feast of St. Simeon (September 1) will be *siromašno* (poor); one born on Christmas Eve will be unhappy; and one born on Christmas Day will be even more unhappy. It is said that, in the old days, such children were burned as they were considered unworthy to live. Mothers would keep secret that a child was born at Christmas. A child born at Easter was a forewarning of the imminent death of his parents (Trebješanin 1991, pp. 99–100).

Holidays are attributed human properties and actions. They can get angry, take offense at people and punish them for not following the prohibitions and regulations. In Polesie, they say: "We celebrate Mikhail, so that the thunder does not beat the hut. On this day, we do not chop wood, do not wash clothes, do not pick up a knife, do not weave a cross. So as not to offend Mikhail" (PA n.d., Velikiye Avtyuki, Gomel region). Another Polesian story personifying a holiday recounts: "One woman was weaving on the minor holiday (*prisviatok*) of *Warm Aleksey* (March 17). She was told: "What are you weaving, because today is a holiday!" She replied scornfully: "What kind of holiday is this!" So, at night someone came with a big stick. The woman got scared and died. It was this holiday that came and scared her" (PA n.d., Chudel, Rivne region).

In Polesie, stories are popular about the autumn holiday Miracle (the day of remembrance of the miracle of Archangel Michael, September 6/19) and how dangerous and insidious it can be. The peasant was going to plough the field on the Miracle holiday, and was punished by it. According to the stories, he and his two oxen were petrified in the field or nothing came up in the field or the oxen died. In another story, a hostess released a cow to graze on the Miracle, and the Miracle punished her by causing her cow to bloat, and then she died, and, in another, a woman began to whitewash the house that day and went blind. Other Slavs have similar beliefs. The Serbs believe that if they had not celebrated the Thursday before Trinity, the river would have risen and the flood would have begun (Zečević 1981, p. 34).

However, holidays are attributed not only punitive, but also positive magical functions. For example, according to Polesian beliefs, in order to avoid meeting with a snake in the forest in the summer, it is necessary to say that the Annunciation was on such and such a day of the week (PA n.d., Kobrin district, Brest region). According to Ukrainian beliefs, if a person gets lost in the forest, he should remember on what day of the week Christmas fell. Then he should take some soil from under his feet and sprinkle it on his head. Then he would be able to find his way home (Nomis 1864, No. 280). In Ukrainian and Belarusian lucky charms, people turn to the holidays for help in the same way as to the saints, to God and the Mother of God. For example: "Annual holidays, great, small, please start to help me..." or in the Belarusian charm: "Holy bright day (Easter), get up to help! Holy Mikola, get up to help! Holy Friday, get up to help! Holy Ascension, get up to help! Saint Peter-Paul, get up to help! Saint Elijah, get up to help! Saint *Spas* (Transfiguration), get up to help! Saint Cover (Intercession), get up to help! ..." (Zamovy 1992, No. 1044).

The holidays are called "God's saints and quick helpers": "Let us pray and worship the Lord God, the Holy Spirit, the Holy Trinity, the One and Most Pure Mother of God, and the holy Resurrection of Christ, the honest, the merciful, and all the holy holidays, the annual holidays–from Kiev, Pechera and Rome, Jerusalem, Isaak and Jacob, and Anthony, and Theodosius, and Saint Yuri and Georgy, and Holy Father Nikola. Holy Father Nicholas, you are a saint of God, you are a quick helper, your prayers are powerful you quickly help, you drive away *volos* (a disease imagined in the form of a long and thin worm that bites into a person's body) and send it to mosses and swamps" (Ibid., No. 749).

One old woman from the village of Sporovo, Brest region, talking about her difficult and joyless life, admitted that the holiday helped her survive: "You know, I still think that this holiday helped me. In our language, it is called Onuphrius. [There was a church dedicated to St. Onuphrius in the village.] He turned my life around. Thanks to him, the pigs began to breed in twos and threes and the cow gave milk, and there was more sour cream and milk. I remember this holiday, and it seems to me that he helped me" (PA n.d.).

A holiday, like a saint, can be considered the patron of a family, clan or native land. Interesting in this respect is the famous South Slavic custom *Slava*, known mainly in Serbia, Montenegro, Macedonia, Bosnia and Herzegovina and named *slava* (glory), *krsno ime* (Christian name), *svetac* (saint), *svechari* (saints) or *praznik* (holy day). A family or clan praises one saint (or two saints) on the day(s) of their memory. The most popular of these patrons are Saints George, Nikola, Sava, Jovan and Luka. However, it is not only saints that can be glorified, but also holidays that do not have their own saint: such as the Exaltation (*Krstovdan*), the Intercession (*Pokrovitsa*), the Ascension (*Spasovdan*), as well as the Candlemas of the Lord (*Sretenie*), the New Year, the Assumption of the Virgin and others (Nedeljković 1990, p. 317). In essence, the same meaning is given to votive holidays among Russians, only they are not generic, but individual or rural and sometimes occasional in nature.

On the further path of personification of the holidays, new "saints" may even be created, who are venerated to the same degree as Christian saints. This is how the Bulgarians acquired a saint named *Mladen*, who owes his origin to the feast of the Forty Martyrs

(March 9), which in some Bulgarian regions is called *Mladenci* (Young men). People turn to the *Mladen* as well as to the traditional saints with prayers and requests: "Saint Mladen, rejuvenate us with God's help". Among Bulgarians, however, the most venerated saint is St. Theodor (who united the cult of Theodor Stratilat and Theodor Tyron), and he is a figure as sacred as he is demonic. St. Theodor's Memorial Day (*Todor*) is considered the first Saturday of Lent. Researchers of folk tradition pay attention to the fact that this saint, with the usual male image, is sometimes (in the northern Bulgarian regions) given the female name *Todoritsa*, *Tudurichka*, *Baba Tuduritsa* and is revered as a female saint. In this we can see traces of the androgyny of the pre-Christian pagan deity (Popov 1994, p. 84). I believe that there may rather be calendar and linguistic reasons for turning a male *Todor* into a female *Todorica*. In the folk calendar, the Slavic nouns for the first week of Lent (*Todorova nedelja*) and the Saturday of that week (*Todorova subota*) are feminine in gender (*nedelja, subota*). Thus, the names *Tudorica* and *Baba Tudorica* simply personalise these feminine-gender units of calendar time.

The same can be said about such calendar characters as *Baba Marta* (personification of the first day of spring, March 1, or the entire month of March) and *Baba Korizma* among the Southern Slavs. In Slavic Catholic traditions *Post* is the personification of the period of Great Lent. Among the Bulgarians, *Baba Marta* is thought to be a wayward and capricious old woman who does not tolerate other old women, who, in consequence, must try not to leave the house on her Saint's Day. Her character is explained by the changeable weather at this season of the year. *Baba Martha*'s relatives are considered to be the personified months of January (*Golyam Sechko*) and February (*Malak Sechko*). In relation to them she is either a sister or a spouse. As a punishment, she can send strong cold winds or blizzards (BM 1994, pp. 17–18). In Serbian calendar beliefs, *Baba Korizma* is endowed with all the features of a mythological (demonic) character: she is an old, skinny, evil and cruel old woman who menacingly drags chains and carries seven sticks on her shoulder, symbolizing the seven weeks of Great Lent (Nedeljković 1990, p. 8).

Another striking example of the personification and anthropomorphising of holidays in the Slavic folk calendar is the so-called semantic model of kinship. In popular ideas reflected in calendar terminology, paremiology and folklore, holidays and calendar days are related to each other: father or mother, brother, niece, etc. In terms of such a family-generic model, the semantic association between calendar units that are chronologically contiguous or symmetrical in the annual cycle becomes one of the ways to structure annual time. (For more information, see Zaykovsky 1994; Tolstaya 2005, p. 12). Family relations link not only one holiday to another, but other units of calendar time (months, seasons): for example, in Belorussian folklore "November is the grandson of September and the son of October and winter's own father" (Lozka 1993, p. 176).

Thus, the personification of holidays and other parts of the annual cycle derive from two different sources. First, it is the book-written, ecclesiastical tradition of designating and interpreting holidays in connection with the names and images of Christian saints which gives the calendar characters and personified holidays a sacred status and makes them objects of veneration together with the canonical saints. Secondly, it is an oral, folklore tradition that involves calendar time in the sphere of mythological representations. This tradition gives them demonological or semi-demonological features which brings them closer to characters of a completely different type—to representatives of lower mythology. However, the Christian saints themselves are also subject to mythologization in the oral tradition see (Tolstaya 1995).

2. Sacred and Magical in the Folk Cult of Saints

It has been said and written many times that in the popular consciousness of the Slavs, Christian saints replaced pagan gods, taking their functions and their place in the pantheon. For example, St. Elijah replaced the god of thunder Perun, St. Nicholas replaced the animal god Veles-Volos and St. Paraskeva took the place of the goddess Mokosh. That such replacements were possible seemed to compensate for the loss of polytheism and smoothed the sharpness of the transition from one cultural model to another. Over time, all the Christian saints were more or less included in the system of folk ideas, beliefs and rituals. They adapted to it and were ultimately mastered by it. In the process of this adaptation, the way in which their images were presented in the lives of the saints and other apocryphal texts changed radically. To a certain extent, the images of God, the Virgin and other evangelical characters seen in folk Christianity have also undergone significant changes.

The events and deeds that justified the canonization of the saints and the circumstances of their lives and of their death for the faith live on only in church texts or similar genres. In the oral peasant culture, the popular stories, legends and songs about these saints ceased to be significant and disappeared from memory even though their names still figured in the national calendar. The saints themselves are often relegated to the level of mythological, semi-demonic or completely demonic characters, and the sacred texts of prayers often mix with and dissolve into primitive magical spells.

The main direction of change and transformation that Christian saints have undergone in oral folk culture has been their mythologization and their endowment with supernatural magical abilities that give them power over natural phenomena and human life.

Following the pagan model, saints have undergone a specialization: each of them is assigned certain areas of nature such as rain, hail, wind, animals and birds or certain human activities such as breeding and farming, beekeeping, spinning and weaving and folk medicine. The archaic worldview of the ancient Slavs was based on the obvious desacralization of the saints, their separation from the higher divine sphere and their joining with the true holders of power over earthly life—the ancestors and dead relatives. The saints are perceived as belonging to the "other world" joined with the deified ancestors and with the characters of lower mythology (evil spirits).

At the same time, Saints descend to the ground, walk through fields and roads, enter houses, and punish people for violating norms and prohibitions. For example, Paraskeva-Friday spins at night on an abandoned spinning wheel, mixes up the yarn or breaks the tools of any woman seen spinning at an untimely time or on a holiday. The saints are asked for help in all aspects of life through prayer or with charms. People make them vows, offer them sacrifices and gifts and take offense with them when their requests are not granted. They even punish the saints; for example, by taking their icon and scratching out the eyes, turning it to the wall, breaking it or dropping it in water. Communications with the saints are not much different from those with mythological characters such as a brownie, a house goblin, a water-sprite or mermaid.

All collections of charms contain direct appeals to the saints for help: for example, in the Serbian charm "Lord, help, and Saints Kuzma and Demian, help, and Saint Ilya, and Saint Basil" (Radenković 1982, p. 209, No. 343), or in the Northern Russian Archangel charm: "Kuzma-Demyan, the god of artisans and selfless adherent of Christ, come to me and my little sister and help stop the blood, so that the 'meat' (body) does not hurt and the bones do not ache. From now on and forever and ever. Amen" (LLZ 1992, p. 23, No. 20). However, more often the figure of the saint acts as a wonderful fairy helper and a defender from evil forces. For example, in the Archangelsk charm against fever, "Saints Kuzma and Demyan meet twelve shaggy, hairy and girdle-less virgins, take out sharp sabres and prepare to take their heads off" (LLZ 1992, p. 42, No. 72). Compared to the Serbian charm: "Saint Tomas, bring the golden cart, burn the *vilas* and *veshtitsas* and all the other wicked" (Radenković 1982, p. 46, No. 56).

The saints serve as sacral prototypes, so that, when magical actions are performed, people turn to the saints to give their actions a sacred status and make them successful. For example, "Mother Most Holy Virgin, Queen of heaven, you washed your son in a steam bath and I (name) brought you water" (LLZ 1992, p. 39, No. 66). Finally, the names of the saints in the charms can curse evil power and disease: for example, "Come out, I conjure you by Saint Jacob, God's brother" (Radenković 1982, p. 211). Saints are also often presented with gifts and sacrifices; for example, there is a widely known custom of leaving ears of grain on the field at the end of the harvest. This is called the beard of Christ or Volos, or Ilya, or Nikola or Kuzma-Demyan. In various folklore formulas (ritual sentences, incantations, etc.), saints often replace each other. For example, when girls seek for their fortunes "Andrew, Andrew, I'm sowing hemp on you (i.e., on your day). Let me know who I will be sleeping with". The name of Andrew can be substituted with other saints who are venerated in the calendar such as St. Basil, Alex and Makovey (often depending on the day of divination). However, there are many cases when sacred characters or saints are united with characters of a completely different level, with lower-level mythological characters or demons. For example, invitations to Christmas dinner may be addressed not only to God, the Mother of God, All Saints and St. John the Baptist, Nikola, Herman, etc., but also to the personified forces of nature (frost, wind, cloud). This example is found in Polesie: "Frost, frost, go to eat *kutia* (a Christmas dish)!" Invitations can also be offered to animals and birds, dead ancestors and even evil spirits (the witch, the brownie, the devil or demons who control the clouds) (Vinogradova and Tolstaya 2005).

In the texts of Serbian incantations directed against hail clouds, the names of the saints are invoked equally with the names of the dead (hanged or drowned), who, according to popular beliefs, control the clouds: for example, "Don't you dare come here, Saint John! Don't you dare, I beg you!" (Tolstoy and Tolstaya 1981, p. 64). In another prayer-incantation, there is an appeal to Saint Sava, who has become seen as the lord of the heavenly cattle (the hail clouds): "Saint Sava, drive away your cows from our village..." (ibid.). The name of a real drowned or hanged person from a given village can also be used in a prayer-incantation for the same purpose: for example. "Oh, Radojka Zimonia! Drive away your cows, keep them out of our fields. Don't you dare, don't you dare, don't you dare!" (ibid., p. 74) or: "Milija, drowned woman, burn your cattle, drive them away! Let them go to the mountains and to the water, where nothing grows" (ibid., p. 77); and "Drowned and hanged, drive your cattle there, do not come here" (Radenković 1982, p. 363, No. 588). This kind of parallelism, the isofunctionality of the characters of the sacred and the mythological, confirms the way in which popular culture assimilates Christian motifs and images—the process of mythologization.

In Russia, the cult of St. Nicholas (commonly called *Nikola*) provides a striking example of mythologization. As Uspensky (1982) showed, the veneration of Nikola can be attributed to the cult of the pagan god Veles–the patron saint of cattle and the opponent of the thunder god Perun. In his book, there are many examples of Nikola being seen not only as a pagan god, but also as a character of lower mythology, revered at the same level as the goblin or even the bear. A similar transformation takes place among the Serbs in the widely revered St. Sava, who in popular belief becomes the patron saint of cattle, their protector from wolves and, at the same time, the wolf master. On his day in December, St. Sava climbs on a pear tree, summons the wolves, entertains them and indicates whose cattle will be their prey for the next year (SMR et al. 1998, p. 263).

Another example of this kind is the cult of St. Herman among the Balkan Slavs, especially in Bulgaria and Macedonia. Saint Herman is considered as the lord of clouds, thunderclouds, rain and hail. In many areas, he was called *gradushkar* (from the word *grad* 'hail' or the *cloud*). This character has already lost the features of a Christian saint and has acquired the status of a natural spirit or a deity of the elements. The only feature that connects him with the Christian belief system is the timing of the rituals, which are performed on St. Herman's Day (May 12). However, these rites are already absolutely pagan in nature. Participants make a clay doll named Herman, then symbolically bury it and mourn. The ritual is perfected during a drought to cause rain, or prophylactically on St. Herman's Day to prevent a summer drought. When people see that a thunderstorm or hail cloud is approaching, they go out of the house and address the cloud, calling it Herman, with the words: "Go away, Herman, there is no place for you here! Go to the desert, Herman, where the meadows are not mowed, Herman, where the wheat is not gathered from the field, Herman, where the vineyard is not harvested, Herman, where the young men are not married, Herman!" (Radenković 1982, p. 368, No. 608).

Thus, in the perception of Christian saints, there is often no trace left of the veneration for either the traditional religious or moral aspects of their holiness. It seems that the cult of saints can be transformed through reduction and desacralization with the ideal of Holiness eliminated, but it can also remain part of a dual faith, in which one aspect is a dominant Christian culture, where the ideal of Holiness has not only been retained but further developed in both the religious and moral senses (Toporov 1995).

Life-cycle rites and everyday practical activities were noticeably less influenced by Christian culture. Here, Christian elements were used almost exclusively as a means of sacralising various ritual acts of a magical (productive or protective) nature and most often did not influence their deep (mythological) content. Examples of this include priests leading processions with icons and banners around fields, saying prayers, sprinkling water to bring on rain and praying for the harvest. Similarly, the objects of Christian worship and even the clergy themselves often became the object or instrument of magical actions; for example, when people girded the church with a long ribbon of cloth, drank water from icons, poured water on the priest during a drought or forced the priest to roll around in the field to make the crops grow better.

3. Folk Ideology and Axiology

Understanding Christian concepts was often facilitated by the presence in the Slavic languages and collective consciousness of similar concepts and values inherited from the pre-Christian era. For example, the concept of holiness, which we perceive exclusively as an element of the Christian worldview and the Christian value system, was actually a most important pre-Christian concept. As V. N. Toporov showed in his works, many texts of Ancient Russian literature, especially those about the first Russian saints Boris and Gleb and Theodosius of Pechersk, contain a pre-Christian substratum of holiness as the highest spiritual value. Etymological analysis of words with the root *svęt-* in Slavic languages and their comparison with related words of other Indo-European languages (above all with the Sanskrit word *śvanta*, found in the Rig Veda and derived from a verb meaning 'to swell, or grow') shows that the original, pre-Christian definition of this word was associated with a fairly specific meaning of swelling, growth or fertility, which was then abstracted and extended to the spiritual sphere and sacralised. This semantic "elevation" determined the further evolution of this word towards such new meanings as righteousness, purity and virtue (Toporov 1987).

We can also be confident that another key concept of Christian culture—the concept of sin—also has ancient pre-Christian origins. Of course, the content of these conceptual and axiological categories of folk culture could not stay the same and would have changed under the influence of the Christian faith. At the same time, the Christian concept was not only assimilated, but also underwent serious semantic transformations.

Sin is one of the main concepts of Christian ethics. The theory of sin is a specially developed component of the Christian doctrine with its own concepts, arguments, motives and texts, etc. In the oral folk tradition, there are no texts discussing the concept of sin. Such texts are found only in the so-called spiritual verses, and this genre derives from a mixture of folk and church sources.

This, however, does not mean that the concept of sin is generally alien to the conceptual sphere of folk culture. It is not just the result of a mechanical assimilation of the Christian concept or a simple cultural borrowing. There is every reason to believe that, as in the case of holiness, there was a certain "correlate" of the Christian concept of sin in the pre-Christian world view. It is clear that it was one of the most important concepts of pre-Christian culture, which, after the adoption of Christianity, underwent a significant transformation. In the same way, the Christian concept underwent profound changes as the adaptation to folk beliefs developed.

Thus, what can be called the folk concept of sin is a complex mix of religious (Christian) and mythological (pre-Christian, pagan) ideas, elements of oral and book culture, postulates of popular law and of traditional value systems. The main parameters and coordinates of this concept are not given to us in pure form, they can only be reconstructed from a variety of indirect data: from the evidence of language (vocabulary, phraseology, paremiology); from customs, rituals, prohibitions and prescriptions (and especially from explanations of the reasons behind them); from stories, legends, songs and narratives about the consequences of sins that tell of the posthumous fate of sinners and the righteous and about the atonement of sins; and from other diverse folklore texts that discuss the motives of sin and sinners.

While the Christian interpretation of sin is "any, deviation by act or word and even thought from the commandments of God (either free and conscious or forced and unconscious) and a violation of the law of God" (Khristianstvo 1993, p. 430), the semantic boundaries of the Slavic word of sin in its literary and dialectal use are much wider. Sin is the violation of any law or rules or regulations, whether it be the law of God, the law of nature or popular laws or behavioural norms (Tolstaya 2000).

According to popular beliefs, every action or act of a person receives a certain "response" from the world, nature and the cosmos, and every act or action that violates the "norm" (i.e., the existing unspoken prohibitions) also violates the balance of the world and threatens negative consequences for the world and for man. In these parameters, the folk concept of sin is formed. Sin is anything that contradicts the norm and triggers a punishing reaction by nature (or a higher power). In the folk tradition, the composition of sins and their ranking in strength differs from measures of the severity of sin in Christian thought. Note, for example: the notion of mortal sin; the notion of the seven deadly sins; and punishment, in accordance with the severity of the offense often set out in quantitative terms (such as the obligation to bow down to the ground forty or a hundred times and observe a certain number days of fasting). According to the Polish researcher of Polesie, Czesław Pietkiewicz, "sin in the concept of a resident of Polesie is not a simple question. He keeps in mind an extensive list of sins, but he ranks them himself: a great sin, simple sin and small sin. He considers many acts sinful, including some that are not recognized by the church. Among the great sins he includes: murder for the purpose of robbery or for the sake of revenge, mutilation, disrespect for the cross, perjury, arson and the theft of bees. At the same time, the murder of a horse thief, a bee thief, or an arsonist is not considered a great sin, not even a small sin, which is the term used to define the most minor offenses" (Pietkiewicz 1938, pp. 271–73).

According to folk ethics, almost every sin can lead to the most serious consequences for a person, for society, for the whole world or for the cosmos. For example, hail can be a punishment for what we would see as incomparable crimes: on the one hand, the birth of a bastard or its murder and, on the other, the chopping of firewood at Christmas. However, in some folklore texts, there is an idea of sin that is particularly close to the tradition of the Church. So, in spiritual verses, according to G. P. Fedotov, the most terrible sins are considered: 1. sins against mother earth and the religion of genus; 2. sins against the ritual law of the church; and 3. sins against the Christian law of love (Fedotov 1991, pp. 84–86).

In those areas of culture that have not experienced the direct influence of Christian ideology, the gravest sins are the crimes against the "law of genus", such as murder and disrespect of parents, incest and marriage between godchildren and adoptive parents. In Slavic folklore, the motif of incest of a brother and sister and the disastrous consequences of this terrible sin is widely known. This is either the death of the sinners themselves or the "punishment" of nature: for example, rivers can dry up, forests can wither and animals run away.

Often in this respect, the folk tradition is stricter than the church tradition. For example, in some areas of Serbia, marriages are prohibited between members of the family of up to the sixth generation. They are also sometimes prohibited between families bearing the same surname or having the same patron saint, between persons having the same godfather, between persons related by a twinship and their descendants, between foster brothers and sisters, etc. In fact, disregard of ritual kinship can be considered as an even more serious sin than incest. The consequences of violating these prohibitions may concern not only the perpetrators themselves (their marriage will not be successful, they will face death, they will have sick or freak children, etc.), but also their entire house and family (the family may die out, the house will burn down from lightning, etc.).

The bride who lost her virginity before marriage is seen as a great danger to society and nature. All Slavs know a variety of ritual forms that establish the "honesty" or "dishonesty" of the bride during the wedding night, as well as a system of punishments for the non-virginal bride such as a ritual scolding that is usually addressed to the bride's mother. In addition, measures are taken to protect the fields from the eyes of such a bride. When she is taken out of the house, she is blindfolded, so that her "sinful" look does not make the fields barren. The voice of such a bride is capable of killing cattle and so she is forbidden to use her voice outside of the house (Tolstaya 1996).

The most common and strictly observed regulations among the Slavs are temporal (calendar, daily, lunar) in nature: notably, bans on working on holidays, especially at Christmas, Easter and the Annunciation. The severity of these prohibitions was explained not so much by the veneration of these holidays as by the fear of them. According to popular beliefs, the penalty for violating such prohibitions could be the birth of ugly offspring in both humans and cattle. Ukrainians believed that freaks would be born when a husband and wife had sexual intercourse on solemn holidays. People were sure that if a pregnant woman painted a hut before the holiday, her child would be born blind, as she will "paint over his eyes". If the husband cut wood on a holiday, the child would be born a cripple (Chubinsky 1872, p. 18). In Polesie (Oniskovichi Brest region), children conceived on memorial days would be born deaf (PA n.d.).

An interesting feature of the folk normative system is the idea that a sin can be atoned for by special magical actions that will enable the avoidance or mitigation of the prescribed punishment. For example, if during the Christmas holidays a woman had to sew or spin despite the ban on doing so, then she could "transfer" this sin to an old broom and after the holidays she should cut the broom into pieces on the doorstep or on the stove door with the words: "What you knitted—let be unknitted, what you sewed, let be unsewed, what you whipped—let be unwhipped" (PA n.d.).

While in customary law the punishing force is society (often through its authoritative representatives or the elders of the family), in Christian law, the punishment comes from God. As the examples cited above have shown, in folk religion, the punitive functions are attributed to an unpersonalised higher power or to nature itself. The object of punishment in folk religion is not the sinner, as in common law, but society as a whole or even the whole world. Of course, the criminal can be punished but the punishment in mythological law is very different to the one in the common law. Mythological punishment is always understood as the intervention of otherworldly forces as is evidenced in the following examples.

The South Slavs adopted various forms of social punishment for those who committed grave sins such as murder, theft, arson, violence and adultery. In these cases of serious crime, the Serbs would gather at the edge of the village and call a curse on the head of the guilty person, regardless of whether he was known or not. For this action, a holiday was chosen, and the more significant the holiday, the greater the effect of the ritual was expected to be. While uttering curses and incantations, people would throw stones, clods of earth and sticks to make a large pile and everyone passing by would have to throw their stone on it and utter a curse. This rite was called anathema or *prokletije* or *kamenovanje* (SD 1995–2012, vol. 1, pp. 106–7).

Punishment in the mythological law would be quite different. There is a well-known belief among Slavs that someone who has committed a grave sin can turn into stone. There are many stories about some specific stone boulders formed from people turned into stone for their sins. According to a Belarusian legend, in the beginning of the world brothers-in-law lived with each other, sisters lived with brothers, daughters with fathers, mothers with sons, godfathers with godmothers not knowing that it was a sin. Then God began to turn them into stones as a punishment. Once, a godfather and a godmother were traveling on epiphany and were tempted to sin along the way, and at that moment they were turned into stones. From that time on, people began to listen to God and began to marry strangers with strangers (Federowski 1897, vol. 209, N 861). This legend is interesting because it explains the very origin of the concept of sin, and, as we see, the punishment for it precedes the realization of sin.

In Polesie, a drought might be seen as a punishment if one of the villagers had touched the ground before the Annunciation (for example, in digging holes for fence posts), or a woman had baked bread. To stop the drought and bring on the rain, it was necessary to break the fence set at the "wrong" time or pour water on the woman who had violated the ban on working. Another reason for the drought was often considered to be that "unclean" dead (suicides or especially a hanged person) had been buried in "clean", consecrated land in the cemetery. Therefore, to bring on rain, it was necessary to destroy the grave of this person or at least tear up a cross from it and sometimes even dig up the deceased and throw the corpse into the water.

Thus, mythological morality, in contrast to social and religious morality, is based on the idea that there is a direct connection between human behaviour and the state of the cosmos. The mechanism for regulating the balance between man and the world is a system of prohibitions, the violation of which is recognized as sin and is punished by natural disasters, death, disease and human misfortunes. The arbiter of judgment and punishment is nature itself or the higher powers. A person can only appeal to these forces, uttering curses and calling for punishment on the head of the guilty. Curses, although uttered by humans, are a mythological mechanism, as evidenced by widespread legends about the miraculous and literal execution of curses (for example, a mother in her heart says to a child: "May the goblin take you!", and the goblin takes the child).

4. Christian Motifs in the Oral Tradition

The Slavic folk tradition adapted not only individual elements, structures, and semantic categories of Christianity, but also ready-made "texts" (whole or parts), plots, motifs and themes developed in various folklore genres. Intermediaries connecting the canonical Christian text and the folklore text included the apocryphal literary tradition and a certain fund of representations, plots and images already prepared in the spirit of the folk tradition and adapted to it. This forms the "textual" (narrative) basis of what is sometimes called folk Christianity (or folk Orthodoxy).

In its full form, of course, this "textual basis" is not recorded anywhere. After a long break, however, attempts are once again being made to reconstruct it from the material of different Slavic traditions. Consequently, the subject of special (including field) research is the topic of the Slavic "folklore bible" (Bulgarian, Polesian, Russian, Belarusian) (Badalanova 1999; Kuznetsova 1998; Ot Bytiya k Iskhodu 1998; Belova 2004; BNB 2010).

It is from this "secondary" source (folk Christianity) that various folklore genres draw specific biblical (evangelical) plots and motifs in the creation of new folklore texts. Different genres deviate to varying degrees from the direct presentation of biblical events. Whereas spiritual poems tend to preserve both the event outline and the ideological meaning of the source, legends handle it much more freely. Other genre forms, charms, for example, only incorporate individual Christian motifs and symbols.

Once in the folklore context, the Christian element is included in the semantic field of another cultural and symbolic language and inevitably undergoes changes, losing some meanings and connotations and acquiring others. This is well illustrated by the example of a folk song based on the Bible story of the meeting of Christ with a girl at a well. Christ asks for water to drink and the girl refuses him, saying that her water is unclean. More than a century ago, A. N. Veselovsky wrote about this song as a reworking of the gospel story about the Samaritan Woman (John 4). Hepointed to the Slavic versions known to him and to some European parallels based on the same the gospel text, but nevertheless very far from the Slavic versions (Veselovsky 1877).

To date, this story has been recorded among the Eastern Slavs (mainly in Belarus and in Polesie) and in the Carpathians. Some examples have been found among the Western Slavs in southeastern Poland and some isolated versions in other Polish regions and in Moravia and Lusatia (about fifty versions in total). No traces of the song have been found in South Slavic folklore. The text is practically unknown in the territory of Greater Russia.

Developing this gospel theme, Slavic folklore follows the Bible story of the meeting of the Lord with a woman at the well quite closely. However, while their conversation stays close to the Bible story, its meaning is very different. The parable of the "living water" of faith turns into a ballad about the punishment of a sinful girl who killed her illegitimate children. In some versions, we find some development of the gospel text and the account of the Samaritan woman's illegitimate husbands: "Jesus said to her, 'You are right when you say you have no husband. The fact is, you have had five husbands, and the man you now have is not your husband" (John 4: pp. 17–18). The ballad also develops the motif of the Samaritan woman's recognition of God: "The woman says to Him, 'Lord! 'I can see that you are a prophet'." John 4: p. 19 (see Tolstaya 1999).

A comparison of the Belarusian, Polesian, Polish, Moravian and Lusatian versions reveals many similarities and differences. They relate primarily to additional motifs and storylines related to such topics as sin, atonement for sin, punishment and salvation of the soul. As for the nuclear part of the text, the variation is much less noticeable. The development of this theme in the Polesian song fully corresponds to the oral folklore tradition, in which all sin (and infanticide in particular) is understood as a mythological and "human" category, and punishment as a reaction of "cosmic force". There is a widespread belief among the Slavs that an illegitimate and murdered child causes torrential rains, hail and natural disasters.

In summary, we can say that the pre-Christian folk tradition of the Slavs did not disappear under the influence of the stronger system of Christian culture. It was able to assimilate many Christian concepts, symbols and texts, translate them into its own language and fill them with its own content. However, in relation to some most important ideas, the folk tradition turned out to be extremely stable and closed. In fact, the cult of the dead and the lower mythology, which form the ideological basis of the mythological (mythopoetic) cultural paradigm, remained untouched by Christian influence.

Funding: This research received no external funding.

Institutional Review Board Statement: Not applicable.

Informed Consent Statement: Not applicable.

Acknowledgments: I would like to express my deep gratitude to Michael Blackwell for his valuable comments and editing of the English version of the article.

Conflicts of Interest: The author declares no conflict of interest.

Notes

[1] Apple Spas is The Transfiguration of the Lord (August 6/19).

[2] Semik is the Thursday before Trinity.

[3] Polesie–border region between Belarus, Ukraine and Russia.

References
Archival Sources

Бадаланова 1999–*Бадаланова Ф.* Болгарская фольклорная Библия//Живая старина. 1999. № 2. С. 5–7.

Белова 2004–«Народная Библия": Восточнославянские этиологические легенды/Составление и комментарии О.В. Беловой. Москва, 2004.

БМ–Българска митология. Енциклопедичен речник/СъставителА. Стойнев. София, 1994.

БНБ–Беларуская«народная» Бiблiя» ў сучасных запiсах/Уступ. артыкул, уклад. i камент. А.М. Боганевай. Мiнск, 2010.

Веселовский1877–*ВеселовскийА*.Н. Опыты по истории развития христианской легенды. II. Берта, Анастасия иПятница // Журнал Министерства народного просвещения. С.-Петербург. С. 186–252.

Виноградова, Толстая2005–*Виноградова* Л.Н., *Толстая*С.М. Ритуальные приглашения мифологических персонажей на рождественский ужин. Формула иобряд//ТолстаяС.М. Полесский народный календарь. Москва, 2005. С. 443–500.

ВП – Валачобныя песні. Мінск, 1980.

Зайковский1994–*Зайковский*В.Б. Народный календарь восточных славян//Этнографическоеобозрение. Москва, 1994. № 4. С. 53–65.

Зализняк1995–*Зализняк*А.А. Древненовгородский диалект. Москва, 1995.

Замовы–Замовы. Мінск, 1992 [Серия: Беларуская народная творчасць].

Зечевић1981–*Зечевић*С. Митска бића српских предања. Београд, 1981.

Кузнецова 1998–*Кузнецова* В.С. Дуалистические легендыо сотворении мира в восточнославянской фольклорной традиции. Новосибирск, 1998.

Лозка 1993–*Лозка* А. Беларускі народны каляндар. Мінск, 1993.

ЛЛЗ–Встану я благославясь … Лечебные и любовные заговоры, записанные в частиАрхангельскойобл. / Изд. подготовлено Ю.И. Смирновым иВ.Н. Ильинской. Москва, 1992.

Мороз1998–*Мороз*А.Б. Семантика, символика и структура сербскихобрядовых песен календарного цикла. Кандидат. диссертация. Москва, 1998.

Недељковић1990–*Недељковић*М. Годишњиобичаји уСрба. Београд, 1990.

Номис1864–Українські приказки, прислівья и таке инше/Спорудив М. Номис. Санкт-Петербург, 1864.

От Бытия к Исходу1998 – От Бытия к Исходу. Отражение библейских сюжетов в славянской и еврейской народной культуре. Москва, 1998.

ПА–Полесскийархив Института славяноведения РАН(Москва).

Попов1994–*Попов* Р. Светци демони// Етнографски проблеми на народната духовна култура. София, 1994. Т.2. С. 78–100.

Раденковић1982–*Раденковић*Љ. Народне басме и бајања. Ниш; Приштина; Крагујевац, 1982.

СД–Славянские древности. Этнолингвистический словарь/ Подобщей ред. Н.И. Толстого. Москва, 1995–2012. Т. 1–5.

СМР–*Кулишић*Ш., *Пантелић*Н., *Петровић*П.Ж. Српски митолошки речник. Београд, 1998. Изд. 2.

Срезневский1893–1903 – Срезневский И.И. Материалы для словаря древнерусского языка. Санкт-Петербург. Т. 1–3.

Толстая1995–*Толстая*С.М. Сакральное и магическое в народном культе святых//Folklor–Sacrum–Religia/ Red. J. Bartmiński i M. Jasińska-Wojtkowska. Lublin, 1995. pp. 38–46.

Толстая1996 –*Толстая*С.М. Символика девственности в полесском свадебномобряде// Секс и эротика в русской традиционной культуре/ Ред. А Л.Топорков. Москва: Ладомир, С. 192–206.

Толстая1997–*Толстая*С.М. Мифология иаксиология времени в славянской народной культуре//Культура и история. славянский мир. Москва, 1997. С. 62–79.

Толстая1999–*Толстая*С.М. Самарянка: баллада о грешной девушке в восточно- и западнославянской традиции//W zwierciadle języka i kultury/Red. J. Adamowski i S. Niebrzegowska. Lublin, 1999. pp. 58–81.

Толстая2000–*Толстая*С.М. Грех в свете славянской мифологии//Идея греха в славянской и еврейской культурной традиции. Москва, 2000. С. 9–43.

Толстая2005–*Толстая*С.М. Полесский народный календарь. Москва, 2005.

Толстая2011–Сакральное и магическое в славянском народном календаре//Annales Universitatis Paedagogicae Cracoviensis. Studia Russologica IV. Europa Słowian w świetle socjo- i etnolingwistyki. Przeszłość–teraźniejszość/Red. E. Książek i M. Wojtyła-Świerzowska. Kraków, 2011. pp. 252–58.

Толстые1981–*Толстые* Н.И. иС.М. Заметки по славянскому язычеству. 5. Защита от града в Драгачево и других сербских зонах//Славянский и балканский фольклор. Обряд. Текст. Москва, 1981. С. 44–120.

Топоров1987–*Топоров*В.Н. Ободномархаическом индоевропейском элементе в древнерусской духовной культуре–SVĘT//Языки культуры и проблемы переводимости. Москва, 1987. С. 184–252.

Топоров1995–*В.Н.Топоров*. Святость и святые в русской духовной культуре. Москва, 1995. Т. 1.

Требјешанин1991–Требјешанин Ж. Представа о детету у српској култури. Београд, 1991.

Успенский1982–Успенский Б.*А.* Филологические разыскания вобласти славянских древностей. Москва, 1982.

Федотов1991–*Федотов*Г. Стихи духовные. Русская народная вера по духовным стихам. Москва, 1991.

Христианство 1993–Христианство. Энциклопедический словарь. Москва, 1993. Т. 1.

Чубинский1872–Труды этнографическо-статистической экспедиции в Западно-русский край, собранныеП.П.Чубинским. Санкт-Петербург, 1872. Т. 4.

Federowski 1897–*Federowski* M. Lud białoruski na Rusi Litewskiej. Kraków, 1897. T. 1.

Pietkiewicz 1938–*Pietkiewicz* Cz. Kultura duchowa Polesia Rzeczyckiego. Materiały etnograficzne. Warszawa, 1938.

Published Sources

Badalanova, Florentina. 1999. Bolgarskaya fol'klornaya bibliya. *Zhivaya starina* 2: 5–7.

Belova. 2004. Narodnaya bibliya. In *Vostochnoslavyanskie Etiologocheskie Legendy*. Edited by O. V. Belova. Moskva: Indrik.

BM. 1994. Bulgarska mitologiya. In *Enciklopedichen rechnik*. Edited by A. Stoynev. Sofia: Logis.

BNB. 2010. *Belaruskaya narodnaya bibliya u suchasnykh zapisakh*. Edited by A. Boganeva. Minsk: Bel. navuka.

Chubinsky. 1872. *Trudy Etnografichesko-Statisticheskoy Ekspedicii v Zapadno-Russkiy kray, Sobrannye P.P. Chubinskim*. Sankt-Peterburg: RGO.

Federowski, Michał. 1897. *Lud białoruski na Rusi Litewskiej*. Kraków: Akad. Umiejetnosci. Kraków. T. 1.

Fedotov, Georgy. 1991. Stikhi dukhovnye. In *Russkaya narodnaya vera po dukhovnym stikham*. Moskva: Gnosis.

Khristianstvo. 1993. *Enciklopedicheskiy slovar'*. Moskva Ros: Enciklopedia. T. 1.

Kuznetsova, Vera S. 1998. *Dualisticheskie legendy o sotvorenii mira v vostochnoslavyanskoy folklornoy tradicii*. Novosibirsk: RAN.

LLZ. 1992. *Vstanu ya blagoslavvyas' . . . Lechebnye i lyubovnye zagovory, zapisannye v chasti Arkhangel'skoy obl. Izd. Yu.I. Smirnov, V.N. Il'inskaya*. Moskva: Russkiy mir.

Lozka, Ales. 1993. *Belaruski narodny kalyandar*. Minsk: Polymya.

Moroz, Andrey B. 1998. Semantika, simvolika i struktura serbskikh obryadovykh pesen kalendarnogo cikla. In *Dissertaciya*. Moskva: MGU.

Nedeljković, Mile. 1990. *Godišni običaji u Srba*. Beograd: Vuk Karadic.

Nomis, Matviy. 1864. *Ukrajin'ski prikazki, prisliv'ja i take inshe*. Sankt-Peterburg.

Ot Bytiya k Iskhodu. 1998. Ot Bytiya k Iskhodu. In *Otrazhenie biblejskikh siuzhetov v slavianskoy i evrejskoy narodnoy kulture*. Moskva: Institute for Slavic Studies RAN.

PA. n.d. *Polesskiy arkhiv Instituta slectavyanovedeniya RAN*. Moskva: Indrik.

Pietkiewicz, Czesław Z. 1938. Kultura duchowa Polesia Rzeczyckiego. In *Materia y etnograficzne*. Warszawa: Polska Akademia Umiejetnosci.

Popov, Rachko. 1994. Svetci demoni. In *Etnografski problemi na narodnata dukhovna kultura*. Sofia: Klub. T. 2, pp. 78–100.

Radenković, Ljubinko. 1982. *Narodne basme i bajania*. Niš–Priština–Kragujevac: Gradina-Jedinstvo-Svetlost.

SD. 1995–2012. *Slavyanskie drevnosti 1–5. Slavyanskie drevnosti. Etnolingvisticheskiy slovar'*. Edited by N. I. Tolstoy. Moskva: Mezhdunarodnye otnosheniya. T. 1–5.

SMR, Kulišić, N. Pantelić, and P. Ž Petrović. 1998. *Srpski mitološki rečnik*. Beograd: Interprint. Izd. 2.

Tolstaya, Svetlana M. 1995. Sakral'noe i magicheskoe v narodnom kulte svyatykh. In *Folklor–Sacrum–Religia. Red*. J. Bartmiński i M. Jasińska-Wojtkowska. Lublin: UMCS. S, pp. 38–46.

Tolstaya, Svetlana M. 1996. Simvolika devstvennosti v polesskom svadebnom obryade. In *Seks i erotika v russkoy tradicionnoy kul'ture*. Moskva: Ladomir, pp. 192–206.

Tolstaya, Svetlana M. 1997. Mifologiya i aksiologiya vremeni v slavyanskoy narodnoy culture. In *Kultura i istoriya. Slavyanskiy mir*. Moskva: Indrik, pp. 62–79.

Tolstaya, Svetlana M. 1999. Samaryanka: Ballada o greshnoy devushke v vostochno- i zapadnoslavyanskoy tradicii. In *W zwierciadle języka i kultury/Red*. J. Adamowski i S. Niebrzegowska. Lublin: UMCS, pp. 58–81.

Tolstaya, Svetlana M. 2000. Grekh v svete slavyanskoy mifologii. In *Ideya grekha v slavyanskoy i evreyskoy kul'turnoy tradicii*. Moskva: Institute for Slavic Studies RAN, pp. 9–43.

Tolstaya, Svetlana M. 2005. *Polesskiy narodnyj kalendar'*. Moskva: Indrik.

Tolstaya, Svetlana M. 2011. Sakral'noe i magicheskoe v slavyanskom narodnom kalendare. In *Annales Universitatis Paedagogicae Cracoviensis. Studia Russologica IV. Europa Słowian w świetle socjo-i etnolingwistyki. Przeszłość–teraźniejszość/Red*. Kraków: E. Książek i M. Wojtyła-Świerzowska, pp. 252–58.

Tolstoy, Nikita I., and Svetlana M. Tolstaya. 1981. Zametki po slavyanskomu yazychestvu. 5. Zashchita ot grada v Dragachevo i drugikh serbskikh zonakh. In *Slavyanskiy i balkanskiy folklore. Obryad. Tekst*. Moskva: Nauka, pp. 44–120.

Toporov, Vladimir N. 1987. Ob odnom arkhaicheskom indoevropeyskom elemente v drevnerusskoy dukhovnoy kul'ture SVĘT. In *Yazyki kul'tury i problemy perevodimosti*. Moskva: Nauka, pp. 184–252.

Toporov, Vladimir N. 1995. *Svyatost' i svyatye v russkoy dukhovnoy kul'ture*. Moskva: Yazyki russkoy kul'tury. T. 1.

Trebješanin, Žarko. 1991. *Predstava o detetu u srpskoj kulturi*. Beograd: Srpska knjizevna zadruga.

Uspensky, Boris A. 1982. *Filologicheskie razyskaniya v oblasti slavyanskikh drevnostey*. Moskva: MGU.

Veselovsky, Aleksandr N. 1877. Opyty po istorii razvitiya khriistianskoy legendy. II. Berta, Anastasiya i Pyatnica. In *Zhurnal Ministerstva narodnogo prosveshcheniya*. S.-Peterburg, pp. 186–252.

Vinogradova, Ludmila N., and Svetlana M. Tolstaya. 2005. Ritual'nye priglashheniya mifologicheskich personazhey na rozhdestvenskiy uzhin. Formula i obriad. In *Tolstaya S.M. Polesskiy narodnyj kalendar'*. Moskva: Indrik, pp. 443–500.

VP. 1980. *Valachobnya pesni*. Minsk: Navuka i tekhnika.

Zaliznyak, Andrey A. 1995. *Drevnenovgorodskiy dialekt*. Moskva: Yazyki slavyanskoy kul'tury.

Zamovy. 1992. *Charms*. Minsk: Navuka i tekhnika.

Zaykovsky, Vitaly B. 1994. Narodnyj kalendar' vostochnykh slavian. In *Etnograficheskoe obozrenie*. Moskva: Nauka. N 4, pp. 53–65.

Zečević, Slobodan. 1981. *Mitska bića srpskih predanja*. Beograd: Vuk Karadic, Etnografski muzej.

Article

The Making of a Marian Geography of Grace for Greek Catholics in the Polish Crownlands of the 17th–18th Centuries

Dieter Stern

Faculty of Arts and Philosophy, University of Ghent, 9000 Ghent, Belgium; dieter.stern@ugent.be

Abstract: This article explores the ways in which the newly founded and highly contested Christian confession of the Greek Catholics or Uniates employed strategies of mass mobilization to establish and maintain their position within a contested confessional terrain. The Greek Catholic clerics, above all monks of the Basilian order fostered an active policy of acquiring, founding and promoting Marian places of grace in order to create and invigorate a sense of belonging among their flock. The article argues that folk ideological notions concerning the spatial and physical conditions for the working of miracles were seized upon by the Greek Catholic faithful to establish a mental map of grace of their own. Especially, the Basilian order took particular care to organize mass events (annual pilgrimages, coronation celebrations for miraculous images) and promote Marian devotion through miracle reports and icon songs in an attempt to define what it means to be a Greek Catholic in terms of sacred territoriality.

Keywords: Greek Catholics; Church Union; Ruthenia; popular baroque piety; Marian devotion; miraculous images; devotional songs; pilgrimages

check for **updates**

Citation: Stern, Dieter. 2021. The Making of a Marian Geography of Grace for Greek Catholics in the Polish Crownlands of the 17th–18th Centuries. *Religions* 12: 446. https://doi.org/10.3390/rel12060446

Academic Editor: Dennis Ioffe

Received: 12 May 2021
Accepted: 9 June 2021
Published: 16 June 2021

Publisher's Note: MDPI stays neutral with regard to jurisdictional claims in published maps and institutional affiliations.

1. Introduction

The veneration of images[1] of the Mother of God and the belief in their miraculous powers have been key components of Christian spiritual practice since time immemorial, but they came to experience a particular boost in the age of the Counter-Reformation, when all kinds of devotional activities were fostered and reinvigorated in what looks like an effort of making a counteroffer of a more material religion to look at, touch and interact with as opposed to the bookish immateriality of the Reformation. It was meant to be an offer to the broad masses, as Counter-Reformation relied on mass mobilization to win back lost terrain. The entrance of the sacred into the material world of ordinary people required downsizing the sacred from its frightening cosmic oversize to more humane dimensions of ordinary sorrows and feelings of compassion. The Mother of God remodeled into a human impersonation of true motherhood would be the natural choice to win over the masses and reinvigorate trust in the church and state authorities. Piotr Hiacynt Pruszcz, a contemporary witness of the rise of Marian devotion and the proliferation of ever new Marian places of grace throughout the Polish–Lithuanian Commonwealth comments accordingly on the situation:

> Zaprawdę, w Polszcze nászey nie mász Miásta, rzadka Wieś z Kośćiołem, w którymby się Obraz PANNY Nayświętszey nie ználazł cudowny. (Pruszcz 1662, p. 142)

> Truely, in our Poland there is no town, and hardly a village with a Church of its own, in which there could not be found a miraculous image of Our holiest Lady.

What is true of Poland is no less true for the more narrow setting that we have chosen to have a closer look at, i.e., the Eastern, formerly a predominantly Orthodox part of the Polish Crown, that was claimed by the Greek Catholic Uniate church.[2] In the confessional borderlands of the Polish–Lithuanian Commonwealth, the momentum of the counter-reformation was taken advantage of so as to also confront Byzantine-Slavic Orthodoxy. The Catholic attempt at winning over the Orthodox faithful by meeting them halfway with the promise of leaving orthodox spirituality and ritual traditions untouched brought forth Greek Catholicism as a confessional hybrid, which came to lead a life of its own through its efforts at asserting itself in a contested confessional terrain. In a parallel movement, both Roman and Greek Catholics made active efforts to secure their share of the confessional terrain quite literally in terms of spatial relations by construing geographies of grace for their respective flocks. Miraculous images were particularly suited to inspire a spatial sense of what it means to be a Greek or Roman Catholic by marking out specific places of grace, which through a complex array of ordered activities, supported by diverse kinds of hierotopic events (annual processions and pilgrimages, coronations), provided devotional objects (votive tablets, print reproductions of icons, commemorative medals) and minor texts distributed in print or handwritten copies (prayers, songs, litanies) for a lived experience of sacred space. This rematerialization of religious experience of folk baroque Catholicism could in the area in question capitalize on the deeply rooted traditional Orthodox conceptualization of the Church as "heaven on earth", in which "the Divinity is accessible through matter, that of sacraments and sacred objects", which "can be not only seen, but even smelled, tasted, kissed" (Fedotov 1946, p. 33). Though both Orthodox and Catholic traditions put an equally strong emphasis on the veneration of the Mother of God, the active reinforcement of the cult of miraculous Marian images in the Polish–Lithuanian Commonwealth in the 17th and 18th century ought to be seen as a particularly Catholic concern, which seems to owe much of its force and fervor to prior Protestant iconoclasm and its Hussitic precursor (Kruk 2011, p. 154).[3] In times of confessional strife and competition, the potentials for securing one's flock by providing them with a sanctified geographical space to feel at home in could not be overlooked and was actively seized upon by Greek Catholics.[4] Capitalizing on physical space and territoriality naturally implies linking up the sacred to the political, and indeed many practices and events around miraculous images openly reflect upon political and historical events and aim at making Marian devotion a foundational part of political power.[5] In this paper we will outline how Greek Catholics organized a network of local, regional and supraregional Marian places of grace in order to gain a foothold in a contested confessional terrain in the face of their specific lack of a long-standing tradition of their own. Where Roman Catholics and Orthodox could lean back on an already established network of places of grace, Greek Catholics had to start from scratch (Jakovenko 2011–14). The acquisition and creation of places of grace as well as their successful advertisement were key to promote the loyalty of the community.[6] Marian places of grace formed indeed the key constitutive element in the foundation of a Greek Catholic sacred space (Al'mes 2018, p. 266). On an as yet more fundamental level, a geography of grace founded on miracles helped also boost Uniate self-confidence in the face of external challenges to its right of existence, as the Uniate metropolitan Gabriel Kolenda clearly put it into words:

> Evidentius argumentum probandi Unionem esse salutarem, et laudabilem non requiritur aliud, quam quotidiana miracula Beatae Virginis Zyroviciensis (. . .). (EMKC n.d., p. 322)

> There is no need for a more evident argument that proves the Union to be salutary and laudable, than the daily miracles of the Blessed Virgin of Žyrovyci.

The argument is quite straightforward. If the Church Union was against the will of God, he would hardly allow his graces to show up at Uniate places. Being able to boast an increasing number of miraculous sites could not fail to convince the faithful of the rightfulness and legitimacy of the Holy Union. Actively establishing a sacred geography is obviously a matter of acquiring and obtaining control over as many Marian sites as possible and successfully spreading the word of their miraculous powers. In a less obvious manner it is also about exploiting and reinforcing preexisting folk ideological notions about the relationship that holds between physical space and the sacred. It is our intention to demonstrate how interpretations of physical spaces as sanctified and thus meaningful places supported and orchestrated the rise of Marian devotion in a way as to help a specifically Uniate geography of grace to take root within the Uniate community. The underlying folk ideological notions of space and geography are most explicitly articulated in small text genres related to Marian devotion, in particular in miracle reports and Marian devotional songs in praise of miraculous images. We will build up our basic argument drawing primarily on these small text genres.

2. Marian Places of Grace in the Ruthenian Lands

Knowledge about Marian places of grace in the Ruthenian lands of the 17th–18th centuries is highly fragmentary and must be pieced together from a variety of sources. Despite the existence of a number of fairly extensive lists of Marian places (Barącz 1891; Fridrich 1904; Lužnyc'kyj 1984; Rožko 2002), it appears difficult to arrive at a complete list of all Marian places of grace that were known and applied to by the faithful throughout the period in question. In many cases, knowledge of the prior existence of a particular place of grace can only be derived from one source, which more often than not amounts to nothing more than a song in its honour.[7] Lužnyc'kyj (1984), whose list is perhaps the most extensive, registers 114 miraculous images for the Ruthenian lands. He distinguishes miraculous[8] from blessed images, the latter of which appear too numerous and knowledge about them too hard to come by to attempt an exhaustive list of them.[9] It appears that the majority of registered images seem indeed to have gained public recognition for working miracles only within the period in question, while only few have a track record of working miracles since the middle ages.

In some cases, as in Pidkamin' or Boruny,[10] the ground for establishing a place of grace was prepared through a grass roots tradition of reporting Marian visions that were unanimously interpreted as a sign to have a specific place of worship erected. When the place was established it was again the pilgrims who by simply coming in large numbers to the place and applying for help initiated the following step of calling in a commission to have the image officially recognized as miraculous. In the case of Pidkamin' it looks like it was the church authorities that were driven by the public to establish a place of grace, but the place would possibly never have acquired its fame without the commitment of the local Dominicans who applied to the bishop of Łuck, Andrzej Gembicki, to call in a commission in 1647, and it was the Dominican brother Szymon Okolski who painstakingly compiled the miracle reports obtained through the commission into a book that was published in 1648 (Rok 2005, p. 143). Often enough, as soon as a religious order or brotherhood took over, the reputation of an image's healing powers underwent a significant boost, as in the case of Krem'janec', which started to attract pilgrims only when the Augustinian Brotherhood of Our Mother of Consolation (*Bractwo Najświętszej Maryi Panny Matki Pocieszenia*) came in charge of the place in 1733 (Fridrich 1904, II, p. 382). The same can be observed anywhere, where Uniate Basilian monks took over.[11] It can be reasonably assumed that Cholm, Počajiv, Terebovlja, Krechiv, Hošiv and Zarvanycja would have never risen to places of mass pilgrimage, were it not for the efforts taken by the Uniate church in the 18th century (Levyc'ka 2017–18). In the case of Žyrovyči, all started as early as 1621 at the founding congregation of the Uniate Basilian order. In 1622 a history of the miraculous image was published by Teodozy Borowik (Jakovenko 2011–14; Wereda 2018). In the course of the 18th century, Basilians had managed to provide in a targeted way for each church province a center of Marian devotion through applying for papal coronations: Žyrovyči for Lithuania, Cholm for the Polish Crown territory and Počajiv for Volhynia (Wereda 2018, p. 69).

The future fame of a miraculous image can, however, be initiated by a very modest first step indeed, as in the case of the exceptionally popular miraculous Mother of God at Sambir, which on being reported to the chaplain of the local parish church for shedding tears of blood in 1727, was immediately acquired by the priest from its private owner and put into place at his church to grant public accessibility (Fridrich 1904, II, p. 420). Within a very short time the image attracted pilgrims in significant numbers and within a year the Greek Catholic bishop of Przemyśl, Hieronim Ustrzycki, formed a commission to have the image officially recognized as miraculous. The tremendous speed of the events are suggestive of the energetic commitment of chaplain Andrzej behind it all, who appears to have been wholly bent upon establishing the fame of this image. Andrzej meticulously registered the weeping activity of the image for 175 successive days from September 1727 until May 1728, and he is also the one who composed a first person witness report on his discovery of the image (Fridrich 1904, II, p. 422). There are still other cases, where the solitary commitment of one or other private person sets things going, and in many of these cases, among them Sambir, one wonders if the image would have ever acquired any fame at all without the personal initiatives of these highly committed individuals.

Indulgences, Pilgrimages and Pilgrims

On the face of it, it should be obvious that news about miracles worked by images would spread fast among the population and attract in an uncoordinated and unorganized grass-roots process the afflicted and the suffering, yielding a steady flow of pilgrims over the course of a year. However, this is only part of the story. As a matter of fact, within the Roman Catholic tradition pilgrimages often combine two strains of intent, which though being somehow related, ought to be kept apart. Relief from actual suffering is the first aspect, while forgiveness of sins according to the Roman Catholic practice of indulgences, readily became an associated aspect. Wherever an image acquired the reputation of working miracles, the order in charge of the place was likely to channel the uncoordinated stream of supplicants into well-structured annual events by linking supplication to confession of sins and organizing regular mass events at fixed dates of the year for processions and pilgrimages of forgiveness (Pol. *odpusty*, Ukr. *vidpusty*). By extending the service profile of miraculous images to encompass indulgences, the entirety of the Catholic communities—Roman and Greek alike—were expected to visit at least once a year a miraculous image of their choice to obtain forgiveness for their sins. This practice was in an obvious manner designed to increase and reinforce public recognition and extend the range and significance of the place from the merely local into something bigger. This process of channeling supplicants into regular pilgrimages required official approval by authorities higher up in the ecclesiastic hierarchy thus putting the incipient and fluid spiritual grass-roots movement on a firm institutional footing in order to secure its long-term establishment.

A further side-effect of institutionalization is turning the private activity of individual supplication into part of a public event.[12] The choice of the Basilian monastery of Žyrovyči to establish an institutionalized practice of pilgrimages about their miraculous image from 1625–26 onwards was by no means accidental. The place was singled out by Josip Veljamyn Ruts'kyj to become a major center of pilgrimage for Greek Catholics. This was achieved by obtaining an exceptional papal approval that pilgrims to Žyrovyči were granted indulgence on occasion of the Holy Year in 1625, which privilege was even extended to the following year 1626 (Senyk 1984, pp. 263–64). The monastery at Žyrovyči happened also to have been headed by Josaphat Kuncewicz, the only martyr of the Greek Catholic church, who was murdered in 1623 only three years prior to the establishment of the annual pilgrimage. There were plans, which became never realized, though, to transfer the relics of the martyr saint from their grave at Biała to the monastery at Žyrovyči, thus transforming the inconspicuous place of grace into the very geographical and spiritual heart of Greek Catholic confessional identity (Wereda 2018, p. 70).

Within the Greek Catholic confessional space the appointment of fixed days of the church calendar[13] for common pilgrimages of forgiveness is attested for a significant number of miraculous images throughout the 18th century.[14] Making places of grace part of the practices of confession and forgiveness, which every believer had to undergo on a regular basis, could not fail to generate a mental map of grace within the consciousness of each and every Greek Catholic, which would comprise all places with officially approved pilgrimages and processions as its stable and immutable core, among which the coronated images would shine out as the most desirable destinations. For the Uniate metropolitan bishop Raphael Korssak it was a plain truth that a Marian geography of grace of one's own was a precondition on the emergence of a Uniate identity which he linked to Žyrovyči:

> Capitaneatus et Districtus Civitatis regiae Slonim totus unitissimus, et si ubi, ibi habet S. Unio consolationem propter ecclesiam Zyrovicensem Beatissimae Virginis miraculosam, ubi sunt Monachi uniti. (EMKC n.d., p. 120)

> The district and territory of the royal town of Slonim is totally Uniate, and if there is one place where the state of the holy Union is truly consoling, this is it, all on account of the church in Žyrovyči with its miraculous image of the Blessed Virgin, where there are Uniate monks. (translation quoted from Senyk 1984, p. 263)

Raphael Korssak's statement may be taken to be representative of the general train of thought among Uniate clerics. Confessional consolidation was made dependent on defining a sanctified territory, that might overlap with others (especially the Roman Catholic geography of grace), but which should also specifically cater towards the Greek Catholic community. Though on the face of it, miraculous powers and territoriality appear not to be intrinsically related, the specific practice of obtaining grace through miraculous powers as well as the legendary reports that frame them, are founded on folk notions of the miraculous that approximate physical materialism and are suggestive of miraculous powers being dependent on physical proximity and fixed places. The following section will provide an overview of the types of Marian miracles in order to be able to look into this issue more closely.

3. Miracles and Physical Space

There is a basic distinction to be made between (1) unsolicited miracles establishing the fame of the miraculous image by drawing for the first time common attention to its powers, i.e., founder miracles, and (2) miracles worked on a daily basis upon solicitation, i.e., through prior acts of supplication. Founder miracles do as a rule not predicate the type of service miracles that the image will work, once its miraculous powers have been established. Service miracles of type (2) can be further differentiated into (a) minor personal miracles (healing illnesses (*Mater remedii*), delivering the possessed by demons (*salus infirmorum*), support in personal crisis or distress (*Mater auxiliatrix*)) that require always personal supplication, (b) major miracles of politico-historical dimension (such as fending off military invaders) that combine both service and consolidation of fame. Miracles of type (2b) do not necessarily require solicitation, though prior instigation through mass supplication seems a common feature. Service miracles of politico-historical dimension may also serve to establish miraculous powers for the first time among believers, thus being at the same time founder and service miracles. Miracles of politico-historical significance are as a rule modeled on the famous Byzantine palladium miracle of the Blacherniotissa at Constantinople. Their prevalence on the Ottoman frontier of the Polish–Lithuanian Commonwealth throughout the 17th century thus draws on Byzantine-Slavic Orthodox traditions (Trajdos 1984, pp. 128–31).

It should be noted that founder miracles primarily focus on the image as an object of veneration for its own sake and are not meant to provide useful services to anyone. Thus, many founder miracles are not about healing and helping, but about stunning an unsuspecting observer by occurrences that go against the laws of nature. Quite common is the shedding of tears, which more often than not also serves as a comment on the deplorable state of affairs of the region in question.[15] Typically, the threat to the integrity of orthodoxy[16] through military attacks of heterodox forces will cause an image to weep. Perhaps it is no coincidence that weeping images made their first appearance on Ruthenian soil during the troublesome second half of the 17th century.[17] In the case of Počajiv the shedding of tears for a period of four weeks one year prior to the Turkish attack on the monastery in 1675 also brings into evidence the prophetic powers of the image (Przesławna gora poczaiowska Dawnością Cudow Przenayczystszey Bogarodzicy Panny od cudownego Jey Obrazu wynikających [1787] 1801, fol. 22r).

In some cases, unusual occurrences of a mystical and supernatural character serve as founder miracle. Thus, there are images which out of the blue make their first appearance in a tree (Žyrovyči, Kup'jatyči),[18] in the woods (Staroduba) or on a river bank (Lin'kiv on the river Desna)[19] or just in the grass of a meadow (Ochtyrci). Their sudden appearance was likely to be taken as a sign of their supernatural origin as 'not being made by hand' (*nerukotvornaja*, ἀχειροποίητος).[20] Others are witnessed to change their color (Kam'janka-Strumylova) or to emit an unearthly light (Dubovyči, Ochtyrci, Počajiv). Still others publicly manifest their powers through being found unscathed by devastating fires (Nestanyč; Onyškivci; Zelenycja (later Rudky); see also Kruk 2011, pp. 173–80). These specific founder miracles may be assumed to serve as marks of distinction among the mass of miraculous images of local and supralocal significance. They moreover give testimony to the magical and spiritual texture of the natural universe of which the image forms part.

Founder miracles are not only meant to provide evidence of miraculous powers and to draw attention to a particular object of worship. They are also meant to establish authority, and doubting authority converts the healing powers into powers of destruction and punishment. The faithful are therefore not only to be comforted and kept in trust by an endless number of healing narratives, but they have at the same time to be warned and kept in awe of the supernatural powers. Rare but exemplary cases of renegades, heretics and unbelievers being severely punished for desecrating the place of grace and doubting its powers are therefore included in any miracle register. A particularly interesting case of a punishment is that of the miraculous image at the Basilian monastery of Puhinki, which raises a key issue of sacred geography. When the monastery at Puhinki was dismantled in 1840 following the abolition of the Union in those Ruthenian territories that came under Imperial Russian control, and the image had to be moved to another place, three villagers who attempted to remove it from its old place were struck by blindness, palsy and madness (Rožko 2002, p. 67). The punishment is obviously not so much about moving the image to another place, but rather about the destruction of its sanctuary and, accordingly, the destruction of its specific relation to space. It looks like miraculous images have a sense of physical belonging to a particular place. Thus images may be moved from their place in times of war and crisis, but this is only acceptable under condition that they will be moved back sometime later. If there is no intention of moving the image back to the place it has chosen for its miracles, the image may even stop its miraculous activity, as in the case of the image of Počajiv that did not show any signs of miraculous powers during its period of sequestration at the hands of Firlej Bełzki from 1617 to 1647. When the image was restored to its original place, it started working miracles again (Przesławna gora poczaiowska Dawnością Cudow Przenayczystszey Bogarodzicy Panny od cudownego Jey Obrazu wynikających [1787] 1801, fol. 9r–11r). The movement of images to better equipped places in the near surroundings poses usually no problem, though. The image of Pidhirci professed its first miracle at Holubycja in 1692, from where it was moved to the nearby Basilian monastery at Pidhirci in 1694.[21] Likewise, the miraculous image of Rudnja was moved to the Ascension monastery in Kiev to give it greater prominence, without this movement impairing her miraculous powers in any way (Senyk 1984, p. 265). On the contrary, the new abode at the Kievan Cave monastery was expected to provide for a better environment which could not fail to enhance even the image's miraculous powers, as an early song in praise of this image states:

> V obiteli panjanskoj, pri cerkvi Pečarskoj,
> V chramě svoemъživetъlěpšъvesi Rudjanskoj.
> (ms. ASP 135, fol. 63v, v. Hruševs'kyj 1897, p. 49)

> In the nunnery, close to the Cathedral of the Cave,
> in a chapel of her own she [scil. the image] lives better than at the village of Rudnia.

Miraculous images seem to choose the place where they work; their miracles are not random. When the village of Nestanyč (Galicia) was burnt down in a Tatar raid, the image at the local church which was found unscathed by the fire and was ordered to be brought to the church at nearby Stanyn, but the cart-horses would not move and when forced to do so, would return to Nestanyč (Lužnyc'kyj 1984, p. 177).[22] Though places are not chosen randomly, it is only rarely that the actual reason for selecting a particular place is made known. When Franciscans, who had heard of the miracles worked by the image at the Basilian monastery at Žyrovyči, made plans to acquire the image and move it to their own monastery, God explained to one of the Franciscan brethren in a dream vision, why the image must not be moved:

> (…) et intellexit hunc locum pro defensione et promotione unionis a Deo excitatum atque ablato unitis loco miracula cessatura.
>
> (Harasiewicz 1862, p. 342)

> (…) and he understood that this place had been chosen by God for the defense and promotion of the Union, and that if [the image] were to be taken away from that Uniate place, the miracles would stop.

It is perhaps not by accident that an explicit motivation is provided in this particular case, where confessional politics of place come into play. As we have seen above, Žyrovyči was indeed intended early on to become the very center of Uniate identity, and losing its miraculous image must have thwarted the Uniate politics of building up a confessional territory of its own.

The principally indissoluble link between image and place implies also a particular loyalty of the image to its place, which is iconically expressed in narratives of defending places of grace against military invaders, as most prototypically represented by the Hodegetria of Constantinople (Blacherniotissa) spreading the palladium (*omofor*) over her wards. It is through these miraculous appearances that images are awarded the name of the place as their specific cognomen, yielding titles typically found in the rubrications of icon songs such as *Bogorodica čudotvorna počajevska* and the like. Kruk (2011, p. 166) observes that through the palladial function of the Mother of God as *mater misericordiae* the according naming practice of local identities was created, reinforced and maintained.[23] The intimate and indissoluble relation to a place of its own seems to be expressed as a territorial ownership of sorts in the workings of miraculous images, especially frontier icons that on the face of it appear to defend the local populations against military invaders, but ultimately aim at maintaining the integrity of the sanctuary which hosts the image and delimits its spiritual dominion, which includes the local population as its integral part. Perhaps, the footprints left by the Mother of God in her visionary appearances at Počajiv and Pidkamin' should accordingly be read as ownership marks. There could not possibly be a more lasting material testimony to the strong territorial bond into which the Mother of God has entered. Počajiv and Pidkamin' appear to be marked out by these footprints as being held in greater esteem by the Mother of God than any other place she owns. The working of miracles through images turns the arbitrariness of physical space into the uniqueness of a meaningful place. By relating the sacred to the physical, the foundation for territorial identity is laid. As each miraculous image has a place of its own, everyone who enters into a relation with it becomes somehow part of the geographical space it defines.

The Materiality of the Sacred, Communication and Personhood

The paradox of the manifestation of the sacred which is not of this world, in objects of this world has been remarked and commented upon by Mircea Eliade (1957, pp. 8–9). It becomes particularly apparent in the practices and beliefs growing around the working of miracles, which bring forth and are supported through material objects, such as images. In an obvious way, it is unavoidable that the sacred turns into material objects in a world of physical matter by the simple act of referring to it. Eliade's amazement is, however, not so much about the semiotic relationship between the material *signe* of sacred objects and the immateriality of the sacred as *signifié*. In fact miraculous images are not conceived of as signs (be they symbolic or iconic in the Peircean sense), but they seem to have been believed to actually be what they represent: "The saints were really present in their relics as well as in their icons" (Fedotov 1946, p. 33).[24] Mere representations could not be hurt and bleed and weep like miraculous images do. As a matter of fact, in a song in praise of the Mother of God at Kam'janka-Strumylova the miraculous image is being depicted as an actual face that reacts to the supplicant:

> Tvarьbo izměnjaetъ, slezy izlivaetъ,
> sklanjaetъušesa, priemletъslovesa.
> (BG 2016, no. 131)

> Her face changes, she starts weeping,
> she lends her ear and takes in the words.

Once, however, personhood is assigned to images, the sacred tends to become locked in into the world of matter and thereby acquires a spatial dimension. Physical space thus turns into a meaningful place through its association with sanctity. As a matter of fact, the close, ownership-like link between place and image, discussed above, bears directly on the issue of personhood. By assigning an image to a fixed place, the image may up to a certain degree achieve independence of the higher abstract personhood of the biblical figure of the Mother of God as such. Having a face of its own that can weep, a body that can bleed, and being indissolubly related to an individual place in an ownership relation of sorts are the basic ingredients of individuality and personhood. Through the proliferation of miraculous images, the mother of God seems to turn into a multiplicity of Mothers of God, the more so where images enter into competition with each other (on which see below).[25] The Mother of God of Počajiv would then in a very immediate sense not be the same person as the biblical Mother of God, whatever else their relationship may be. This folk ideological interpretation could not fail to make theologians feel uneasy about the cult of miraculous images and to ultimately inspire iconoclastic thinking. There is, of course, no way of knowing if individual worshippers perceived images as true persons, but there is at least one Marian song that favors Surožskij's interpretation of the Mother of God being present and acting through the icon without becoming the icon:

> Inši kto žъtvoi čudovny obrazy
> zličiti može panien'ko bez zmazy,
> v kotorychъesъvel'mi južesъsja vslavila
> i rozmaitychъlaskъljudъnabavila.
> (Kamjanskij Bogoglasnik 1734; cf. Žeňuch 2006, p. 908)

> And who is capable of counting
> all the other miraculous images,
> in which you have acquired fame
> and proffered variegated graces to the people.

This stanza may possibly reflect an attempt to maintain the official high church interpretation in the face of problematic grass roots tendencies.

Miraculous images grant unmediated access to the sacred for the supplicant who can speak to the sacred and touch it—especially by kissing as a mark of respect—but also to induce the sacred to flow over to the supplicant upon the touching of the image. The latter aspect becomes yet more apparent in less common forms of the material manifestation of sacred powers, as in the case of healing water (Počajiv, Pidkamin', Univ, as well as a number of minor places of grace in Ruthenia, such as Nastašiv). For healing water to become effective, physical application to the sufferer is indispensable. As we have seen, restrictions on physical space also apply to the working of miracles. Miraculous images are bound to a particular place through their founder miracles, and the same is even more obviously true of healing water. In order to overcome the restrictions of physical space techniques of replication and transfer have to be applied. Thus the healing water at Počajiv was taken home by pilgrims in small bottles (Rožko 2002, p. 82), in order to be applied whenever and wherever needed. The same holds for images, which can be made miraculous through the simple technique of rubbing them against an already miraculous image, as in the case of the image of Nastašiv which is said to have acquired its healing powers through rubbing it against the Black Madonna of Częstochowa (Barącz 1891, pp. 189–90; Fridrich 1904, II, p. 263; Stern 2000, p. 36).[26] It may be assumed that the sanctification through the physical act of rubbing was quite common and also applied to minor objects as print reproductions of famous images and small figurines of the Mother of God, which could be taken along wherever one went.[27] This all looks as if the sacred was conceptualized as a kind of invisible, but ultimately material property of this world, as a kind of electric energy of sorts that prefers particular configurations of the material world to manifest itself as effectively as possible. There is here a general danger of the profanation and ultimate dissolution of the sacred into an utterly materialist perspective. This is, though somewhat lamely, counteracted by the requirement that successful treatment depends on a spiritual attitude which creates a personal relationship between the supplicant and the sacred through the medium of material objects.

Imagined pilgrimages, as suggested by Chociszewski (1882, p. 106) are suggestive of an immaterial and placeless, apparently purely spiritual interpersonal, communicative relationship, though true physical experience still looms large in this exceptional practice. Moreover, the interpersonal relationship rather than counterbalancing the materialism of the physical experience of pilgrimage and acts of supplication, is itself liable to bring about profanation, especially where the sacred becomes impersonated, as is most obviously true of the veneration of the Mother of God. Marian devotional songs show a common strong tendency to emphasize the role of the *mediatrix* who can be applied to like any person of influence of the material world. Among Marian songs, icon songs carry this tendency to the extreme. The impending profanation of the interpersonal relationship is counteracted not so much by the sanctity of the *mediatrix*, but by the sanctity of the suffering brought to her witness. As soon as the *mediatrix* is applied to for minor causes where suffering in the commonly accepted sense of the word is not involved, the sanctity dissolves altogether and the supplication turns into a grotesque.[28]

In most cases the procedure and basic mechanism of obtaining a miracle would be quite obvious. Miraculous images were addressed as directly as possible by either praying in front of them or, if possible, even touching them. It looks as if some kind of face-to-face communication has to happen before the Mother of God will take up the case for the supplicant. The image appears to serve the function of a communication device. Possibly it is even considered as a personal living being in its own right, as is suggested by the lines quoted above about the Mother of God of Kam'janka-Strumylova. These lines reinterpret the topical miracles of shedding tears and changes of appearance, reported for so many images as founder miracles, as actual facial expressions. Where the sacred is conceived of as personal, there is a direct pathway to its being made present in the material world. Relating to the sacred means entering into an interpersonal relation, which implies communication. All communication needs a medium, which in this world must needs be physical in kind. In order to say your prayers, words should be audibly spoken, and ideally there ought to be someone visibly present to be spoken to. Thus, images allow for a greater approximation of natural face-to-face interaction, but still communication appears to remain a one way road of sorts, where the image usually will respond only indirectly through performing or refusing to perform a miracle. It is only through weeping, as the lines about the Mother of God of Kam'janka-Strumylova so vividly show, that the sacred can take directly perceptible, physical action that can be interpreted in terms of a communicative act in the sense of Hymes (1972). In fact, weeping will be regularly interpreted as a comment on recent deplorable events, mostly of a politico-historical dimension, such as the frequent Tatar and Ottoman raids and military expeditions. It can be more generally taken as an admonition to the faithful to reconsider their moral behaviour and redirect their steps, as in Wespazjan Kochowski's lines:

> Te oczy płaczą Matki naszej drogi
> Dla upomnienia nam i dla przestrogi. (quoted from Kruk 2011, p. 267)

> Those dear eyes of our Mother keep weeping,
> for us to remember and as a warning.

However, contrary to what the lines from Kam'janka-Strumylova suggest, this is still a far cry from face-to-face interaction. Sacred communication in physical space still remains a monological one-way road, though it may change directions at times.

From the long row of miracle reports from Počajiv it can be gathered that ideally the diseased should be present in person in front of the image, even if he or she is incapable of offering a prayer and needs someone doing it in his stead. Thus, several miracle reports provide details on how even the fatally ill were transported over long distances in order to place them in a lying position in front of the miraculous image. If the place was known for highly potent healing powers, even people who were believed to be dead were presented in this way.[29] This looks like the Mother of God would require some visual evidence before granting her support and that she can perceive this evidence only through the miraculous image, as if it were her eyes on earth. Whatever conceptualization of the mechanical laws and requirements of miracles may stand behind this, supplicants must have believed that effective help was less reliably obtained through an intermediary with the person in need of help being absent. In contrast to what appears to be the common practice at Počajiv, many supplicants at Pidkamin', however, followed a less strict and troublesome procedure. If the needy person in his disease was unable to visit the place of grace in person, he would send a donation in order to apply for help and would visit the place afterwards to bring his votive in person to the image.[30] The reasons for this marked difference in practice can only be guessed at. On the face of it, the practice observed at Pidkamin' contradicts to some degree our assumption that the working of magic is bound to a particular place and requires face-to-face interaction. The practice does, however, not downright deny the precondition of the uniqueness of place, but makes only allowances for a more supple and practicable interpretation, by postponing face-to-face interaction to a later point in time. The offering of votive tablets in these particular cases is then not only a sign of gratitude, but restores the spatial order of magic.

General instructions for the supplicant on how s/he ought to physically approach a miraculous image is also provided in some song texts, as in the following stanza:

> Podźmy przed obraz Maryjej,
> Tej najśliczniejszej lilijej,
> Poklęknąwszy, rączki słożmy,
> Do jej się nożek ucieczmy.
> (quoted from Kruk 2011, p. 214)

> We come to stand in front of the image of Mary,
> that most beautiful lily,
> and, once we have kneeled down, we fold our hands,
> and we seek shelter below her feet.

These instructional verses can be taken to depict a standard manner of approaching an image, possibly even a recommended form of behavior that is meant to prevent the masses of supplicants from engaging in too intimate a manner (such as touching and kissing) with the miraculous image that must inevitably cause wear marks in the long run. Apart from these more practical worries, an overly intimate relation allowing for direct physical contact would also run counter to the idea of a celestial hierarchy, which according to Catholic notions would turn the Mother of God into the Queen of Heaven. The imposition of a modicum of spatial distance would then serve to maintain the hierarchy and dividing line between the sacred and the profane. As Erazm Gliczner's report suggests, pilgrims seem indeed to have been instructed by chaplains of what to do (and possibly also what rather not to do) in front of a miraculous image, which in Gliczner's protestant eyes is already too much:

> I zwłaszcza że liud prostiy z nauki Kapłanow Kościoła Rzimskiego nauczeł się y uczy przed Obrázmi ták ritemi bałwany táko y málowánimi kłaniać y klękać y klęcząc swe modlitw przed nimi mowić szapke zeimować y insze nabożne pokłony czynić, swieczky przed nimi zápalene stawiać, Krzyżem padać, lieżeć { ... } (Gliczner 1598, fol. 51v)

> And the common people were indeed instructed from the knowledge of the Roman Catholic chaplains and learned to bow and kneel down in front of the images, be it sculpted figures or paintings, and once they kneeled down, to say their prayers in front of them and to take off their hats and do other kinds of pious exercises, and to put burning candles in front of them, and to prostrate and stay lying in the form of the cross (...).

All actions listed restrict the interaction between supplicant and image to the level of mere perception over a certain physical distance. This distance was often imposed by the spatial arrangements of the sanctuary. As a matter of fact, the more famous an image became, the more prominent a location it was assigned in its sanctuary. The more famous images were, as a rule, put up in a shrine behind the main altar (Kruk 2011, p. 216), which effectively precluded any form of intimate interaction, thus thwarting the folk ideological idea of physical contact as a precondition for miraculous powers to become effectively operative. Putting images in side altars and specifically designed chapels allowed for a more private and secluded form of supplication, to be sure, but it would still not allow for physically intimate forms of interaction.[31]

The modest offerings of votive tablets for individual help combine with monumental celebrations, the building of chapels and altars and the donation of rich adornments to create a space of worship and devotion that fuses both acts of individual gratitude and the display of political power through its link to the sacred to yield a meaningful place that ultimately transcends the division between the private and the public. The latter aspect becomes particularly evident in the practice of coronations. Coronations of miraculous images and the pompous ceremonies associated with them can be seen as the culmination and ultimate completion of the general trend for visualizing and sensualizing spiritual experience within the framework of fully developed 18th century Catholic baroque culture.[32] Moreover, the very imagery of a coronation clearly implies a parallelization of sacred and secular power. The coronated icon is quite obviously the visible representation on earth of the *regina coeli*, who enters into a specific political territorial alliance with the Polish–Lithuanian Commonwealth as the Most Holy Virgin Mary, Queen of Poland (*NMP Królowa Polski*).[33] Coronations of miraculous images became a fashionable competition in Ruthenia (Levyc'ka 2017–18), which was set going by the coronation in 1717 of the Black Madonna of Częstochowa. Only a decade later two more sites applied almost simultaneously for papal coronations, one being Basilian (Žyrovyči 1726/1730) and the other one Dominican (Pidkamin' 1727). The celebrations were usually organized as ecumenical events (addressing Roman, Greek and Armenian Catholics) and attracted tens of thousands of faithful. They were put on stage with great pomp, including the declamation of elaborate sermons on Marian topics, public disputations on theological issues, artful fireworks, that lasted for hours and depicted various pertinent symbols and images, the erection of triumphal arches, the artful illumination of buildings, concerts, military parades, canons firing salutes, the exposition of historical paintings commissioned for the occasion, the issuing of commemorative medals by the thousands. The celebrations were in an obvious manner meant to excite all senses and to turn Marian devotion into an overwhelming experience of joy that did not shun forms of entertainment that must strike a modern observer as carnivalesque and almost profane, but which were described in all detail with obvious pride in commemorative volumes that were laboriously compiled and issued years after the celebrations.[34] Enthusiastic mass excitement and mobilization was the key objective of these celebrations, which is most vividly illustrated by the manner in which the commemorative medals were distributed among the attending masses. Instead of giving the medals individually to anyone who wanted one, they were thrown by the handful among the masses (Rok 2005, p. 146). The multi-layered structure of these events testify to a monumental enactment of the power of a self-confident unified Catholic church that is firmly anchored in this world and does not shy away from publicly displaying its close links to secular power. The sacred alliance of clerical and secular power is visibly reflected in the patronage of coronation events, which are represented in portraits on the scene of the celebration. In the course of the coronation celebrations at Počajiv 1773 on the walls inside the monastery cathedral paintings of the Polish king Stanisław August Poniatowski and the local patron and donor Mikołaj Potocki were opposed to paintings of pope Clemens XIV and bishop Sylwestr Rudnicki (Wereda 2018, p. 74). The unity of sacred and secular power is further underlined by paintings commemorating historical events within the confines of the monastery, as in the case of the coronation at Cholm. The presence of a tremendous number of ordinary citizens gives proof to the general alignment of popular, political and clerical will. It goes without saying that an event of this kind and size could not fail to kindle ambitions among all monasteries throughout the Ruthenian lands. Počajiv which vied for regional primacy with Pidkamin', with which it shares the prominent feature of combining a miraculous image with an additional pond of healing water, could not fail to do its utmost to have a coronation of equal dimensions as the earlier one at the Roman Catholic (Dominican) monastery at Pidkamin'.[35] It is not known, how many faithful attended, but the expense made was similar. In one point Počajiv managed to outscore Pidkamin' by having 5 instead of 4 triumphal arches at its coronation festivity.

4. The Geography of Grace According to Ruthenian Devotional Songs in Praise of Images

Minor texts such as witness reports on miracles as well as devotional songs in praise of individual images contributed in an essential manner to the construal of a specifically Greek Catholic Marian landscape, but they did so in different ways.[36] Both text genres conveyed information about miracles and the places where they occurred in sufficient detail so as to build up and mnemonically maintain the collective consciousness of sanctified space throughout the Greek Catholic community. This referential function, important as it is by itself, was, however, implemented and expanded in widely different ways. Witness reports may be supposed to have started as orally transmitted tales that spread through the untraceable channels of gossip and rumor in an undirected and uncoordinated manner. At this early stage, there may have been little enough political intent at play to help confessional identity building. With the documentation, collection and publication of these tales in written form things take a more political turn that clearly aims at engrossing the as yet confessionally unmarked miracles for the benefit of one's confession, but is is doubtful that miracle reports in written form had a direct impact on the collective memory and consciousness of the community. Songs, however, are likely to have been composed right from the start with the intent to draw public attention to the place in question and enhance its significance within the community. Their communal role is played out by making them part of the lived bodily experience of sanctified space through singing them or just listening to them on one's way to the miraculous image as well as on the spot, in group or individually.[37]

Icon songs dealt in various ways with their objects of praise. If there was not much to be said or little known about the image and its miraculous history, songs would take the form of a common and unspecific Marian doxology which would be rounded off by an unspecific stanza of general supplication. Most songs, however, try to convey more specific knowledge about the image in question. Two basic types can be observed: (1) narrative songs reporting on founder miracles (this is preferably made use of if the founder miracle is of a politico-historical dimension, such as fending off Ottoman aggression), (2) advertising songs providing lists of ailments that are cured and examples of historically attested spectacular healing successes. It goes without saying that songs can and often do combine both types. Thus, a song in praise of the miraculous image at the Dominican monastery at L'viv offers the following treatments:

> Mertvyja voskrešaešъ,
> Čelověkъot běsovъuvolnjaešъ,
> i ot bědnychъljudej febry otganjaešъ,
> slěpychъprosvěščaešъ. (ms. Maslov 54, fol. 30r; cf. Hnatjuk 2000, p. 197)

> You raise the dead to life again,
> You deliver people from demons,
> you chase away fevers from the poor,
> you make the blind see again.

Generic types of wonders are advertised in a sober, matter of fact fashion, sometimes awkwardly phrased owing to the requirements of rhyme and meter, as in the case of the healing miracles worked by the image of Zahoriv:

> Umarli wstaią na Twe załowanie,
> Dziękuią chromi za wyskakiwanie. (BG 2016, no. 118)

> The deceased stand up at your command,
> the lame thank you for [their] jumping up.

It is remarkable that in many cases only few words are assigned to praising and mentioning the miracles themselves. The larger part of any song usually consists of common doxological set phrases. This nonspecificity is in line with the practice to adapt songs from better known miraculous images.[38] Advertising can also include references to the sheer masses of pilgrims that are being attracted, as well as to the fact that pilgrims come from faraway places and that some of them belong to the higher strata of society (noblemen and kings).

An estimated amount of some 100 to 150 different devotional songs referring to Marian places of grace may have been circulating within the Ruthenian lands (Medvedyk 1999; Medvedyk 2015). The first songs of this kind were composed already in the second half of the 17th century (Medvedyk 2015, p. 372), but the vast majority of them date from the 18th century. Devotional songs were primarily disseminated in private manuscript song collections, and only exceptionally in print. As the area of Greek Catholic activities was by and large bilingual, songs in praise of miraculous images of that region may be either in Ruthenian, which was the common majority language, or to a somewhat lesser extent in Polish, as the language of learning and of the higher classes (and increasingly of the Basilian order). Some images were honored by songs in both languages. Though it is almost impossible to pin down the exact origin for any Ruthenian devotional song, it is generally assumed that many if not most of them were composed by Basilian authors.[39] As for icon songs, there is an obvious bias for Greek, and to a lesser extent Roman catholic sites. Famous orthodox miraculous images like at the Kievan Cave monastery or Černihiv attract fairly little attention of song writers, though, it has to be admitted that one of the earliest icon songs was composed in praise of the image of Vyšhorod, that was translated to Kiev in 1662.[40] As a matter of fact, only Basilian monks cared to prepare print editions of Ruthenian devotional songs, most notably the Bogoglasnik of Počajiv of 1791 (BG 2016). Moreover, a closer look at the places of grace represented in the repertoire of Ruthenian devotional songs shows a very pronounced bias for Basilian sites. Thus, it can be safely assumed that Ruthenian devotional songs, though they are in no way openly confessional in their contents, convey a Uniate, or more specifically a Basilian perspective.

Among the 114 images registered by Lužnyc'kyj (1984) for the Ruthenian lands, less than a quarter (26) are specially honored by songs. Our own estimate[41] of the number of images being praised in Ruthenian devotional songs yielded 39 locations.[42] To these may be added a number of 9 images for which devotional songs in Polish instead of Ruthenian have been composed.[43] It goes without saying that the more popular an image, the more likely it will be honored by a song, with the most popular sites such as Počajiv being honoured by a fairly large number of songs.[44] As for the lesser known images there is no way of telling what made a song composer select them, but it can be assumed that Basilian interests stood often behind the decision to have a song composed for a particular image.[45] In the case of the icon songs assembled in the Bogoglasnik there is no getting around the impression that the icon songs for Basilian sites authored by Basilian authors were at least in part specifically composed for the edition of the Bogoglasnik in order to bring into focus the wealth of the Basilian geography of grace.[46] At least some of the songs in praise of the select few coronated icons may be assumed to have been composed on the occasion of a coronation with the intention to sing them publicly at the festivity (Kruk 2011, p. 268).[47] Especially, songs that make their first appearance in the years after the coronation may be rightly assumed to have in fact been composed for the occasion. It can be assumed that songs were composed to serve the needs of regular ritual devotional celebrations of a paraliturgical kind, such as annual processions, rosary devotions and the like. In fact, any miraculous image that came under the patronage of a brotherhood or a religious order was likely to become the object of organized worship (Kruk 2011, p. 269), which would include prayers and songs. On occasion, songs seem to have been composed by individuals to commemorate a plight which the author experienced personally and through which the author desires to express feelings of gratitude for having survived the plight. This is apparently the case with the two songs in praise of the Mother of God at Sambir, composed by Andrzej Jakubiński in October 1770 right after the epidemic of that same year (Ščurat 1908b, pp. 181–85). Other songs were composed on the occasion of a translation to a newly built sanctuary, as in the case of the song *Drogi klejnocie naroda ruskiego* (Al'mes 2018, p. 272).

For songs to be able to shape the Greek Catholic geography of grace they should ideally be disseminated in large numbers, and in addition dissemination should also be directed and coordinated in order to avoid undesirable outcomes. However, up to the end of the 18th century, when the *Bogoglasnik*, a large print collection of Ruthenian (and a smaller number of Polish) devotional songs, was published at the Uniate monastery of Počajiv, devotional songs were almost exclusively disseminated in highly individualized manuscript collections, some of them possibly compiled and arranged by Basilian monks,[48] but most of them appear to be private collections that were randomly compiled. Prior to the publication of the Bogoglasnik, there were only few editions of commemorative volumes in praise of individual images that contained a small number of icon songs.[49] A further possible source of dissemination, which is for the time being difficult to assess, are print reproductions (wood cuts, etchings) of miraculous images and sites that seem to have been provided with texts of devotional songs, akathists or other liturgical hymns, at least in a number of cases (Levyc'ka 2017–18).

Song collectors included in their private collections not only icon songs from their home region, but often display an interest in Marian places of grace from all over the Ruthenian lands. Thus, song collections originating from Transcarpathia regularly register songs from other Ruthenian regions of the Polish–Lithuanian Commonwealth, such as Galicia, Volhynia, Podolia, Polesia and Kiev, and occasionally even include places from geographically as well as politically quite remote Slobozhanshchyna, such as Kaplunivka (Medvedyk 2015, p. 371), thus creating a sense of cultural-geographic belonging beyond the confines of one's immediate home region.[50] On the whole, Ruthenian collections of devotional songs display, however, a clear focus on Galicia and Volhynia, identifying these as the heartlands of Greek Catholic Marian devotion. The vast majority of song collections register only icon songs for one or two sites, with a clear bias on Pidkamin', Počajiv and (Novyj) Sambir, which three places figure also prominently in the remainder of the manuscripts that register a larger number of sites.

It is not quite clear if the choices made in song collections reflect the personal history of pilgrimages of song collectors, as is insinuated by Pap (1971, p. 119). There is at least one observation that would speak against such an assumption. Longer lists, of up to 7 different Marian sites are usually found in collections of a larger size. A strong correlation with the overall size of collections would be suggestive of the number of sites reflecting the collecting activity rather than the actual pilgrimage history of the collector.

If we assume that the choice of icon songs in private manuscript song collections somehow reflects the common popularity of Marian sites of pilgrimage, then the most often attested sites would represent the core sites on the Greek Catholic mental map of grace. Interestingly enough, the list of the sites most often included in song collections is headed by the Roman Catholic (Dominican) sanctuary at Pidkamin', followed by the Basilian sites at Počajiv, Sambir and Bar.[51] Most remarkably the declared center of Uniate Marian piety Žyrovyči, as well as the famous and time-honored Marian site at Cholm do not show up at all in song collections, despite the tremendous investments and efforts of the Basilian order to popularize these sites. The assumption that no songs were composed and published for these images is highly unlikely in view of the fact that for both sites coronation celebrations were held, which implies the composition of songs of praise.[52] The question, however, remains what has become of these songs? Why did they not become sufficiently popular to survive into our age? There can be no doubt, though, that both sites attracted huge numbers of pilgrims each year, so that, though the existence of icon songs and their dissemination in song collections may tell us something about the main sites on the Greek Catholic mental map of grace, the lack or absence of icon songs tells us little enough about the relative significance and publicity of any Marian sanctuary. The reconstruction of the mental map of grace of 18th century Greek Catholics would thus necessarily remain incomplete, were it to rely exclusively on the presence of icon songs in song manuscripts. There are still other famous Uniate Marian sites, such as the Lavra at Univ, for which no songs can be identified in manuscript song collections. The particular circumstances and reasons why no songs for these obvious candidates of praise have come down to us can only be guessed at. The simple and straightforward assumption of a documentary gap, i.e., the assumption that there might have been dozens of copies of the respective songs which somehow did not survive into our age, cannot offer a convincing answer. If there ever were songs in praise of these images, they must be assumed to never have attracted the common interest of song collectors. Since we know little enough about what guided collectors of devotional songs in selecting and compiling their personal song treasuries, we are unable to provide an answer for this intriguing question, and it looks like we will be never able to come up with an answer.[53]

4.1. Geographical Hierarchies

All icon songs that we analyzed reveal a common pattern of relating grace to geographical space, which reproduces patterns of hierarchically nested geographical entities. The highest level is defined by the universal *orbis christianus*. On the next lower level the Polish Crown is regularly identified. Interestingly enough, hardly any songs make reference to the Polish–Lithuanian Commonwealth. Instead, some songs refer to the cultural area of 'Russia' (*Rossija* or *ruskij kraj* in the songs), i.e., the Ruthenian lands, which form part of the Polish Crown. On the next lower level, the administrative region (e.g., Volhynia, Podolia) or even district (Polish *powiat*) is referred to. Though the regional level is usually identified in terms of administrative units, it can exceptionally be identified in more volatile and diffuse terms of a specific plight. Most prominently the most heavily afflicted places on the Ottoman frontier are conceived of as a region in their own right.[54] Regions can be defined even more ephemerically in terms of having been afflicted by an epidemic in the more recent past.[55]

The universal dimension of the *totus orbis christianus* is employed only very sparingly to mark off those places of grace that have gained the highest degree of celebrity. The capacity of an image to attract pilgrims from far beyond the immediate local or regional surroundings was made a requirement for a miraculous image to qualify as a candidate for a papal coronation (Levyc'ka 2017–18). Reference to the world at large is therefore likely to be found in songs for icons that have already been coronated as an epithet of honour and entitlement. In the songs we investigated it was mainly the songs in praise of Počajiv (e.g., BG 2016, nos. 111–14) that regularly alluded to the Christian world at large in such expressions as *vselennyja vsja strany* 'all countries of the inhabited world', *pojušče vsja strany* 'all countries sing', *svidětelstvuet . . . blagodatь tvoju i po cělom světě* 'your grace is borne witness to all over the world'. Only few other images are praised for attracting pilgrims from all directions, and it can only be surmised that reference to pilgrims from far away was possibly introduced into song texts with the aim of preparing a coronation procedure.[56] The politico-historical dimension of miracles is commonly addressed by referring to the national and immediately subnational level of geography. Thus, with an obvious reference to the Marian epithet of Queen of Poland (*NMP Królowa Polski*), established by king Jan Kazimierz in 1656, many icon songs relate miraculous action to the lot of the Polish Crown, as in one of the songs in praise of the Mother of God of Počajiv *Veselo spěvajte, vsi čelom vdarjajte*, which depicts the intervention during the Ottoman crisis of 1675 as an act meant to save the entire Polish nation:

> V čas Zbaražskoj vojny bjachu nespokojny
> Hrady vsja ot běsurmanov { . . . }
> Ktož vo zemnom troně i v Polskoj koroně
> Sčirym serdcem nepriznaet,
> iž děva čista
> mati boha ista
> vsěch ot vrahov zastupaet. (BG 2016, no. 112)

> In the times of the Zbarazh war [of 1675]
> the towns were unsettled because of the Muslims { . . . }
> Whoever on the worldly throne and in the Crown of Poland
> will not admit with an open heart
> that the chaste virgin
> the true Mother of God
> defended us all from those enemies.

More frequently, however the relation of miraculous images to the Polish Crown is framed in an indirect manner. Many songs close with a stanza of general supplication, which is not directed specifically to the image in question, but to the Mother of God more generally. In these supplications standard formulas asking for the protection of the Polish Crown are regularly included: *Błogosław Polskiey Koronie* 'Bless the Polish Crown' (BG 2016, no. 122, Bilostok), *Da dast v Polskoj koroně/mirno vsěm nam žiti* 'Grant to everyone of us living under the Polish Crown a life in peace' (BG 2016, no. 128, Tyvriv). Some of the songs also specifically refer to the Ruthenian lands of the Polish Crown, thus establishing a prestabilized politico-geographical harmony in which the Ruthenian lands are imagined as a specific manifestation of the Polish Crown. Though the famous first line of the song in praise of the Mother of God of Pidkamin' *Prečistaja děvo, mati Ruskaho kraju* 'Most chaste virgin, mother of the Ruthenian land' has been repeatedly referred to as a Ruthenian or rather Ukrainian national anthem of sorts, this interpretation is proven wrong by the last stanza of this song, which indeed closes on the words *Budi vsěm nam oborona/Prosit tja Polska korona jako carici* 'The Polish crown asks you as her queen, to serve as a defense to everyone of us' (BG 2016, no. 134).

There is a general tendency for songs in praise of lesser known local images to use geographical references that identify the next higher administrative unit to which the place belongs—probably for practical considerations of making the place identifiable for the recipient (listener, collector)—or to drop the geographical dimension altogether. In the remarkable case of the song *Preběhaem k tebě, nebesna carice* in honour of the Mother of God of Povča, there is indeed no specific reference to the place and the deeds of the image at all. The song would look like an ordinary song of general supplication, were it not for the refrain line, which makes explicit reference to the image (Žeňuch 2006, pp. 910–11).

The texts of icon songs seem to contribute little enough to the establishment of a specific geography of grace beyond putting it in place within the common framework of sociopolitical geography. The role of these songs seems rather to lend structure to the totality of all places of grace by singling out the more deserving places from their vast mass and thus creating a hierarchy of places for particular needs against places of everyday spiritual counsel, support and consolation. One possible reason for why songs would not contribute in a direct manner through their text narratives to establishing a commonly shared Greek Catholic geography of grace might be the competitive attitude prevalent in most songs, which adhered to the golden rule of keeping silent about any other competing places of grace, even those which belonged to one's own confessional network. Even songs in praise of second to third order miraculous images would phrase their praise in a way, as if there was no other miraculous image around. Thus, the song *Pomozi nam Christe Bože* in praise of the Mother of God of Piddubci depicts its object of praise as a unique blessing of the heavens offered to the whole of Volhynia:

> Ježeli zvažišъpričinu,
> ČemъJu za matku jedinu,
> Na Volynju namъNebo dalo? (BG 2016, no. 119)

> If you wonder about the reason,
> why the Heavens gave her
> as the one and only Mother to Volhynia?

In a likewise competitive attitude the song in praise of the Mother of God of Počajiv *Nyně proslavi sja Počajevska skala* (BG 2016, no. 116) makes a claim of having defeated the epidemic of 1770 even at Pidhirci, Kam'janec'-Podil's'kyj and Krem'janec', which can boast well-known miraculous images of their own, thus denying their healing powers by hushing up their existence.[57] There is only one icon song to break this rule of silence,[58] i.e., the song *Przed wieki Niebios i Ziemi Królowa* in praise of the image of Puhinki, which somewhat too frankly states that there are other places such as Puhinki where grace can be obtained:

> Przed wieki Niebios i Ziemi Królowa,
> U którey pomoc dla wszystkich gotowa,
> Którey iak w wielu mieyscach tak w Puhinie,
> Łaska z znacznemi dobrodzieystwy płynie. (BG 2016, no. 126)

> Pre-eternal Queen of the Heavens and Earth,
> who offers help to everybody,
> whose grace flows forth with considerable benefits in Puhinki,
> as in so many other places.

These lines make Puhinki look an almost random choice, which would make any pilgrim wonder why he should take the trouble to specifically go there. It is here that a quite different perspective on Uniate places of grace is granted. In these lines, places of grace are not so much seen as something exceptional and unique for one specific place. The many places alluded to may be reasonably assumed to be Basilian places, such as Puhinki. Puhinki is thus being advertised as part of a dense supply network, that provided each and every believer with conveniently accessible local grace. Individual places of grace are being framed here as part of a supply chain for graces. It is here that grace turns from something supernatural and inexplicable into an ordinary, everyday service provision. Making miracles a feature of mass consumption is liable to ultimately break down the dividing line between the sacred and the profane.

With this one notable exception, it is not so much songs in praise of specific images, but rather common Marian songs that put the networks of images into a broader perspective of a geography of grace. A case in point is the song *Radujsja Marie, nebesna carice* (Stern 2000, pp. 613–14) which establishes the idea of a chain of frontier icons defending the Polish Commonwealth against the impending Ottoman invasion. The song depicts the simultaneous weeping activities of images at three frontier sites (Terebovlja, Kam'janec'-Podil's'kyj, Chotin) as a coordinated action to fend off the aggressors (Stern 2000, p. 36). Another example, which is even more obviously meant to convey the idea that grace is ubiquitous and evenly distributed, is *Čistaja děvo, nebesna* carice, first attested in the Kamjanskij Bogoglasnik of 1734 (Žeňuch 2006, pp. 907–8). This song lists in apparently random succession miraculous images at Roman and Greek Catholic sanctuaries throughout Poland and Lithuania (Sokal', Częstochowa, Borek, Jaroslav, L'viv, Pidkamin', Žyrovyči, Boruny, Počajiv). By and large, it looks like icon songs focus exclusively on the local image, as if they were meant to advertise it against competing images around. Greek catholic icon songs testify to the conflicting forces of territorial competition, which on the one hand aim at establishing a common and shared Uniate geography of grace, but on the other hand evolved into an internal competition among Uniate monasteries vying for supremacy, which to some extent was liable to counteract the emergence of a collective Uniate geographical identity.

4.2. Lop-Sided Competition

Establishing a collective Uniate geographical identity appears to be at the heart of Basilian efforts at promoting places of grace. The ultimate aim was to create a secluded mental territory within which Greek Catholic identity would be allowed to grow without being affected by the dreaded pull of Orthodox tradition. Luckily for the Uniates, the remaining Orthodox strongholds of centuries-old Marian devotion were few, and, moreover, Orthodox monasteries and clerics did not enter into a comparable competitive activity. It looks, like Orthodox clerics took a quite different attitude to confessional competition, which appears to reflect the difference between an old confession (orthodoxy) that has lost its political backing and support and owing to its loss of cultural significance has drifted into a state of fatalistic inaction,[59] and a new confession that, though it was confronted by prejudice from all sides, still had the fervor of a young and dynamic movement that was ambitiously struggling for its place (to be taken quite literally) in this world. The Orthodox-Uniate competition thus turns out to be asymmetrical, with competitive action being basically restricted to the Uniate side.[60] Uniate competitive action did, however, nowhere, except perhaps in the case of the eparchy of Cholm (Jakovenko 2011–14), take the form of active encroachment upon the remaining Orthodox territory.[61] The Uniate method of choice to secure one's territory was complete ignorance and erasure from the mental map of one's confessional community of anything Orthodox in order to create a hermetically closed parallel confessional world. The success of this attempt may, however, be doubted. Though, by and large, pilgrims stuck to their local or regional places of Marian devotion, many of the more renowned places seem to have been frequented by supplicants across the confessional divides.[62] Whether these occasional crossings of confessional boundaries sufficiently substantiate the assumption of a common Uniate-Orthodox hierophantic macrospace (Jakovenko 2011–14) remains doubtful, though. Especially the introduction of the ultimately Roman Catholic practices of annual pilgrimages of forgiveness, as well as other forms of mass mobilization (the setting up of commissions for public hearings of witnesses, coronation procedures and celebrations, propagation of knowledge about places of grace through the proliferation in print of miracle reports, songs and print reproductions of images, etc.) were liable to erase within a relatively short time span the hereditary Orthodox map of grace and write it over with the new Greek Catholic map.[63]

Thus, as it comes to advertising and promoting places of grace, there is a slight confessional bias in Greek Catholic, especially Basilian efforts to get a foothold on the market of devotion and grace, which, however, never amounts to more than keeping silent about the great places of grace of Orthodoxy nearby and farther off. Only occasionally are there reports—all of them Roman Catholic in fact—that stress the superiority of Catholicism by referring to the loss of miraculous powers whenever a place of grace falls into the hands of Orthodoxy (e.g., Počajiv in 1839). The most outspoken case of confessional shaming is the legend of the icon of Smolensk turning against its Russian Orthodox flock on the approach of the Polish troops of Władysław IV (Kruk 2011, p. 166). For the rest, competition is largely silent and implicit in actions and choices. The ultimate cause for the mildness of competition can possibly be seen in the vividly felt unity of worshippers of miraculous images against the backdrop of Protestant and Muslim iconoclasts, which at the grass roots level resulted in a solidarity of pilgrims from all competing confessions applying to the same Marian images of supraregional fame, such as the Mothers of God at Rudki, Sokal and the Dominican monasteries at L'viv and Pidkamin', and last but not least at the Basilian monastery at Počajiv (Kruk 2011, p. 179).

4.3. Closing the Ranks: Frontier Icons at the Antemurale Christianitatis

The rise of Marian devotion in the 17th–18th century has been repeatedly explained and rationalized as a reaction to the political and military insecurity of the age caused by Tatar and Ottoman Turkish raids which affected the larger part of the Ruthenian lands (Medvedyk 2015, p. 371). There can be no doubt that the insecurity of the situation helped the rise of Marian devotion, but the propagation of Marian places of grace was rather a general project of the counter-reformation all over Europe. The military confrontation offered, however, a convenient opportunity to heighten the sense of spiritual territoriality by closing the ranks against the one and only true adversary of faith and traditions alike. Miraculous images came to be imagined as part of the Christian defense system of the *antemurale christianitatis*. Within the Polish–Lithuanian Commonwealth the term came to be applied to the chain of frontier towns and fortresses of Chotim, Terebovlja, and most notably of Kam'janec'-Podil's'kyj (Tazbir 2017, p. 78). All of these frontier towns hosted Marian places of grace which in a way served as spiritual fortresses, thus becoming part of a war which was not only about physical conquest. It is here that the notion of sacred space is extended once again into the public and political domain, as in the case of coronation celebrations. Moreover, it is again devotional songs that make this notion of miraculous images as part of the military defense system explicit. Songs in praise of icons that can boast a victory over Ottoman troops, often take the form of a fairly detailed narrative account of the historical event, as most prominently in the case of the song *Veselo spěvajte, čelom udarjajte* (BG 2016, no. 112) that celebrates the defeat of the Turkish troops during the siege of Počajiv on 20 July 1675. The authors of the Bogoglasnik made a point of explicitly referring to this event in the title heading to that song: Ѡ преславной побѣдѣ еѧ надъ Турками бывшей Року Бжіа АХОЕ Мⷰа Іюла к҃ дна ("On the glorious victory over the Turks that took place in the year 1675, in the month of July, on the 20th day"). The song provides a vivid description of how the Ottoman troops were frightened off by a visionary appearance of the Mother of God and a legion of angels above the monastery, protecting her wards with her cloak (*omofor*) in imitation of its most famous model, the Blacherniotissa (Trajdos 1984, no. 128). In fact the imagery of the *Pokrov* miracle experienced a repopularization in the 17th–18th centuries in paintings of the period, but perhaps much more so in devotional songs in which reference to the *omofor* in connection with the defense from the Turks became a common and almost ubiquitous topos (Senyk 1984, pp. 268–69).

Likewise, songs in praise of less prominent images that provided their service on the immediate Ottoman frontier, tend to renarrate the military events, as in the song *Prijdĕte vsi činy, aggelstii lici* (ms. Maslov 48, fol. 116v–117r; cf. Hnatjuk 2000, pp. 193–94) that details in two stanzas the unsuccessful Ottoman attempts to blow up the fortifying walls of Terebovlja. As a matter of fact, the event by itself does not qualify as particularly miraculous. There is nothing obviously supernatural in the failed attempt of detonating an explosive charge. However, explaining the happy turn as a simple stroke of good luck would do injustice to the presence of a miraculous image nearby in a way as to almost verge on blasphemy. So, in this situation of imminent danger even the most inconspicuous event turns into a miracle and is thus made part of the larger history of divine intervention in favor of Christianity and the Polish Crown. The panegyrical narration of these historical events in songs turns the respective places of grace into symbolic places of collective historical memory, and thus links the sacred to the political.

It looks like Marian images formed part of the Polish defense system against Ottoman attacks. Where military means failed, local images were applied to as a last resort. Marian images are therefore likely to have been perceived as a necessary and indispensible piece of equipment of any fortified frontier town. As a matter of fact in some accounts miraculous images are in fact wielded almost like weapons. Thus, the Carmelite monastery at Berdyčiv is reported to have survived an Ottoman attack and siege owing to the courageous action of a nun who confronted the Ottoman troops on the monastery wall wielding the miraculous image as a shield to deflect the Ottoman arrows and send them back on the enemies who were thus killed in significant numbers (Chociszewski 1882, p. 263).[64] The most extreme interpretation of miraculous images as weapons of defence is represented in the notion and associated practice of the war icon, which accompanies troops heading for battle. In 1673 the Mother of God of Cholm was indeed taken to the battleground in king Michał Korybut Wiśniowiecki's campaign against the Ottomans (Wereda 2018, p. 71).

Besides proving potent weapons against troops of infidels, icons seem also to be indestructible and do not give way to the brute force of attackers. Thus, at least in several cases, miraculous images are reported to have survived unscathed devastating fires layed by Tatar raiders (e.g., Nestanyč, Onyškivci). Individual attackers, however, may succeed in inflicting wounds to miraculous images, which leave scars, as in the case of the icon at Rudnja, which was whipped by heretics (Lužnyc'kyj 1984, p. 181).[65] These scars, however, add to the glory of the image and are kept as a sign of suffering, not unlike the stigmata of Christ, which, as in the case of the icon at Pidhirci, resist attempts at being painted over (Barącz 1891, p. 219).[66] Stories which elaborate on the invincibility and indestructability of miraculous images ought to be seen against the backdrop of profanations of religious artefacts, especially images and crosses by Ottoman invaders. Especially, the profanation of icons at Kam'janec'-Podil's'kyj upon the order of Mahomet IV in 1672 (Kruk 2011, pp. 155, 317) as well as the destruction of the church and the desecration of the icon of the Mother of God at Cholm (Kruk 2011, p. 154) must have left a traumatic imprint on the collective memory of the Ruthenian population of all denominations alike, which the legends of indestructability seem to try to make amends for by glossing the spiritual wound over by narratives of perseverance. In fact, in this particular period of crisis, rather than offering protection to the populace, miraculous images were in need of protection and care themselves. Thus, the miraculous Mother of God of Kam'janec'-Podil's'kyj was brought into safety at L'viv in 1672 by Armenian catholics, only to return to its original place in 1699 after the Ottomans were defeated for good.

In relating the sacred to the suffering in a world of historical and political upheaval, miraculous images turn not only into weapons against the enemy, but are also invoked to serve the consolidation of the political power of the state as means to preserve peace and stability: "Matko pokoiu zrządź Polskiey swobodę" ('Mother, provide for the peace of Polish freedom', BG 2016, no. 118). It is here that the individual quest for salvation and protection is transcended into a national concern, but it should also be noted, as Kruk (2011, p. 166) rightly pointed out, that the basis for collective identity formation through miraculous images is essentially local, i.e., it remains ultimately restricted to the experiential domain of the individual. It is not so much the abstract idea of a Polish nation which is at stake here, but the conditions of everyday life, of which spirituality forms an integral part in a very specific place of its own, the "world of the frontier fortresses" of Podolia, Volhynia and Galicia with their lived experience of "incessant battles against the perennial Turkish and Tatar enemy", where a particular environment of popular baroque culture emerged (Tazbir 2017, p. 73), in which miracles became enlarged into almost cosmic dimensions, that ultimately would even transcend merely national concerns. We have developed earlier the idea of the sacred engaging with physical space through images by establishing tight relations of inalienable ownership of place. This very idea is fully played out on the politico-social level of frontier and war icons, which, bluntly speaking, do nothing but defend their own place. Miraculous images thus become part of the overall power structure which defines itself in terms of territorial ownership.

5. Conclusions

The core of religious identity can reasonably be assumed to consist basically in shared beliefs about the sacred and spiritual dimensions of the world we inhabit and the shared practices associated with these beliefs. However, what if very similar beliefs are shared by competing religious communities to a degree that it becomes hard to tell one from the other and if it is, moreover, felt in one of these communities inopportune to overstress differences against common features, as was the case with Greek and Roman Catholics? The Greek Catholics of the 17th and 18th century were encumbered with a twofold identity problem. First, theirs was a confession of recent creation which had to create cohesion out of a mix of the centrifugal forces of initial indifference and traditionalist antagonism. Moreover, theirs was about to become a hybrid confession at best, with very little which it could call its own in terms of exclusive ritual or dogmatic assets. Relying on exclusively shared beliefs or age-old family allegiances was no option. Where the conceptual approach to the world of the sacred through spiritual and metaphysical ideation and dogma failed, the immediacy of physical experience of the spiritual world offered an attractive alternative to reach out to one's flock in an attempt to create communal cohesion. Following the counter-reformatory Catholic model of popular mass mobilization through the objectification and commodification of the sacred in magic practices enacted and made sense of in popular texts, objects and events, Greek Catholic clerics set out to define what it means to be a Greek Catholic in terms of physical space being imbued with spirituality. Uniate spirituality manifested itself in soundscapes of songs, prayers, litanies and coronation concerts as well as in the visual landscapes of pilgrimage routes and the events and facilities offered at the places of grace as their final destinations, thus turning Uniate identity into an embodied experience. Claiming patronage and ownership over this geography of grace and designing it according to one's own hierarchical patterns, thereby embedding it into the structure of the state as a whole, proved the way for Greek Catholicism to take firm root among their envisaged community by demonstrating its sacred and secular legitimacy. The adoption of mass pilgrimages of forgiveness according to the Catholic model of indulgences as well as establishing a hierarchy of sacred places through papal coronations welded the flock to their confession, but also the confession to the Polish Crown. Territoriality became instrumental in the integration of the sacred, the political and the communal. Whenever, though, this narrowly defined territory would become subject to disintegration, Greek Catholic confessional cohesion was likely to lose its grip.

Funding: This research received no external funding.

Institutional Review Board Statement: Not applicable.

Informed Consent Statement: Not applicable.

Conflicts of Interest: The author declares no conflict of interest.

Notes

[1] In this article the terms 'icon' and 'image' are used somewhat interchangeably at times. We are well aware that 'icon' is usually understood in the quite specific art historical sense of a painted sacred image of Byzantine origin and/or style, distinguishing it from sacred images in Western styles (in Poland mostly baroque) or even statuettes. As a matter of fact, the vast majority of miraculous images mentioned and discussed in this paper are indeed icons in the stricter sense of paintings in the Byzantine style, but some of them are not. In order to avoid confusion, we use '(miraculous) image' as a default term, employing 'icon' only sparingly in those cases where the object referred to is indeed a true Byzantine style icon. Out of stylistic considerations, we also took the liberty to use 'icon' in the more general sense of 'image' in compounds such as 'icon song' or 'frontier icon'.

[2] For convenience's sake we will call this fuzzily defined terrain the Ruthenian lands with reference to the historical, ethnic, cultural and linguistic heritage of its majority population. Most miraculous Marian images on Ruthenian soil made their first appearance in the 17th–18th c. with a clear peak in the 18th c. See the lists in Šćurat (1910, pp. 10, 14). The spread of miraculous images of the

Mother of God forms, of course, part of the general rise of Marian devotion throughout Poland and the Ruthenian lands, which for the latter territories can also be gleaned from the increase of the relative share of parish churches bearing the name of the Mother of God (Jakovenko 2011–14). See also Senyk (1984). As for the Lithuanian part of the Polish–Lithuanian Commonwealth with its predominantly Ruthenian population that likeweise had to accept the Church Union, similar forms of Marian devotion can be observed, though to a notably lesser extent. We will therefore not specifically look at that region and only include it where it adds to our understanding of Marian devotion in the Ruthenian lands in general. As a matter of fact, the Ruthenian part of the Polish Crownlands (especially Galicia and Volhynia), where the Uniate Chruch came to grow particularly strong, clearly emerge as the active epicenter for the adoption and diffusion of devotional practices that follow the Western, i.e., Roman Catholic models of their time, such as the foundation and active maintenance of Marian places of grace, the establishment of annual pilgrimages, composing and collecting of devotional songs, etc. An apparent reason for this may be that in the Polish Crownlands the presence of Roman Catholic culture was much more imminent and formed part of the lived experience of the entire population, Roman and Greek Catholics alike, so that the ground for cultural confessional bricolage was better prepared here than in Lithuania.

3   The key event of iconoclastic aggression in the Polish–Lithuanian Commonwealth is evidently the 1430 Hussitic raid on Jasna Góra and the damaging and defilement of the icon of the Black Madonna. Moreover, the abortive Swedish siege of Częstochowa in 1655, which was ascribed to the intervention of the Black Madonna and led to the Mother of God being pronounced as the Queen of Poland, must have once again conjured up the iconoclastic threat (Niendorf 2010, p. 159). To these can be added minor acts of Protestant iconoclastic agression throughout Poland in the 16th century (Kruk 2011, pp. 152–54). On the whole, however, outspoken anti-Marian iconoclastic attitudes and actions were a rare exception among Polish protestants (Tazbir 1984, p. 226). For the pertinent topos of damaged and desecrated images in general cf. Kretzenbacher (1977). It must be added here that for Greek Catholics protestant iconoclasm was much less of an issue. Their territory being largely situated in the endangered frontier zone of the Ottoman empire, the role of desecrating aggressor was naturally taken over by Muslim military invaders (on which see below). On alternative explanatory accounts of the rise of Marian devotion in the 16th–17th century throughout Europe, see Jakovenko (2011–14) and Rok (2005, p. 137).

4   Levyc'ka (2017–18) identifies the 18th century bishop of L'viv, Iosif Šumljans'kyj (1700–1708) and the Uniate metropolitan bishop Atanasij Šeptyc'kyj (1729–1746) as the most active supporters of the spread of Marian worship among the Greek Catholic community through the establishment and propagation of ever new Marian places of grace. For a recent case of promoting confessional identity through pilgrimages and apparition sites among the Greek Catholics of Eastern Slovakia see Halemba (2008).

5   The political dimension of the rise of Marian devotion in the Polish–Lithuanian Commonwealth has already been noticed and commented upon by Ščurat (1910, p. 4), but we do not agree with his interpretation of Marian devotion being employed to serve the Ukrainian national cause in an attempt to fend off Polish advances into Ukrainian religious domains.

6   It is not clear whether our ethnographic approach to places of grace with its focus on interpretation and sense-making of geographical space at large would be covered by Lidov's (2006) notion of hierotopy, which is about the organisation of sacred space, to be sure, but focusses rather on aspects of immediate sensual experience especially of the ritual space as well as the art of translating mystical into sensual experience, where our approach is rather about the creation of group cohesion and corporate identity employing hierotopic features to induce interpretations of geographical space in relation to community.

7   Examples are the miraculous Marian images at Pan'kivci (Brody region, L'viv county, Galicia), Benevo (Turkiv region, L'viv county, Galicia), Lysovci (Zališčyky region, Ternopil' county, Podolia), Prjaživ (Žytomyr county), and Jatwiesk (Volkovysko area, Podlachia).

8   Lužnyc'kyj's (1984) distinction between miraculous (*čudotvorni*) and blessed (*blahodatni*) images, which seems to follow common practice, is somewhat fuzzy, so that assigning one of both labels to any one image may in some cases verge on the arbitrary. Blessed images are usually recognized only locally, and they are rather known for offering solace to the supplicant and answering prayers in a less conspicuous manner. The difference is certainly one of degrees, but it may possibly be given a sharper outline by looking at it from a more technical vantage point. Miraculous images produce eyewitness evidence that qualifies for being checked and evaluated by an episcopal committee, whereas blessed images usually rely for their fairly modest fame on the inner subjective experience of individual supplicants that as a rule will not be able to produce any kind of convincing evidence of intercession. As a matter of fact, for an image to become eligible for papal coronation it had to be officially recognized as miraculous by an episcopal committee. The recognition procedure would require a public hearing of witnesses and included also the publication of miracle reports in printed form (Levyc'ka 2017–18).

9   Though, Rožko (2002) makes an attempt at including also all images of the *blahodatna* type in his survey of the places of grace of Volhynia.

10   On Boruny v. Senyk (1984, p. 265).

11   The Uniate Basilian order (*Ordo Sancti Basilii Magni*) was founded in 1631 with the clear objective to promote the Church Union among the as yet Orthodox faithful of the Polish–Lithuanian Commonwealth. In addressing their task the Basilian Order adopted the organizational forms and patterns as well as the educational programme and policy of the Jesuit order. Like the Jesuit order, the Basilian order laid within its domain of action the foundation for a school system consisting of a dense network of monasteries and associated seminaries.

12   For the Greek Catholic Basilian order utilizing this pathway to establish a stable geography of grace of their own meant to adopt Roman Catholic ecclesiological positions on forgiveness and indulgences (Wereda 2018, pp. 69–70).

13 Mostly on Marian feast days after the harvesting season, when the rural population was free to move. Cf. e.g., Žyrovyči, Zahoriv—September 8 (Nativity of the Blessed Virgin Mary), Cholm—September 15 (Our Lady of Sorrows). Cf. Wereda (2018, p. 75).

14 Apart from the coronated images, for which the establishment of approved pilgrimages of forgiveness was a prerequisite, these are: Povča, Novyj Sambir, Krasnyj Brod, Zarvanycja, Zahoriv, Puhinki, Bilostok, and possibly still others, which may be lacking from this list owing to the fragmentariness and incompleteness of much of the available historical information on Ruthenian places of grace.

15 For Polish examples see also Kruk (2011, p. 187), who also refers to an exceptional case of tears shed by an image being interpreted as tears of joy over the imminent victory of Władysław IV.

16 To be here understood in opposition to heterodox, i.e., to be understood as orthodox in the broad sense of any Christian confession that is conservatively oriented towards tradition and accordingly allows for the worship of images (Greek or Roman Catholic, or Russian Orthodox). The according Slavic term would be the broader *pravověrnyj* rather than the specifically confessional *pravoslavnyj*. As a matter of fact, despite their often embittered rejection of each other, proponents of Russian Orthodoxy as well as Greek and Roman Catholicism tend to form a unified cultural block in the face of any form of belief that does not embrace miracles worked through saintly images as part of its spirituality. The fear of Muslims seems to be largely inspired by a fear of the destruction of sacred images and the concomitant loss of the sense of living in a sanctified space. The same would hold in principle for Protestants, who are indeed in some Ruthenian devotional songs portrayed as a menace to the proper faith of no lesser evil than muslims. There is one notable case of a Marian devotional song which comments on the Ottoman inroads into Podolia in the 1670s, where Muslims (commonly adressed in songs as *bisurmany, aharjany* or *pohani*) are addressed by the term 'heretic' which is usually reserved for Protestants: *Jeretičeskaja ruka vsěchъnasъogortujetъ* 'the hand of the heretic has a firm grip on us' (Stern 2000, pp. 44–45). An as yet more explicit identification of Muslims with Protestants, who are contrasted to all orthodox Christians (in the broad sense outlined above), is found in the following lines taken from a song in praise of the Mother of God of Tyvriv: *Smiri glavy agarjanъ/Postydi sonmъLjuteranъ/Voznesi rogъpravověrnychъchristianъ*" ['humble the heads of the sons of Hagar (i.e., Muslims)/Put shame on the assembly of Lutherans/Increase the power of all Christians in the right faith'] (BG 2016, no. 128). For the parallelisation of Muslim and Protestant iconoclasm specifically directed against Marian images throughout South Central Europe, cf. also Kretzenbacher (1977, pp. 94–105).

17 The following is a list of weeping images of the Ruthenian lands in the chronological order of their first miraculous manifestation: Ilins'ko-Černyhivs'ka ikona 1662, Terebovlja 1663, Klokočiv 1670, Pidhirci 1692, Povča 1696, Verchrata (Krechiv) 17th c., Vicyn end of the 17th c., Horodok beginning of the 18th c., Nastašiv after 1701, Balykyni 1711, Sambir 1727, Bucniv after 1729, Rohizno 1734, Vyšnivčyk 1742, Ternopil' 1st half of the 18th c., Tartakiv 1765, Dalešiv 1781. The rise of reports about weeping images should, however, not exclusively be accounted for in terms of the heterodox iconoclast threat. The particular emphasis on strong emotions of compassion possibly reflects the general folk baroque Catholic trend to foreground the human aspects of the relation of the Mother of God to her son, which found its most outspoken expression in the depiction of the Mother of God as fainting from pain and woe under the Cross, a topic which is also repeatedly found in Ruthenian Marian songs on the passion of Christ, as in the following lines taken from the song *Vse stvorenie po umeršem paně: Mati Christova pod krestom stojala/Z žalju tjažkago srodze omlěvala* ['The Mother of Christ stood under the Cross, she fainted severely from her heavy woe'] (Stern 2000, pp. 155–56, 533). The particular emphasis on emotions of compassion is ultimately grounded in Franciscan mysticism and its conceptualisation of the Mother of God as *mater dolorosa* being pierced by the sword of pain (*gladius doloris*). Later on, similar notions were elaborated upon by the prominent theologist of Russian orthodoxy, Gregory Camblak in his Homily on the Dormition of the Mother of God (Trajdos 1984, p. 134). Among the Greek Catholic community the cult of the *Mater dolorosa* spread, however, via Western *pietà* iconography from the mid 17th century onwards (Senyk 1984, p. 273).

18 See Kruk (2011, p. 174) on Žyrovyči and parallel cases in Poland.

19 For the origins and other examples of this topos see Kruk (2011, p. 173).

20 Vasil' Surožskij stresses in his treatise *O jedinoj istinnoj prawoslawnoj věrě* (Ostrog 1588), accordingly, that the miraculous powers of icons derive from their being created through the assistance of God (RIB 7 1882, col. 929–30).

21 It goes without saying that reports and legends about miraculous images reflect the perspective and interests of its authors. Thus, wherever the author of a legend is likely to be a Basilian, as in the case of Pidhirci, movement to a Basilian sanctuary is unlikely to be reported as having caused averse reactions by the image being moved.

22 This report, in fact, reflects a very common topos of long-standing, which in the Polish–Lithuanian Commonwealth was strongly reinforced by a similar legend about the Black Madonna of Częstochowa (Kruk 2011, pp. 180–83).

23 Only exceptionally and under very specific circumstances will miraculous images swap loyalties, as in the case of the Mother of God of Smolensk, which is reported by a Polish chronicler to have turned on their own Orthodox wards when being attacked by the troops of the son of the Polish king Władysław IV (Kruk 2011, p. 166). Similarly, an indeterminate number of smaller 'Muscovite' icons seem to have been brought to Poland by the same invading troops during the times of trouble without losing their miraculous powers (Kruk 2011, p. 62). Thus, even the strict local loyalty of images seems to have its limits where issues of the right confession are concerned, as when an Orthodox surrounding can be exchanged for a Catholic one. Here, again, the confessional background of the report is crucial.

24 A key contemporary text which elaborates from an Orthodox perspective on the relationship between icons and the sacred, is Surožskij's *O edinoj istinnoj pravoslavnoj věrě* (1588)—especially the closing chapter on images and icons (*O obrazochъi ikonachъ*).

Surožskij admits on the one hand that icons and other holy objects are created by men out of dead matter, but that God chooses to dwell in them and engages through them with this world. Thus, the icon is for Surožskij at the same time representational of the sacred as well as imbued by the sacred. It holds the middle ground between conceptual sign (col. 921 and col. 929–30) and material medium cum chosen place for God to be in this world (col. 921, 923–24, 926–27). According to Orthodox notions, miracles are not worked immediately through icons, but through the grace of God who thus reacts to icons being applied to for help (Levyc'ka 2017–18). Where according to this officious Orthodox conceptualization icons are not to be identified with the sacred and cannot therefore acquire personal features, under the influence of Catholicism this boundary becomes increasingly porous through phenomena such as weeping and bleeding images, which strongly suggest a true animacy and personhood for images (Niedźwiedź 2005, p. 95; Kruk 2011, pp. 184–85). See also Kretzenbacher (1977, pp. 7–8) who refers to this as empsychosis and ensomatosis. It is through weeping that communication between the faithful and the sacred turns from a one way road of mere supplication into a mutual, though not truely dialogic exchange of personal perspectives and intentions by means of interpretable communicative behaviour. When in her attempt to deny this material aspect of Marian worship Oleksandra Hnatjuk concludes that in the praise of icons "нівелюється не лише світський час, але й простір" (Hnatjuk 1994, p. 118) she falls victim to a disembodied idealism that is not borne out by the facts, as was already remarked by Medvedyk (1999, p. 76).

[25] See also Niendorf (2010, pp. 158–59).

[26] Copies of the Black Madonna of Częstochowa that had been rubbed against the original were held in highest esteem, and some of them became famous destinations for pilgrimages themselves, such as the icon of Mydlów (1748) and Solec (1763) (Kruk 2011, p. 161). See also Rožko (2002, p. 86) for the miraculous icon of Počajiv, copies of which, that were made miraculous in the way described, were disseminated over Ukraine and Russia.

[27] Evidence for the Ruthenian lands is hard to come by Chociszewski (1882, pp. 263–64) mentions an anecdote of Juliusz Słowacki's being sent by his mother a print repoduction of the Mother of God of Berdyčiv, which served him as support in many hopeless situations. Print reproductions (wood cuts and copper etchings) were made of the miraculous images at Pidkamin', Sokal', Berdyčiv, Počajiv, Hošiv and Zarvanycja (Levyc'ka 2017–18; Wereda 2018, p. 76). It is, however, not known whether these reproductions were being sanctified by rubbing them against their originals. It can, however, be surmised that this practice which was quite commonly applied to small print reproductions of the Black Madonna of Częstochowa (Kruk 2011, p. 161) was also known and accordingly employed among the Greek Catholics of Ruthenia. It seems, however, that a more common practice at Częstochowa consisted in rubbing kerchiefs and rozaries against the image (Tazbir 1984, p. 227). See also Kürzeder (2005, pp. 52–54) for a detailed discussion of similar practices in the Southern German-speaking lands.

[28] This point is neatly illustrated in Saltykov-Ščedrins novel *Gospoda Golovlevy*, where the major protagonist Arina Petrovna, head of the Golovlev family who is obsessively bent on the acquisition of material riches, orders a service of supplication (*moleben*) to the famous miraculous icon at the Iviron monastery on Mount Athos to help her purchase by auction a village of 1000 souls (Saltykov-Ščedrin 1979, p. 39). The *Iverskaja bogomati* was indeed one of the most popular icons among all the Eastern Slavs and had many replicas at various places throughout the Ruthenian lands (Lužnyc'kyj 1984, p. 167). Citing this biting satire is not meant to suggest that Arina Petrovna's might reflect a common attitude. It serves only to highlight the inherent tendencies in supplication and its desanctifying potentials. By the by, her application to a far away icon of world wide fame instead of to a local icon nearby tells a lot about the competitive hierarchy that holds between places of grace which translates into the branding of efficacy and the emergence of a system of exchange values on the market of miracles.

[29] As in the case of the resurrection of a child on 15 August 1710 (Przesławna gora poczaiowska Dawnością Cudow Przenayczystszey Bogarodzicy Panny od cudownego Jey Obrazu wynikających [1787] 1801).

[30] Several of the reports on the early healing miracles at Pidkamin' describe this procedure with the prefabricated formula "ofiarował się do Obrázu Naświętszey Pánny mieyscá Podkamienieckiego" ['he offered to the Image of the most holy Lady at Pidkamin"]. When the healing was granted and health fully recovered, the supplicant would set out to Pidkamin' in person ("y nieco po chrobie wypocząwszy reddendi voti causa był w Podkamieniu", Okolski 1648, fol. 27v).

[31] A desire for the physical representation of miraculous powers and their visible manifestation can also be gleaned from the common practice of adorning miraculous images with gold or silver frames, studding them with precious stones, and the like, as well as hanging votive tablets around them. These practices were common all over Catholic Europe and the Roman and Greek Catholics of the Polish–Lithuanian Commonwealth did definitely not lag behind this common trend. For Roman Catholic Poland see Kruk (2011, pp. 214–17, 245–58). Among the Greek Catholic places of grace Počajiv and Pidhirci stand out for their wealth and abundance of the votive tablets they were adorned with. This basically Catholic practice became also quite popular among the remaining Orthodox communities (Senyk 1984, pp. 263–64). It goes without saying that the number of votive tablets accumulated served also as visible evidence of the efficacy of the image in question and would raise its esteem and value among pilgrims.

[32] On coronations of miraculous icons as a key hierotopic element of Greek Catholic efforts to stage the sacred foundations of state and society, see Levyc'ka (2017–18) and Wereda (2018). The coronation of miraculous images came to be a common practice under Pope Urban III from 1630 onwards. It was in the beginning restricted to Italy, especially Rome, and spread abroad with the coronation of the Black Madonna of Częstochowa on 8 September 1717. Only those miraculous images that could be shown to be ancient, which preferentially implied a Greek (Byzantine) origin, were eligible for coronation. Cf. Kruk (2011, pp. 241–45) and Levyc'ka (2017–18). Among the images that constitute the Greek Catholic geography of grace, the following were coronated in the 18th century: Žyrovyči (Basilian, 1726/1730), Pidkamin' (Dominican, 15 VIII 1727), Bucniv (Basilian, 1738), L'viv (Dominican, 1751), Berdyčiv (Carmelite, 16 VII 1753), Cholm (Basilian, 1765), Počajiv (Basilian, 8 IX 1773). The inclusion of the Roman

Catholic sites Pidkamin' and Berdyčiv into the Greek Catholic geography of grace can be additionally gleaned from the fact that the coronation celebrations at both sites were designed as ecumenical (including Roman, Greek and Armenian Catholics) (Levyc'ka 2017–18).

[33] We are not suggesting here, though, that the 1656 Lwów Oath (*śluby Jana Kazimierza*), when the Polish king Jan Kazimierz declared the Mother of God Queen of Poland, served as a direct source of inspiration for the coronation procedures of the 18th century, making it thus a prerequisite for this later practice. Although the metaphorical link is quite obvious, a direct causal link between the Oath and later coronations would still have to be made evident by further research into the matter. There is, by the way, little direct evidence that the idea of the Mother of God as the Queen of Poland was actually extended to the Lithuanian part of the Commonwealth. See Niendorf (2010, p. 158), who cites at least one contemporary source in support of this assumption.

[34] An impressive example of this kind of commemorative volume is Trześniewski (1767).

[35] The ambitions of Počajiv were not restricted to the immediate regional level, but aspired to catch up with the famous Kievan Cave monastery, with which it came to share the honorary title of *lavra* (Trajdos 1984, p. 134).

[36] See Nastalska-Wiśnicka (2015) for an overview of text genres associated with Marian sanctuaries in 16th–18th century Poland.

[37] For the role of music and singing in creating a soundscape of sanctity and spirituality, see Fisher (2014), and with particular reference to the Greek Catholic church Lorens (2019).

[38] Especially the famous song in praise of the Mother of God at Pidkamin' *Prečistaja děvo, mati ruskago kraju* has been repeatedly used as a model for songs in praise of images at other places (Počajiv, Tyvriv, Chotyn). Likewise the songs in praise of the miraculous image of Vicyn *Děvo mati preblagaja, tys' carice nebesnaja* (>Počajiv, Kiev, Povča) and the miraculous image of Sambir *Divna tvoja tajna, čistaja, javisja* (>Bar, Povča) were appropriated and remodeled for the sake of other places of grace.

[39] In fact, the majority of icon songs in the Bogoglasnik of Počajiv (BG 2016, nos. 110–11, 114–25) are identified as being authored by Basilian monks (творение инока чинуС. Василїа В. or *Dzieło zakonnika Bazyliana*). In some cases explicit reference to the Basilian order within the song points to a Basilian authorship (e.g., in the 5th stanza of the song in praise of the icon of Jatwiesk, cf. Stern 2000, p. 648). The Polish song *Drogi klejnocie naroda ruskiego* in praise of the Mother of God of Vicyn was also authored by a Basilian (Al'mes 2018, p. 272).

[40] Incipit '*Poběditel'naja vsěm kievskaja strana*', attested in ms. ASP 233 and the Kam'jans'kyj BG of 1734 (cf. Medvedyk 1999, pp. 72–73).

[41] Being based on Stern (2000), Żeňuch (2006), Medvedyk (2015) and (BG 2016).

[42] **Transcarpathia**: (1) Klokočev—(2) Krasnyj Brod—(3) Povča. **Ciscarpathia**: Kuty (4). **Galicia**: (5) Hošiv—(6) Kam'janka-Strumylova—(7) Pan'kivci—(8) Pidhirci—(9) Pidkamin'—(10) (Novyj) Sambir—(11) Ulaškivci—(12) Zarvanycja—(13) Benevo—(14) Tartakiv. **Volhynia**: (15) Krystynopil'—(16) Piddubci—(17) Počajiv—(18) Zymne—(19) Radechiv. **Podolia**: (20) Bucniv—(21) Jarka—(22) Kam'janec'-Podil's'kyj—(23) Nastašiv—(24) Ternopil'—(25) Tyvriv—(26) Lysovci. **Žytomyr**: (27) Berdyčiv—(28) Prjaživ—(29) Tryhory. **Černihiv**: (30) Lin'kiv—(31) Tupyčiv. **Bukovina**: (32) Chotyn—(33) Terebovlja. **Podlachia**: (34) Dalešiv—(35) Jatwiesk—(36) Žyrovyči. **Mohyliv**: (37) Bar. **Peremyšlja**: (38) Peremyšlja—(39) Vicyn.

[43] On the face of it, the language choice would be suggestive of an ongoing language shift from Ruthenian to Polish among the local population. However, none of the places in question is likely to have undergone a majority shift to Polish. **Galicia**: (1) Bilostok—(2) Bil'šivci. **Volhynia**: (3) Puhinki—(4) Zahoriv. **Podolia**: (5) Krem'janec'—(6) Zahajci. **Polissia**: (7) Jurewicze. Language choice might perhaps rather reflect the linguistic preferences within the ranks of the Basilian order, which has long been known for increasing Polonizing tendencies. There are several places (Počajiv, Pidkamin' and Zarvanycja) for which songs in both languages have been composed. Given the supraregional fame of these places of grace, bilingual usage can best be explained as an effort to cater for the needs of pilgrims from all over the Commonwealth. We are quite confident that our count comes quite close to the actual number of images for which songs have ever been composed, but an extended search that would include a larger number of the countless manuscript collections of Ruthenian devotional songs might reveal yet a few more items of less widely circulated songs in praise of other miraculous images.

[44] For Počajiv an impressive number of 11 different songs could be identified: (1) *Witay cudowna matko w Poczaiowie*, (2) *K tebě, božija mati, priběgaem, Obremenenni grěchami vzyvaem*, (3) *Mnogimi usty glasi ispusti*, (4) *Veselo spěvajte, vsi čelom udarjajte pred Matkoju Christovoju (cesarevoju)*, (5) *Vselennyja vsja strany, zemljane*, (6) *Děvo mati preblagaja, Tys' carice nebesnaja*, (7) *Prečistaja děvo mati v Volynskom kraju*, (8) *Nyně proslavitsja Počaevska skala*, (9) *Zlatozarnye zori počavskie*, (10) *Izyjděte dvory so sobory*, (11) *Pasli pastyri ovci na gorě*.

[45] Basilian places of grace honoured in songs are: Bar (former Dominican monastery), Bilostok, Bucniv, Dalešiv, Hošiv, Jatwiesk, Krasnyj Brod, Krem'janec', Krystynopil', Piddubci, Pidhirci, Počajiv, Puhinki, Pušni, Tryhory, Ulaškivci, Vicyn, Zahajci, Zahoriv, Zarvanycja, Zymne—Dominican places: Pidkamin', Tyvriv—Carmelite places: Berdyčiv—Jesuit places: Nastašiv.

[46] In a similar vein, Ščurat (1908a, p. 48) argues that the icon songs in praise of diverse Basilian Marian places of grace were commissioned for the edition of the Bogoglasnik. See also Medvedyk (2006, pp. 26–27).

[47] This seems to be true of the song in praise of the Mother of God at the Dominican monastery at L'viv *Izyjděte dščery ierusalimskija* (ms. Maslov 54, Hnatjuk 2000, pp. 197–98), which mentions the coronation as an ongoing event happening simultaneously to the singing of the song (*dnes' ukoronovana*). Cf. also the small brochure for the coronation of the icon of Počajiv 1773 (Medvedyk 2000a), which contains a number of songs in praise of the icon that are attested in this very brochure for the first time and may therefore be assumed to have been specifically composed for the compilation of the brochure, i.e., for the coronation feast itself.

[48] Such as the Suprasl' collection F 19–233 of the 1730s–40s; v. Stern (2000).

49 The earliest booklet of this kind appeared in 1734 in honour of the icon of Rohizno, which in that very same year had manifested her miraculous powers (Medvedyk 1999, p. 76; Medvedyk 2001, p. 92). The most prominent of this kind of occasional print brochures is certainly the *Gora počajevskaja stopoju i obrazom čudotvornym Presvjatoj Děvy Bogorodicy počtenna* published in 1742, and then again in 1755, 1772 and 1793. Around 1772–73 appeared a small print edition of songs and prayers on the occasion of the coronation of that same icon (for an edition see Medvedyk 2000a). A polish language volume in praise of the same icon, that contained one Ruthenian icon song, was published in 1765, and again in 1778 (Kasparowicz 1765). A small print edition of songs in praise of the miraculous icon at the Dominican monastery at L'viv was published on the occasion of the icon's coronation in 1751 (Pieśni o Nayświętszey Maryi Pannie w Cudotwornym Obrazie Lwowskim; cf. Medvedyk 2006, pp. 120–21). The commemorative volume *Obrona Polskiey korony ot granic* Ukrainskich (1760) in praise of the miraculous image of Berdyčiv contains one Ruthenian icon song, which, however, with one exception never found its way into manuscript song collections (Medvedyk 2006, pp. 131–32). This volume was later on significantly enlarged by Trześniewski (1767).

50 For instance, the selection of Maslov 45 (cf. Medvedyk 2000b) offers a representative choice of all major Ruthenian regions: Pidkamin' (Galicia), Počajiv (Volhynia), Vicyn (Peremyšlja), Lin'kiv (Černihivščyna), Bzov and Rudnja (Kyjivščyna). In contrast to this, the Ivanovce song manuscript (cf. Javorskij 1934, pp. 111–27) has a more narrowly defined regional profile, with a clear focus on Galicia: Pidkamin', Sambir, Zarvanycja, Kam'janka-Strumylova (all Galicia), Počajiv (Volhynia), Bar (Podolia), Povča (Transcarpathia).

51 An analysis of 86 manuscript collections of Ruthenian devotional songs from the 18th–19th c. yielded the following list for Marian sites being included in more than 5 manuscripts: Pidkamin' (44 mss.), Počajiv (33), Sambir (29), Bar (24), Kam'janka-Strumylova (20), Zarvanycja (13), Povča (10), Vicyn (9), Lin'kiv (8).

52 Žyrovyči in 1727 and Cholm in 1765.

53 On collectors of Ruthenian devotional songs and the emergence of a popular collectors' culture see Stern (2021).

54 A fine example of this is the song in praise of the Mother of God at Počajiv *Veselo spěvajte, čelom udarjajte* (BG 2016, no. 112) that offers a doxological narrative in verse of the Ottoman siege of the larger region in 1675. In the lines *Vъ časъZbaražskoj vojny bjachu nespokojny/Grady vsja ot běsurmanovъ* the idea of the Ottoman frontier as a chain of towns forming the *Antemurale Christianitatis* is neatly evoked.

55 The song in praise of the Mother of God at Počajiv *Nyně proslavi sja Počajevska skala* (BG 2016, no. 116) capitalizes on the healing powers of the image by reporting on the regional epidemic (*mor*) of 1770. It lists stanza by stanza the places that were afflicted and granted deliverance from the plague by the icon of Počajiv. By listing these places the author seems to rely on a common cognizance shared by his addressees of which localities are most intimately linked to Počajiv as its spiritual point of reference and thus constitute a geography of grace, which the author construes as a space of shared embodied experience, in contrast to the largely disembodied administrative and political conceptions of space that predominate in other icon songs.

56 Tyvriv (BG 2016, no. 128: *Soglasno krikněte, christiane sja sniděte, so liky, timpany, ot vseja strany* 'Call out in harmony, join in, Christians from all countries, with the choirs and the tympana'), Novyj Sambir (Ščurat 1908b, p. 183: *Dneś wo wsia strany uže razsianny/Ko tebi Maty! od siudu wołaiem* 'Today all countries which are spread over the world call to you from all directions').

57 This Počajevan partisanship is countered by two songs having been authored by one Andrzej Jakubiński in october 1770, who gives credit to the miraculous image of Sambir for having ultimately saved the region from the same epidemic (Ščurat 1908b, pp. 181–85).

58 The rule of silence, which we saw being applied internally to competing places of grace of one's own confessional domain, was with even greater naturalness applied to the great competitor in the field, the Orthodox church which in the places, where it still remained, could often boast miraculous images of centuries-old tradition. Greek Catholic songs would attempt to obliterate these from the consciousness of the Greek Catholic pilgrims. In the case of Zymne the Basilian monks actively promoted the cult of their recently introduced miraculous image by means of an icon song (BG 2016, no. 125) in order to outshine its much older and more dignified counterpart at the nearby Orthodox monastery (Lužnyc'kyj 1984, p. 167). By perfidiously ignoring the existence of the latter it seems as if the Basilian monks hoped for their own new miraculous image to profit from the fame of the latter through the confusion of both by the unsuspecting pilgrim.

59 See Trajdos' (1984, p. 131) assessment of Marian worship among the Orthodox faithful ever since the Polish–Lithuanian Union of 1386 and the subsequent catholization of almost the entire political class of the Grand Duchy of Lithuania. Where up to this point the Ruthenian lands partook in the expanding common East-Slavic cult of palladium icons as symbols of the union of the sacred and the political power, which typically were located at the centers of political power (Kiev, Černihiv, Smolensk, Wolodymir Volyns'kyj, Luc'k, etc.), the loss of the political power of Orthodoxy brought the establishment of new Marian places of grace to a sudden halt. The appearance of the Greek Catholic Basilians with their strong emphasis on Marian devotion in the 17th and 18th century could thus capitalize on the general decline of Orthodox culture by filling in the gap left by centuries of lethargy to set up an entirely remodeled geography of grace for their flock, which focussed more on the as yet Ruthenian speaking countryside than on the heavily polonized and catholicized towns and cities.

60 If we apply the imagery of the religious marketplace, as did Mahieu and Naumescu (2008, pp. 14–16) with respect to the present situation of the Uniate Church, to the 17th–18th centuries, there was in fact not much of a market place offering free choices to the individual believer. Yet, the steps taken by the Uniates in promoting Marian devotion clearly bespeak a strategy applicable in a situation of open competition. Uniate competitive behabviour does, in fact, not directly address a competitor, but is directed at the believers' competing loyalties that oscillate between whole-heartedly accepting the legally imposed Church Union and sticking to the tradition of defeated Orthodoxy.

[61] Jakovenko (2011–14) aptly calls this 'territorial correctness'.

[62] See among others the often repeated remark in Barącz (1891) that the image being described is being visited by all confessions alike. See also Jakovenko (2011–14).

[63] Writing over the traditional Orthodox map of grace smacks of cultural domination, and, as a matter of fact, Uniates, in particular Basilian monks in their efforts to follow the Jesuit model of confessional cultural re-education significantly contributed to an active policy of Western, i.e., Catholic reorganization and domination of the Orthodox cultural sphere. Yet, Uniate action lacked the open assertiveness of their Roman Catholic models so much so that the notion of dominance does not impose itself rightaway on the observer. It seems that, shunning any form of open confrontation on the discourse level of e.g., confessional polemics, Uniates were forced to resort to positive action, in which they took great care not to incur the accusation of encroachment (v. above on the Uniate policy of 'territorial correctness'). This restrained self-assertion may be taken as a symptomatic feature of the Uniates' role as a "bridge between two cultural spheres" (Mahieu and Naumescu 2008, p. 2).

[64] This again is a variation of the topos of the Blacherniotissa. The Marian visions reported for Počajiv (1675) and Terebovlja (1673) reflect in a more direct manner its Constantinopolitan prototype.

[65] According to the folk ideological notion of ensomatosis (Kretzenbacher 1977, p. 7), wounds inflicted to miraculous images are true physical wounds, which ought to be bleeding, accordingly.

[66] The ultimate model for this specific topos of the injured image which resists attempts at repair, is the Black Madonna of Częstochowa. The interpretation as stigmata that testify to the act of redemption through sacrifice and suffering can be gleaned from Rotter (1756): "To dziw u wszystkich sprawuje Przecudowna Matka, że ran zostawionych na Twojej twarzy, żaden sposób ludzki zgładzić nie może. Darmo do tego końca i największej doskonałości malarze w farbach moczą. Ale domyślam się, dlatego chcesz je mieć niezgładzone { . . . }, abyś dała znak jawny niewygasłej ku nam miłości Twojej, dla której tak ciężkie przepuściłaś sobie rany" ('This miracle professed the miraculous Mother in the presence of all, that the wounds inflicted on Your face, cannot be removed by any human art. To this end even the most artful painters apply their colours in vain. But I surmise that it is for this reason that you refuse to have your wounds removed { . . . } that you might give a visible sign of your inextinguishable love to us, for which you suffered to have these heavy wounds inflicted upon you.'). See also Kretzenbacher (1977, pp. 50–52).

References

Al'mes, Ivan. 2018. «Drogi klejnocie naroda ruskiego»: Духовна піснясерединиXVIII століття на честьікони божої матері Віцинського Василіянського монастиря. *Калофонія* 2018: 265–77.

Barącz, Sadok. 1891. *Cudowne obrazy Matki najświętszej w Polsce.* Lwów: Nakładem autora.

BG. 2016. *Bogoglasnik—Pěsni blagogověnyja (1790/1791). Eine Sammlung geistlicher Lieder aus der Ukraine.* Edited by Hans Rothe and Jurij Medvedyk. 2 vols, Cologne, Weimar and Vienna: Böhlau.

Chociszewski, J. 1882. *Pielgrzymka do ważniejszych miejsc w ziemiach polskich wsławionych cudownemi obrazami N. Maryi Panny.* Poznań: G. Jalkowski w Grudziądzu.

Eliade, Mircea. 1957. *Das Heilige und das Profane. Vom Wesen des Religiösen.* Reinbek bei Hamburg: Rowohlt.

EMKC. n.d. *Epistolae metropolitarum Kioviensium catholicorum (Raphaelis Korsak, Antonii Sielava, Gabrielis Kolenda, 1637–1674).* Edited by Athanasius G. Welykyj OSBM. Rome: PP. Basiliani.

Fedotov, George P. 1946. *The Russian Religious Mind. Kievan Christianity, the Tenth to the Thirteenth Centuries.* New York, Evanston and London: Harper & Row.

Fisher, Alexander J. 2014. *Music, Piety, and Propaganda: The Soundscapes of Counter-Reformation Bavaria.* Oxford: OUP.

Fridrich, Alojzy. 1904. *Historye cudownych obrazów najświętszej Maryi Panny w Polsce.* Kraków: Wydawnictwo Tow. Jez.

Gliczner, Erazm. 1598. *Appellatia ktorą sie popiera z znowu wywodźi obrona dołozna Confederatiey krolestwa polskiego, z okażaniem pewnym, ze euangelicy Auspurskiey confessiey, tu w Polszcze, w Litwie, w Prusiech y wszędzie w państwie korony Polskiey, w miástách koronnych, stołecznych, i innych { . . . }.* Królewiec: Dziedzicy Georga Osterbergera.

Halemba, Agnieszka. 2008. Greek Catholics of Zemplin: Dilemmas of Contemporary Identity Politics. In *Churches In-between. Greek Catholic Churches in Postsocialist Europe.* Edited by S. Mahieu and V. Naumescu. Berlin: LIT Verlag, pp. 299–318.

Harasiewicz, Michael. 1862. *Annales Ecclesiae Ruthenae.* Leopolis: Institutum Ruthenum Stauropegianum.

Hnatjuk, Oleksandra. 1994. *Українська духовна барокова пісня.* Warsaw-Kiev: Pereval.

Hnatjuk, Oleksandra. 2000. *Барокові духовні пісні зрукописних співаниківXVIII ст. Лемківщини.* L'viv: Misioner.

Hruševs'kyj, Mychajlo. 1897. Сьпіванник з початку XVIII в. *ЗапискиНаукового Товариства імени Шевченка* 15 & 17: 1–98.

Hymes, Dell. 1972. Models of Interaction of Language and Social Life. In *Directions in Sociolinguistics: Ethnography of Communication.* Edited by J. J. Gumperz and D. Hymes. New York: Holt, Rinehart & Winston, pp. 35–71.

Jakovenko, Natal'ja. 2011–14. «Битва за душі»: Конкуренція богородичних чуд міжуніятами та православнимиу 17 ст. (відТеодозія Боровика до Йоаникія Ґалятовського). *Harvard Ukrainian Studies* 32–33: 807–25.

Javorskij, Julian A. 1934. *Материалы для историистаринной песенной литературы в подкарпатскойРуси.* Prague: Orbis.

Kasparowicz, Gabryel Andrzej. 1765. *Zrodło ogrodów, studnia wod żywych ktore płyną impetem z Libanu [. . .] to iest Officium albo godzinki Nayś. Maryi Pannie na Górze iasney Poczaiowskiey [. . .].* L'viv: Drukarnia Collegii.

Kretzenbacher, Leopold. 1977. *Das verletzte Kultbild. Voraussetzungen, Zeitschichten und Aussagewandel eines abendländischen Legendentypus.* München: Bayerische Akademie der Wissenschaften.

Kruk, Mirosław Piotr. 2011. *Ikony-obrazy w świątyniach rzymsko-katolickich dawnej Rzeczypospolitej.* Kraków: Collegium Columbinum.

Kürzeder, Christoph. 2005. *Als die Dinge heilig waren. Gelebte Frömmigkeit im Zeitalter des Barock*. Regensburg: Schnell & Steiner.

Levyc'ka, Mar'jana. 2017–18. Короновані ікони Богородиці вукраїнськійунійній традиціїXVIII-XIX ст. (істориографіяі зразки). *Карпати: людина, етнос, цивілізація* 7–8: 270–84.

Lidov, Aleksey. 2006. Иеротопия. Создание сакральных пространств как вид творчества и предмет исторического исследования. In *Hierotopy. Creation of Sacred Spaces in Byzantium and Medieval Russia*. Edited by A. Lidov. Moscow: Progress-Tradition, pp. 9–31.

Lorens, Beata. 2019. Muzyka w działalności duszpasterskiej Bazylianów w Rzeczypospolitej w XVIII wieku. *Muzyka* 64: 49–69. [CrossRef]

Lužnyc'kyj, Hryhor. 1984. Словник чудотворних богородичних ікон України. In *Intrepido Pastori. Науковий збірник на честь блаженнійшого патріярха Йосифа в40-ліття вступлення на Галицький престіл, 1.11.1944*. Rome: Universitas Catholica Ucrainorum S. Clementis Papae, pp. 153–88.

Mahieu, Stéphanie, and Vlad Naumescu. 2008. Introduction: Churches In-between. In *Churches In-Between. Greek Catholic Churches in Postsocialist Europe*. Edited by S. Mahieu and V. Naumescu. Berlin: LIT Verlag, pp. 1–32.

Medvedyk, Jurij Jevhenovyč. 1999. Українські богородичні канти другої половиниXVII-XVIII ст. *Народна творчість та етнографія* 269: 71–80.

Medvedyk, Jurij Jevhenovyč. 2000a. *Пісні до Почаївської богородиці*. L'viv: Misioner.

Medvedyk, Jurij Jevhenovyč. 2000b. Нижньо-тварожськийспіваниксередини30-тих роківXVIII ст.: його місце вукраїнській духовнокантовій культурі та рукописній традиції жанру. In *Slovensko-rusínsko-ukrajinské vzťahy od obrodenia po súčasnosť*. Edited by Ján Doruľa. Bratislava: Slavistický kabinet SAV, pp. 350–70.

Medvedyk, Jurij Jevhenovyč. 2001. Духовні канти в честь чудотворних храмових святиньГаличини. *Українське музикознавство* 30: 90–103.

Medvedyk, Jurij Jevhenovyč. 2006. *Українська духовна пісняXVII-XVIII століть*. L'viv: Ukrajins'kyj katolyc'kyj universytet.

Medvedyk, Jurij Jevhenovyč. 2015. Іконославильні пісні Полісся, Поділля, Волині та Галичини вспівочій культурі Мукачівської єпархіїXVIII-XIX ст. In *Ad fontes: ЗісторіїукраїнськоїмузикиXVII—початку XX ст*. L'viv: KNU imeni Ivana Franka, pp. 371–90.

Nastalska-Wiśnicka, Joanna. 2015. Staropolskie piśmiennictwo sanktuaryjne jako źródło do badań nad kultem Maryjnym na ziemiach Rzeczypospolitej. *Textus et Studia* 4: 47–70.

Niedźwiedź, Anna. 2005. *Obraz i postać. Znaczenia wizerunku Matki Boskiej Częstochowskiej*. Kraków: Uniwersytet Jagielloński.

Niendorf, Mathias. 2010. *Das Großfürstentum Litauen*. Wiesbaden: Harrassowitz.

Okolski, Szymon. 1648. *Góra święta Najśw. Panny Różańca św. w Łuckiem biskupstwie na Wołyniu nad miastem Podkamień, przedziwnemi cudami, y stopkami Panieńskiemi, Dekretem Pasterskiem y wielu pielgrzymowaniem wsławiona*. Kraków: Łukasz Kupisz.

Pap, Stefan. 1971. Духовна пісня на Закарпатті. *Analecta Ordinis S. Basilii Magni* 7: 114–42.

Pruszcz, Piotr Hiacynt. 1662. *Morze Łaski Bożej, które Pan Bog w Koronie Polskiey po różnych mieyscách, przy Obrázách Chrystusa Páná, y Matki iego Przenayświętszey ná Sercá ludzi pobożnych, y w potrzebách rátunku żądaiących, z głębokości miłosierdzia swego, nieprzebránego, co dzień obficie Wylewa*. Kraków: Drukarnia dźiedźików Stanisława Lenczewskiego Bertut.

Przesławna gora poczaiowska Dawnością Cudow Przenayczystszey Bogarodzicy Panny od cudownego Jey Obrazu wynikających. 1801. Poczajów: Drukarnia WW. OO. Bazylianów. First published 1787.

RIB 7. 1882. Vasil' Surožskij, О единой истинной православной вѣрѣ и о святой соборной апостолской церкви, откуда начало приняла, и како повсюду распростреся. Сочиненіе Острожскаго священника Василія 1588 года. "О образохъ рекше о инконахъ". *Русская историческая библиотека, издаваемая археографическою коммиссиею* 7: 918–38.

Rok, Bogdan. 2005. Sanktuaria Maryjne ziem ruskich w kulturze Rzeczypospolitej XVIII wieku. In *Czasy nowożytne. Studia poświęcone pamięci prof. Władysława Eugeniusza Czaplińskiego w 100 rocznicę urodzin*. Wrocław: Instytut historyczny, Uniwersytet Wrocławski, pp. 137–47.

Rotter, o. Ksawery. 1756. *Ucieczka grzeszników albo nabożny zasmuconych recurs do dzielney cudami Częstochowskiey Matki*. Częstochowa: Drukarnia Jasney gory Częstochowskiey.

Rožko, Volodymyr. 2002. *Чудотворні ікониВолині і Полісся*. Luc'k: Media.

Saltykov-Ščedrin, M. E. 1979. *Господа Головлевы. Сказки*. Moscow: Chudožestvennaja literatura.

Senyk, Sophia. 1984. Marian cult in the Kievan metropolitanate XVII-XVIII centuries. In *Intrepido Pastori. Науковий збірник на честь блаженнійшого патріярха Йосифа в40-ліття вступлення на Галицький престіл, 1.11.1944*. Rome: Universitas Catholica Ucrainorum S. Clementis Papae, pp. 261–75.

Stern, Dieter Hubert. 2000. *Die Liederhandschrift F 19–233 (15) der Bibliothek der Litauischen Akademie der Wissenschaften*. Cologne, Weimar and Vienna: Böhlau.

Stern, Dieter. 2021. Ruthenian devotional songs as collectors' items? *East-West: Journal of Ukrainian Studies* 8. in press.

Ščurat, Vasyl'. 1908a. *Iзстудій над почаївським Богогласником*. L'viv: Nyva.

Ščurat, Vasyl'. 1908b. Iзстарої галицької літератури. *ЗапискиНаукового Товариства* ім. Шевченка 81: 181–88.

Ščurat, Vasyl'. 1910. *Мариинський культ на українських землях давної польської держави*. L'viv: Nyva.

Tazbir, Janusz. 1984. Różnowiercy polscy wobec kultu maryjnego. *Studia Claromontana* 5: 224–46.

Tazbir, Janusz. 2017. From Antemurale to Przedmurze, the History of the Term. *Odrodzenie i reformacja w Polsce* 61: 67–87. [CrossRef]

Trajdos, Tadeusz M. 1984. Kult wizerunków maryjnych na ziemiach ruskich Korony i Litwy drugiej połowy XIV i pierwszej połowy XV w. w społecznościach katolickiej i prawosławnej. *Studia Claramontana* 5: 127–47.

Trześniewski, Grzegorz. 1767. *Ozdoba y oborona ukraińskich krajów, przecudowna w Berdyczowskim obrazie Maryi watykańskiemi koronami od Benedykta XIV papieża z pobożney hoyności własnym Jego kosztem sporządzonemi, przez ręce { . . . } Kajetana Ignacego Sołtyka, krakowskiego, na ten czas kijowskiego biskupa { . . . } w roku 1756, dnia 16 miesiąca lipca ukoronowana, to jest opisanie teyże odprawioney koronacji na dwie części rozdzieloney.* Berdyczów: Drukarnia Karmelu Fortecy NMP.

Wereda, Dorota. 2018. Koronacje wizerunków maryjnych w Cerkwi unickiej. In *Koronacje wizerunku Matki Boskiej na przestrzeni dziejów.* Edited by Ewelina Dziewońska-Chudy and Maciej Trąbski. Częstochowa and Warsaw: Uniwersytet Humanistyczno-Przyrodniczy im. Jana Długosza w Częstochowie, pp. 67–79.

Žeňuch, Peter. 2006. *Kyrillische paraliturgische Lieder. Edition des handschriftlichen Liedguts im ehemaligen Bistum von Mukačevo im 18. und 19. Jahrhundert.* Cologne, Weimar and Vienna: Böhlau.

Article

Christian Saints in Russian Incantations

Aleksey Yudin

Department of Languages and Cultures, Ghent University, 9000 Ghent, Belgium; Oleksiy.Yudin@UGent.be

Abstract: This article discusses the Christian saints who are most often mentioned in Russian incantations: Sts. George, Nicholas, Florus and Laurus, Kossma and Damian, Zosima and Savvaty of Solovki, as well as the semi-apocryphal saints Sisinius and Solomonia. The first six are among the most popular saints of Russian folk Orthodoxy. The article presents the naming conventions of saints, and their attributes and functions in Russian folk magic. Depending on their magical function, the protagonists of the incantations can act as helpers, protectors, and healers. They assist in various practical areas of life, and protect against real and magical dangers, in addition to helping healing from diseases and wounds.

Keywords: Russian; Orthodox; Christianity; saints; folklore; magic; incantations

Citation: Yudin, Aleksey. 2021. Christian Saints in Russian Incantations. *Religions* 12: 556. https://doi.org/10.3390/rel12080556

Academic Editor: Dennis Ioffe

Received: 7 June 2021
Accepted: 5 July 2021
Published: 21 July 2021

Publisher's Note: MDPI stays neutral with regard to jurisdictional claims in published maps and institutional affiliations.

1. Introduction

Popular Orthodox Christianity is widely represented in various genres of Russian folklore: songs, legends, spiritual poems, and incantations. However, the images of saints represented in folklore genres differ significantly. This article is devoted exclusively to their images and functions in Russian incantations. It presents in English the results of my research previously outlined in (Yudin 1997, 2010) and other publications.

From the point of view of methodology, this is a study of the onomastics of East Slavic folk magic. The study was carried out by means of a continuous scan of the corpus of published texts of incantations and a complete extraction of data on the names, attributes, and magical functions of saintly figures. In the lexicon (Yudin 1997), I also described with which figures each figure appears (in a group), and which figures he is opposed to. In this short article, I will limit myself to describing the forms of names of saintly figures, sets of their magical functions, and associated motifs.

Incantations are oral or written magical texts that were intended to guarantee the achievement of some desired state: security, health, love, success in the hunt, etc. In the Russian tradition, they could be short incantatory formulas, but they could also be long apocryphal prayers mentioning the names of many Christian figures. A large number of important Christian figures are mentioned in Russian incantations, including around 200 names of saints. Many of them are mentioned only once or twice, but certain saints are particularly popular in folk magic. The two most frequently invoked figures are Saint George and Saint Nicholas.

2. Saint George (*Georgii*)

The image of one of the most popular saints, *Georgii-Egorii-Iurii*, in Slavic folk culture is one of the best studied. However, most of the fundamental research about him in the 19th–21st centuries was published in Russian. Here, I will only draw upon one book (Senderovich 2002). With regard to publications in Western languages, I would like to name the following works: (Loorits 1955; Oinas 1982; Stangé-Zhirovova 1981).

The image of George varies significantly in different folklore genres. The East Slavic Yuri (*iur'evskie*) songs are most closely associated with him. They regularly contain motifs of the invocation of spring, the awakening of nature, the unlocking of the earth (the spring awakening of nature is, in *iur'evskie* songs, presented as unlocking of the earth with a key),

Religions **2021**, *12*, 556. https://doi.org/10.3390/rel12080556   https://www.mdpi.com/journal/religions

and, in general, the motif of keys. Additionally, common are the motifs of dew, grass, cattle protection, and improving the harvest. The object (attribute) most associated with George-Yuri in these songs is a horse. In spiritual poems he slays the dragon and thus saves *Elizaveta* the Beautiful (cf. Sokolov 1995). We also see this in the saga about Saint George the Brave (Barsov 1903, p. 18). Within the framework of the East Slavic syncretism of Christianity and paganism, called in Russian double faith (*dvoeverie*), he had a whole series of functions. According to the list of "The Graces Bestowed by God to Saints and when to appeal to them"[1] (Shchapov 1863, p. 64), which goes back to Russian icons of the same name[2], St. George protects livestock from wild animals. As A.N. Afanas'ev noted, he is "revered as the patron saint of livestock and chief of the wolves, telling them where to feed and what to eat" (Afanas'ev 1914, p. 195); see: (Toporkov 2013). In the folklore, wild animals are represented as "the flock of God, but as their nearest chief and protector God has appointed Saint George" (Ermolov 1905, p. 4; Tokarev 1957, pp. 47, 115). People prayed to George everywhere on George's Day (*Iur'ev den'*), when the cattle were first taken out to pasture. In the Vologda region, people prayed to George for their horses (Adon'eva and Ovchinnikova 1993, p. 11). At the same time, he is seen as the patron saint of warriors (Ganchenko 1995, p. 258). Subsequently, I will analyze his image in Russian incantations.

I found approximately 200 mentions of St. George in Russian incantations. This is too much to refer to all sources here. As a representative corpus of references to St. George, I will indicate here only the pages in the most important publication of Russian incantations: (Toporkov 2010, pp. 101–2, 115–16, 124–25, 136, 141, 143–44, 316, 336, 347, 353, 358, 435–36, 439, 459, 468–69, 480–81, 508, 510, 512, 514, 522, 525, 540–41, 544, 681, 702, 707, 713, 729). Saint George is found in texts from the following provinces of the Russian Empire or regions of the USSR: Amur[3], Arkhangelsk (*Arkhangel'sk*), Karelia (*Kareliia*), Kostroma, Mangazeya (*Mangazeia*), Minusinsk, Nizhny Novgorod (*Nizhnii Novgorod*), Novgorod, Karelia (*Kareliia*), Perm (*Perm'*), Saratov, Simbisk, Smolensk, South Siberia (*Iuzhnaia Sibir'*), Tobolsk (*Tobol'sk*), Vologda, Vyatka (*Viatka*), Yeniseysk (*Eniseisk*). The saint's name is presented in the following forms: *Georgii, Egorii, Egorei, Egor, Egorushka, Grigorei*. The name has permanent epithets: George the Brave (*Georgii Khrabryi*[4]), George the Victorious (*Georgii Pobedonosets*), George the passion bearer (*Georgii strastoterpets*), the great martyr George (*velikomuchenik Georgii*{XE "Георгий"}), George, Gracious Saint George (*Georgii Svet Milostivyi*); the Great George (*Velikii Egorei*), etc. He rides on an *Egor'evskii* horse. His functions in Russian folk magic are various. However, like the functions of other figures in Russian incantations, these functions can be grouped into three broad categories: helper, protector, healer.[5]

Helper. George helps to conjure up someone's love and assists with the hunt, more specifically the hunt for foxes, ermines and other fur-bearing animals, as well as hares and bears. He also helps find lost livestock and with keeping the livestock at home and by calming the cow, so that it does not kick, thus increasing the cow's milk production.

Protector. He protects people from sorcerers, from witches, the evil eye, and diseases, as well as from enemies in general; against rulers; against any weapon; against dogs (except rabid dogs); he protects wedding processions from sorcerers, evil, witches, and diseases; protects livestock from all harm; protects livestock from bears, wolves, etc.; ensures that a partridge that falls into a trap is not eaten by ravens.

Healer. He heals people (including infants) and livestock from all diseases, including from ailments that come about by magical means (the evil eye, etc.). Upon intercession he feeds and herds the livestock when they are first taken out to pasture in the spring. He can also deliver people from the fire dragon that visits young women. In the latter case, among the Eastern and Western Slavs, it was believed that unmarried girls or married women could be visited by flying fire dragons, which shed their wings on land and took the appearance of a woman's husband or lover, or simply turned into beautiful young men and entered into a love affair with these girls and women, who subsequently lost weight, became ill, and often died. The dragon sucked milk or blood from the breasts of his victim and finally killed her. The dragon itself was considered a variant of the walking dead.

Copulation with the dragon resulted in the birth of demons and freaks. It was considered a widespread, serious problem, so much so that the people invented a whole cycle of incantations against the fire dragon. The same dragon, associated with the devil, could befriend the master of the house and bring money into the house. For more information on this figure, see: (Levkievskaia 2009).

Sometimes George is portrayed as a rider dressed in white (or gold) on a white (gray, silver) horse, with "a sharp spear and a holy knife". This image goes back to the traditional depiction of St. George on icons that represent a motif from his apocryphal life of the saint, namely as a rider on a white horse taming or defeating a dragon. St. George can also descend from heaven on a golden ladder and send down "three hundred golden daggers, three hundred tightly stretched bows" or "one hundred arrows and one hundred swords". He is often mentioned together with St. Michael and is connected with the motifs of thunder arrows, of shooting down diseases with a bow and arrow, stabbing them with a spear, and of chasing them away with a staff and whip. He is also a rider who has authority over wolves. He is the superior to the forest, earth, and water kings and the father of the two archer brothers who shoot down diseases. In the incantation against cataracts, he walks across an iron bridge, while three dogs—one gray, one white, and one black—run after him and lick his cataracts away. In the incantation for inducing cows to produce milk, there is the motif of the three-pound stone which St. George carries and places on the cattle's "lower vertebrae".

3. Saint Nicholas (*Nikolai*)

St. Nicholas of Myra in Lycia is the most popular saint of the folk Christian cult of the eastern Slavs, and his image is one of the most researched. In Russian folk Christianity, the functions of St. Nicholas are almost universal. He was especially revered as the patron saint of sailors. According to the folk list of "The Graces Bestowed by God to Saints", he delivers from drowning, disaster, and sorrow (Shchapov 1863, p. 64). "He is prayed to for help in all kinds of mischief and is worshiped as the one who feeds widows and orphans ..." (Ganchenko 1995, p. 194). In the Vologda area, he was thought to assist travelers (Adon'eva and Ovchinnikova 1993, p. 11). Nicholas in folk performances is the patron saint and savior of sailors, and the patron of girls awaiting marriage and children. Russians are raised with the legend of Nicholas as the lord of mice and rats (which is a version of the famous "Pied Piper") (Klinger 1931, p. 76).

For more details about his image and functions in folk culture, see: (Uspenskii 1982). St. Nicholas is provided in magical texts—and in the Russian popular Christianity in general—with numerous epithets from ecclesiastical practice. He is the (great) wonder-worker of Myra (Mirlikiiskii chudotvorets), the holy man, Christ's saint, the great keeper, and a speedy helper (velikij khranitel, skoryi pomoshchnik[6]). Furthermore, the personified icon of Nicholas of Mozhaisk is mentioned. The name of St. Nicholas appears in several canonical and folk variants, including *Nikolai*, *Nikola*, *Mikolai*, and *Mikola*. It is found in texts from the following provinces of the Russian Empire or regions of the USSR: Amur (Azadovskii 1914, p. 9), Arkhangelsk (*Arkhangel'sk*), (Efimenko 1878, pp. 149–50, 166–72, 185–86, 191; Maikov 1869, pp. 501–2, 553–55; Ivanova 1994b, p. 44), Kostroma (Vinogradov 1907, pp. 53–56, 58, 74; Vinogradov 1909, pp. 10–11, 44–45, 88), Kursk (Maikov 1869, pp. 486, 489), Minusinsk (Putilov 1984, p. 52), Novgorod (Maikov 1869, pp. 530–39), Karelia (*Kareliia*) (Sreznevskii 1913, pp. 502, 507), Oryol (*Orël*) (Maikov 1869, p. 488), Perm (*Perm'*) (Maikov 1869, p. 498, Saratov (Maikov 1869, pp. 505–8), Vologda (Popov 1903, p. 229), (Sreznevskii 1903, p. 148), (Sheremetev 1902, pp. 51–53), Voronezh (Selivanov 1886, p. 91; Maikov 1869, pp. 503–4; Lobanova 1993, p. 14), Vyatka (*Viatka*) (Ivanova 1994a, pp. 21, 93; Osokin 1856, p. 4), Yaroslavl (*Iaroslavl'*) (Balov 1893, p. 427; Sokolova 1982, p. 15), unspecified location (Buslaev 1861, pp. 1511–13; Eleonskaia 1917, p. 48; Tikhonravov 1863, pp. 352–53; Maikov 1869, pp. 457, 493, 534; Zenbitskii 1907, p. 1; Rybnikov 1867, pp. 254–55).

St. Nicholas has the following functions in Russian incantations:

Helper. He helps travelers along their way and assists with hunting, especially hunting foxes, stoats, and other fur-coated animals. He enhances fishing and helps summon and stop rain. Furthermore, he is the helper in the grazing of cattle (incantation at the first pasture) and reinforces the words of the incantation.

Protector. He protects people from sorcerers, from witchcraft, the evil eye, diseases and the like (his dominant function), and protects people from enemies and rulers and from any weapon. He protects wedding processions from wizards, witchcraft, and other mischief, and protects cattle from all misfortune (this is also his dominant function).

Healer. St. Nicholas heals people from all diseases, especially from inguinal and umbilical hernias, ulcers, and all kinds of growths in infants as well as in adults. He heals adults of hemorrhages; of the twelve (also seven or seventy-seven) shivering fevers and the twelve demonic spirits; of erysipelas; of inflammation of internal organs; stabbing pain including in the ears; and general pain in the body. He heals children from convulsions, adults from snake bites, enchantments, witches, the evil eye, *perepoloch* (dismay, an illness resulting from a child being severely frightened by someone, such as the night demons of childhood insomnia), and the like. He provides assistance during birth. He also heals cattle of all diseases. Like St. George, St. Nicholas saves people from the fire dragon that visits young women.

In the incantation texts, St. Nicholas can sit on a golden throne on the *"Ocean sea"* and use his bow to defend against witchcraft and other harmful magical acts[7]. He is also found in long lists of saints called for help. He rolls away diseases with his bow and spears into the *"blue sea"*, *"the sandy sea"*, in *"dark forests and treacherous swamps"*. In an invocation text, he sails across the golden sea in a golden ship (probably as the patron saint of sailors and seafaring).

4. Saints Florus and Laurus (*Flor* and *Lavr*)

The feast day of the martyrs Florus and Laurus, August 18, O.S, is known in popular culture as a "horse holiday" (Ermolov 1905, p. 423), and the saints themselves are the patrons of horses. In the spiritual verse, Florus and Laurus graze horses (Shchapov 1863, p. 49); according to the list of '"he Graces Bestowed by God to Saints", they deliver horses from death as a result of an epidemic (ibid, p. 64; Ganchenko 1995, p. 302). Their names are found in the following regions: Amur (Azadovskii 1914, p. 11), Kostroma (Vinogradov 1907, pp. 53–56), Novgorod (Maikov 1869, pp. 530–39), Karelia (*Kareliia*) (Sreznevskii 1913, pp. 496, 512), Smolensk (Dobrovol'skii 1891, p. 209), Tobolsk (*Tobol'sk*), (Gorodtsov 1916, pp. 43, 52), Vologda (Sheremetev 1902, p. 49; Maikov 1869, p. 528), unspecified location (Maikov 1869, pp. 492, 534; Zenbitskii 1907, p. 5).

These saints are always mentioned together as a pair. Sometimes they are seen as one indivisible figure. They also appear in spiritual folk verses that mention horse grazing (Shchapov 1863, p. 49). According to the list of "The Graces Bestowed by God to Saints", they deliver from horse mortality (ibid, p. 64; Ganchenko 1995, p. 302).

Widespread forms of their names are: *Flor* and *Lavr*; fathers *Flory* and *Lavry*; *Flora* and *Lavr*; *Flor* and *Laver*; *Frol* and *Lavr*; *Frol* and *Laver*; father *Frol-Laver* the horse herder; *Frël* and *Lavër*; *Khrol* and *Lavër*, great martyrs and wonderworkers. Their magic functions are limited, they are mainly associated with the feast day of the martyrs Florus and Laurus, the so-called "horse feast" (Ermolov 1905, p. 423). In this way, they were regarded as the "horse saints", the protectors of horses. In the incantations, they contribute to improving the harvest of the oats so much desired by horses. They protect livestock against all harm (this being their dominant function) and horses against bears, wolves, and other wild animals. They cure livestock—especially horses—of all diseases, particularly eyelid hardening. They can also cure people—including infants—of diseases that have arisen by magical means: enchantment, bewitching, the evil eye, *perepolokh*, and others (motifs include shooting down diseases with a bow and an iron fence that is placed around the object of the incantation to protect it).

5. Saints Cosmas and Damian (*Kuz'ma* and *Dem'ian*)

Another pair of saints that are almost always mentioned together, in accordance with the worship in the Church and with the iconography, are the saints Cosmas and Damian. According to the list of "The Graces Bestowed by God to Saints", they clarify the mind for the purpose of learning to read and write (Shchapov 1863, p. 64). According to folklore, they are the patrons of calligraphy (cf. ibid., p. 69). They are mentioned in the "tooth prayer" of the Holy Martyr Antipa (Tikhonravov 1863, p. 356).

In Russian folk culture, some pairs of saints bearing this name unite into one.

The day of commemoration of the unmercenary (*bessrebrenniki*) wonderworkers and martyrs of Rome, Cosmas and Damian, is 1 July; that of the unmercenary martyrs Cosmas and Damian of Cilicia is 17 October; that of the unmercenary wonderworkers Cosmas and Damian of Asia, sons of Theosodia, 1 November. "In the scientific literature, three holy pairs of brothers are apparently distinguished with these names, according to the place where they were buried: resp. Cosmas and Damian of Asia (feast day 1 November), of Arabia (17 October) and of Rome (1 July). Common to all three couples are their family ties and their activities: they are brothers and act as unmercenary healers" (Likhachev 1987, p. 154).

There are also different forms of their names: besides the ecclesiastical names, folk variants from spoken language were used. First of all, there is the old canonical form of the name (*Kozma*). There is also a Russian literary variant (*Kuz'ma*) and various colloquial forms. The contemporary Russian canonical form (*Kosma*) is only found in the spelling variant *Kossma*.

The following forms of the names of the saints were found in the sources: the holy "lords" wonderworkers *Kozma* and *Damian*; the saints *Kuzma* and *Dem'ian*; *Kuz'ma-Dem'ian* "the god of this place"; *Kuz'ma* and *Dem'ian* (*Damian*) the unmercenary saint of Christ; *Kuzma* and *Dem'ian*, the craft god and unmercenary saint of Christ; the honorable saints *Kuz'ma* and *Dem'ian*; *Koz'ma* and *Dem'ian*, saints of Christ; *Kaz'ma* and *Damien*; *Kossma* and *Damien*, holy men; *Kozma* and *Dem'ian* (*Dam'ian*, *Damiian*, *Domiian*); *Kos'ma* and *Damiian*; *Kuz'ma-Dev'ian* (sic!); *Kuzma*, *Demian*, great helpers (*velikie pomoshchniki*); Saint *Kozma Demiian*; *Kuz'ma* and *Damian*, the unmercenary saints; *Kozma* and *Damian*, holy unmercenary wonderworkers; *Kozma* and *Damian*, unmercenary holy lords, teachers, masters, leaders, intercessors and guardians; holy honorable *Kozma* and *Dam'ian*, wise masters; *Kos'ma* and *Dem'ian*; teachers and punishers and mentors and protectors and preservers (*uchiteli i nakazateli i nastavniki i zastupniki i sokhraniteli*). United in one figure, they can act as a third helping angel (along with *Michael* and Gabriel) in the incantation against disease.

Their names are found in the following regions: Amur (Azadovskii 1914, p. 13), Arkhangelsk (*Arkhangel'sk*) (Efimenko 1878, pp. 149–50, 169–70, 214; Smirnov and Il'inskaia 1992, p. 23), { ХЕ "ХриСТОС" }Kostroma (Vinogradov 1907, pp. 40–43, 73–74; Vinogradov 1909, pp. 39–40, 44–45, 80), Nizhny Novgorod (*Nizhnii Novgorod*) (Popov 1903, p. 226); Novgorod (Maikov 1869, pp. 504–5, 507; Maikov 1869, pp. 530–39), Karelia (*Kareliia*) (Maikov 1869, p. 471), (Sreznevskii 1913, pp. 488, 496, 504–5, 507), Oryol (Orël) (Popov 1903, p. 237; Skalozubov 1905, p. 10), Priangarye (*Priangar'e*) (Kliaus 1990, p. 8), Smolensk (Dobrovol'skii 1891, p. 209), Transbaikalia (*Zabaikal'e*) (Maikov 1869, p. 469), Tula (Maikov 1869, p. 747), Vologda (Vinogradov 1907, p. 14; Popov 1903, p. 229; Sreznevskii 1903, p. 155), Yaroslavl (*Iaroslavl'*) (Balov 1893, p. 427), unspecified location (Vinogradov 1907, pp. 22–25), (Dal' 1989, p. 37; Efimenko 1878, p. 205; Maikov 1869, pp. 466, 492–93; Rybnikov 1867, pp. 254–55; Shchurov 1867, p. 164; Anikin 1998, pp. 202, 220, 232, 265, 274, 294, 339, 349–50). Uniting into one figure, they can act as a third helper angel (+ Michael and Gabriel) in incantations against diseases (Popov 1903, p. 226).

These saints help make someone fall in love with the person who reads the incantation, or make a loving couple lose love; further, they aid in the hunt, and they reinforce the words of the incantation. They protect people against wizards, witchcraft, the evil eye, diseases, etc. (using the motif of an iron fence that is placed around the object of the incantation to protect it), against demonic spirits (the devil), arrows, and any weapon. They protect bridal processions from wizards, witchcraft, and other mischief and cattle from wizards, demonic forces, witchcraft, disease, etc. They also protect against the traps of hunters. They cure people of all diseases, including angina, scarlet fever, scrofula, toothache (their dominant function), (groin and navel) fractures, hemorrhages, and the twelve (also seven or seventy-seven) shivering fevers; and protect from snakebites, enchantment witches, the evil eye, and others. They save the cattle from hardening of the eyelid and from colic.

In the incantation against bleeding, one also encounters their (or rather "his", because the saints are often perceived as one figure) younger sister, with whom they help stop bleeding together. They can hang from or stand in a tree.

6. Saints Zosimas and Sabbathius

This pair of Russian saints is associated with *Solovetskii* Monastery and they are widely venerated as the patron saints of beekeeping. In the *Solovetskii* Monastery, in the Russian north, bees were bred, of which Saint Zosima (died in 1478, holidays 17 April and 18 August) was said to be the first beekeeper. Zosima was both the superior and the founder of the monastery. Saint Sabbathius (died in 1435, feast days 8 August and 27 September) is known for erecting a cross on Solovki Island in 1429. In the spiritual poems, Zosima of Solovki loves bees (Shchapov 1863, p. 50). According to folklore, Zosima and Sabbathius help bring back lost horses (Shchapov 1863, p. 68). Their names are found in the following regions: Karelia (*Kareliia*), (Anikin 1998, p. 208), Kostroma (Vinogradov 1909, p. 77), Novgorod (Maikov 1869, pp. 530–39), Perm (*Perm'*) (Maikov 1869, p. 498), Tula (Maikov 1869, p. 747), Vologda (Popov 1903, p. 229), (Sheremetev 1902, pp. 51–53), Vyatka (*Viatka*) (Maikov 1869, pp. 540–41), unspecified location (Shchapov 1863, pp. 52–54), (Toporkov 2010, pp. 426, 698, 726).

The forms of these saints' names also vary in the texts. They are called: *Zosima* and *Savvatii* (*Savvati*); the wonderworkers of Solovki; the saints *Zosima* and *Savvatii* of Solovki full of grace; *Zosima* and *Savatei*, wonderworkers of Solovki; the holy fathers *Zosima* and *Savvatii* of Solovki; *Izosim* and *Savvatii*, wonderworkers of Solovki; the saints of God *Izosim* and *Savvathii*; *Izosima* and *Savatel'*.

These saints also help in the hunt, and they protect people against wizards, witchcraft, the evil eye, diseases, etc. (using the motif of an iron fence iron fence that is placed around the object of the incantation to protect it). They protect cattle from all disasters, and bees from danger (their dominant function). They heal people of the twelve (also seven or seventy-seven) shivering fevers, of enchantment, witchcraft, the evil eye, *perepoloch*, and others (sporadic motif of shooting of diseases with a bow).

7. Saint Sisinius (*Sisinii*)

This is a rather enigmatic saint, who appears in only one category of texts, viz. in the apocryphal invocation against the twelve fevers widely distributed in Russia. As a noted researcher of Russian folk culture wrote, "Sisinius is one of the forty saints celebrated on March 9 according to the ecclesiastical calendar. In the ecclesiastical tradition about Sisinius there is nothing to be found about his ability to drive away diseases. Nevertheless, in folk tradition his name is invariably associated with the cure of fever" (Gromyko 1975, p. 94). Incidentally, as Liatskii already noted, nine Sisiniuses are known, five of whom are saints. According to him, three of them influenced the tradition of incantations, while the Sisinius of the incantations is said to be a contamination of Sisinius the bishop of Laodicea, who lived in the early fourth century, at the time of Diocletian, and—indirectly—from Sisinius of Cyzicus (Liatskii 1893, pp. 133–34). In connection with the presumed sources of the texts about Sisinius—the magical treatise "Testamentum Salomonis" and the Byzantine legend about Gilo (Gilu, Γιλλυ, Γυλου), the infanticide who was conquered by the brothers Sisinius and Sisinodorus—see: (Mansvetov 1881; Sokolov 1888; Miller 1896; Cherepanova 1977; Toporkov 2017). Toporov believes that the name and representation of the saint ultimately go back to the Sisinius who was an heir of Mani, the founder of Manichaeism

(Toporov 1993, pp. 102–3). "False prayers" against fever, with reference to Sisinius, were mentioned in the ancient indexes of forbidden books. On "The Saint Sisinius Legende" in Christian culture, see the fundamental collective monograph (Toporkov 2017).

His name forms are: Saint Sisinius (*Sisinii*); the holy father *Sisinei*; the St. *Sisenii*; *Sosinii*; Saint Apostle *Sizinii*; the holy martyr *Sisimi*; the saint father *Sosentii*. His name was found in texts from the following regions: Arkhangelsk (*Arkhangel'sk*) (Efimenko 1878, p. 205; Efimenko 1878, pp. 205–6; Efimenko 1878, pp. 207–8), Kostroma (Vinogradov 1909, pp. 6–7), (Vinogradov 1909, pp. 7–8), Novgorod (Popov 1903, p. 239), Karelia (*Kareliia*) (Toporkov 2010, p. 125), Siberia (Novombergskii 1907, p. 223; Gromyko 1975, p. 92), South Siberia (*Iuzhnaia Sibir'*), (Buslaev 1861, pp. 1596–97), Yaroslavl (*Iaroslavl'*) (Balov 1893, pp. 425–26), unspecified location (Maikov 1869, pp. 461–64; Vokhin 1878, p. 490; Tikhonravov 1863, pp. 351–52; Toporkov 2010, p. 527).

The usual tale associated with Saint Sisinius is that he encounters fevers that come from a lake and take the guise of multicolored, often naked girls. The saint asks them who they are and where they are going, whereupon they usually say their (multiple, often obscure) names and say that they are going to Russia to torment the people, each in their own way. Then, the saint exorcises and expels these demons from the person.

8. Saint Solomonia

Another Christian apocryphal figure who had made it into East Slavic folklore was Grandmother Solomonia. According Adon'eva and Ovchinnikova is "Grandmother Solomida"—a frequent figure in children's incantations—an obvious cross-contamination of two figures. The apocryphal proto-gospel of Jacob tells of how a midwife was invited by Joseph and Mary to the birth of Christ and told a passerby named Salomé that "the virgin had given birth and yet kept her virginity" < … >. Separate portions of this text (second–third centuries AD) were translated from Greek in the fifteenth century at the latest and appear in various documents of Old Russian literature (for example, in Makarii's menologion on September 8). Another possible source for the appearance of Solomonida or Solomida in oral tradition is icon painting. The icon of the Nativity shows the "Apocryphal midwife" (Adon'eva and Ovchinnikova 1993, pp. 167–68). This figure can also be compared with Solomonia, who was, according to ecclesiastical tradition, the mother of the seven brothers and martyrs of the Maccabean family. The day on which they are all remembered is August 1. The name Solomonia does not appear in the Holy Scripture (cf.: Soliarskii 1884, pp. 56–57). On the possible connection between the image of Solomonia-Salamanida and that of King Solomon cf.: (Toporov 1993, p. 102). For the reflection of the name of Solomonia in Russian folk names of plants (phytonyms), see: (Berezovich and Osipova 2018, p. 129).

There are many variants of her name: midwife (*baba*, *babushka*) or mother (*matushka*) *Solomoniia, Solomoneia, Soloman'ia, Solomon'iushka, Solomoneiushka, Solomonida, Salmanida, Solomonida, Solomonidushka, babushka Salamanidushka, Solomat'iushka, Soloveia Mikitishna, Solomida, Salama, baba Sov'ia*, grandmother *Solomonidushka*, midwife of Christ (*babushka Solomonidushka, Khristova povivalushka*). Her name was found in texts from the following regions: Arkhangelsk (Arkhangel'sk) (Adon'eva and Ovchinnikova 1993, p. 89; Zagovory 1895, pp. 240–41; Cherepanova 1977, p. 81; Efimenko 1878, pp. 201, 212; Maikov 1869, p. 471), Kostroma (Vinogradov 1907, p. 62), Olonetsk (Sreznevskii 1913, p. 508), Oryol (Orël) (Popov 1903, p. 228), Saratov (Shchapov 1863, p. 59), Smolensk (Dobrovol'skii 1891, p. 194), Vologda (Popov 1903, p. 224), (Sheremetev 1902, p. 49), Voronezh (Maikov 1869, p. 484), Vyatka (Viatka) (Maikov 1869, p. 447; Ivanova 1994a, p. 19; Ivanova 1994a, p. 82), unspecified location (Zenbitskii 1907, p. 3; Anikin 1998, p. 340).

She is the protectress of infants and, in general, a healer. She cures little children of all kinds of diseases, including insomnia, and protects them against the evil eye and other harmful magical acts. She helps with childbirth, heals people from bleeding, cures horses from all diseases, and makes sure that cows give milk. There is a permanent motif associated with her: she was the midwife at the birth of Jesus Christ, wrapping him in a diaper, washing him, and, according to some sources, she even gave birth to him herself (!). She also carries the keys with which she opens a cow's udders, or three iron bars which she swings to ward off witches (the evil eye, harmful magical acts).

This concise overview is intended to describe the images and functions of saints in Russian folk magic. From a genetic point of view, Russian folk Christianity is a complex amalgamation of Byzantine orthodoxy (along with elements of Mediterranean paganism and folk culture brought by this orthodoxy) on the one hand, and substratic or borrowed pre-Christian elements of Slavic folk culture on the other. These factors have been the subject of scientific analysis for more than a century, but an exhaustive interpretation is still a matter for the future.

Funding: This research received no external funding.

Institutional Review Board Statement: Not applicable.

Informed Consent Statement: Not applicable.

Data Availability Statement: The study does not report any data other than those extracted from the sources cited in the bibliography.

Conflicts of Interest: The author declares no conflict of interest.

Notes

[1]   *"Skazanie, kiim sviatym kakovyia blagodati ot Boga dany i kogda pamjat' ikh"*.

[2]   These Old Believer icons depict about 50–60 figures of popular saints, with signatures like the following: "Saint Nicholas the Wonderworker—(pray to him) for deliverance from drowning; Hieromartyr Cyprian and Holy Martyr Justina—for protection from evil spells; Holy Great Martyr George the Victorious—to save livestock from animals; Saints John the Warrior and Theodore Tiron—to find stolen things." (see image: http://varvar.ru/arhiv/slovo/skazanie.html (accessed on 7 July 2021)).

[3]   I render geographic names in the transliteration accepted on Google maps. In case it does not match the transliteration of the Library of Congress, the latter is shown in parentheses.

[4]   Permanent ecclesiastical or folk epithets of saints are capitalized, while the designations of classes of saints (martyrs, passion-bearers ...) are written with a small letter.

[5]   In order to save space, references to sources are not given for each name, motif or function.

[6]   Stable poetic descriptions of the saints that go back to prayer formulas are written with a small letter.

[7]   Saints in incantations often shoot away the evil eye, *uroki* (an extraneous harmful magical effect through words, eyes, objects) and illness. The saints shoot their arrows at the evil eye and diseases to make them disappear.

References

Adon'eva, Svetlana Borisovna, and Ol'ga Ovchinnikova. 1993. *Traditsionnaia Russkaia Magiia v Zapisiakh Kontsa XX Veka*. Sankt-Peterburg: Frendlikh-Taf.

Afanas'ev, Aleksandr Nikolaevich. 1914. *Narodnye Russkie Legendy*. Kazan': Molodye Sily, vol. 1.

Anikin, Vladimir Prokop'evich. 1998. *Russkie Zagovory i Zaklinania: Materialy Fol'klornykh Ėkspeditsii 1953–1993 gg.* Edited by Vladimir Prokop'evich Anikin. Moskva: Izdatel'stvo Moskovskogo universiteta.

Azadovskii, Mark. 1914. Zagovory amurskikh kazakov. *Zhivaia Starina* 3–4: 5–15.

Balov, Aleksei. 1893. Molitvy, zagovory i zaklinaniia, zapisannye v Poshekhonskom uezde Iaroslavskoi gubernii. *Zhivaia Starina* 3: 425–31.

Barsov, Elpidifor Vasil'evich. 1903. *O russkikh Narodnykh Pesnopeniiakh*. Moskva: Universitetskaia Tipografiia.

Berezovich, Elena L'vovna, and Kseniia Viktorovna Osipova. 2018. Sud'ba imeni v svete podkhoda Wörter und Sachen: Solomon v russkikh govorakh i prostorechii. *Russkii Iazyk v Nauchnom Osveshchenii* 2: 123–47.

Buslaev, Fiodor Ivanovich I. 1861. *Istoricheskaia Khrestomatiia*. Moskva: Universitetskaia Tipografiia.

Cherepanova, Ol'ga Aleksandrovna. 1977. Tipologiia i genezis nazvanii likhoradok-triasavits v russkikh narodnykh zagovorakh i zaklinaniiakh. In *Iazyk Zhanrov Russkogo Fol'klora: Mezhvuz. Nauch. sb.*. Petrozavodsk: Petrozavodskii Gosudarstvennyi Universitet, pp. 44–57.

Dal', Vladimir Ivanovich. 1989. *Tolkovyi Slovar' Zhivogo Velikorusskogo Iazyka*. Moskva: Russkii iazyk, vol. 2.

Dobrovol'skii, Vladimir Nikolaevich. 1891. *Smolenskii Ėtnograficheskii Sbornik*. Sankt-Peterburg: Tip. Evdokimova, vol. 1.

Efimenko, Piotr Savvich. 1878. Materialy po ėtnografii russkogo naseleniia Arkhangel'skoi gubernii. Ch. 2. Narodnaia slovesnost'. In *Izvestiia Ėtnograficheskogo Otdela Imp. Obshchestva Liubitelei Estestvoznaniia, Antropologii i Ėtnografii pri Moskovskom Universitete*. Moskva: Tipo-litografiia Arkhipova i Ko, vol. 30, Book V. Part II.

Eleonskaia, Elena. 1917. *K Izucheniiu Zagovora i Koldovstva v Rossii*. Moskva: Tipografiia Shamordinskoi Pustyni, Part 1.

Ermolov, Aleksei Sergeevich. 1905. *Narodnaia Sel'skokhoziaistvennaia Mudrost' v Poslovitsakh, Pogovorkakh i Primetakh*. Sankt-Peterburg: Tip. A.S. Suvorina, vol. 3, Zhivotnyi mir v vozzreniiakh naroda.

Ganchenko, Oleg. 1995. *Chudotvornyi Dukh Istseleniia. Chudotvornye Ikony, Zhitiia Sviatykh, Molitvy*. Sost. o. Oleg (Ganchenko). Khar'kov: Grif.

Gorodtsov, Piotr Alekseevich. 1916. Prazdniki i obriady krest'ian Tiumenskogo uezda. In *Ezhegodnik Tobol'skogo Gubernskogo Muzeia*. Tobol'sk: Tipografiia Eparkhial'nogo Bratstva, Part 26. pp. 1–65.

Gromyko, Marina Mikhailovna. 1975. Dokhristianskie verovaniia v bytu sibirskikh krest'ian XVIII–XIX vekov. In *Iz istorii sem'i i byta sibirskogo krest'ianstva v XVII nachale XX v*. Sbornik nauchnykh trudov Novosibirsk: Nauka SO, pp. 71–109.

Ivanova, Anna Aleksandrovna, ed. 1994a. *Viatskii fol'klor. Zagovornoe Iskusstvo*. Kotel'nich: Kotel'nichskaia tipografiia.

Ivanova, Anna Aleksandrovna, ed. 1994b. *Zagovory i zaklinaniia Pinezh'ia*. Karpogory: Pinezhskaia tipografiia.

Kliaus, Vladimir Leonidovich, ed. 1990. *Lechebnye «nagovory» Priangarskogo Kraia iz Sobraniia A. A. Savel'eva*. Moskva: Dom Russkoi Kosmetiki.

Klinger, Witold. 1931. *Doroczne Święta Ludowe a Tradycje Grecko-Rzymskie*. Kraków: Gebethner i Wolff.

Levkievskaia, Elena Evgen'evna. 2009. Zmei Ognennyi. In *Slavianskie Drevnosti. Etnolingvisticheskii Slovar'*. Edited by Nikita Il'ich Tolstoi. Moskva: Mezhdunarodnye otnosheniia, vol. 2, pp. 332–33.

Liatskii, Evgenii Aleksandrovich. 1893. K voprosu o zagovorakh ot triasavits. *Etnograficheskoe Obozrenie* 4: 121–36.

Likhachev, Dmitrii Sergeevich, ed. 1987. *Slovar' Knizhnikov i Knizhnosti Drevnei Rusi*. Leningrad: Nauka, Part 1.

Lobanova, Irina. 1993. Zagovornoe iskusstvo Kotel'nichskogo raiona. Zagovory. *Proselki. Literaturno-kraevedcheskii al'manakh* 11: 6–21.

Loorits, Oskar. 1955. Der heilige Georg in der russischen Volksüberlieferungen Estlands. In *Veröffentlichungen der Abteilung für Slavische Sprachen und Literaturen des Osteuropa-Instituts (Slavisches Seminar) an der Freien Universität Berlin*. Wiesbaden: Harrassowitz, Berlin: Osteuropa-Institut, pp. 11–29.

Maikov, Leonid Nikolaevich. 1869. Velikorusskie zaklinaniia. In *Zapiski Imperatorskogo Russkogo Geograficheskogo Obshchestva po Otdeleniiu Etnografii*. Sankt-Peterburg: [s.n.], vol. 2, pp. 417–580.

Mansvetov, Ivan Danilovich. 1881. "Vizantiiskii material dlia skazaniia o dvenadtsati triasavitsakh" Drevnosti. Tr. Imp. Moskovskogo arkheologicheskogo obshchestva. *Moskva* 9: 24–36.

Miller, Vsevolod Fiodorovich. 1896. Assiriiskie zaklinaniia i russkie narodnye zagovory. *Russkaia mysl'* 7: 66–89.

Novombergskii, Nikolai Iakovlevich. 1907. *Materialy po Istorii Meditsiny v Rossii*. Tomsk: Iakovlev, vol. 4.

Oinas, Felix Johannes. 1982. St. George as Forest Spirit. *International Journal of Slavic Linguistics and Poetics* 25/26: 313–17.

Osokin, S. (Mikhail Ionovich.). 1856. Narodnyi byt v Severo-Vostochnoi Rossii. Zapiski o Malmyzhskom uezde (v Viatskoi gubernii). Stat'ia vtoraia. *Sovremennik* 11: 1–40.

Popov, Gavriil Ivanovich. 1903. *Russkaia Narodno-Bytovaia Meditsina*. Sankt-Peterburg: Suvorin.

Putilov, Boris Nikolaevich, ed. 1984. *Russkaia Svadebnaia Poėziia SIBIRI*. Novosibirsk: Nauka.

Rybnikov, Pavel Nikolaevich. 1867. *Pesni, Sobrannye P. N. Rybnikovym*. Ch. 4. Narodnye byliny, stariny, pobyval'shchiny, pesni, skazki, poveriia, sueveriia, zagovory i t. p. Sankt-Peterburg: Kozhanchikov.

Selivanov, Aleksei Ivanovich. 1886. Etnograficheskie ocherki Voronezhskoi gubernii. In *Voronezhskii Iubileinyi Sbornik*. Voronezh: Izdanie Voronezhskogo Gubernskogo Statisticheskogo Komiteta, vol. 2, pp. 69–115.

Senderovich, Savelii. 2002. *Georgii Pobedonosets v Russkoi Kul'ture: Stranitsy Istorii*. Moskva: AGRAF.

Shchapov, Afanasij Prokof'evich. 1863. *Istoricheskie Ocherki Narodnogo Mirosozertsaniia (Pravoslavnogo i Staroobriadcheskogo)*. Sankt-Peterburg: [s.n.], vol. 1.

Shchurov, Ivan. 1867. Znakharstvo na Rusi. In *Chteniia v Imperatorskom Obshchestve Istorii i Drevnostei Rossiiskikh pri Moskovskom Universitete*. Moscow: V universitetskoi tipografii Publ., pp. 143–74.

Sheremetev, Pavel Sergeevich. 1902. *Zimniaia Poezdka v Belozerskii Krai*. Moskva: Sinodal'naia tipografiia.

Skalozubov, Nikolai Lukich. 1905. Materialy k voprosu o narodnoi meditsine. Narodnaia meditsina v Tobol'skoi gubernii. In *Ezhegodnik Tobol'skogo Gubernskogo Muzeia*. Tobol'sk: Tipografiia Eparkhial'nogo Bratstva, pp. 1–30.

Smirnov, Igor', and Vera. Il'inskaia, eds. 1992. *Vstanu ia Blagoslovias' . . . Lechebnye i Liubovnye Zagovory, Zapisannye v Chasti Arkhangel'skoi Oblasti*. Moskva: Russkii mir.

Sokolov, Matvei Ivanovich. 1888. *Materialy i Zametki po Starinnoi Slavianskoi Literature*. Moskva: Universitetskaia tipografiia, Part 1.

Sokolov, Boris Matveevich. 1995. *Bol'shoi Stikh o Egorii Khrabrom: Issledovaniia i Materialy*. Edited by Valentina Aleksandrovna Bakhtina. Moskva: Nasledie.

Sokolova, Vera Konstantinovna. 1982. Zaklinaniia i prigovory v kalendarnykh obriadakh. In *Obriady i Obriadovyi fol'klor*. Moskva: Nauka, pp. 11–25.

Soliarskii, Pavel Fiodorovich. 1884. *Opyt Bibleiskogo Slovaria Sobstvennykh Imen: 5 Vol*. Sankt-Peterburg: Tipografiia Eleonskogo, vol. 4.

Sreznevskii, Vsevolod Izmailovich. 1903. *Otchet Otdeleniiu Russk!ogo Iazyka i Slovesnosti Imp. Akademii Nauk o Poezdke v Vologodskuiu Guberniiu (mai iiun' 1901 goda)*. Sankt-Peterburg: Tipografiia IAN, Part 2.

Sreznevskii, Vsevolod Izmailovich. 1913. *Sreznevskii V. I. Opisanie Rukopisei i Knig, Sobrannykh Dlia Imp. Akademii Nauk v Olonetskom krae*. Sankt-Peterburg: Tipografiia IAN.

Stangé-Zhirovova, Nadia. 1981. Le cult de loup chez les slaves orientaux d'après les sources folkloristiques et ethnographiques. *Symposium Internationale et Pluridisciplinaire sur le Paganisme slave, Slavica Gandensia* 7/8: 181–91.

Tikhonravov, Nikolaj Savvich, ed. 1863. *Pamiatniki Otrechennoi Russkoi Literatury*. Moskva: Universitetskaia tipografiia, vol. 2.

Tokarev, Sergej Aleksandrovich. 1957. *Religioznye Verovaniia Vostochnoslavianskikh Narodov XX Nachala XX v.* Moskva-Leningrad: Izdatel'stvo AN SSSR.

Toporkov, Andrei L'vovich, ed. 2010. *Russkie Zagovory iz Rukopisnykh Istochnikov XVII—Pervoi Poloviny XIXv.* Moskva: Indrik.

Toporkov, Andrei L'vovich. 2013. Siuzhet o Egorii Xrabrom—Volch'em pastyre v slavianskom fol'klore i russkoi literature pervoi treti XX v. In *Pis'mennost', Literatura, Fol'klor Slavianskikh Narodov. Istoriia Slavistiki. XV Mezhdunarodnyi s"ezd Slavistov. Doklady rossiiskoi delegatsii*. Moskva: Institut russkogo iazykoznaniia RAN, pp. 529–57.

Toporkov, Andrei L'vovich, ed. 2017. *Sisinieva Legenda v Fol'klornykh i Rukopisnykh Traditsiiakh Blizhnego Vostoka, Balkan i Vostochnoi Evropy*. Moskva: Indrik.

Toporov, Vladimir Nikolaevich. 1993. Ob indoevropeiskoi zagovornoi traditsii (izbrannye glavy). In *Issledovaniia v Oblasti Balto-slavianskoi Dukhovnoi kul'tury. Zagovor*. Moskva: Nauka, pp. 3–103.

Uspenskii, Boris Andreevich. 1982. *Filologicheskie Razyskaniia v Oblasti Slavianskikh Drevnostei*. Moskva: Izd. MGU.

Vinogradov, Nikolai. 1907. *Zagovory, Oberegi, Spasitel'nye Molitvy i Proch.* Sankt-Peterburg: Tipografiia MPS, Part 1.

Vinogradov, Nikolai. 1909. *Zagovory, Oberegi, Spasitel'nye Molitvy i Proch. (Po starinnym Rukopisiam i Sovremennym Zapisiam)*. Sankt-Peterburg: Tipografiia MPS, Part 2.

Vokhin, Nikolai Nikolaevich. 1878. Narodnye zagovory protiv boleznei nachala XVIII-go veka. *Russkaia Starina* 22: 490–92.

Yudin, Aleksey Valerievich. 1997. *Onomastikon Russkikh Zagovorov. Imena Sobstvennye v Russkom Magicheskom Fol'klore*. Moskva: MONF.

Yudin, Aleksey. 2010. Orthodoxe heiligen in de Russische volksmagie. In *De drie Romes. Heiligenlevens, Vormen van Verering en Intellectuele Debatten in de Westerse Middeleeuwen, in Byzantium en in de Slavische Tradities*. Onder redactie van Danny Praet, met de medewerking van Nel Grillaert. Gent: Academia Press, pp. 221–34.

Zagovory. 1895. Zagovory Arkhangel'skoi gubernii Shenkurskogo uezda sela Kurgomen'. *Zhivaia Starina* 2: 240–41.

Zenbitskii, Piotr Nikolaevich. 1907. Zagovory (kontsa XVII veka). *Zhivaia Starina* 1: 1–6.

Article

Sacred and Profane: Tabooing in Russian Magical Manuscripts of the 17th–18th Centuries (Incantations and Herbals)

Aleksandra B. Ippolitova

A.M. Gorky Institute of World Literature of the Russian Academy of Sciences, 121069 Moscow, Russia; alhip@yandex.ru

Abstract: Linguistic taboos (euphemisms, omissions, and other) are an essential part of Slavic verbal and written culture. In this article, we analyze cryptography as a form of tabooing in the magical texts of the grassroots manuscript tradition of the 17th and 18th centuries (handwritten incantations and herbals). Our main objective is trying to see a system behind separate examples and define which kinds of texts are usually tabooed in incantations and herbals, their topics, and messages. We have managed to find out that the function of keeping secrecy is not relevant for the magical tradition; rather, encryption was used to emphasize the elements that are of special importance. In the book of incantations called the Olonets Codex, dating back to the 17th century, ciphering was used for the names and titles of sacred and demonological characters, antagonists, descriptions of certain rituals, closing phrases for the incantations (amen, "key"), etc. We hypothesize that the encryption is used in the Olonets Codex as a means of retaining the magical strength of all the texts in the manuscripts, protecting from hostile beings, sacralizing where necessary, tabooing what was considered sinful for religious reasons, accentuating the main meanings of the incantations, etc. In the herbals, cryptography is basically used for tabooing of "sinful" or trappy topics (love magic, magic used against courts and authorities, some contexts concerning sorcery, jinx, and "secret" knowledge), and in the texts that had to bear sacral meaning (incantations and prayers).

Keywords: magic; folklore; Slavic studies; Russian studies; herbals; incantations; manuscript studies

Citation: Ippolitova, Aleksandra B. 2021. Sacred and Profane: Tabooing in Russian Magical Manuscripts of the 17th–18th Centuries (Incantations and Herbals). *Religions* 12: 482. https://doi.org/10.3390/rel12070482

Academic Editor: Dennis Ioffe

Received: 24 May 2021
Accepted: 24 June 2021
Published: 28 June 2021

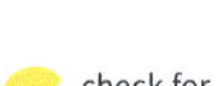

1. Introduction

The phenomenon of taboo is one of the cultural universals common to both archaic and modern societies. The term *taboo* comes from the Polynesian word *tabu* or *tapu*. It became known to Europeans through Captain James Cook's journals of 1777 (Allan 2018a, p. 3). English social anthropologist A.R. Radcliffe-Brown so described this phenomenon:

In the languages of Polynesia, the word means simply "to forbid", "forbidden", and can be applied to any sort of prohibition. A rule of etiquette, an order issued by a chief, an injunction to children not to meddle with the possessions of their elders, may all be expressed by the use of the word tabu (Radcliffe-Brown 1939, pp. 5, 6).

An important role in the history of research into the taboo concept belongs to Sir James G. Fraser, who discovered that taboos are characteristic not only for the cultures in the Pacific area, but also for other ancient and modern peoples (Frazer 1888). He examined the phenomenon of taboo in detail in his book "Golden Bough" (3rd ed., Frazer 1911) and proposed classification of taboos into tabooed acts, persons, things, and words.

The book by Dmitry K. Zelenin, a Russian ethnographer, summarizes a lot of information about linguistic taboos in various Eurasian cultures (Zelenin 1929, 1930). The researcher divided the prohibitions into two large groups: those related to trades (hunting, fishing, beekeeping, etc.) and prohibitions in domestic life.

Modern approaches to the problem of linguistic taboo and the corresponding literature are discussed in *The Oxford Handbook of Taboo Words and Language* (Allan 2018b).

Linguistic taboos are an essential part of Slavic oral and written culture. Russian ethnolinguists Yelena L. Berezovich and Svetlana M. Tolstaya define linguistic taboo as "evading calling objects by their principal names or mentioning them in certain situations" (Berezovich and Tolstaya 2012, pp. 224, 225). In folklife culture, it is done, on the one hand, to protect oneself from the danger allegedly coming from the object in question (e.g., a disease, a wild animal, etc.), and on the other hand, to protect something very valuable or sacred (God, a child, a domestic animal, etc.). Additionally, words connected with the body, uncleanness, sexuality and fertility, vices and crimes are also tabooed in some situations (Berezovich and Tolstaya 2012, p. 225).

In oral tradition, taboos can take the form of euphemisms or omissions. Written tradition offers broader opportunities for tabooing. In manuscripts, for example, the presence of taboos can be recognized through edits in the initial text or corrections made in a text subject to translation. Thus, in Russian books dating back to the cusp between the Middle and the Modern Ages, tabooing becomes an issue when dealing with another cultural context, e.g., when European scientific works are translated into Russian. In the 17th-century copies of the Gaerde der Suntheit medical essay (Luebeck 1492), which was translated into Russian in 1534, the aetiological legend of the Morsus diaboli (Succisa pratensis Moench) plant was eliminated because of some mentions of the devil and sorcery (Ippolitova 2018). The late 16th century's Russian version of Liber de arte distillandi (the book about the art of distillation) by the Strassbourg doctor Hieronymus Brunschwig (first ed. 1500) was deliberately missing the medicines based on human blood and feces, angleworms, as well as astrological and esoteric advice (Sapozhnikova 2016, 2019).

Another sign of tabooing in manuscripts can consist in cryptography: certain words or texts are encoded using various ciphers. The main work dedicated to Cyrillic cryptography was written by Mikhail N. Speransky in 1929, offering a plethora of facts about South Slavic and Russian cryptography, the methods of ciphering and deciphering, and some of their functions (1. hiding sensitive information from prying eyes; 2. hiding the author's name out of Christian humility; 3. just for fun or as a trick[1]).

As A. Arkhipov further defines the list of functions suggested by Speransky, he mentions that cryptographic tradition is not that rich in texts that hide some sensitive information, and supposes that the main function of cryptography is to emphasize the importance of key points in the text and underscore the "sacral value of personal names or other important elements". Therefore, its use "contained some kind of sacral, sometimes even magical force." (Arkhipov 1980, pp. 85, 86).

In a recent study, D. Bulanin supports Arkhipov's view, mentioning that the array of medieval cryptographic texts does not actually contain anything really "secret"; these are names of sacral concepts and objects, those of writers and other persons, references to the circumstances under which the writing was done, reverential quotes, invocations, and some elements marking the structure of the book. So far, studies of cryptography have mostly been sporadic (enough to say that there has not been any monograph on this topic after Speransky's), with most attention being paid to the deciphering techniques. Although it may seem that cryptographic sources actually yield little information, they are "priceless as a means to take a look on the spiritual world of the early Slavs". That said, Bulanin suggests focusing on the functions of cryptography in written Slavic culture, its intended use, objectives, and the kind of texts that used to be ciphered (Bulanin 2020, pp. 58, 59). This approach is also the methodological basis for our work.

In this article, we intend to analyze cryptography as a form of tabooing in the magical texts of the grassroots manuscript tradition of the 17th and 18th centuries: handwritten incantations and herbals. In those genres, the functioning of cryptography has not been studied in a close manner so far[2]. Our main objective is trying to see a system behind separate examples and define which kind of texts are usually tabooed in incantations and herbals, their topics, and messages.

In our search for incantations, we studied the most extensive collection of texts pertaining to the period in question, A.L. Toporkov's "Russian Incantations from Manuscripts of the period between the 17th and early 19th century" (RZRI 2010). Out of the 36 manuscripts included in this study, four turned out to have encrypted text; however, it was only complete in one of them, the Olonets Codex of Incantations[3] dating back to the second quarter of the 17th century. So, we decided to focus our analysis on this collection.

As for the herbals, we have used a significant pool of texts from 17th- and 18th-century manuscripts collected over many years from archives and libraries of Moscow, St. Petersburg, and other Russian cities.

2. Cryptography in the Olonets Codex of Incantations, 17th Century

The Olonets Codex, henceforward the OC, is unique in many ways against the background of handwritten incantation tradition, "unequaled in volume, variety of topics and plots, not only as of the 17th century, but, apparently, in the entire Russian manuscript tradition" (RZRI 2010, p. 37). This is one of the earliest extant incantation books, coming from the Russian North (surroundings of the Onega Lake), and is written in two languages, Russian and Karelian-Veps[4]. It was introduced into research as early as in the second half of the 19th and early in the 20th century (Malinovsky, 1876; Sreznevsky 1913, pp. 481–512), but was only published in full, accompanied by A. Toporkov's extensive comments, in 2010 (RZRI 2010, pp. 37–310). Several linguistic works on the OC, significantly supplementing the ideas about the manuscript, have been written recently by A.S. Alekseeva and A.A. Gippius (Alekseeva 2017, 2018, 2019, 2020; Alekseeva and Gippius 2019).

According to A.L. Toporkov, the OC is written by two persons (texts No. 1–35 and No. 36–125) (RZRI 2010, p. 62). In the first part, only No. 32 is ciphered, while there are six more (No. 48, 58, 104, 107, 109, 124) in the second one[5]. Often, the word «amen» (Russ. *аминь*) is also ciphered (about 20 cases, all in the second part), as well as the word "key" (Russ. *ключ*) in the zakrepka[6] (once, No. 88) (RZRI 2010, pp. 63, 64). All encrypted texts are written in Russian. Both writers used the so-called simple lithorea (Russ. *простая литорея*), a cipher where each one of the ten consonants of the first half of the alphabet was replaced with the corresponding letter from the other ten consonants but going in reverse order, from the end of the alphabet to its beginning[7] (Speransky 1929, pp. 97, 98, see Table 1).

Table 1. Simple lithorea cipher (after Speransky 1929, p. 98).

б	в	г	д	ж	з	к	л	м	н
щ	ш	ч	ц	х	ф	т	с	р	п

According to A.L. Toporkov, cryptography in the OC "is used in the most doubtful parts, which could have caused much trouble if read by anyone unauthorized" (RZRI 2010, pp. 63, 64). A.A. Turilov and A.V. Chernetsov are also inclined to see cryptography in the incantations as a means to avoid displeasure of the authorities[8] (for example, researchers believe that the name of Tsar Mikhail Fyodorovich is written with simple lithorea in the OC, "for secrecy") (Turilov and Chernetsov 2002a, p. 71).

As for the Karelian-Veps texts in the Codex, A.A. Turilov and A.V. Chernetsov suggest interpreting them as a secret language of sorcery, a kind of cryptography, too (Turilov and Chernetsov 2002a, p. 71). A.L. Toporkov believes that this hypothesis needs more substantial rationale (RZRI 2010, p. 82).

We believe that the functions of cryptography in the OC are not limited to secrecy (if any at all) but are more complicated and diverse than has been supposed so far.

The functionality of ciphered texts in the OC is various enough. They include incantations about tsars and the authorities (No. 32), cattle release (No. 58), fistfights (No. 104), love spells (No. 107), incantations against snake bites (No. 109), against trees and foes (No. 124), and an incantation to scare mice away from stacked rye[9] (No. 48). Thus, the incantations can be classified into groups based on their functions and topics: social and interpersonal (32, 104, 107), protective (124), housekeeping (48, 58), and curative (107). This classification is partly conventional, because, say, No. 124 can also be considered a text about social interactions (with foes). The curative incantation (against snakes) stands somewhat apart because it is of Karelian-Veps origin, only its title being written in Russian using lithorea[10]. It is important that the distribution of cryptographic texts by function and topic does not correlate with the general distribution of incantations by those groups in the Codex in general[11] (see Table 2 and Figure 1).

Table 2. Statistic of ciphered Russian incantations in the OC.

	Protective	Curative	Social/Interpersonal	Housekeeping	Total
Overall number of Russian incantations in the OC	35 (38.5%)	33 (36.3%)	15 (16.5%)	8 (8.8%)	91
Number of ciphered incantations in the OC	1 (16.6 %)	–	3 (50%)	2 (33.3%)	6
Proportion of ciphered incantations in each group	1 of 35 (2.9%)	–	3 of 15 (20%)	2 of 8 (25%)	6 of 91 (6.6%)

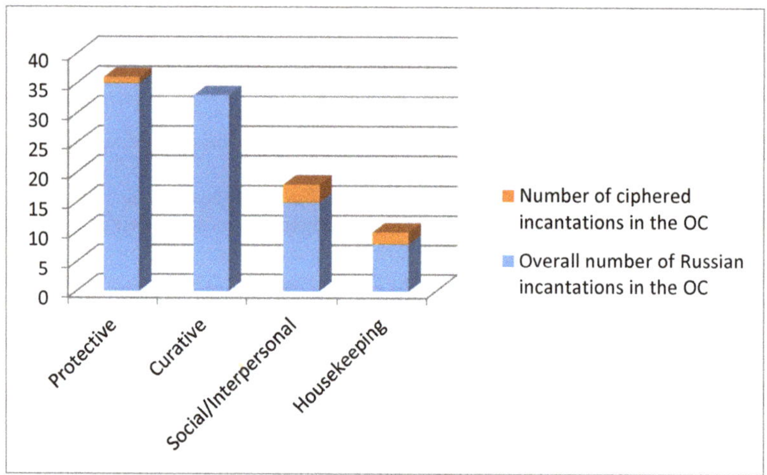

Figure 1. Distribution of ciphered Russian incantations in the OC.

According to the table, half of the incantations containing cryptography belong to the social/interpersonal group (and even two thirds of them if we include No. 124, too), while the remaining third contains household-related incantations. Meanwhile, within the Codex in general these groups are on the third and fourth positions in terms of occurrence. Accordingly, one fifth of the social/interpersonal incantations contains cryptography, as does somewhat about a quarter of the household-related ones.

The texts are encrypted not in their entirety, only some words or phrases therein. There is cryptography both in the titles of the incantations (No. 107, "Слова *к женкам*"[12] ("Words *to women*"), No. 109 "Слова *змеиные*" ("*Snake* words")) and in the instructions to them (No. 48, 58, 104, 124) as well as in the incantations themselves (No. 32, 48, 58, 104, 107, 124, i.e., all of them except for the "snake" one). That means that the incantation about the tsar and the government (No. 32) only has cryptography in the text itself; the snake one (No. 109), in the title only, the love one (No. 107), in the title and in the body, and all the rest (No. 48, 58, 104, and 124) have it both in the texts and in the instructions.

The incantation No. 32 is intended to make all people and authorities like the caster. The name and title of Tsar Mikhail Fyodorovich are mentioned twice, ciphered. First, it goes on about the beauty of the Tsar, which the caster wants to associate with, then about the caster's desire to be liked by the Tsar, as well as by the authorities, the higher-ups, and the noblemen.

И как свет сей блистаетца и Христос наряжаетца, свет настал и Христос народис(я)[13], и как падут утренные и вечерние росы на землю и на во(ду), також бы пала красота и лепота *царя и господаря и великого князя Михаила Федоровича всея Руси*[14]. < … > и как любы Господу Исусу Христу свои Б(о)жии престолы, також бы и яз люб был, раб Б(о)жий имярек, кому говориш(ь) имярек, *царю, господарю и великому князю* имярек *всея Руси*[15] и всем властем, и началом, и велльможам. (RZRI 2010, p. 106).

And like this light shines and Christ is adorned, light has shone and Christ is born, and like morning and evening dews fall upon earth and water, so may fall the beauty of *Mikhail Fyodorovich, the Tsar and Lord and the Grand Prince of all Russia*. < … > and like Lord Jesus Christ loves His Godly thrones, may also me, servant of God (name), be loved by the one called by the name (name), and also by the *Tsar, Lord and Grand Prince* (name) *of all Russia* and by all authorities, higher-ups, and noblemen. (RZRI 2010, p. 106).

The love incantation (No. 107), apart from the title «Слова к женкам» (Words *to women*), ciphers the name of the mythological Tsar, *Zhazhda* (Rus. *жажда* "thirst"), the phrase «*во уста*» (meaning to the lips of the woman to whom the magic is addressed; a river of fire is meant to enter her lips), and the phrase «*сердце у той рабе*» (the heart of that woman, God's servant):

Слова к женкам[16].

На зоре на утренней пойду яз < … > и увижу яз царя *Жажду*[17]. Цар(ь) *Жажда*[18], об(ъ)яви мне огненную реку < … > пади ты, огненная река, имярек *во уста*[19]. И как та огненная река горит, так бы горело *сердце у той рабе*[20] имярек < … >. (RZRI 2010, p. 132).

Words *to women*.

In the early dawn will I go < … > and see Tsar *Zhazhda*. Oh Tsar *Zhazhda*, show me a river of fire < … > may you river of fire fall *into the lips* of (name of woman). And like that river of fire burns, may *the heart of that woman*, [God's] *servant* (name) burn, too < … >. (RZRI 2010, p. 132).

Thus, in addition to the name of Tsar Zhazhda, everything that has to do with the woman (references to her, like *женка* "woman", *раба* "(God's) servant", and her body parts, like *уста* "lips", *сердце* "heart") is ciphered here.

In the fistfight incantation (No. 104), the names of demonological characters (forest and water spirits, leshy and vodyanoy respectively), whom the caster summons for helping him win, are also ciphered, as is the designation of the opponent, *the fistfighter*, and all the words referring to the harm done to him: (arms and legs) *weakened*, (eyes) *turned blind*, (arm) *raise*, *beat* (my enemy). The caster's *fists* are also mentioned ciphered (evidently as a means of doing harm to the enemy) as well as the pronoun "my" in the phrase "*моему супостатаю*" («for *my* enemy»), maybe in order to make him stand out from among the other enemies.

Се яз, раб имярек< … > призываю к себе на помощ(ь) *из леса лешех, из воды водяных*[21]; и вы, *из леса лешые, из воды водяные*[22], подите ко мне на помощ(ь) противо моего супостатая, *кулачног(о) бойца*[23], и пособите вы мне *побит(ь)*[24] моего супостатая, *кулачного бойца*[25] имярек, своими *кулаками*[26]. И вы, *из леса лешие, из воды водяные*[27], возмите(у) сег(о) имярек *мертвеца*[28], камен(ь) и положыте моему супостатаю, *кулачному бойцу*[29] имярек, на руки, или на ноги, или на главу; кол(ь) есть сему *мертвецу*[30] тяжело от земли и от камени, стол(ь) бы тяжело было *моему*[31] супостатаю, *кулачному бойцу*[32] имярек, противо меня руки *поднят(ь)*[33], чтоб у моего супостатая *кулачног(о) бойца*[34] имярек руки и ноги *ослабли*[35] и в глазах *слепота*[36] от моего приговора и до моег(о) отговору. (RZRI 2010, pp. 130, 131).

Herewith do I, God's servant (name) < … > summon for help *leshys from the forest and vodyanoys from the waters*; so may you, *leshys from the forest and vodyanoys from the waters*, come to help me against my enemy, *fistfighter*, and help me *beat* my enemy, *fistfighter* (name), with my *fists*. And you, *leshys from the forest and vodyanoys from the waters*, take from this *dead man* (name) a stone and put it onto my enemy's, *fistfighter* (name)'s, arms, or legs, or head; just like earth and stone are heavy for the *dead man*, so may it also be heavy for *my* enemy, *fistfighter* (name), trying to *raise* his arms against me, and may my enemy, *fistfighter* (name)'s arms and legs *weaken* and his eyes turn *blind* from [the moment of] my incantation and up to [the moment] when I cancel it. (RZRI 2010, pp. 130, 131).

There is also another figure in this incantation, the dead man, whose objective is to make the enemy weaker. The reference to the dead man is ciphered both in the incantation and in the instructions that follow, describing the rite that has to be performed on the dead man's tomb: this is where the incantation is to be read thrice; then the caster bows unto the dead man and takes the stone, putting it into his breeches or his sleeve. Apart from the dead man, the following is also ciphered here: the location (*могила* "tomb"), the denomination of the incantation itself (*слова* "words"), the references to the caster's ritual actions (the thing that he takes off (*К*)[37]), the repetition of the incantation (*в другой ряд* "again"), the bow (*поклонися* "bow"; *да поклонися мер(т)вому до земли* "and bow unto the dead man to the ground"), the cardinal point where the caster is looking *(на север* "northward")*, and the place where the caster hides the object *(в портки, за рукав* "into your breeches or behind your sleeve"):

Пришед *к могилы*[38], да соими *крест*[39] да проговори *слова*[40] ряд *да поклонися мер(т)вому до земли*[41], а сам зри *на север*[42], да *в другой ряд*[43] проговор(и) тож, *поклонися*[44], а в третей ряд проговор(и) *да поклонися*[45], да возми [к]*амень*[46] невелик, да положи *в портки или за рукав*[47]. (RZRI 2010, p. 131).

As you come to *the tomb*, take off your *cross* and say the *words* [spell] once and *bow unto the dead man to the ground*, while looking *northward*, then *again* say it, *bow,* and then the third time say it *and bow*, and take a small *stone*, and put it *into your breeches* or *behind your sleeve*. (RZRI 2010, p. 131).

The cattle release incantation (No. 58) is meant for protecting the cattle during the forest pasturage season. Such texts used to consist of "three parts. (1) an address to the forest tsar, his wife and children; (2) mentioning of a gift or treat that the caster has brought; (3) the ask to care for the cattle throughout the pasturage season" (RZRI 2010, p. 231). In the instructions for this one, cryptography covers ritual elements, ciphering the location (forest, a hidden place, *остров* (a higher ground in the forest), *выскить* (an uprooted tree)) and time (early in the morning before sunrise) of the ritual treatment, as well as the imperative of the action (*неси* "bring"):

Аще хощеш(ь) скота отпущат(ь) мног(о), и ты щуку свежую да 3 яйца да 3 бочки меду розсыти, да *неси в лес на остров (в)* укромное место *на выскит(ь),* а *неси поутру рано до солнца*[48]. (RZRI 2010, p. 116).

If you want to release a lot of cattle for pasture, then [take] a fresh pike and 3 eggs and 3 barrels of liquid honey and *bring it all to a hidden high ground in the forest next to an uprooted tree,* and *do it early in the morning before sunrise.* (RZRI 2010, p. 116).

The incantation No. 58 itself ciphers the address to the lords of the forest, Tsar Gongoy and Tsarina Gogeya, and their children (but their servants are written as usual). Then, certain significant words are also ciphered (*обед* "dinner"), i.e., the ritual treat, *лебеди* "swans", which is the metaphorical name for the eggs the caster has brought, and the ask to accept the treat *в чести* "in honor":

Государь *царь Гонгой и царица Гогея, и сынове, и дочери*[49], и слуги, вам *обед*[50], щука свежая, да три белые *лебеди*[51], да три бочки меду; а приймете *в чести*[52], а за то стерегите и берегите и пасите вес(ь) мой скот< . . . > (RZRI 2010, p. 116).

Oh you *Tsar Gongoy and Tsarina Gogeya, and their sons and daughters* and servants, here is a *meal* for you, a fresh pike and three white *swans*, and three barrels of honey, too; please accept this *in honor*, and as a reward protect and observe and care for all my cattle < . . . >. (RZRI 2010, p. 116).

In the incantation No. 124 (against trees and foes), the caster addresses a mythological figure, Tsar Shustiya, asking him to turn his body into stone, his bones into bulat (steel) and his ribs into copper, so that he becomes invulnerable to any foes or falling trees. In this case, only some parts of the zakrepka are ciphered:

... слова(м) клю(ч), а древа(м) тле(н), *а телу моему креп*[53]. Во имя О(т)ца и Сына и Святого Духа. Сим слова(м) *ами(н)*[54]. (RZRI 2010, p. 143).

... key to the words, and rot to the trees, *and strength to my body*. In the name of the Father, the Son, and the Holy Spirit. To these words, *amen*. (RZRI 2010, p. 143).

The instruction No. 124 says that the caster should read the incantation while standing on iron, and the name of the metal is ciphered (however in the second instance it is written in the regular way):

Гово(р) 3(ж), стоя *на желизе*[55] да на камени, а в рука(х) де(р)жа(т) железо да каме(н), да оче(р)ти(т) собя 3(ж) о(т) утра да до вечера. (RZRI 2010, p. 143).

Tell this three times, standing *on iron* and on stone, with iron and stone in your hands, and circumscribe yourself, three times from morning to evening. (RZRI 2010, p. 143).

The instruction to the incantation against mice (no. 48) encrypts the action to be done with harvested rye: *класть в стог* ("stack"). The incantation itself encrypts mice themselves[56]:

Аще кто хощет рож *класти в (с)тог*[57], и ты возми 9 каменей да 9 жеребейков ол(ь)ховых да положы наперед, а молви: *Мышем*[58] камен(ь) то им хлеб, а древо то им волога, за колико лет обновляется круг земный. (RZRI 2010, p. 112).

Should someone want to *stack* rye, then take nine stones and nine alder rods and put [them] first [i.e., before stacking the rye], and then say: *To mice*, stone is bread and wood is soup [for so many years], as the earth renews (RZRI 2010, p. 112).

Thus, we have seen that only a few texts (6.6%) of the OC actually use cryptography for separate words and phrases. It is worth noticing that in other parts of the same manuscript, the same words or phrases are written in the regular manner. This could be a powerful argument against the earlier version stating that the encryption in the OC (and in other handwritten incantation books, too) was a means to avoid persecution from the authorities. In our mind, should the authors of the OC actually have intended to do so, the encryption would have been more consistent. For instance, all the names of mythical creatures and fragments of "black" incantations would have been ciphered, too.

Therefore, it appears to make more sense that cryptography in the OC was used in approximately the same way as it was in other genres of written texts. That is, as D.M. Bulanin puts it, to emphasize important parts of the text rather than to hide anything "secret" (from intruders or authorities) (Bulanin 2020, pp. 59, 60).

As we have seen, there is an array of characters that appears in the encrypted parts of the OC. These are Tsar Mikhail Fyodorovich (full name and title), some mythical creatures, such as Tsar Zhazhda, Tsar Gongoy, and Tsarina Gogeya and their children, forest and water spirits, a dead man; also, people having to do with the caster: the woman (addressee of the love spell), the fistfighter (enemy of the caster); also, animals (mice or snakes doing harm to the caster).

We believe that Tsar Mikhail Fyodorovich's name is ciphered as a sacred word, which is actually confirmed by various studies (Toporkov 2007, pp. 69–72, 78; Toporkov 2018, pp. 117, 118). We should note that this name also appears in the Veliky Ustyug incantation book, created at the same time as the OC, and is not ciphered there (Turilov and Chernetsov 2002b, p. 212). The Tsar's name could also have been ciphered in the OC for being perceived as a kind of talisman, so that no one could do harm to the Tsar[59].

The names of mythical creatures, such as leshys, vodyanoys, the dead man, Tsar Zhazhda, Tsar Gongoy, and Tsarina Gongeya, could have been tabooed out of fear that they could do harm to the caster. Additionally, tsars, even mythical, could be perceived as sacred beings, too[60].

The names of the caster's "enemies", such as the fistfighter or mice or snakes could have been tabooed so that they would never know what was being planned against them. The ciphered description of such a common thing as stacking rye could also be something of this kind: to prevent the mice from knowing. The mentions of women were evidently tabooed in some contexts that had to do with love magic, since it was considered a sin.

Additionally, cryptography in the OC is used in the cases of using magic against other people, like debilitating a fistfighter or attracting a woman; in both cases, the addressees' body parts are mentioned.

The descriptions of rituals were also encrypted. These are, first, the ritual treats to the lords of the forest: the place and time, the actions themselves, and the treats themselves (*обед* ("meal"), *лебеди* ("swans", metaphorical name for eggs), and the ask to receive the meal *в чести* ("in honor")). Second, the ritual at the dead man's tomb: the ciphering is used for the location, the actions (bowing, facing northward), verbal elements (reference to the incantation (*слова* ("words"), and their repetitions)), and certain things (breeches, sleeve, cross).

In some cases, cryptography serves to emphasize the basic meaning of the text. For example, in the incantation No. 124 all the encrypted words appear to contribute to one single objective: making the caster invulnerable. They are the zakrepka *телу моему креп* ("strength to my body"), the second zakrepka with the encrypted *аминь* ("amen"), and the word *железо* ("iron") which appears encrypted in the instruction.

3. Cryptography in the Herbals of the 17th–18th Centuries

Russian handwritten herbals as a tradition existed from the early 17th to the early 20th century, and this tradition was heterogeneous. The herbals describe plants both real and fictional, intended for use for household, cures, or magic; some plants were said to have supernatural properties (see Ippolitova 2008 for more details). The texts about plants in the herbals have a rather consistent structure: the plant's name, description, where and when it grows, time and ritual of collection, functions, and ways of using.

In the herbals of the 17th–18th centuries, there is cryptography in the texts about 12 plants, which is not that many (about 2%) for an array of about 500 plant descriptions, and it is used sporadically as a rule (usually about 1–3 texts for one plant)[61].

Here, only some parts of the text are encrypted, from one word to several phrases. In the structure of the articles, the use of cryptography is aligned with the descriptions of the plants' functions (how and what for they are used), to-wit: love magic (5 plants), various kinds of other magic (6 plants), and cure (1 plant). Aside from that, one incantation, two prayers and an unclear fragment (possibly incantation or plant picking ritual) are encoded.

Out of these 12 plants, the majority (9 of 12) are said to have completely magical functions, while at least 5 are charmful (have an anthropomorphic root; have flowers that burn at night like candles; the plant can destroy iron objects; has 4 or 12 different flowers; the one who finds the herb will get lost and go mad).

There are basically two types of cryptography in the 17th–18th century herbals, popular in Russia back then: simple lithorea and using the initial letters of the words[62] (sometimes with a complicating twist). There is also an example of ciphering using Latin letters (GIM. Museum coll. No 1226; Speransky 1929, pp. 70–72). For some texts, we failed to detect the type of cryptography and, therefore, to decipher them.

3.1. The Case of the Semitar Plant

The texts about the magical herb of Semitar[63] are interesting in terms of both encryption methods and the specific vocabulary being tabooed. In the 17th–18th century herbals and home cure books in question, there are about fifteen variations of texts about this plant, which has an anthropomorphic root and four flowers of various colors (a Russian mandrake, in its way). The names for Semitar can vary in different manuscripts, e.g., *семитар, симтарм, сирман, царь трава сам парамонть,* etc. There are encrypted fragments in two texts about Semitar.

In a manuscript dating back to the second quarter of the 18th century (IRLI. Velichko coll. No. 26), there is one of the lengthiest descriptions of Semitar (over 300 words), involving the phytonyms of *sirman, pokrik*[64]. This one contains a detailed description of the "man-root"'s anatomy and explains how it should be used in magic. For instance, it was believed that the root's "head" could help a man get to love his wife. The "right hand" prevented both husband and wife from adultery. The "liver", when cooked in milk, was said to be able to cure sterility. Two fragments of the text dedicated to the use of the magic root are written using simple lithorea.

- The root's "chest" is cut in two and the "heart" is taken out. Then, the instructions to the love magic appear ciphered:

И в том корени велико нутр, как в человеке. И ты сердце вынемши< ... >, и да пить *девице*[65], *коея восхощеши, или жа[н]ке, и по тебе учнет тужить*[66] (IRLI. Velichko coll. No. 26. Fol. 297. No. 73).

And in that root, there are many entrails, just like in a man. So you take out the heart < ... > and give it to drink *to the girl you might want, or a woman, so [she] will yearn for you.* (IRLI. Velichko coll. No. 26. Fol. 297. No. 73).

- It was also suggested to use the magic root for success in trials and with people in general:

Коли хочешь *тягатися* или *на суд*[67], тот корень держать при себе, *виноват* не будешь *ничем, властелином*[68] будешь, учнет тебя *любеть, и вси люды* учнет *теба любить зело*[69]. (IRLI. Velichko coll. No. 26. Fol. 297v. No. 73).

Should you want to go to *court*, take this root with you so you will not be *guilty of nothing*, you will be *governor*, and also *loved, and all the people will love you* a lot. (IRLI. Velichko coll. No. 26. Fol. 297v. No. 73).

Thus, the *sirman* text taboos phrases related to love magic, as well as the magical impact of the caster over other people's feelings and over court, too. At the same time, the words related to marriage magic (to make a husband love his wife, to prevent adultery) and to the magical ability for a woman to become pregnant using the "liver" of the root remain unciphered in the same text.

In the text about the *simtarm* herb, titled "Should one want to get married" and included into a herbal of the late 17th century, cryptography is used in about the same manner as in the *sirman* example. A.A. Turilov, the first publisher of this text, suggested a partial reading of this encryption (Turilov 1998) and noticed that it employed various ciphering methods (Turilov 2002, p. 375).

И выняти из того человека сердце: есть бо нутр в том человеке, как в жи[во]м же. И половину того же сердца или третюю часть истерти мизиным пальцом и дать кому ни буди пити *ли дѣ ко а х. ш з с б*[70] *ино д+ и ж к* возждеют хотя б буди *c^r* или *knzna* и они *б*[71] твои, аще испиеть ис твоих рук. (GIM. Museum coll. No. 1226. Fol. 204r-204v).

Indeed, there is a mixed-up ciphering technique here, based on one or two first letters of a word (sometimes also the last letter and the one prior to it), and based on the Latin alphabet. Based on A.A. Turilov's findings and on our own collation of the text with the original of the manuscript, we suggest our own version of a deciphered fragment.

... и дать кому ни буди пити *[женке и]ли де[вке], ко[е]а х[още]ш з[а] с[е]б[я]*, ино *де[вка]* и *ж[ен]к[а]* возжде[ле]ют, хотя б буди *ц[а]р[евна]* или *кн[я]жна*, и они *б[удут]* твои, аще испиеть ис твоих рук.

And take the heart out of that man, for he has entrails, just like a real man does. Then crumble a half or a third of that heart with your little finger and give someone to drink, [*to a woman*] or *a girl you want to possess*, then *the girl and the woman* will yearn, no matter if they are *a tsarevna* or *a princess*, and they *will be* yours if only they drink [the potion] from your hands. (GIM. Museum coll. No. 1226. Fol. 204r-204v)

In this fragment, encryption is used to refer to the women (woman, girl, tsarevna (tsar's daughter), princess), for the phrase that describes which women is it about (" ... you want to possess") and the verb "will be" which confirms that the magical procedure does work.

It is notable that in the texts about Semitar ciphered fragments are not repeated in various texts, i.e., they were ciphered independently, at the discretion of the specific writer.

There is also a special version of the Semitar text that contains no encryption but has appeared as a result of the "profane" being tabooed. It has no mentions of love or marriage magic (except for childbirth); the folk chrononym of *Иванов день* (Ivan's day) is replaced with the canonical *Рождество Иоанна Предтечи* (The Nativity of John the Forerunner); the description of the collection ritual requires the caster to read a prayer and to be clean from any filthiness; the magical function involving the use of the "heart" is replaced with the curing one: to have a wash with the "heart" of the root in order to cure one's own heart. In other words, this text was edited in the tideway of the Christian tradition and religious motives (handwritten cure book of the 17th century, GIM. Uvarov coll. No. 114. Fol. 15v-16r).

3.2. Love Magic

In a 1703 text about the *Khorobrets* herb (Figure 2), there are three words ciphered by simple lithorea and also three letters under titlos in the end:

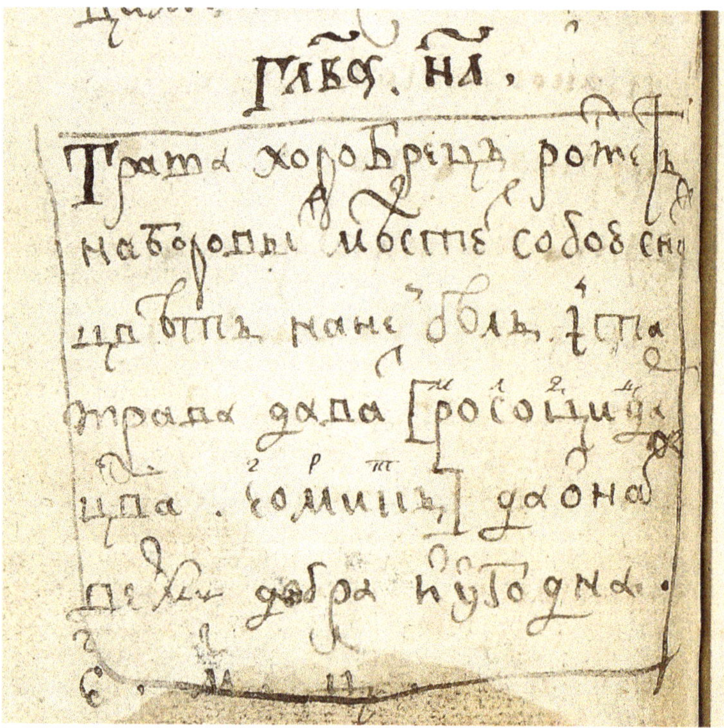

Figure 2. Text about *Khorobrets* plant with cryptogram, 1703 (RGB. Undolsky coll. No. 1072. Fol. 28v. No. 51).

Трава хоробрец< . . . > И та трава давать {р^{м}ос^{л}оц^{д}ид^{ц}а^{р} цва ч^{г}ом^{р}ик^{т}}, да она ж велми добра и угодна е̃. м̃. п̃. (RGB. Undolsky coll. No. 1072. Fol. 28v. No. 51).

The herb Khorobrets < . . . > Is to be given {р^{м}ос^{л}оц^{д}ид^{ц}а^{р} цва ч^{г}ом^{р}ик^{т}}. And it is very good and useful indeed. е̃. м̃. п̃.

The meanings of the consonants are inscribed over them in the first cryptogram, perhaps by a reader from the 18th century. When deciphered, the text looks as follows:

И та трава давать *молодицам*[72]—*душа горит.*

And that herb be given to *young women—soul burns.*

Here, encryption is added at the discretion of the writer, since the same phrase does not appear encrypted in other texts about the same plant. This text is about using magic to make a woman love the caster[73].

The three letters with titlos, «ē. м̄. п̄.», in the end of the *Khorobrets* text, are the initial letters of the encrypted text's words. In such cases, deciphering is made easier based on versions of the same text in different manuscripts, which sometimes contain unciphered words. One case of this is the text about the *Peresyanka* plant (18th century), where references to women and girls appear both ciphered and not (see Table 3).

Table 3. Texts about the *Peresyanka* plant.

GIM. Uvarov coll. No. 705. Fol. 26. No 79	BAN. 45.8.175. Fol. 12r. No 76
Кто ея корень носит при себе, и того человека любят {ж и д}.	Тот корень кто носит при себе, и того человека любят женки и девки.
The one who wears the root [of this plant] is loved by {ж и д}.	The one who wears the root is loved by women and girls.

Another trouble when deciphering this kind of encryption is that the scribes sometimes misinterpreted the texts they did not understand, replacing some letters with other ones (usually looking similar).

We again meet references to *молодые молодицы* (young moloditsas) and encryption in three texts from the 18th century dedicated to the Gnida plant: black, not tall, with a black flower, known to protect from bears (who neither attack nor roar) and from dogs (who don't bark):

Трава гнида < ... > Да угодна и молодым молодицам (д) (ж) (ц) (к) (с) (RGB. Museum coll. No. 9530. Fol. 29. No. 76).

Gnida herb < ... > Also good for young moloditsas (д) (ж) (ц) (к) (с).

We can suppose that the first letters of the encrypted text, *Д* and *Ж*, can have the close meaning as *молодицы*, i.e., girls and women. In all probability, this text is about love magic, too.

3.3. Other Magical Functions

The rest of the examples speaking about plants' functions is encrypted by their first letters and usually not deciphered, the context being reconstructed based on the surrounding words. Let us take a brief look at some examples.

In the text about the Palochnik plant (*Typha latifolia* L.), encryption "(м) (п) (е)" closes the statement that it is good for millers at their mills (RGB. Museum coll. No. 9530. Fol. 28. No. 72). Meanwhile, the encryption in the text about the magical Sova 1 plant is probably related to the destruction of a mill:

Кто на ту траву наидет — заблудитца, а кто вырвет — без ума будет, а в мелницу кинет — мелницу рознесет *п д б з е р ж ч* железо *д*. (IRLI. V.M. Perets coll. No. 489. Fol. 4v. No. 38).

The one who steps on this herb, will get lost, and the one who pulls it by the roots, will go mad, and when thrown into a mill, it destroys the mill *п д б з е р ж ч* iron *д*. (IRLI. V.M. Perets coll. No. 489. Fol. 4v. No 38).

Among peasants, millers were reputed sorcerers, hobnobbing with vodyanoys (water spirits) and leshys (forest spirits). In case of a conflict with a miller, peasants tried to destroy the stanch or even the mill itself (Shchepanskaia 2001, pp. 17–19).

In the copy of the herbal from a 1770 investigation report, there is a text about the magic herb of Muravey (Rus. *муравей* "ant") that breaks scythes, frees horses from hopples and opens locks. Thanks to the abundance of texts about Muravey, we had no problem deciphering the П: У: fragment: these letters mean *петля удавить* ("loop strangle"), i.e., this plant could save someone sentenced to hanging from being strangled with a rope (see Table 4).

Table 4. Texts about the Muravey plant.

(Mihajlova 2003, p. 272) (RGIA. F. 796. Op. 51. D. 322)	BAN. 33.14.11. Fol. 35v. No 78
А в рот положить—не может ево: причем под титлами поставлены в клетках два слова П: У:	А кто в рот положит, и петля удавить не может.
Put [this herb] into your mouth and can't him: there are two П: У: words under the titlos in the cells.	Put [this herb] into your mouth and no loop will strangle you.

In several texts about the Myshka (Rus. *мышка* "mouse") plant, there is encryption looking like this: *д. в. х. к./д в х/Ж: I: Е: Д: В: Х:*, which correlates with the words *хмель* (hop plant or drunkenness) and *пьет* (drinks) (RGB. Dolgov coll. No. 111. Fol. 1; RNB. Q.VI.18. Fol. 38v. No 68; Mihajlova 2003, p. 272: RGIA. F.796. Op. 51. D. 322). Other versions of texts about Myshka make us think that this plant was helpful for quitting drinking.

The magic plant Levuppa, as the herbals put it, cannot be found during the day, but at night its flowers burn like candles, and the other plants bow unto it. Levuppa was said to have some charmful properties: its owner was given honors, it also protected from jinx and foes, at weddings and feasts, and also helped in hunting. In some manuscripts, the plant's functions are not listed in detail but mentioned as a whole: "ко всячине добра" ("[it] is good for everything"), followed by a cryptogram of 4–5 letters, e.g., *з р с т* (BAN. 33.14.11. Fol. 32v, No. 70), possibly standing for a description of some specific properties of the plant.

3.4. Incantations and Prayers in the Herbals

We can only be sure about the presence of an encrypted incantation in the herbals in the case of one plant, Ulik. There are four texts from the 18th century about it, two of them containing encryption by initial letters of the words; in two others, the entire text of the incantation for the collection of that plant is encrypted. It should be said that the letters in the two ciphers only coincide partially, and the compliance between the incantation and its encryptions is obviously incomplete. Perhaps both encrypted texts have been corrupted during copying. The incantation had to be read while collecting the plant, and its effect consisted in restoring conjugal love. Let us quote the encrypted and the regular versions (see Table 5).

Table 5. Texts about the Ulik plant.

Kizhi Museum. KP-4281/1. Fol. 25. No. 78.	GIM. A.S. Uvarov coll. No. 705. Fol. 28v. No. 86.
Как ты т: г: п: так п т к или м: к своей главу с в д: и р: с д ни от х: в: п д: д:.	Как ты, трава, приклонила главу свою в землю, так бы приклонили оне ж меж себя главы свои всею душеи и ретивым сердцем, думою и мыслию хотно век по веку до гробныя доски.
Just like you т: г: п: so п т к or м: to own head с в д: и р: с д ни от х: в: п д: д:.	Just like you, the herb, have inclined your head toward the ground, let also them incline their heads toward one another, with all their souls and passionate hearts, reflections and thoughts, readily, forever as long as they live.

In one ciphered incantation about Ulik, the plant's function is different: instead of marital magic, it should be used for attracting maids and women (DTYuS 1998, p. 434).

In some texts about Solneshnik plant, there is a rather large encrypted passage, of about 30 symbols, and it has not been deciphered so far. The text appears to have been ciphered by first letters or using a mixed-up method, by the numeric values of the letters. Some letters can differ from manuscript to manuscript, however there are coincidences, too. The encrypted part appears following the phrase that extracting this plant is "not easy"; this latter expression has a special meaning in the herbals, to-wit, having special magic skills; here, those skills consist in knowing which ritual to perform while collecting the plant.

Есть трава солнешник< . . . > А копать ея не просто: м.н.о.а.а.о.п.р.р.х. | | .л.л.н.о.п.ж. р.р.р.а.а.s.а. дух. ч.н.н.о. И носить на себе, тот будет силен велми храбр воин. А копать в маие месяце в 9-м числе. А как пропустишь, и ты на Иванов день Купалницы, и никому не сказывай, то велми богат будешь. (RNB. Q.VI.18. Fol. 35v. No. 50).

There is a herb called Solneshnik. < . . . > And it is not easy to extract: м.н.о.а.а.о.п.р.р.х. | | .л.л.н.о.п.ж.р.р.р.а.а.s.а. spirit. ч.н.н.о. And to be carried with yourself, to become a very brave and valiant warrior. Dig it out on the ninth day of the month of May. If you miss that, do <dig it out> on St. John's day, [after St. Agrippina's day], and tell no one, then you will be very rich. (RNB. Q.VI.18. Fol. 35v. No. 50).

One could suppose that the cryptogram contains an incantation to be read while digging the plant out, or just a description of the collecting ritual. It should be noted that in an 18th century manuscript from RGB, the cryptogram and the words *непросто и* ("not easy and"), directly preceding it, were deliberately crossed out from the text (RGB. Museum coll. No. 9530. Fol. 26r. No. 65). This is a particular situation when something tabooed is tabooed additionally.

There are also two cases when prayers are also written with encryption. An 18th century manuscript advised to say the Trisagion prayer, ciphered by its first letters, while collecting the above-mentioned herb of Sova 1:

. . . а рвать непростому *С(вятый) Б(оже) С(вятый) К(репкий) С(вятый) Б(ессмертный) П(омилуй) Н(ас) ав* (GIM. Barsov coll. No 2257; quoted in: Strakhova 1988, p. 42).

. . . and collecting is not easy: *Holy God, Holy Mighty, Holy Immortal, have mercy on us (thrice)*. (GIM. Barsov coll. No 2257; quoted in: Strakhova 1988, p. 42).

While collecting the Bogorodichnaya herb (i.e., belonging to Virgin Mary), a prayer to Virgin Mary should be said:

Есть трава Богородичная . . . та трава добра человеком и скотом и у которыя жонки болят груди или титки парь да хлебай да говори *Б(огородице) Д(ево)* . . . (GIM. Barsov coll. No 2257; quoted in: Strakhova 1988, p. 42).

There is also the Bogorodichnaya plant, which is good for both people and cattle, and if a woman's breasts ache, [apply it hot] and drink it and say *the Virgin Mary* [prayer] . . . (GIM. Barsov coll. No 2257; quoted in: Strakhova 1988, p. 42).

Thus, there are actually not that many encrypted texts in the herbals (just about 2% of the plants) but they are not distributed uniformly. Ciphering was used for fragments of texts about the plants, the incantations and prayers included therein; there are various types of encryption; there are also entries that can and cannot be deciphered; the encryption is mainly used in some specific texts about one plant (apparently at the discretion of the writer of each manuscript in particular).

For 7 of 12 plants, encrypted texts have been deciphered successfully. The majority (4) of them have to do with love magic. In such texts, the encryption covers references to women (*женка* "woman", *девка* "girl", *княжна* "princess", *царевна* "tsarevna", *молодица* "young woman"), verbs and phrases that express love desires and emotions (*восхотеть* and *хотеть* "start yearning for", *тужить* "long for", *душа горит* "soul burns"), and the text of the love incantation is ciphered almost in full.

As for social magic, i.e., related to courts, authorities, and people in general, there is only one deciphered fragment (Semitar). This is somewhat related to delinquent, secret knowledge, expressed in the fragment about the rope that will not strangle you (Muravey).

As for the four plants with undeciphered descriptions, we can barely make guesses about their meanings based on the adjacent phrases. As far as we can understand, these parts keep discussing love magic (Gnida), secret knowledge and sorcery (the mill and its destruction, Palochnik and Sova 1), plus the motif of quitting drinking (Myshka) and the multi-purpose plant (Levuppa).

The use of encryption in incantations and prayers (three plants deciphered and one not) apparently marked them as sacral texts, simultaneously making them stronger. Importantly, unlike in the incantation books, herbals include such entries encrypted in their entirety, i.e., they are emphasized against the background of the regular text. This could have something in common with the prohibition on saying the incantations aloud in the presence of outsiders, or giving them to outsiders, to prevent these texts from losing their powers (Tolstaya 1999, pp. 240, 241).

4. Conclusions

In this article, we have studied the practice of tabooing in handwritten magical texts of the 17th–18th centuries (herbals and incantation books), performed by means of cryptography. Our main idea is that there is some kind of system behind the seemingly unrelated fragments of the manuscripts, even if such a system is not clearly seen now. Let us share some insights.

We believe that in the materials we have studied cryptography is not used for the purpose of secrecy[74] but rather to emphasize some parts of the texts that appear to be of special value. The most complicated thing here is to understand the nature of this value.

The encryption in the OC is more complicated and diverse than in the herbals, and has a wider functionality, too. In the OC, encryption has to do with the relationships between the caster and various aspects of reality and the magical world. The caster is surrounded and affected by real persons (men and women, authorities, foes), animals (mice, snakes), and demons; they also perform ritual actions in special, hidden places (forest, tomb) using various objects. In each specific case, encryption in the OC has its own function: acting as a talisman for the caster and their possessions, protecting them from foes; as a means to sacralize the figure of the tsar; as a means of tabooing of sinful passions, etc. At the same time, the presence of encryption was meant to make the incantation stronger. Thus, encryption is an instrument to make the incantations multidimensional and complicated, strengthening the existing narratives and adding new ones.

The incantations from the OC were definitely intended for verbal use, i.e., for being read out (whispering or hushfully, but read out anyway, not silently). How did one have to handle the ciphered parts, then? Most probably, the cryptograms were intended for written texts only; the words were deciphered before being read out. Thus, the encryption served as a kind of container for words of special value, to prevent them from "escaping" ahead of time and thereby losing their strength, as it can happen with incantations that are disclosed to outsiders. Let us recall that the word *amen* is ciphered about 20 times in different parts of the OC, plus the word *key* one more time. We could suppose that these words served as a kind of zakrepka for the entire book, holding together the magical force of the incantations contained therein[75]. This is similar to the evidence of a sorceress from the Pomorye (Northern Russia) area. In the 1930s, she told a folklore researcher the text of an incantation but not the *ключевые слова* (Rus. "key words", i.e., zakrepka), to prevent the incantation from losing its effect:

. . . then come some key words, but the sorceress did not tell them to me. She said, 'I can only disclose those key words in my dying hour, to the one who would take up my trade'. (RGALI. F. 1489. Op. 1. D. 24. Fol. 80).

The same function was probably ascribed not only to the amens but also to all the encrypted texts in the OC. It could well be that the owner of the OC had to teach their successors in trade to read simple lithorea, in order to be able to make good use of the incantations.

Understanding the encrypted parts as the "key words" of the entire manuscript can explain the sporadic use of ciphers in the incantations (and not only there), which appears strange to a modern person. This includes the situations when some important word is only ciphered once or twice, and then is written in the regular manner further on (e.g., dead man, tsars of different names, iron). Apparently, it was enough to encrypt the word once to give it all the necessary sacral and magical functions.

The functions of ciphers in herbals appear to be less complicated. In this case, they are mostly used for tabooing sensitive topics (love magic, magic related to courts and authorities, some contexts related to sorcery, black magic, "secret" knowledge), and to mark verbal texts that should have sacral effects (incantations and prayers). It could well be that in the latter case, just like in the OC, cryptography was simply believed to strengthen a text's magical effect and at the same time to protect it from outsiders. Still, almost a half of the encrypted parts of the herbals still remains ciphered; hence, the complete picture could well look less complicated than it really was.

We believe we could further focus on studying the functions of the ciphers, involving more and more sources (published and archived incantation books[76] and handwritten cure books) and also by expanding their chronology toward the 19th and the early 20th century, a period of transformation in the magical handwritten tradition (Ippolitova 2008, p. 16; RZRI 2010, p. 21). This would mean a very different nature of the encryptions (e.g., herbals include encrypted jokes and phytonyms). Such a study could enable us to perform comparative research of similar genres in other Slavic and, widely, European manuscript traditions. (Important observations in the field of functional study of cryptography from the Middle Ages and early modern times have been made recently, in particular for the German and Hungarian traditions, which opens up the possibility for cross-cultural research: Müller 2014; Láng 2015, 2018). Then, we could study other methods of tabooing in magical manuscripts, such as omissions, euphemisms, and traces of tabooing in edited texts.

Funding: This work was funded by RUSSIAN FOUNDATION FOR BASIC RESEARCH, grant number 20–012–00117.

Institutional Review Board Statement: Not applicable.

Informed Consent Statement: Not applicable.

Acknowledgments: The author is sincerely grateful to this Special Issue guest editor, Denis Ioffe, for inviting to contribute to it.

Conflicts of Interest: The author declares no conflict of interest.

Abbreviations

BAN	Manuscript Department of the Russian Academy of Sciences Library in St Petersburg
GIM	Manuscript Department of the State Historical Museum in Moscow
Kizhi Museum	Kizhi State Open Air Museum of History, Architecture and Ethnography in Petrozavodsk
IRLI	Drevlekhranilishche im. V.I. Malyshev of the Institute of Russian Literature of Russian Academy of Sciences in St Petersburg
RGALI	Russian State Archive for Literature and Art in Moscow
RGB	Russian State Library
RGIA	Russian State Historical Archive
RNB	Manuscript Department of the National Library of Russia in St Petersburg

Manuscripts
BAN. 33.14.11. Herbal from the late 18th century. 47 ff.
BAN. 45.8.175. Herbal from the middle of the 18th century. 21 ff.
GIM. A.S. Uvarov coll. No 114. Cure book from the second half of 17th century.
GIM. A.S. Uvarov coll. No. 705. Herbal from the middle of the 18th century. 37 ff.
GIM. E.V. Barsov coll. No 2257. Herbal from the late 18th century. 40 ff.
GIM. Museum coll. No 1226. Herbal from the late 18th century. 307 ff.
IRLI. V.M. Perets coll. No 489. Herbal from the late 18th century. 18 ff.
IRLI. V.V. Velichko coll. No 26. Herbal from the second quarter of the 18th century. 326 ff.
Kizhi Museum. KP-4281/1. Herbal from the late 18th century. 48 ff.
RGALI. F. 1489. Op. 1. D. 24. Murmansk folklore expedition of 1932.
RGB. Dolgov coll. No 111. Herbal from the early 18th century. 30 ff.
RGB. Museum coll. No. 9530. Herbal from the 18th century. 133 ff.
RGB. V.M. Undolsky coll. No 1072. Herbal and medical book from 1703, 1705. 160 ff.
RGIA. F. 796. Op. 51. D. 322. Investigation report from 1770 about a priest called Jacob from Kolomna.
RNB. Q.VI.18. Herbal from the second half of the 18th century. 39 ff.

Notes

[1] Apart from that, Speransky mentions letter strings in incantations as a kind of cryptography (Speransky 1929, p. 3).

[2] A.A. Arkhipov also planned to take on this question (Arkhipov 1980, p. 86), however, in the next edition of his work he limited himself to emphasizing the need for such a study (Arkhipov 1995, p. 145).

[3] Today it belongs to the Manuscript Department of the Russian Academy of Sciences Library in St Petersburg (no. 21.9.10).

[4] According to A.L. Toporkov's calculations, the OC contains 130 texts, including 90 Russian and 9 Karelian-Veps incantations, as well as "prigovors" (short ritual-magical texts), non-canonical prayers, and texts without verbal magic (RZRI 2010, pp. 48, 75–77).

[5] A.S. Alekseeva and A.A. Gippius have supposed that the words *вомр* and *вомра* in the love incantation No. 122 are written, "using a special cipher where letters are replaced with the ones most similar to them in shape: Б for В, Ж for М, Ь for Р, and therefore mean божий, божья (God's). < ... > The attribute preposition, unusual for the phrase раб Божий (God's servant), could also be a form of distortion, a kind of "anti-behavior", according to B.A. Uspensky, caused by the fact that using God's name in an erotic text would have been a blasphemy" (Alekseeva and Gippius 2019, pp. 153, 154). We will not talk about this text herein, since deciphering is still hypothetical.

[6] Zakrepka (from Russ. *закреплять* "to anchor, to fix")—Russian term for final element of incantation, magically "locking" the action of the entire previous text, that gives the words power and neutralizing possible mistakes made in it (Yudin 1997, p. 9).

[7] As A.A. Arkhipov puts it, this kind of cipher is borrowed from the Jewish tradition (Arkhipov 1980).

[8] In the 17th and 18th centuries, sorcery and incantations in Russia were strictly persecuted by church and state, with legal action taken against people owning "incantation books" (Lavrov 2000; Smilianskaya 2003; Kivelson 2013; Mihajlova 2018).

[9] In the 2010 publication, the text is called "so that mice don't eat the straw" (RZRI 2010, p. 112). However, the text itself has not a single reference to straw (and mice don't actually eat it), so we have corrected the name.

[10] The statistics regarding the functional distribution of Karelian-Veps texts is not given in RZRI 2010 (maybe because it is often hard to define).

[11] The information on the frequency of Russian incantations in the Olonets Codex is taken from (RZRI 2010, p. 80) but we have also added the mice one (No. 48) to the statistics.

[12] Hereinafter, the decipher of encrypted text is given in italic within the quotes from the sources.

[13] In the original text: *надодис.*

[14] In the original text: *дамя и чолноцамя и шесикочо тпяфа Рижаиса Зецомошига шлея Мули.*

[15] In the original text: *д(а)рю, чолноцаму и шелитору тпяфю имяреквлея Мули.*

[16] In the original text: *т хептар.*

[17] In the original text: *Хахцу.*

[18] In the original text: *Хахца.*

[19] In the original text: *шо улка.*

[20] In the original text: *лемцде у кой маще.*

[21] In the original text: *иф села свеж, иф шоцы шоцацы.*

[22] In the original text: *иф села свые, иф шоцы шоцяпые.*

In the original text: *тусагпоч(о) щойда.*

In the original text: *нощик.*

In the original text: *тусагпоч(о) щоида.*

In the original text: *ту сатари.*

In the original text: *иф села севые, иф шоцы шоцяпые.*

In the original text: *ремкшеда.*

In the original text: *тусачпору щойду.*

In the original text: *ремкшеду.*

In the original text: *роеру.*

In the original text: *тусагпою щойду.*

In the original text: *ноцпяк.*

In the original text: *тусачпочо щойда.*

In the original text: *олсащли.*

In the original text: *лсенока.*

A.L. Toporkov believes that it could be a garment (cap, caftan, etc.) (RZRI 2010, p. 131). A. Alekseeva and A.A. Gippius believe that it is about the baptismal cross (Alekseeva and Gippius 2019, p. 152).

In the original text: *т рочисы.*

In the original text: *к.*

In the original text: *лсоша.*

In the original text: *ца нотсопиля ремшору цо ферси.*

In the original text: *па лешем.*

In the original text: *ш цмучой мяц.*

In the original text: *нотсопиля.*

In the original text: *ца нотсопися.*

In the original text: *[т]апеп* (reading according to: (Alekseeva and Gippius 2019, p. 152)).

In the original text: *в номкти или фа мутав.*

In the original text: *пели ш сел па олкмош в утморпое место па шылтик, а пели ноукму мапо цо лоспда.*

In the original text: *дапь Чопчой и дамида Чочея, и лыпоше, и цогеми.*

In the original text: *ощец.*

In the original text: *сещеци.*

In the original text: *ш гелки.*

In the original text: *кесу роеру тмен.*

In the original text: *ари(п).*

In the original text: *па хесифе.*

Strictly speaking, the word *мышем* cannot be the beginning of the incantation but instead the last word of the instruction. See also: (Alekseeva and Gippius 2019, p. 146).

In the original text: *тсалки ш коч.*

In the original text: *рывер.*

Compare D.M. Bulanin's supposition that the encrypted text on the Zvenigorod bell could have included the name of Tsar Aleksey Mikhailovich so that no devilish forces could harm the Tsar (Bulanin 2020, p. 60; concerning the inscription, see: (Speransky 1929, pp. 94–97, 112, 113)). Regarding the Tsar as a sacred figure, see (Uspensky 1998; Lukin 2000; Toporkov 2007).

However, there are other Tsars mentioned in the OC (Tsar Parfey, Tsar Shustiya, etc.) and their names remain unencrypted. (See (Toporkov 2007, 2018) for more information).

Only in the text about the Solneshnik plant is cryptography indeed used in most texts.

In the absence of additional information, such texts usually cannot be deciphered (Speransky 1929, pp. 41, 42).

Hereinafter, the herbal's phytonym written in regular will refer to the entire array of textually similar texts about the same plant (referred to using one of the most frequent phytonyms that appears in the aforesaid array). The phytonym written in italic will refer to the name of a plant that appears in a specific manuscript.

Compare Polish *pokrzyk* for "Atropa belladonna L.".

Corrected. Otherwise, it is written *цувеце* when deciphered, probably because of the writer's mistake.

In the original text: *душеде тоея шолкобеви иси хате, и но кеще угпек кухикь.*

[67] In the original text: *кячакиля или па луц.*

[68] Corrected. Otherwise, it is written *бастелином* when deciphered, probably because of the writer's mistake.

[69] In the original text: *шипошак не будешь нимет, щалкесипор будешь, учнет тебя сющекь и шли сюды учнет неща сющикь зе.*

[70] Could also be read as *e* or *ѣ.*

[71] Letter *Б* appears ringed.

[72] In the 18th century, the word *молодица (moloditsa)* was used to refer to a young woman, usually married (SRYa 18 veka 2003, p. 10).

[73] Such formulas with the "love burning" motif are popular in Russian incantations and literary tradition, too (Toporkov 2005, pp. 24–45).

[74] It should be said that encrypted manuscripts often attracted the attention of criminal investigators. Thus, a herbal was confiscated in 1770 from a priest called Jacob and some parts of it were copied into the investigation report, including the ones that were ciphered by the initial letters. The investigators left the following note next to one of the encrypted fragments: " . . . there are certain letters written from the cells, apparently as explanation for that [text] or some kind of incantation, but the meaning thereof remains unknown" (Mihajlova 2003, p. 270; RGIA. F.796. Op. 51. D.322).

[75] About magical power of zakrepka, see also (Levkievskaya 2002, p. 245).

[76] There are also a lot of encrypted texts in an 18th century manuscript containing some incantations and a herbal (Vinogradov 1909, pp. 28–41), and in the incantations from a 17th century cure book (Pushkarev 1977, pp. 87, 103, 104, 111, 114, 121).

References

Alekseeva, Alina S. 2017. O dvukh zagadkakh Oloneckogo sbornika [About two riddles of the Olonets codex]. In *Antropologiya. Fol'kloristika. Sociolingvistika (Sankt-Peterburg, 23–25 marta 2017)*. Sankt-Peterburg: EUSpb, pp. 4–7.

Alekseeva, Alina S. 2018. Tri kommentariya k Oloneckomu sborniku [Three comments to the Olonets codex]. In *Problemy Istorii i Kul'tury Srednevekovogo Obshchestva. Materialy XXXVII Vserossijskoj Mezhvuzovskoj Konferencii Studentov, Aspirantov i Molodyh Uchenyh «Kurbatovskie Chteniya» (27–30 noyabrya 2017 g.* Sankt-Peterburg: Skifiya-print SPB, pp. 139–44.

Alekseeva, Alina S. 2019. O protografe Oloneckogo sbornika [About the protograph of the Olonets codex]. In *Tekstologiya i Istoriko-Literaturnyj Process*. Moscow: Buki Vedi, vol. 7, pp. 11–21.

Alekseeva, Alina S. 2020. Russkoyazychnye teksty Oloneckogo sbornika: Fonetika i grammatika [Russian texts of the Olonets codex: Phonetics and grammatics]. *Russkij Yazyk v Nauchnom Osveshchenii* 2: 128–50.

Alekseeva, Alina S., and Aleksei A. Gippius. 2019. Nabliudeniia nad tekstom Olonetskogo sbornika zagovorov. [Some observations on text of olonetskii sbornik]. *Drevniaia Rus'. Voprosy Medievistiki* 2: 141–57.

Allan, Keith. 2018a. Taboo words and language: An overview. In *The Oxford Handbook of Taboo Words and Language*. Edited by Keith Allan. Oxford: Oxford University Press, pp. 1–27.

Allan, Keith, ed. 2018b. *The Oxford Handbook of Taboo Words and Language*. Oxford: Oxford University Press.

Arkhipov, Andrei A. 1980. O proishozhdenii drevneslavyanskoy taynopisi. [About the origin of ancient Slavic cryptography]. *Sovetskoye Slavyanovedeniye* 6: 79–86.

Arkhipov, Andrei A. 1995. *Po tu Storonu Sambationa: Etiudy o Russko-Evreiskikh Kul'turnykh, Iazykovykh i Literaturnykh Kontaktakh v X–XVI Vekakh*. [Beyond the Sambation: Studies of Russian-Jewish cultural, linguistic and literary contacts in the X-XVI centuries]. Oakland: Berkeley Slavic Specialties.

Berezovich, Yelena L., and Svetlana M. Tolstaya. 2012. Tabu yazykovoe. [Tabu of words]. In *Slavyanskie drevnosti: Etnolingvisticheskij slovar' v 5 t*. Edited by N. I. Tolstoy. Moscow: Mezhdunarodnye otnosheniya, vol. 5, pp. 224–28.

Bulanin, Dmitrii M. 2020. Tajnopisanie i inopisanie (Novacii russkoj pis'mennosti v poru Vtorogo yuzhnoslavyanskogo vliyaniya). [Cryptography and alternative writing systems (inopisanie) (Innovations of Russian Literature at the Time of the Second South Slavic Influence)]. *Palaeobulgarica/Starobulgaristika* 3: 58–80.

DTYuS. 1998. Dejstvuyushchij travnik iz Yuzhnoj Sibiri. [Acting herbal from South Siberia]. Podgot. teksta, predisl., slovnik V.A. Lipinskoj, G.A. Leont'evoj. In *Tradicionnyj opyt Prirodopol'zovaniya v Rossii*. Edited by L. V. Danilova and A. K. Sokolov. Moscow: Nauka, pp. 414–479.

Frazer, Sir James G. 1888. Taboo. In *Encyclopaedia Britannica*, 9th ed. Edinburgh: A. and C. Black, vol. 23, pp. 15–18.

Frazer, Sir James G. 1911. *The Golden Bough. Pt. 2: Taboo and The Perils of the Soul*, 3rd ed. London: Macmillan.

Ippolitova, Aleksandra B. 2008. *Russkie Rukopisnye Travniki XVII–XVIII Vekov: Issledovanie Fol'klora i Etnobotaniki [Russian Herbal Manuscripts of the Seventeenth–Eighteenth Centuries: Research in Folklore and Ethnobotany]*. Moscow: Indrik.

Ippolitova, Aleksandra B. 2018. Legenda o trave s otkushennym kornem v russkoj rukopisnoj tradicii XVI–XIX vv. [The Legend of Grass with a Bitten Root in the Russian Manuscript Tradition of the XVI–XX centuries]. *Zhivaya Starina* 1: 6–10.

Kivelson, Valerie. 2013. *Desperate Magic: The Moral Economy of Witchcraft in Seventeenth-Century Russia*. Ithaca and London: Cornell University Press.

Láng, Benedek. 2015. Ciphers in Magic: Techniques of Revelation and Concealment. *Magic, Ritual, and Witchcraft* 10: 125–41. [CrossRef]

Láng, Benedek. 2018. *Real Life Cryptology*. Amsterdam: Amsterdam University Press. [CrossRef]

Lavrov, Aleksandr S. 2000. *Koldovstvo i Religiia v Rossii: 1700–1740 gg. [Witchcraft and Religion in Russia: 1700–1740].* Moscow: Drevlekhranilishche.

Levkievskaya, Yelena Ye. 2002. *Slavianskiy Obereg. Semantika i Struktura. [Slavic Protective Magic. Semantics and Structure].* Moscow: Indrik.

Lukin, Pavel V. 2000. *Narodnye Predstavleniya o Gosudarstvennoj Vlasti v Rossii XVII veka. [The Popular Political Beliefs in the 17th Century Russia].* Moscow: Nauka.

Malinovsky, Lev L. 1875–1876. Zagovory i Slova, po Rukopisi XVIII Veka. [Charms and Words, According to the Manuscript of the XVIII Century]. In *Oloneckij sbornik*. Petrozavodsk: Izd. Olonec. Gubern. Stat. Komiteta, vol. 1, pp. 69–92.

Malinovsky, Lev L. 1876. Zagovory i slova, po rukopisi XVIII veka. [Charms and words, according to the manuscript of the XVIII century]. *Oloneckie gubernskie vedomosti*. 1876. 28 fevr. No 15. pp. 161–63; 3 marta. No 16. pp. 172–73; 10 marta. No 18. pp. 192–95. Available online: https://ogv.karelia.ru/catalog.shtml?year=1876 (accessed on 2 May 2021).

Mihajlova, Tatiana V. 2003. *Koldovskie Processy v Rossii: Oficial'naya Ideologiya i Praktiki Narodnoj Religioznosti (1740–1801 gg.). [Witchcraft Processes in Russia: The Official Ideology and Practices of Folk Religiosity (1740–1801).].* Diss . . . kand. istorich. nauk. Sankt-Petersburg: Evropejskij Universitet.

Mihajlova, Tatiana V. 2018. *Ot Kolduna do Sharlatana. Koldovskie Processy v Rossijskoj Imperii XVIII veka (1740–1800). [From Sorcerer to Charlatan. Witchcraft Processes in the Russian Empire of the XVII Century (1740–1800)].* Sankt-Petersburg: Evropejskij Universitet.

Müller, Stephan. 2014. Warum mittelalterliche Geheimschriften keine Geheimschriften sind. In *Schriftträger—Textträger*. Edited by Annette Kehnel and Diamantis Panagiotopoulos. Berlin, München and Boston: De Gruyter, pp. 169–78. [CrossRef]

Pushkarev, Lev N. 1977. Drevnerusskij lechebnik. [Old Russian medical book]. In *Redkie Istochniki po Istorii Rossii*. Edited by A. A. Novosel'ski and L. N. Pushkarev. Moscow: Iinstitut istorii SSSR AN SSSR.

Radcliffe-Brown, Alfred R. 1939. *Taboo*. Cambridge: Cambridge University Press.

RZRI. 2010. *Russkie Zagovory iz Rukopisnykh Istochnikov XVII—Pervoi Poloviny XIX v. [Russian Charms from Handwritten Sources of the XVII—The First Half XIX Century].* Edited by A. L. Toporkov. Moscow: Indrik.

Sapozhnikova, Olga S. 2016. Neizvestnyj russkij perevod XVI veka iz mediko-astrologicheskogo traktata «Kniga o zhizni» ital'yanskogo gumanista Marsilio Fichino. [Unknown Russian translation of the XVI century from the medical-astrological treatise "The Book of Life" by the Italian humanist Marsilio Ficino]. *Peterburgskaya Bibliotechnaya Shkola*. 3: 12–24.

Sapozhnikova, Olga S. 2019. Evropejskoe izdanie epohi Vozrozhdeniya (Liber de arte distillandi): Prichiny izbiratel'nosti russkogo perevoda [Renaissance European edition (Liber de arte distillandi): Reasons of the selectivity of the Russian translation]. *Drevniaia Rus'. Voprosy Medievistiki* 4: 175–93.

Shchepanskaia, Tatiana B. 2001. Muzhskaya magiya i status specialista (po materialam russkoj derevni konca XIX—XX vv). [Male magic and the status of a specialist (based on materials from a Russian village at the end of the 19th–20th centuries)]. In *Muzhskoj sbornik*. Edited by I. A. Morozov and S. P. Bushkevich. Moscow: Labirint, vol. 1, pp. 9–27.

Smilianskaya, Elena B. 2003. *Volshebniki. Bogohul'niki. Eretiki. Narodnaya Religioznost' i «Duhovnye Prestupleniya» v Rossii XVIII v. [Magicians. Blasphemers. Heretics. Popular Religiosity and "Spiritual Crimes" in Russia XVIII c.].* Moscow: Indrik.

Speransky, Mikhail N. 1929. *Taynopis' v Yugo-Slavyanskih i Russkih Pamyatnikah pis'ma (=Entsiklopedia Slavyanskoy Filologii, 4.3). [Cryptography in South Slavic and Russian Monuments of Writing (=Encyclopedia of Slavic Philology, 4.3].* Leningrad: Izdatelstvo AN SSSR.

Sreznevsky, V. I. 1913. *Opisanie Rukopisei i Knig, Sobrannykh v Olonetskom krae. [Description of Manuscripts and Books Collected in the Olonets Region].* Sankt-Petersburg: Tipografiia Imperatorskoi akademii nauk.

SRYa 18 veka. 2003. *Slovar' Russkogo Yazyka 18 veka. [Dictionary of Russian Language of 18 Century].* 13. Sankt-Petersburg: Nauka.

Strakhova, Olga B. 1988. Fragmenty zagovorov i molitv v travnikah. [Fragments of charms and prayers in herbals]. In *Etnolingvistika Teksta. Semiotika Malyh form Fol'klora. Tezisy i Predvaritel'nye Materialy k Simpoziumu*. Moscow: Nauka, pp. 40–42.

Tolstaya, Svetlana M. 1999. Zagovory. [Incantations]. In *Slavyanskie Drevnosti: Etnolingvisticheskij Slovar' v 5 t*. Edited by N. I. Tolstoy. Moscow: Mezhdunarodnye Otnosheniya, vol. 2, pp. 239–44.

Toporkov, Andrei L. 2005. *Zagovory v Russkoj Rukopisnoj Tradicii XV-XIX vv.: Istoriya, Simvolika, Poetika. [Charms in the Russian Handwritten Tradition of the XV-XX Centuries: History, Symbolism, Poetics].* Moscow: Indrik.

Toporkov, Andrei L. 2007. Nominaciya «car'» i personazhi-«cari» v russkih zagovorah XVII–XIX vv. [Nomination "tsar" (king) and characters-"tsars" (kings) in the Russian conspiracies of the XVII-XX centuries.]. *Voprosy Onomastiki* 4: 67–82.

Toporkov, Andrei L. 2018. Onomastikon Oloneckogo sbornika zagovorov XVII veka. [Proper Names in the 17th Century Olonets Codex of Verbal Charms]. *Voprosy Onomastiki* 2018: 115–33.

Turilov, Anatolii A. 1998. Narodnye pover'ya v russkih lechebnikah. [Folk beliefs in Russian medical books]. *Zhivaya Starina* 3: 33–36.

Turilov, Anatolii A. 2002. Narodnye pover'ya v russkih lechebnikah. [Folk beliefs in Russian medical books]. In *Otrechennoe Chtenie v Rossii XVII–XVIII Vekov*. (Tradicionnaya duhovnaya kul'tura slavyan. Publikaciya tekstov). Edited by A. L. Toporkov and A. A. Turilov. Moscow: Indrik, pp. 367–75.

Turilov, Anatolii A., and Alexei V. Chernetsov. 2002a. Otrechennye verovaniya v russkoj rukopisnoj tradicii. [Renounced Beliefs in Russian Manuscript Tradition]. In *Otrechennoe Chtenie v Rossii XVII–XVIII Vekov*. (Tradicionnaya duhovnaya kul'tura slavyan. Publikaciya tekstov). Edited by A. L. Toporkov and A. A. Turilov. Moscow: Indrik, pp. 8–72.

Turilov, Anatolii A., and Alexei V. Chernetsov. 2002b. Velikoustyuzhskij sbornik XVII v. [Veliky Ustyug charms collection of the XVII c.]. (Predislovie i publikaciya A.A. Turilova i A.V. Chernecova). In *Otrechennoe Chtenie v Rossii XVII–XVIII Vekov*. (Tradicionnaya duhovnaya kul'tura slavyan. Publikaciya tekstov). Edited by A. L. Toporkov and A. A. Turilov. Moscow: Indrik, pp. 177–224.

Uspensky, Boris A. 1998. *Tsar' i Patriarch: Kharizma Vlasti v Rossii. [Tsar and Patriarch: The Charisma of Power in Russia]*. Moscow: Yazyki slavianskoy kultury.

Vinogradov, Nikolai. 1909. *Zagovory, Oberegi, Spasitel'nye Molitvy i Proch. [Verbal Charms, Saving Prayers, and So On]*. Sankt-Petersburg: Tipografiya MPS, vol. 2.

Yudin, Aleksei V. 1997. *Onomastikon Russkikh Zagvorov. [Onomasticon of Russian Incantations]*. Moscow: MONF.

Zelenin, Dmitrii K. 1929. *Tabu Slov u Narodov Vostochnoj Evropy i Severnoj Azii. [Taboos of Words among the Peoples of Eastern Europe and North Asia]*. Leningrad: Izdatelstvo AN SSSR, vol. 1.

Zelenin, Dmitrii K. 1930. *Tabu Slov u Narodov Vostochnoj Evropy i Severnoj Azii. [Taboos of Words among the Peoples of Eastern Europe and North Asia]*. Leningrad: Izdatelstvo AN SSSR, vol. 2.

Article

Catherine's Icon: Pavel Filonov and the Orthodox World

Nicoletta Misler [1] and John E. Bowlt [2,*]

[1] Dipartimento di Slavistica, Istituto Universitario "L'Orientale", 80133 Naples, Italy; mislern@gmail.com
[2] Department of Slavic Languages, University of Southern California, Los Angeles, CA 90007, USA
* Correspondence: bowlt@usc.edu

Abstract: The authors discuss the Orthodox icon which Pavel Filonov (1883–1941) painted in 1908 or 1909 for his sister, Ekaterina, placing it within the broader context of his oeuvre, his family and his understanding of 'religiosity'. Making reference to Filonov's system of Analytical Art and to what he called 'madness', the authors focus on the particular technical devices which he used in the icon and on the *podlinnik* (or primer) from which he copied the main elements. Reference is also made to other religious motifs in Filonov's art such as the Magi, Flight into Egypt and Crucifixion.

Keywords: Pavel Filonov and his family; icons of St. Catherine the Martyr; Orthodoxy; Old Believers; Imperial Academy of Arts; Crucifixion; Natal'ia Goncharova; Kazimir Malevich; Vladimir Tatlin; Cathedral of the Intercession at Rogozhskaia sloboda (Moscow)

1. Preamble

The artist Pavel Nikolaevich Filonov (1883–1941), a primary member of the Russian avant-garde, is remembered for the highly enigmatic paintings and drawings which he executed according to his system of Analytical Art or Madeness.[1] Although occasionally investigating abstraction, Filonov tended to concentrate on intensely crowded compositions showing numerous heads and figures often languishing in a mysterious and ominous city, compositions striking in their dense stratification and almost microscopic detail. Although, by way of his mysterious subject-matter and 'Expressionist' style, Filonov stands apart from his more familiar colleagues such as Kazimir Malevich and Vladimir Tatlin, there is a single—and unexpected—common denominator which interconnects, i.e., an appreciable debt to the Russian Orthodox or, rather, Christian, tradition. This denominator, however, often manifested itself indirectly or metaphorically, rather than in immediate extrapolations in the sense that Malevich, for example, described the early phase of his artistic evolution as 'iconic', although he never painted icons, Tatlin examined Mediaeval icons such as *The Virgin Mary of Vladimir* from a purely formal perspective, reducing their compositions to geometric schemes, Natal'ia Goncharova painted saints and apostles, making recourse to Neo-Primitivist and Cubo-Futurist styles and, of course, Vasilii Kandinsky looked to the Old Testament for images of the Flood and the Apocalypse. Still, other salient artists of the avant-garde such as Aleksandra Ekster, Ivan Kliun, Liubov' Popova and Aleksandr Rodchenko seem to have ignored the Christian legacy altogether. In contrast, and more than any other of his radical colleagues, Filonov did focus on the Biblical story, at least during the early part of his career, producing numerous, highly personal interpretations of key episodes such as *Flight into Egypt* (1918, Figure 1), *The Magi* (1913, Figure 2), *Easter* (1912–13, private collection, St. Petersburg) and *Holy Family* (1913–14, Figure 3).[2] In particular, in 1908 or 1909 Filonov painted an "image of Catherine the Great Martyr from a figural icon primer" (Figure 4)[3] for his sister, Ekaterina (perhaps as a wedding present)—or, more probably, from an original which he may have seen in the Cathedral of the Intercession (Pokrovskii sobor) in the Rogozhskaia sloboda, Moscow (Figure 5), and which he seems to have copied almost literally, "*toch' v toch'*".[4] The focus of this article is on the wider, familial context of this icon, its accommodation within Filonov's *oeuvre* and its ultimate

Citation: Misler, Nicoletta, and John E. Bowlt. 2021. Catherine's Icon: Pavel Filonov and the Orthodox World. *Religions* 12: 502. https://doi.org/10.3390/rel12070502

Academic Editor: Dennis Ioffe

Received: 3 June 2021
Accepted: 29 June 2021
Published: 6 July 2021

Publisher's Note: MDPI stays neutral with regard to jurisdictional claims in published maps and institutional affiliations.

destiny—and, in general, the difficult issue of Filonov's religious beliefs.[5] The result is at once a detective story, an archaeological expedition and an art-historical evaluation.[6]

Figure 1. Flight into Egypt (1918, oil on canvas, 71.1 × 88.9, Thomas Whitney Collection, Mead Museum, Amherst College, USA).

Figure 2. The Magi (Adoration of the Magi) (1913, tempera on paper, 35.5 × 45.5, private collection, Switzerland).

Figure 3. Holy Family (Peasant Family) (1913–14, oil on canvas, 160 × 117, State Russian Museum, St. Petersburg; hereafter SRM: Zh 9578).

Figure 4. Icon of St. Catherine of Alexandria, the Great Martyr (1908–9, oil on canvas on wood, 27 × 18, collection of Mikhail Suslov, Russia).

Figure 5. Anon. (Stroganov school): Icon of St. Catherine (17th century, Pokrovskii sobor, Rogozhskaia sloboda, Moscow).

In rendering his illustration of St. Catherine, Filonov was, on the one hand, following the ancient tradition of icon-painting with its hieratic iconography, strict arrangement of figures, colours and perspectives and absence of signature (the icon-painter serving merely as the translator of a divine image and word)—and the fact that Filonov hardly ever signed his own easel paintings might be a gesture to this convention. Obviously,

Filonov was well aware of the particular and intrinsic elements of the traditional Russian icon—treatment of the wooden board, special ground (*levkas*), constant and formulaic composition, instructive purpose to the faithful—but in *St. Catherine* he bypassed or at least emended some of these constituents. Indeed, in his icon Filonov was referring to a comparatively late original, reminiscent not of the Mediaeval canons of Dionysius or Rublev, but of the early 17th century decorative and "worldly" style of Simon Ushakov or Fedor Zubov—an innovation which coincided with the fundamental *raskol* (schism) within the Orthodox church, Patriarch Nikon calling for strong reforms in liturgy and acts of worship. Those who resisted such innovation came to be known as Old Believers (*raskol'niki*), a contingent which was especially notable in Riazan', the seat of the Filonov family.

Filonov, however, in removing the icon of St. Catherine from what should have been a cultic purpose to a profane context (a present for his sister), Filonov seems to be supporting the Orthodox reformists and even the bold trend whereby art historians began to consider the Russian icon as a "work of art" rather than as a sacred image, a trend which culminated in Pavel Muratov's laic exhibition of "Ancient Russian Painting" at the Imperial Archaeological Institute in Moscow in 1913. Presumably, Filonov, too, was drawn to the Stroganov *St. Catherine* more as a "picture" than as an item of devotion and, in any case, he seems not to have been an eager champion of the Orthodox rite with its priesthood, liturgy, iconostasis, holy water, prayers and so on, but rather to have cultivated his own system of religious belief tinged with nuances of pantheism and cosmism. As to whether Filonov or his sister ever entertained the idea of placing his icon in an *oklad* or precious cover is not known.

Filonov's particular attitude towards the icon coincides with the mutation, if not, crisis, which continued to bait Orthodoxy as an intact institution in the late 19th and early 20th centuries. At this moment of question, doubt and counteraction, leading intellectuals such as Vladimir Solov'ev and Viacheslav Ivanov began to explore alternatives to the severe canon of Greek Orthodoxy, encouraging a more universal, more humanist approach to the church service, the holy calendar and festivals and rituals such as the Eucharist, even proposing that the Orthodox church move closer to Catholicism and Protestantism.

Unfortunately, Filonov's personal testimonies, notes and diaries provide scant information on his early artistic development. Consequently, in attempting to reconstruct Filonov's formative years, the researcher is bound to make recourse to ancillary materials so as to cope with the lacunae and self-censorship of the Stalin period. In this sense, the rediscovery and study of Filonov's *Icon of St. Catherine of Alexandria, the Great Martyr* (hereafter: *Icon of St. Catherine*) in 1990 and its inclusion in his retrospective exhibition at the Kunsthalle in Dusseldorf the same year have been fundamental to understanding Filonov's worldview—which, rather than "Christian" or "Orthodox" might be described more broadly as religious, even if, as Nikolai Lozovoi, one of his students, recalled, "Pavel Nikolaevich was not a mystic and was distinguished not by religiosity, but by an overriding belief in the force of the human spirit, in human will—and this was very much part of him. I remember someone saying that, during the course of work paints change their essence beneath the paintbrush. Pavel Nikolaevich supported this strange idea. 'Yes', he said, 'a kind of emanation does take place, but let's not dwell on this, otherwise we'll be accused of vitalism". (Lozovoi 2005, p. 272).

2. Family Values

In 1924, perhaps remembering his painting called *Easter* (1912–13) of a family at table, Filonov painted a group portrait entitled *Family Portrait* (*Easter*) (Figure 6). The importance of the picture lies in its representation not only of Filonov's solid, middle-class relatives, but also of the alliance between a patriarchal Russian family and the most important festival in the Orthodox calendar, i.e., Easter (an association which, in Soviet references to the picture was, of course, lost from view).[7]

In *Family Portrait* Filonov included many members of his immediate family (see caption to Figure 6) who can also be identified in the family tree (Figure 7), although outsider and renegade, observer rather than participant, Filonov himself is noticeable by his absence. The composition begs many questions—the almost total absence of side plates and cutlery, the ten individuals with only three cups and three glasses, the vacuum background, the black, 'cut-out' silhouettes surrounding some of the figures—elements which elicit associations with a funerary repast than with a joyous celebration, or perhaps with an altar offering rather than with food for consumption. In any case, the subject of *Family Portrait* is not simply a group of close individuals, but also a scene common to many of Filonov's paintings and drawings, i.e., of individuals at table, serious and detached, if not, lugubrious, as if observing a venerate ritual, reminiscent of some last and joyless supper as in the celebrated *Feast of the Kings* (1912–13, SRM), for example.

Figure 6. Family Portrait (Easter) (1924, watercolour and pencil on paper, 20.6 × 52; SRM: RS 14833). From left to right: Liubov Aleksandrovna Gue (Goué) (daughter of Aleksandr Gue and husband of Aleksandra Nikolaevna Gue, one of Filonov's sisters); Nikolai Nikolaevich Glebov-Putilevsky, husband of Evdokiia Nikolaevna Glebova (Filonov's youngest sister missing here); Mariia Aleksandrovna Emel'antseva (daughter of the Gues); René Armanovich Aziber (son of Filonov's oldest sister, Ekaterina, and her second husband, Arman Aziber [Azibert]); Ekaterina Fokina-Aziber (Filonova); Vladimir Aleksandrovich Gue (son of the Gues); Aleksandra Aleksandrovna Makokina (daughter of the Gues); Mariia Nikolaevna Filonova (another of Filonov's sisters); Aleksandra Nikolaevna Gue (wife of Aleksandr Andreevich Gue); and Makokina's baby daughter, Galina.

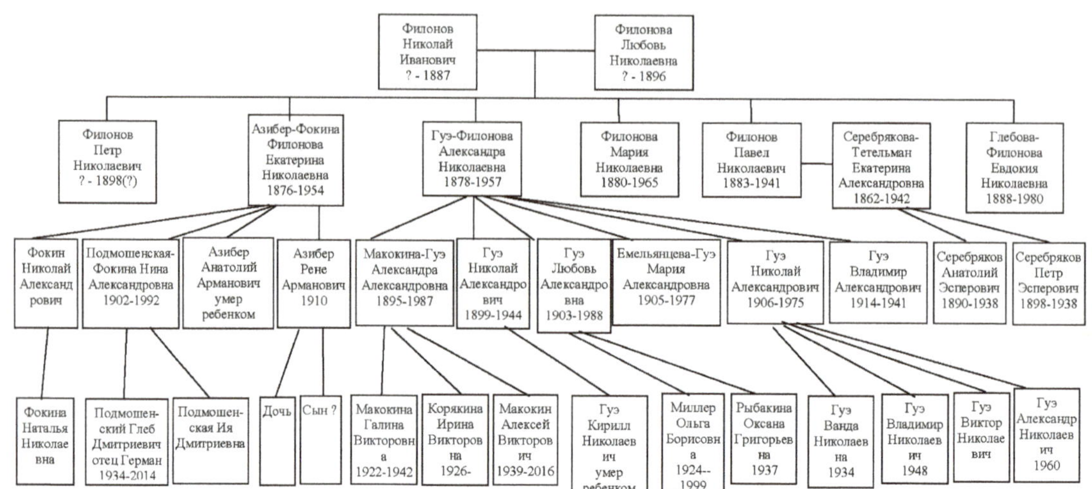

Figure 7. Family tree compiled by Oksana Rybakina, St. Petersburg, 1990.

Of particular relevance to our story is the presence in *Family Portrait* of Filonov's sister, Ekaterina Nikolaevna Filonova-Fokina (1876–954, Figure 8), seated fifth from left, and her son, René (to her immediate right). The eldest of six children (Nikolai, Aleksandra, Mariia, Pavel and Evdokiia),[8] in 1900 (?) Ekaterina married Aleksandr Mikhailovich Fokin (1877–937, brother of the celebrated dancer and choreographer Michel Fokine), to whom she bore two children, Nina (in marriage: Nina Podmoshenskaia; abbreviated to Podmo; St. Petersburg, 1901-Platina, CA, USA, 1994; Figure 9) and Nikolai (1905–after 1937). Judging from verbal and visual information, Ekaterina's marriage to Aleksandr granted her entrée into the upper bourgeoisie of St. Petersburg inasmuch as he was a successful businessman, owning a major bicycle store and also, it would seem, at one point, a car dealership. In any case, Fokin was not foreign to the arts, acquiring and managing the Troitskii Theater of Miniatures in St. Petersburg in 1911 as a gift to his second wife, the ballerina and star of the Maryinsky Theater, Aleksandra Aleksandrovna Fedorova (Alexandra Fedorova, 1884–972), even receiving a gold watch from Tsar Nicholas for his production of an operetta at the Winter Palace.[9] In emigration he continued his theatrical activities, underwriting Fedorova's dancing career in Latvia. That Ekaterina was elegant and solvent is apparent from a photograph of her taken in ca. 1904 and given to us by her grandson (son of Nina Podmo), Abbot Herman (Figure 10), at the conference, "Pavel Filonov: Painter of Metamorphosis", held at the Solomon R. Guggenheim Museum in New York in April, 1983 (Figure 11), a photograph very different from the family photograph taken in Moscow in ca. 1882 where Ekaterina sits dourly on her father's knees (Figure 12).

Figure 8. Anon.: Photograph of Ekaterina Nikolaevna Filonova (1876–954; by first marriage: Fokina; by second marriage: Aziber), St. Petersburg, ca. 1904. SRM: Department of Manuscripts. Call No. f. 156, op. 1, ed. khr. 216, l. 1.

Figure 9. Nicoletta Misler: Photograph of Nina Podmo, Santa Rosa, California, 1989.

Figure 10. Anon.: Photograph of Abbot Herman (Gleb Podmoshensky), California, ca. 2010.

Figure 11. Anon.: Photograph of Lazar Fleishman, John E. Bowlt and Nicoletta Misler at the conference, "Pavel Filonov: Painter of Metamorphosis", held at the Solomon R. Guggenheim Museum, New York, 22–23 April 1983.

Figure 12. Anon.: Photograph of the Filonov family taken in Riazan' in ca. 1882. From left to right: Liubov' Nikolaevna Filonova, Aleksandra Nikolaevna; Petr Nikolaevich (1869–98); Nikolai Alekseevich (1832–87); Ekaterina Nikolaevna.

After Ekaterina and Aleksandr divorced in 1907 or slightly before,[10] Ekaterina married a Frenchman (thereby obtaining French citizenship), Arman Frantsevich Aziber (Armand Azibert, 1878–914; Figures 13 and 14), whose family was well ensconced in St. Petersburg society inasmuch as Arman was the owner and director of an important factory of preserves (fruit, vegetables, meat) which, at one time, supplied the Imperial Army. They had two children, Anatolii (sometimes referred to as Nikolia; died in 1913 or 1914 aged four) and René (1910–after 1970?) who also assumed French citizenship thanks to his father, a status which allowed him and his mother to emigrate to France from Leningrad in 1925. Ekaterina and René, figuring on the left side of the *Family Portrait*, took the icon of St. Catherine, plans and analogous papers related to the Aziber preserve factory in St. Petersburg and, allegedly, a portrait of Ekaterina by Filonov[11] with them when they emigrated. How Ekaterina and the young René made ends meet in France after 1925 is hard to determine, although the husband and father, Arman, had been domiciled not only at 44, Staro-Petergofskii Prospect in St. Petersburg (the site of his preserves factory which functioned until 1914) where, after returning from the front, Filonov also lived for a year, but also (according to his Death Certificate) at 61, rue du Château in Boulogne-Billancourt, an upscale suburb of Paris.[12] This would mean, presumably, that Ekaterina inherited the French apartment after Arman's death in which case she and René might have relocated there after leaving Leningrad. Subsequently, René married a French woman, Giselle, who survived him, inheriting the icon and the documents of the factory. We saw these items when we visited Mme. Azibert in Hauts de Seine, Sèvres, in 1989.

Figure 13. Anon.: Photograph of Arman Frantsevich Aziber (Armand Azibert, 1878–914) and son, Anatolii, 1909–13?), 1913 or 1914 (private collection).

Figure 14. Portrait of Arman Aziber and son, Anatolii (1915; oil on canvas, 115 × 82, SRM: Zh 9570).

Ekaterina's daughter, Nina, remained with her father, Aleksandr, accompanying him and his new wife to Vologda and Tiflis in 1922 and settling in Riga three years later. In 1933 in Riga she married the businessman and paper manufacturer, Dmitrii Ivanovich Podmoshensky (1904–43), settling with him in Pytalovo (also known as Jauntlatgale and then Abrene) near Pskov and bearing two children, Gleb (1934–2014) and Ia (died 2018), both of whom we met at the Guggenheim Filonov conference. In 1940 the Soviets occupied Latvia and Podmoshensky was arrested as an "enemy of the people" and deported to prison camp at Vorkuta where he died three years later. In 1942 Nina and her children moved to Germany and six years later joined Alexandra Fedorova in the USA. In the 1960s Gleb became a Serbian Orthodox monk in Boston and then California, assuming the name of Abbot Herman and co-founding the St. Herman of Alaska Brotherhood, the St. Herman of Alaska Monastery (where he served from 1969 until 2000) and the journal *Russkii palomnik* in Platina, CA, USA (1980 onwards), while his sister Ia moved to London. After her mother's death in 1992, Ia, a devout Orthodox, was also living in California under the surname Schmid (Schmit).

Between 1983 and 1991 we were in correspondence with Nina and in the spring of 1989 took a road trip from Los Angeles to visit her in Napa Valley. Late one morning we arrived at her mobile home just outside the town of Santa Rosa to be greeted by a petite, but vivacious, elderly lady who, immediately, welcomed us in Russian. We spent much of the day listening to her recollections about "Uncle Pania",[13] the Filonov family and also Filonov's *Icon of St. Catherine* which Abbot Herman had mentioned during our meeting in New York and which Nina remembered well. Although she had not seen the icon since the passing of her mother, Ekaterina, in 1954, she assumed that it was still in Paris, in the home of René Azibert's widow, Giselle, but did not have the street address. Determined to track down the icon, one year later we flew to Paris and, after locating 'Azibert, Giselle' in the Paris telephone directory, called Mme. Azibert, scheduled an appointment and took the train out to her home in Hauts de Seine, Sèvres. A spritely and elegant middle-class lady welcomed us, and, after preliminary conversation, produced the *Icon of St. Catherine*, wrapped carefully in fabric, from a drawer in the living-room. Naturally, we were overwhelmed by the rediscovery of an artifact which for us had assumed a mythological dimension, although we soon realized that Giselle herself was unswayed by this strange object in bad condition and without a frame, manifesting much greater interest in her father's Sunday paintings of cherries. Incidentally, Giselle also showed us the documents and photographs pertaining to Arman Aziber's factory in St. Petersburg, although, unlike the icon, the fate of those materials remains unknown. In any case, Giselle agreed to loan the icon to the exhibition, 'Pawel Filonow und sine Schule' organized by the Kunsthalle, Dusseldorf, in September–December, 1990, to which we also contributed as essayists for the catalogue. In 2014 the icon was acquired directly from Giselle by a Russian dealer through a French gallery whence it passed into the hands of the Russian collector, Mikhail Suslov.

Ekaterina's and Aleksandr's second child, Nikolai stayed behind in Soviet Russia, was arrested and then freed in 1937, and lived in Pskov and, according to Nina Podmo, during the 1940s (?) she and her mother sent care packages to him. In any case, there is a strong parallel between Abbot Herman and Pavel Nikolaevich, i.e., between the latter's expulsion from the Imperial Academy of Arts (Figure 15) and then the reinstatement in 1910, and Herman's own iconoclastic activity, because, for all his dedication to the Orthodox cause, in 1988 the Synod of Orthodox Bishops excommunicated him for alleged impropriety, a scandal which drove him into spiritual exile and isolation.

Figure 15. Aleksandr Yagel'sky, photographer: Pavel Filonov as a freshman at the Imperial Academy of Arts, St. Petersburg, 1908.

As for Arman Aziber (Armand Azibert), fresh information has come forth recently regarding his life and work, which helps clarify his position in the St. Petersburg business world and certain issues surrounding the extraordinary oil *Portrait of Arman Aziber and His Son* which Filonov painted in 1915. The assumption that Filonov had painted this double portrait from a photograph after Arman had died in action at Souain Perthes-lès-Hurlus, Marne, on 29 February, 1915 (according to his Death Certificate) was often questioned and, in fact, at least one source still denies the rumour, asserting that the portrait was painted from life[14]—even though Filonov himself mentioned that the portrait was "after a photograph"[15] and the actual photograph was discovered among family papers and reproduced for the first time in the Dusseldorf catalogue of the Filonov retrospective,[16] providing ample testimony not only to Filonov's phenomenal skill—to his 'Photo-Realism'— but also to his delicate pictorial adjustments and impositions as in the facial expressions, the hairstyles and the carpet. The carpet, in fact, recalls the luxurious fabrics of Mikhail Vrubel's *Girl against a Persian Carpet* (1886, Figure 16) where the model holds a pink rose much as the boy does in the Aziber portrait, and of *Fortune-Teller* (1894–96; State Tretiakov

Gallery, Moscow). Here, is the same kind of minute elaboration and organic decoration as in the florid mantle of Filonov's *Icon of St. Catherine* or her ornamental scroll, the script of which is still entirely legible, i.e., Verse 7 from Psalm 139: "Kuda poidu ot Dukha Tvoego, i ot litsa Tvoego kuda ubegu" [Whither can I go to escape Your Spirit? Whither can I flee from Your presence?].

Figure 16. Mikhail Vrubel: Girl against a Persian Carpet (1886, oil on canvas, 104 × 68, Museum of Russian Art, Kiev).

It is almost as if Filonov, expelled from the Academy of Arts in April, 1910, for his "disgusting drawings and studies", but convinced that "I can also work with any technique, coming right down to the natural world"[17] was reminding the establishment that, however "incorrect" and unorthodox his own Analytical art, if need be, he could also execute perfectly academic and "correct" paintings. Indeed, however we judge the *Portrait of Arman Aziber and His Son*, it is a haunting, if not, ominous, *memento mori*, challenging us to accept the sudden death of a flourishing businessman, of a first born, Anatolii (presuming that the boy is, indeed, Anatolii), even of a freshly cut rose. In any case, the photograph must date from before August, 1914, when Arman volunteered for the front and joined the 130th Régiment d'infanterie. That the son here is Anatolii rather than René can be deduced from comparison of the photograph of the young René taken in ca. 1920 (on the left of Figure 17) as well as the photograph of him published on the internet dated 1923 and his portrait in *Family Portrait*, all of which differ markedly from the child in *Portrait of Aziber and His Son*. This conclusion, therefore, contradicts the general assumption that this is René rather than Anatolii.[18]

Figure 17. Anon.: Photograph of members of the Filonov family, Petrograd, ca. 1920. Left to right: Nikolai Aleksandrovich Gue; Liubov' Aleksandrovna Gue; René Aziber (Azibert); Aleksandra Aleksandrovna Gue; Ekaterina Fokina-Aziber; Mariia Aleksandrovna Gue; Vladimir Aleksandrovich Gue; Aleksandra Niklaevna Filonova-Gue (seated); Mariia Nikolaevna Filonova (seated); Nikolai Aleksandrovich Gue; and Evdokiia Nikolaevna Filonova (Glebova-Putilevskaia).

3. Keeping the Faith

The Orthodox connection, in particular, betrays Filonov's upbringing in a strict, Christian family (Podmo recalled that "Filonov belonged to the merchant class, so his parents had brought him up in the Orthodox faith"),[19] with its unfailing celebration of Easter, something which latently, at least, remained a vital point of departure for his aesthetic and philosophical journey, even transmuting itself in some of the politically tendentious and atheist pictures of the Soviet period. In any case, according to his autobiography, Filonov undertook a pilgrimage to the Holy Land in 1907, visiting Constantinople, Mount Athos, and Jerusalem, and the Monastery of St. Catherine in the Sinai Desert,[20] painting souvenir icons en route to cover his expenses—a prelude to master paintings such as *St. George, the Victorious, Mother* and *Holy Family*.[21] Additionally, according to his auto-biography, Filonov copied an engraving of a Crucifixion for monks in Jerusalem, which perhaps looked forward to his *Golgotha* (1912, Figure 18) and two other paintings of the crucifixion, i.e., *Execution (After 1905)* 1913, (Figure 19) and *Execution* (1920–21, Figure 20). In 1914 Filonov even argued that the artistic center of gravity was moving to Russia, "our motherland, creator of marvelous temples, of the art of artisans and of icons".[22]

Figure 18. Golgotha (1912, pencil and watercolour on paper, 33.5 × 32.7, State Tretiakov Gallery, Moscow: R 2918).

Figure 19. Execution (After 1905) (1913, brown ink and pencil on paper, 11.6 × 13.4, SRM: R 58236).

Figure 20. Execution (1920–21, watercolour and mixed media on paper, 27.8 × 44.5, SRM: R 14867).

Some of Filonov's paintings of the Soviet period with their manifest crosses and Madonnas, still carry remnants of the Christian symbology for all their Communist veneer and, after all, Filonov did hang a few Biblical scenes in his one-man exhibition at the State Russian Museum in 1930 (which did not open). For example, there was *Abraham and the Holy Trinity* (1912; Figure 21), *Easter* (1912–13) and *Adoration of the Magi* (1913). Furthermore, in the 1920s and 1930s Filonov continued to refer to icons in his public statements and lectures, praising icon-painters, for example, in his essay on the "Proletarianization of Visual Art"[23] and rejecting the Byzantine influence on Russian icons in his 'Declaration of 'World Flowering".[24] In the programme for the reform of the Petrograd Museum of Painterly Culture which he outlined at the Conference for the Reorganization of Museums in 1923 Filonov included various 'peripheral' disciplines, including the art of children, of the insane, of self-taught artists and also icons—'in covers and without, new and old styles, moving on from the Byzantine and Russian cathedralic canon of Athos and Jerusalem'.[25] Given Filonov's support of the Communist Party's militant atheism, during the Soviet period his religious motifs sometimes become anti-religious ones, even cartoons as in *Last Supper* (1920s–early 1930s, State Russian Museum, St. Petersburg). At the same time Filonov reinforced his new ideological stance by issuing emphatic statements condemning religion both as a hierarchical institution and as a source of artistic inspiration: 'The matter of Russian and global art is, in all its theoretical and ideological interrelationships, the same as in questions of religion, the church, believers and God with all His rituals. And together with all their gods and saints, with their mysteries of creation, with all their prophets, priests, harbingers and wizards of art, all these interrelationships must also be destroyed'.[26]

Figure 21. Abraham and the Holy Trinity (1912; watercolour, Indian ink and pencil on paper, 70.9 × 95.5, SRM).

4. The Icon of St. Catherine

Although, before 1917, Filonov painted a number of major paintings on Evangelical themes, he was not, primarily, a religious painter and, after all, his sister, Ekaterina, may well have suggested the topic of an icon to him and even to have dictated certain prerequisites. Perhaps the fact that the Domestic Church at the Imperial Academy of Arts, Filonov's *alma mater*, was dedicated to St. Catherine, the Martyr, may also have influenced the choice. The extent to which Filonov was following his own theory in the icon vis-à-vis what appears to be its original prototype—a 17th century icon from the Stroganov school in the Cathedral of the Intercession in Rogozhskaia sloboda in Moscow—tells us how faithfully he was keeping to the antique formulae on the one hand, but how innovative he was, on the other.[27]

If we compare Filonov's *Icon of St. Catherine* and the original upon which, apparently, it was based, we can see how closely he followed the iconographic canon, while also diverging. With her royal crown and precious mantle, Catherine stands upright and majestic like a princess, one hand with the two-finger, Old Believer gesture of blessing, the other holding the scroll. The exotic architecture duplicates what we see in the Moscow icon, although perhaps the exclusion of other narrative elements is justified by the fact that Ekaterina, Filonov's sister, may have dictated her own preferences (or they may be missing simply as a result of wear and tear). As for the omission of the martyr's wheel and the palm (signifying Christian victory), Orthodox elders explain that the "wheel [and palm] is a Catholic influence which is why you find it in only the crusader icons at Sinai and in icons which have been copied from them, i.e., this is a 'foreign' influence, which the Old Believers would have avoided."[28] In any case, however canonical the icon may be, the flower at Catherine's feet, even if repeating the accepted model, now merges into a strange animal-vegetable form, just as the intense ornament of the mantle seems to be undergoing an organic metamorphosis. Incidentally, once again, this kind of biological 'degeneration' reminds us of the ornamental promiscuity visible in the late works of Vrubel'—who, a fellow "Decadent", exerted a formative influence on Filonov.[29] With his demonic subjects, crepuscular tones and fragmented surfaces, Vrubel' evoked a nether world of anxiety and despair, a mood identifiable with much of the culture of *fin de siècle*, and although, in his writings, Filonov does not mention Vrubel', he would seem to be sharing the notion that traditional values—of the Academy and of Realism, in particular—had fallen from grace and were to be replaced by new criteria. Indeed, one of the most striking elements in *Icon of St. Catherine* is the mantle with its intense decoration painted in accordance with Filonov's Analytical system, encompassing the "genesis of being, the pulsation of the sphere."[30] His was a sensibility which heeded the living processes of nature right down to the "physiological processes in trees . . . their smell . . . creating phenomena"[31]—which is to say that Filonov perceived form, including artistic form, as a living unit, like a plant or animal.

Of particular interest is the technical structure of the *Icon of St. Catherine* which, in some respects, contravenes the methods of traditional icon-painters, a fact illuminated by the inspections of museum professionals, first at the State Russian Museum, St. Petersburg, in January, 2014, and then at the State Tretiakov Gallery, Moscow, during 2019, the second report being more thorough than the first. The State Russian Museum concluded that the medium of the icon was "mixed" (including temper) and that Filonov's "careful processing of the board, the *pavoloka* [fabric] made up of three different sized segments and the *levkas* are testimony to a conscious application of the devices of icon-painting".[32] On the other hand, the findings of the specialists at the State Tretiakov Gallery, above all, senior restorer, Yulian Khalturin, more comprehensive and more precise, point to the opposite, i.e., that "Filonov was foreign to the technique of the professional icon-painter, witness to which were his application of materials and his methods of execution, for example, how the canvas and its ground (the ground being more typical of oil painting) had been combined with the very thin stratum of the depiction itself, something creating a far from durable structure. Canvas is a material which . . . is subject to contraction and expansion (especially if it had never been attached to the board). In that case, the thin layer of paint will react ... and detach from the ground completely".[33]

In effect, Filonov much preferred oil painting, often applying it with a thin brush even to paper (an application far from secure) so we should not be surprised to see it in his icon, an element separating the artifact from the Orthodox iconic tradition. Furthermore, as Khalturin comments, "Traditional icon-painting technique ensures a stable connection with the wood", whereas Filonov favored a more experimental method, something which, once again, emphasizes the distance between Filonov the studio painter and Filonov the icon painter. The *Icon of St. Catherine*, with its organic embellishments, also brings us to the question of time in the sense that, for the icon-painter, what counts is the actual time spent on preparing the sacred icon, i.e., on creating the various strata and sediments which make for the fast attachment of the image to the board. For Filonov, time meant 'time of execution', something quite different, because the production of the surface image, as opposed to the physical support, had to be slow and measured so that 'Every atom be made ... Think obdurately and accurately over each atom of the work being made'.[34] Such an approach led Filonov to apply a "thin, but bright and variegated, sometimes semi-transparent paint which presumably, was very effective, but not durable. The very presence of a grid tells us that [Filonov] was, probably, copying a specimen ... The original painting is distinguished by a richness of colour, looking almost like watercolour (especially) visible in the architecture and on Catherine's sleeve and shoes."

Khalturin also observes that "We should also remember that there is a thick layer of drying-oil (*olifa*)in this icon and that this can tear the paint away from the surface. We can see that, initially, the artist had made a pencil grid, which helped him reproduce the subject-matter. Using a microscope allows one to see the difference more clearly, since to the untrained eye the impression is erased owing to the dirty varnish which has been removed unevenly and destroyed".[35]

That in his icon Filonov had used a varnish detrimental to the image reminds us that in some of his easel paintings he would apply a varnish which tended to drip down and darken the surface, as, for example, in *Man and Woman* (1912–13, State Russian Museum, St. Petersburg)—which is to say that the issue of Filonov's technical devices, especially during the 1910s and 1920s, is of paramount importance to our understanding of his artistic versatility—as, for example, his use of oil on paper, a difficult and disjointed technique which, thanks to the distinctive, jewel-like quality of his execution, permitted him to attain a pictorial surface of particular brilliance—easily recognizable in the fine detail of St. Catherine's mantle.

Wear and tear have damaged, even cancelled, entire sections in the Filonov icon, including parts of the buildings on the right and left of the Saint, the faithful at her feet and perhaps a scene of her entombment top right. It is also possible that at one point the secondary figure of Catherine, visible lower right, for example, in the Moscow version and other specimens, was also present here, which might also mean that at some point part of the righthand side of the icon was cut. In addition, the vestiges in the lower far left allude to the kind of host of heads and faces with which Filonov populated other pictures of the same period such as *Heads* (1910, State Russian Museum, St. Petersburg, Russia) and he may also have repeated the cameo of Christ in His cruciform halo surrounded by angels, lowering a censer towards Catherine's right palm visible in the Moscow version. The cameo to the right, on the other hand, is an unusual supplement, but may represent a heavenly Jerusalem or Paradise which is often painted with white background and green trees or bushes next to the building, the moreso since the vault behind the circle is dark blue with a personified sun. In any case, for all the missing links and material blemishes, Filonov's icon remains a very close copy of the Moscow original, although the circumstances under which he travelled to the Cathedral (if he did so) and painted his version and how his sister responded are open to conjecture. True, Filonov could have seen black and white reproductions of the Moscow icon in published sources, especially in Nikolai Likhachev's *magnum opus* on the history of icon-painting.[36]

5. Coda

Examination of Filonov's *Icon of St. Catherine* demonstrates that Orthodox does not automatically signify orthodox and that an artist such as Filonov, radical and uncompromising, could find inspiration in one of the most ancient and hieratic traditions of Russian culture—the icon. The confrontation also indicates that, however prodigal and intransigent in behaviour, Filonov was still bound by the canons and conventions of his upbringing—by family, Christianity and the Romantic notion of the artist as witness to the unseen. Even if Filonov interpreted the Biblical story in a very private manner and in later life even replaced it with a Soviet hagiography, he remained faithful to his vision of art as a vehicle of spiritual remedy and transformation. After all, painting is believing.

Author Contributions: All authors contributed equally to this work. All authors have read and agreed to the published version of the manuscript.

Funding: This research received no external funding.

Institutional Review Board Statement: Not applicable.

Informed Consent Statement: Not applicable.

Conflicts of Interest: The authors declare no conflict of interest.

Notes

[1] For more information on Filonov see (Misler and Bowlt 1983; Kovtun 1988; Petrowa and Harten 1990; Gmurzynska et al. 1992; Misler et al. 2005; Bowlt et al. 2006; Petrova 2006, 2008; Pravoverova 2008; Sokolov 2008; Ershov 2015; Pravoverova 2020).

[2] After 1917 Filonov changed the title *Holy Family* to *Peasant Family*, even though, according to Irina Pronina, the picture may not only depict the Holy Family, but also carry a reference to the Massacre of the Innocents and to the Russian pogroms (Pronina 2010). A second so-called *Peasant Family* dated 1912–14 (watercolour; 36 × 46), which would seem to be another interpretation of the Holy Family, is reproduced in (Kriż 1966), plate 6.

[3] P. Filonov: "Avtobiograficheskie teksty" in Bowlt, Misler, Pronina, Sarab'ianov, p. 78. English translation in (Misler and Bowlt 1983, p. 122).

[4] P. Filonov: "Try with all your might to study nature and transmit it exactly [toch' v toch']"—'Pis'mo nachinaiushchemy khudozhniky Baskanchinu" (19 February 1940) in Bowlt, Misler, Pronina, Sarab'ianov, p. 198. English translation in (Misler and Bowlt 1983, p. 291).

[5] This religious or, rather, Christian, aspect of Filonov's *oeuvre* has received relatively little attention, although mention should be made of the following: N. Misler; "Von der Ikonenmalerei zum Fotorealismus" in (Petrowa and Harten 1990), pp. 36–49; Pronina, "Shkola Filonova: neizvestnye stranitsy istorii"; and Elena Fadeeva's dissertation, i.e., "The Judgement Day of Art". Orthodox Iconography, Messianism and Martyrdom in the Art of Pavel Filonov", for Master of Literature at the School of Art History, University of St. Andrews, Scotland, in 2014–15.

[6] Among the many individuals who rendered valuable assistance during our researches for this essay we would like to mention the following: Konstantin Akinsha, Robert Chandler, Donna Decker, Isabelle Fokine, Il'dar Galeev, Lynn Garafola, Musya Glants, Ol'ga Golubeva, Viktor Golubinov, Yulian Khalturin, Viktor Men'shikov, Irina Men'shova, Ol'ga Pal'chevskaia, Wendy Perron, Evgeniia Petrova, Jane Pritchard, Irina Pronina, Ksenia Radchenko, Elena Rybakina, Oksana Rybakina, Ol'ga Selivanova, Sister Gabriela, Wendy Salmond, Yuliia Solonovich, Mikhail Suslov, Mariia Trubacheva, Boris Yavdin, Andrei Vasil'ev, Zel'fira Tregulova and Arlene Yu. The following depositories were also consulted: Military and Historical Society, Moscow; Russian State Archive of Literature and Art, Moscow; Russian State Archive of the Military and Maritime Fleet, St. Petersburg; State Military and Historical Archive, Moscow; State Russian Museum, St. Petersburg; State Tretiakov Gallery, Moscow.

[7] In the catalogue of Filonov's one-man exhibition at the State Russian Museum in 1930, the work is entitled *Family Portrait* without the word 'Easter'. See (Isakov 1930), Introduction, No. 162.

[8] On the Filonov family hailing from Riazan' see Semina (2017), pp. 6–13.

[9] In 1917 director and actor Konstantin Mardzhanov took over management of the Troitskii Theater, but it closed down five years later.

[10] Judging from an affectionate dedication to Ekaterina and Arman, dated 16 April 1907, in a book given to them by the lawyer, Grigorii Osipovich Rozentsveig, i.e., *Iz zaly suda. Sudebnye ocherki i kartinki* (St. Petersburg: Trud, 1900), the couple were already engaged, if not, married. The book was offered at auction by the Moscow auction house, Antikvarium, on 3 December 2015.

[11] See (Glebova 1984): 'Dnevniki Filonova v vospominaniiakh', p. 150. Evdokiia Glebova (Filonova, 1888–1980), married to Nikolai Glebov-Putilovsky, was Filonov's youngest sister. In her monograph of 2008 (*Pavel Filonov, Real'nost' i mif*, p. 237) Pravoverova wrote that the portrait was in the possession of René Azibert in Paris.

2   See http://www.culture.fr/ (accessed on 3 June 2021) at: http://shorturl.at/uwFOY (accessed on 3 June 2021).

3   See N. Podmo: 'Uncle Pania' in Gmurzynska, pp. 173–76.

4   See (Stolbova 2012), p. 11. According to Stolbova, *Ves' Petrograd* for 1915 still listed Aziber as the proprietor of the Preserves Factory on Staro-Petergofskii Prospect, indicating that he was alive at the time of the portrait. We can presume, however, that news of his death reached the editors of *Ves' Petrograd* only after substantial delay.

5   See Isakov, *Filonov*, No. 35.

6   Petrowa, Harten, *Pawel Filonow und seine Schule*, p. 40.

7   P. Filonov: 'Proshenie v Sovet professorov Vysshego Khudozhestvennogo uchilishcha ot vol'noslushatelia, Pavla Nikolaevicha Filonova', 1910 [Petition Addressed to the Council of Porfessors, Higher Art Institute (Imperial Academy of Arts), St. Petersburg from the Audior, P.N. Filonov, 1910]. Manuscript Department, Academy of Arts, St. Petersburg.

8   In conversations with the authors the late Evgenii Kovtun, one-time curator of Soviet graphics at the State Russian Museum in St. Petersburg, maintained that the child was Anatolii and not René, although he did not adduce documentary evidence.

9   Podmo, 'Uncle Pania', p. 173.

10   Podmo affirmed that: 'Filonov, evidently, visited the St. Catherine Monastery in Jerusalem, and, under the impression of the image of the Great Martyr Catherine which he saw, he painted a wonderful icon of St. Catherine and gave it to his elder sister, Ekaterina. I saw this icon. It was amazing, just like an ancient Byzantine one—no bright colors, just brown and green tones.' (Podmo, 'Uncle Pania', p. 175). Additionally, see I. Pronina: 'Zhizneopisanie Pavla Filonova' in Bowlt, Misler, Pronina, Sarab'ianov, p. 27. Filonov visited the Holy Land three times.

11   The State Russian Museum possesses a drawing entitled *Costantinople*, dated 1907. The date of 1918 which Glebova furnishes in 'Dnevniki Filonova' (p. 153) is obviously a lapse of memory or a misprint.

12   See the original source: P. Filonov et al.: *Intimnaia Masterskaia zhivopistsev' i risoval'shchikov 'Sdelannye kartiny'*, St. Petersburg, 1914, unpaginated.

13   P. Filonov: 'Proletarizatsiia izobrazitel'nogo iskusstva' in Bowlt, Misler, Pronina, Sarab'ianov, p. 165. English translation in (Misler and Bowlt 1983, p. 258).

14   P. Filonov: "Deklaratsiia 'Mirovogo rastsveta'" in *Zhizn' iskusstva*, Petrograd, 22 May 1923, No. 20, p. 14.

15   Discussing the establishment of a Museum of Modern and Contemporary Art, Filonov even proposed including a Department of Icons, an idea to which he returned even in the 1930s, See P. Filonov: "Polozhenie ob Institute issledovaniia sovremennogo izobrazitel'nogo iskusstva" in Bowlt, Misler, Pronina, Sarab'ianov, p. 117. English translation in (Misler and Bowlt 1983, p. 181).

16   P. Filonov: 'Ya budu govorit" (ca. 1924) in Bowlt, Misler, Pronina, Sarab'ianov, p. 125. English translation in (Misler and Bowlt 1983, p. 225).

17   We would like to thank Yulia Solonovich of the State Russian Museum, St. Petersburg, for suggesting that Filonov had copied the icon of St. Catherine in the Cathedral of the Intercession in Rogozhskaia sloboda and for finding vintage photographic reproductions of the same.

18   Letter from Sister Gabriela, Orthodox nun and icon-painter, to John E. Bowlt dated 22 February 2021.

19   In his review of the 'Non-Partisan Exhibition' in St. Petersburg, 1913, to which Filonov contributed, Ieronimin Yasinsky referred to him as 'one of the Decadents'. See I. Yasinsky: "Vystavka vnepartiinykh" in *Birzhevye vedomosti*, St. Petersburg, 1913, 16 February, No. 13403, p. 4 (Evening Edition).

30   Filonov, 'Deklaratsiia 'Mirovogo rastsveta', p. 90. English translation in (Misler and Bowlt 1983, p. 170).

31   Filonov, 'Avtobiograficheskie teksty', p. 78. English translation in (Misler and Bowlt 1983, p. 122).

32   Report No. 168/12 dated 23 January, 2014, from the State Russian Museum, St. Petersburg, signed by A.B. Liubimova et al., and relayed to Mikhail Suslov, present owner of the icon.

33   Letters from Yulian Khalturin to Nicoletta Misler and John E. Bowlt from between 9 and 23 February 2021.

34   P. Filonov: 'Ideologiia analiticheskogo iskusstva' in Isakov, *Filonov*, p. 42. English translation in Bowlt (1988), p. 286.

35   Ibid.

36   See, for example, *Snimki s drevnikh ikon, nakhodiashchikhsia v staroobriadcheskom Pokrovskom khrame pri Rogozhskom kladbishche* (Moscow: Sherer and Nabgol'ts, 1899, p. 111). Vol. 2 (plate CCLXXI, item No. 500) of Nikolai Likhachev's fundamental *Materialy dlia istorii russkogo ikonopisaniia* (St. Petersburg: Department for the Preparation of State Papers, 1906) carries a black and white reproduction of the Stroganov icon in question where it is described as "St. Catherine, the Great Martyr, upright, with the depiction of certain events from her life". Other early reproductions can be found in A. Burtsev ed.: *Muraveinik, khudozhestvennyi, bibliograficheskii i etnograficheskii sbornik*, St. Petersburg, 1910, No. 9–10, p. 163; and his *Dosugi. Khudozhestvenno-bibliograficheskii sbornik*, St. Petersburg, 1911, No. 3.

References

Bowlt, John, ed. 1988. *Russian Avant-Garde. Theory and Criticism, 1902–1934*. London: Thames and Hudson.

Bowlt, John, Nicoletta Misler, Irina Pronina, and Andrei Sarab'ianov, eds. 2006. *Filonov. Khudozhnik, issledovatel', uchitel'*. Moscow: Tomesh.

Ershov, Gleb. 2015. *Pavel Filonov. Khudozhnik 'Mirovogo rastsveta'*. St. Petersburg: European University.

Glebova, Evdokiia. 1984. Dnevniki Filonova v vospominaniiakh ego sestry, E.N. Glebovoi. In *SSSR. Vnutrennie Protivorechiia*. New York: Chalidze.

Gmurzynska, Krystyna, Nicoletta Misler, and John E. Bowlt. 1992. *Die Physiologie der Malerei. Pawel Filonow in den 20er Jahren/The Physiology of Painting. Pavel Filonov in the 1920s*. Cologne: Catalog of exhibition at Galerie Gmurzynska.

Isakov, Sergei. 1930. *'Introduction', Filonov*. Leningrad: Catalogue of exhibition at the State Russian Museum.

Kovtun, Evgenii. 1988. *Filonov*. Leningrad: Catalogue of exhibition at the State Russian Museum, Avrora.

Kriż, Jan. 1966. *Pavel Nikolaevič Filonov*. Prague: Nakladetelstvi Československych Vytvarnich Umělcu.

Lozovoi, Nikolai. 2005. 'Vospominaniia o Filonove'. *Experiment* 11: 269–86.

Misler, Nicoletta, and John Bowlt. 1983. *Pavel Filonov. A Hero and His Fate*. Austin: Silvergirl.

Misler, Nicoletta, John Bowlt, and Irina Men'shova, eds. 2005. *Experiment*. Los Angeles: Institute of Modern Russian Culture.

Petrova, Evgeniia. 2006. *Pavel Filonov. Ochevidets Nezrimogo*. St. Petersburg: Palace Editions.

Petrova, Evgeniia, ed. 2008. *Pavel Filonov. Sbornik Statei*. St. Petersburg: Palace Editions.

Petrowa, Ewgenija, and Jurgen Harten, eds. 1990. *Filonow und seine Schule*. Dusseldorf: Catalogue of Exhibition at the Kunsthalle.

Pravoverova, Liudmila. 2008. *Pavel Filonov, Real'nost' i mif*. Moscow: Agraf.

Pravoverova, Liudmila. 2020. *Pavel Filonov. Analiticheskoe Iskusstvo. Sdelannye Kartiny*. Moscow: Akademcheskii proekt.

Pronina, Irina. 2010. 'Shkola Filonova': Neizvestnye stranitsy istorii'. *Russkoe Iskusstvo*, 84–93.

Semina, Maiia. 2017. Pavel Filonov: biografiia i geografiia. *Zolotaia Palitra*, 6–13.

Sokolov, Mikhail. 2008. *Pavel Filonov*. Moscow: Art-Rodnik.

Stolbova, Ekaterina. 2012. Konservnaia imperiia Aziberov. *Yunyi Kraeved*, 6–14.

Article

Poems by Polish Female Poets and the Burning Issue of Religion

Dorota Walczak-Delanois

Faculté de Lettres, Traduction et Communication, Département de Langues et Lettres, Université Libre de Bruxelles, 1050 Bruxelles, Belgium; Dorota.Walczak@ulb.be

Abstract: The aim of this paper is to show the presence of religion and the particular evolution of lyrical matrixes connected to religion in the Polish poems of female poets. There is a particular presence of women in the roots of the Polish literary and lyrical traditions. For centuries, the image of a woman with a pen in her hand was one of the most important *imponderabilia*. Until the 19th century, Polish female poets continued to be rare. Where female poets do appear in the historical record, they are linked to institutions such as monasteries, where female intellectuals were able to find relative liberty and a refuge. Many of the poetic forms they used in the 16th, late 17th, and 18th centuries were typically male in origin and followed established models. In the 19th century, the specific image of the mother as a link to the religious portrait of the Madonna and the Mother of God (the first Polish poem presents *Bogurodzica*, the Virgin Mary, the Mother of Jesus) reinforces women's new presence. From Adam Mickiewicz's poem *Do matki Polki* (To Polish Mother), the term "Polish mother" becomes a separate literary, epistemological, and sociological category. Throughout the 20th century (with some exceptions), the impact of Romanticism and its poetical and religious models remained alive, even if they underwent some modifications. The period of communism, as during the Period of Partitions and the Second World War, privileged established models of lyric, where the image of women reproduced Romantic schema in poetics from the 19th-century canons, which are linked to religion. Religious poetry is the domain of few female author-poets who look for inner freedom and religious engagement (Anna Kamieńska) or for whom religion becomes a form of therapy in a bodily illness (Joanna Pollakówna). This, however, does not constitute an otherness or specificity of the "feminine" in relation to male models. Poets not interested in reproducing the established roles reach for the second type of lyrical expression: replacing the "mother" with the "lover" and "the priestess of love" (the Sappho model) present in the poetry of Maria Pawlikowska-Jasnorzewska. In the 20th century, the "religion" of love in women's work distances them from the problems of the poetry engaged in social and religious disputes and constitutes a return to pagan rituals (*Hymn idolatrous* of Halina Poświatowska) or to the carnality of the body, not necessarily overcoming previous aesthetic ideals (Anna Świrszczyńska). It is only since the 21st century that the lyrical forms of Polish female poets have significantly changed. They are linked to the new place of the Catholic Church in Poland and the new roles of Polish women in society. Four particular models are analysed in this study, which are shown through examples of the poetry of Genowefa Jakubowska-Fijałkowska, Justyna Bargielska, Anna Augustyniak, and Malina Prześluga with the Witches' Choir.

Keywords: religion; Polish mother; Sappho; love poem; religious; Polish female poetry; 21st-century female poetry; lyric matrixes; Polish Catholic Church; feminine models of poetry; engaged poetry; body; Polish society

Citation: Walczak-Delanois, Dorota. 2021. Poems by Polish Female Poets and the Burning Issue of Religion. *Religions* 12: 618. https://doi.org/10.3390/rel12080618

Academic Editors: Dennis Ioffe and Klaus Baumann

Received: 3 June 2021
Accepted: 30 July 2021
Published: 9 August 2021

Publisher's Note: MDPI stays neutral with regard to jurisdictional claims in published maps and institutional affiliations.

1. A Word of Introduction

As part of the Slavic tradition, Polish poetry represents much more than just a lyric form and is more than an ensemble of literary production in verse. It illuminates specific strata of society and otherwise hidden phenomena. Poetry is an artform that is sensitive to and records moods and invisible tensions contained within the human experience.

Furthermore, poetry anticipates the future events of society and can be described as a kind of social prognostic. For example, in the period of Polish Romanticism, poets often played the role of those "who know better", as spiritual guides of the nation, savants, or visionaries. At other times, poets' aims have been projecting a voice of warning as a "guardian", an oppositionist, or a voice of common or individual conscience.

Today, in the era of the predominant medium of the internet, enabling instantaneous conveyance of texts and images in public and private forums, poetry and poets continue to be visibly involved in the life of society, and their work continues to be significant. In this paper, I am especially interested in the contribution of female poets: Is there evidence of interest in poetical discourse on religion[1]? What sorts of matrixes are present in current poems by Polish women? What is happening with their lyric form?

I explore these complex questions and express points of view from a literary–historical perspective across several epochs to explain the most interesting for me—the current situation. A close reading method was applied, especially in the last part of this paper concerning 21st-century poetry, but I also relate some other points of methodological approaches (for example, feminist approaches) when necessary. To understand the present, we must look to the past, which is why two different aspects, the style and time of writing, are considered in this paper. They are articulated in a complementary dialogue in which **poems by female poets** are the most interesting to me, more so than the history of religion, cultural cohabitation in ancient Poland, or important querelles on democracy after 2015. The historical approach examining the timeline of one poet can be built upon by examining the religious aspects of their poem. Both are characterised by one of the most apparent motifs: the iconic model of "Madonna". I examine the poems of female Polish poets who are interested in religion according to their beliefs or their contestations concerning religion. Further, I examine the most popular poetic forms, as well as the origin of the models and their stereotypicality. Finally, I partly examine the impact of the mutual influence between female poets and society. In such a large-scale study, attentively examining the poems is a kind of invitation to further and deeper reflection, a stimulus, and the first step before a future book publication. This transverse study of female Polish poets and the issue of religion will be the first of its kind written, in particular, for non-Polish readers. Prior works of different researchers have presented particular religious topics linked to one poet (Bogalecki 2020), to a singular period (Milewska-Waźbińska 2018), questions of atheism (Bielak 2020), or even non-evident metaphysics (Grądziel-Wójcik 2017).

The first historical part of this paper helps one to understand the subsequent contemporary analysis. The authors and works from the Middle Ages to the present day are only a select—but I hope also representative—choice of texts. Some are well known; others remain virtually unknown. Several difficulties arose in presenting this analysis. For example, the topic touches upon numerous other problematic and rich themes such as women's corporeality and spirituality; their place in Polish society; and their relationships with other women and with men, with education, with politics, and the Catholic Church in Poland, which has fluctuated over the centuries (Weintraub 1971) as it is the only one of the different faiths in ancient Poland that is predominant nowadays. When necessary, the fragments of the historical context appear as a background. As this panorama of works necessitated making choices regarding their inclusion, these choices, unfortunately, led to the omission of many interesting examples.

2. The Particular Presence of Women in the Roots of Literary Tradition

Critical theories of different forms of feminism but also works on history of literature assert that despite the changing position of women in society, during recent centuries, some expressions of beliefs, published in the different forms of literature, still held that a woman is often perceived as uncapable of activities beyond "running a house and raising children" (Clark 1998; Jay 2005; Radzik 2020). The iconic image of a woman with a pen in her hand was one of the most important *imponderabilia* in prior eras. However, there was a kind of ambivalence: women were presented as a "devil's habitat" but also a saintly figure. In *Works and Days* by Hesiod (Hesiod VIII BC 2021, pp. 80, 94, 374–75, 405–6, 700–5); in Greek mythology (Graves 1955); in the works of Pythagoras, who places woman on the side of Chaos (Perictyone 430 BC 2021); and in Aristotle's *Politics* (Aristotle 1984), who sought to prove the natural inferiority of an ordinary woman, her image is seldom presented positively.

As stated by Kirster Stendahl (Stendahl 1974) and Joanna Partyka, in the *Book of Genesis*, the *Book of Proverbs*, and the *Ecclesiastes*, women also appear as Satan's tool, the cause of pain and illness (Partyka 2004). In the Old Testament, out of more than 200 women, only a few are deemed as deserving praise: Sarah, Esther, Judith, Rachel, Rebecca, and Ruth. The appearance of Mary, mother of Jesus, in the New Testament did not create a meaningful change in the ordinary image of women. Mary stays as the unsurpassed ideal and the example to be followed as a woman humbly accepting her lot and motherhood. In the *First Epistle of Paul to the Corinthians* (Saint Paul, ch. 14 vs. 34–35), it is said that a woman should not speak in the Church and participate by herself because it is a shame for her community, and if she wants to learn, she should ask her husband. This tradition continued into the Middle Ages, but balance can be seen to emerge in the 12th and 13th centuries with the development of the *gesta* and in troubadours' lyrics. Here, women are portrayed as beautiful and intelligent, respectful but also inaccessible ladies.

In Poland, the new status of women was built alongside the emerging identity of the new state in the context of a medieval European Christian[2] community, with examples such as Princess Świętosława-Sygryda (born c.a. 960–972–dead after 1016) and Queen Jadwiga-Edwidge (1372–1399), in parallel with new monastic education and the founding of the first schools and renovating Academy of Cracow (1364).

3. From the Ancient Cultural Models to the Beginning of Polish Literature

Many literary and historical sources confirm that the name of the first Polish female poet is **Gertrude (Gertruda) (1025?–1108)**, the daughter of the Polish King Mesco II and the German princess Rycheza. Gertrude is well educated, probably in one of the German abbeys. Gertrude becomes the wife of the oldest son of Jaroslav the Wise (Isaslas) and leads an eventful life. She holds a position of positive notoriety in Polish and Slavic history. Historians of literature associate Gertrude with the prayer book (called *Modlitewnik Gertrudy* (Prayerbook of Gertrude)). She authored prayers in hundreds of volumes, as testified by her characteristic use of the literary language of the epoch and the volume: Latin. The expression of *ego Gertrude* accompanies the stories of the adventures of her family in the prayer book created in the period from early 1075 to late 1086. Specialists of this *libelli precum* write that Gertrude was also able to break with conventional forms to express the authentic individual emotions:

> Beyond the Latin prayers arranged in sublime prose, there is an inner image of a woman torn apart by spiritual contradictions that give her personality the stigma of tragedy. Gertrude felt lonely, alienated from her surroundings and surrounded by enemies "visible" and "invisible". Living in a constant sense of threat, in fear of unfriendly people and of Satan, who "subjects many evils" and "captivates into slavery"—she was looking for support in religion. Her existence was suspended between the unreal, desired good and the almost tactile, almost tangible, surrounding misconduct[3].

(Michałowska 2000, p. 97)

Teresa Michałowska suggested that even if Gertrude and her spiritual depiction were formed by medieval Christian models, Gertrude produces the image of an individual entangled in dramatic events and tormented by violent passions like a person divided between a religious desire for spiritual perfection and love for her son, lust for power, and hatred for her enemies (Michałowska 2000, p. 98). The first literary text composed in the Polish language, the poem *Bogurodzica*, appears at the turn of the 13th or 14th century (the written version is from 15th century) and became immediately popular. The title means "the one who gave birth to God". This great masterpiece of Polish literature from the Piast epoch and "gold autumn" of medieval Poland reveals an exceptional artistry:

> Its stanzas' form indicates that it arose as a *trope* on the acclamation *Kyrie eleison.* What is probably most important, however, is the assertation that the work is wholly original. No Latin source has been uncovered, a fact that is even more noteworthy since the hymn demonstrates an almost perfect structure, a cohesive construction evincing the author's exceptional sense of beauty, manifest in its perfect symmetry and veritably mathematical precision. The work's symmetrical structure, almost faultless rhymes, and parallelisms underscored by rhyme draw our attention to its perfection of form, the perfection of this appear addressed to Mary and via her intercession to Christ, the Son and God. A detailed analysis of this prayer, wherein an entreaty is first made of Mother of God, while the second stanza speaks of the intercession of John the Baptist, allows us to discern the schema of the iconographic theme of *deesis* (Greek for "entreaty" or "prayer"), cultivated in Byzantine culture and assimilated by the culture of Roman Europe in the 11th–13th centuries. In *deesis,* as in our prayer, the central figure is Christ the Majesty, with Mary standing on the right and St. John the Baptist on the left.
>
> (Karpiński 2004, pp. 17–18)

We do not know the name of the anonymous author of this anthem in honour of St. Mary, but we know that the author was a well-educated and talented poet. This image of Mary, from the earliest days of the Roman Church, continues to influence and form the ideal model of the woman.

During the period of the Middle Ages and from numerous countries, works appear by acclaimed female authors involving topics of religious culture. These authors include Christine de Pizan (1364–1430), Hrotsvitha de Gandersheim (935–1002), and Hildegard von Bingen (1098–1179). The writings of women all over Europe, such as Régine Pernoud (Pernoud 1998), show the considerable influence of history and culture. Gertrude, Queen Edwidge (1372–1399), Edwidge of Silesia (1174–1243), and Saint Kinga-Cunegonde (1234–1292) prepare the foundation for future generations of women. Detailed knowledge of their lives remains limited, however. The Renaissance period in Poland produced new models of poetry and new silhouettes of women who expressed themselves in the context of their function or position toward men. For example, the young Ursula, the four-year-old daughter of Jan Kochanowski (1530–1584), was called the Polish "Ronsard". She appears as an emblem of possible feminine creativity. Having died before reaching adulthood, she is the heroine of the elegies written by her famous father, *Treny* (Lamentations) (Kochanowski 1986). As a talented author, she is compared in the sixth elegy to the Greek poet Sappho. Julian Krzyżanowski wrote on the phenomenon of Ursula Kochanowska and the poem about her, one of the lamentations, is linked to the development of lyric form:

> [The poet] made a clear distinction between two kinds of thirteen-syllable verse, when besides the more common kind (7 + 6):

Thou hast made this house for me // a place of emptiness	Wielkieś mi uczyniła // pustki w domu moim,
Ursula, my dearest child // reft by thy absentness	Moja droga Orszulo, // tym zniknieniem swoim.

> He also included in his *Laments* the rare form (8 + 5):

Joyous little songster of mine! // my Slavic Sapho	Ucieszna moja śpiewaczko! // Safo słowieńska,
The heiress to whom not only // my estate will go . . .	Na którą nie tylko moja // cząstka ziemieńska . . .

(Krzyżanowski 1978, p.102)

This important poetic image, that of Sappho, forms and transmits to the future a second strong model (apart from the medieval Madonna). Portraits of women in forms of Renaissance poetry can also be found, especially in the form of *carmen* or *frasca*, which are so cherished in Renaissance literature and beloved as much as the form of elegies. This period also produced an interesting new Polish anonymous text (not obviously present in academic literature courses) and a kind of anti-model of the female personage. This model is exemplified by Barbara Giżanka (1550–1589), the last mistress of the Polish King

Sigismund August. Giżanka was kidnapped for him from a convent because of her physical similarity to his beloved and still mourned late wife (also named Barbara). The poem *O [Barbarze Giżance]* (About Barbara Giżanka) presents Giżanka as a strong woman, who, introduced by the lyrical subject, talks about her colourful life, about her obliged carnal and material compromises, about unequal social hierarchy, and about the consequences of her status (she was from the middle class). This independent and lovable woman, through whose bed "both: priests and lords has passed" (Anonymous 1984, p. 244), defends herself against too-hasty judgement with energy and engagement.

The unfolding of the Renaissance and Baroque eras also produced some rare but clearly identified female poets, although a minority in comparison to men and official authors. Ursula Phillips offers precise descriptions of five women writers in the period of the 16th and 17th centuries (Borkowska et al. 2001). These authors, such as Magdalena Mortęska (1554–1631) and Marianna Marchocka (1603–1652), wrote primarily in prose, in letters, and lived under Carmelite and Benedictine rules. In the 17th century, we find the first mystic lyrics with strong characteristics of love poems:

Dusza strapiona od różnych ciężkości	The soul afflicted by various burdens
Wszystka omdlewa od wielkiej teskności,	All faint from great longing,
Bo się jej ukrył gdzieś Oblubieniec,	For her Bridegroom is hiding somewhere,
Włożywszy na nię poślubny wieniec.	Having placed upon her the wedding wreath.
Zadawszy ranę od strzały miłości,	Having inflicted a wound from the arrow of love,
Ukrył się prędko, a onę we mdłości	He hid himself quickly, and left her in weakness
Zostawił, srodze boleścią ściśnioną,	He left her in very severe pain,
Od wszystkich zmysłów prawie oddaloną.	Out from all her senses, out of her mind.
Myśli nieboga, co ma dalej czynić,	The poor thing thinks what to do next,
Kogo by miała o zranienie winić?	Whom should she blame for her injury?
Nie trzeba długo z myślą tej zabawy;	It does not take long to play with this thought;
Zranił mię, zranił, Oblubieniec krwawy,	I am wounded, I am wounded, by the bloody Bridegroom,
Ale co cięższe, że odszedł zraniwszy,	But more grievously, he went away after,
A mnie zbolałą w żalu zostawiwszy.	And left me in pain and left me in grief.
(Anonymous 1998)	

As we follow the manuscript canticles from women in Polish monasteries in the 17th century, sisters living in various spiritual and religious centres exchange not only letters, but also volumes of poetry and thoughts. In the example above, one can see the amalgam of a religious image of Christ and the beloved human as a "spouse" such as in the erotic, courtyard poetry of the Baroque *concetto* (Hanusiewicz 1998). Notably, some of these women created their own style. In Lviv, in the Polish and Lithuanian Commonwealth, the entire "writers' school" of Magdalena Mortęska appeared with the new generations of female authors. The Carmelite manuscript, which seems to be close in style to the writings of St. Teresa of Avila, records more frivolous lyrics from a convent during recreational periods:

Pódźmyż siostry zaraz po kolędzie	Let's go sisters after carol with visit
zali na nas pan Jezus łaskaw będzie:	Let's see if Jesus will be gracious to us:
Każ nam rekreować, każ nam rekreować Panno Ksieni nasza.	Let us have a leisure, let us have a leisure, our Lady [Priestess.
Którego nam Panna czysta porodziła	Whom the pure Virgin gave birth
I powiwszy w jasełeczkach położyła:	And bore him in a nativity scene:
Każ nam rekreować, każ nam rekreować Panno Ksieni nasza.	Let us have a leisure, let us have a leisure, our Lady [Priestess.
Pódźmyś tedy miłe siostry po kolędzie,	Let us go, then, dear sisters, to visit you,
Z nami też wesoła Panna Ksieni będzie:	The Lady Prelate will be with us:
Każ nam rekreować, każ nam rekreować Panno Ksieni nasza.	Let us have a leisure, let us have a leisure, our Lady [Priestess.
(…)	(…)

Nasza Panna Ksieni (z) siestrzyczkami,	Our Lady Priestess (with) the sisters,
Każe nam wina pełnić kubeczkami:	She makes us fill in wine with cups:
Każ nam rekreować, każ nam rekreować Panno Ksieni nasza.	Let us have a leisure, let us have a leisure, our Lady [Priestess.
(…)	(…)
Jużże teraz zaisadajmy wszystkie stoły,	Let us sit at all the tables now,
Będziemy mieć wesele z apostoły:	We'll have a great joy with apostles:
Każ nam rekreować, każ nam rekreować Panno Ksieni nasza.	Let us have a leisure, let us have a leisure, our Lady [Priestess.
Amen	Amen
(Anonymous 2004)	

By the 17th century, one of the rare figures of feminine poetry appears, presented in the official history of literature—**Elżbieta Drużbacka (c.a. 1695–1765)**. Her most known poem, *Opisanie czterech części roku* (Description of Four Parts of Year), begins:

Bogdajżeś przepadł w piekło z ateistą,	Goddess, you are lost in hell with the atheist,
Ty, który mówisz, że Bóg nie jest Panem,	You who say that God is not the Lord,
Ni Stwórcą rzeczy! A któż oczywistą	Nor the Creator of things! And who the obvious
Machinę świata oblał oceanem,	machinery of the world has surrounded by ocean,
Kto słońce, miesiąc, planety w swym biegu	Who sun, month, planets in their course
Komenderował, kto gwiazdy w szeregu?	Commanded, who put the stars in the row?
(Drużbacka 1903, p. 46)	

With the fever of counter-reformation typical of the Baroque period and already in a classic form of verses, Drużbacka manifested her worldview in a period of transformation. In the beginning of 17th century, Poland was still multi-confessional, with free choice of religion guaranteed by the Compact of Warsaw signed in 1573. This was an exceptional situation in comparison to the rest of Europe (Hernas and Hanusiewicz 1995). Thus, even when the Counter Reformation appeared, it occurred in quite a passive way—without religion wars. In this historical context, another Polish particularity appears: the Polish Enlightenment, initiated by the men of the Catholic Church: Ignacy Krasicki (1735–1801), Hugo Kołłątaj (1750–1812) and Stanisław Staszic (1755–1826).

In the 18th century, **Franciszka Urszula Radziwiłłowa (1705–1753)**, the first female author of threnodies after Jan Kochanowski (16th century), wrote an elegy bearing similarity to Kochanowski's *Treny* after the death of her son. In this literary and social context, her poetic *atelier* is significant (Judkowiak 1992):

> Radziwiłł's poem was not a spontaneous creation—prepared, thought out, subjected to the rules of the genre (or genres), it captured the mother's despair. Everything suggests that no Polish woman before her had ever mourned her child in this way. The Threnody is a genre steeped in antique tradition and philosophical content—for centuries available only to men. In the 16th and 17th centuries, and even at the beginning of the 18th, as can be seen from the example of Franciszka Radziwiłł, women mourning the death of their children took on male roles. This could only be done by those of them, who had acquired "masculine" (read: classical) education. It wasn't until the 17th century in Poland, and a century earlier in France, that women would write in a "feminine" way. The lack of classical erudition would no longer stand in the way of their "literary play".

(Partyka 2004, p. 210)

"Male" erudition, which allows a woman to be a writer, was emerging, but was, at the same time, a barrier for her own creativity and her own "femininity". This obvious conflict stops when what is considered "male" or "female" in society defies rigorous definition. The real changes became visible in the 18th century with the model of the "femme savant", first inspired by developments in the Enlightenment. Even in Poland—though this period of reforms and freethinking was rather short and was a bridge between the long Baroque and still significant Romantic periods—the "women's Enlightenment" and their new role as authors allowed them to appear to be officially at the same level (Jamrożek and Żołądź 1998). Still, they were not as numerous in comparison to male authors and certainly less visible. The scientific discourse on literary work of women from the 18th century is more apparent. Wacław Borowy, in his memorable study about Polish poetry in the 18th century, fully described three of them: the already-mentioned Elżbieta Drużbacka, **Konstancja Benisławska (1747–1806)**, and **Antonina Niemiryczowa (1702–1780)**. In this trio, Drużbacka is perhaps the most known and accomplished poet whose poetry touches upon both metaphysical and religious reflections (Borowy 1978, pp. 32–34).

Until the 19th century, Polish female poets continued to be rare. Where female poets appear in the historical record, they are linked to institutions such as monasteries, where female intellectuals were able to find relative liberty and a refuge. Many of the poetic forms they used in the 16th, late 17th, and 18th centuries were typically male in origin and followed established models. In the whole period of Old Polish literature, there were few female poets writing about religion, and we know little about them. Furthermore, in the 19th century, the specific image of the mother as a link to the religious portrait of Madonna and the mother of God appeared again as a strong and powerful motif.

4. Matrixes of 19th-Century Romanticism

The portrayal of the mother as an iconic and literary fact throughout previous centuries was endowed by the 19th century with an additional denotation. From Adam Mickiewicz's (1798–1855) poem *Do matki Polki* (To Polish Mother), the term "Polish mother" becomes a separate literary, epistemological, and sociological category (Mickiewicz 1993, pp. 320–21). The real success of the poem, beyond its artistic and technical excellence, is based on the historical and political context. Maria Janion, one of the most important Polish scholars and specialists on the history of literature, the history of feminism, and literary critics, writes:

> Mickiewicz was the author of both *Śmierć pułkownika* [Colonel's Death] and the poem *Do matki Polki* [To Polish Mother]. In these poems, he extolled female bravery, which had a powerful influence on the formation of women's self-awareness in Poland. He also created the legend of the self-sacrificing Polish woman, a legend to which Polish women so often conformed themselves later in their life. As a result of Romantic dictates, Polish women became accustomed to carrying the burdens of family and public life in the shade and in silence, as long as the sacrifice was fulfilled.
>
> (Janion 2006, p. 99)

In 1772, 1793 and finally in 1795, the territory of Poland was divided and incorporated into Russia, Austria and Prussia for 123 years. Russian and German language, legal code, education system, and even the religious choices in some historical moments were imposed on Poles for more than a century (Beauvois 2004). After 1830, after the November Uprising, the persecutions and repression of Poles led to new religious interpretations and comparisons, but not only by Poles (Malinowski and Styczyński 2008). These emerging ideas considered Poland to be in the position of a suffering Christ, whereby his mother becomes one of the stylistic figures of the image of the Polish woman and mothers who suffer from the persecution of their children. As Julian Krzyżanowski writes:

> So, whereas for some Messianism was a program of an armed fight for liberation, for others it turned into a quietist attitude, waiting for a miracle that would, without any effort on the part of man, bring freedom to the world, for slavery was in opposition to the notion of divine justice. Still others, taking the principles of Messianism as a basis, maintained that God had saved Poland from taking an active part in rotten political life which was based on the rule of force of the imperialist states, in order to resurrect her at the moment when these states had become just a part of the shameful past. As these ideas were occupied with biblical symbols in such formulations as: Poland, the Messiah of the nations; Poland-the Christ; Poland-Mary Magdalene, etc., the whole of this nebulous mixture gave and still gives the impression of religious and political speculation,

to such an extent that it overshadows the original, simple and noble greatness of the principles, as greatness that was understood by those who, living outside Poland at the time, had read Mickiewicz's brochure, *Books of the Polish Nation,* which was a kind of national bible.

(Krzyżanowski 1978, p. 225)

The Biblical, Messianic comparisons emerge in different variants and they provide a new, but still limited, legacy to the woman considered the most noble: the Polish mother as Christ's mother. In a Romantic landscape, one can also find a woman-lover or a woman-knight (and rarely, Mary Magdalene); they are less present and less powerful in their influence or their strength. Of course, the forms of lament, elegy, and rhapsody are still preferred in poetry and are connected to the religious connotations. The fact that only men—Adam Mickiewicz, Juliusz Słowacki (1809–1949), and Cyprian Kamil Norwid (1821–1883) (or even Zygmunt Krasiński (1812–1859))—could obtain the title of national bard or national guide ("wieszcz") is significant. As was the norm for this period, mothers would educate poets and suffer for them but stand at a distance (as well as their daughters) from the places of the first poetic tribune. However, due to the context of suffering and struggling for independence, women regained a bitter form of equality in accordance with the Tsarist decree of 1863, which introduced equal military courts for offenders of both sexes. Pointing to the similarities between the 19th and 20th centuries, Maria Janion reminds us that:

The imprisonment of most of the male leaders (similar to what happened in the 19th century, which was further depopulated by the military emigration of men after the November Uprising) meant that women took over the leadership and united the shattered movement. Those men who were not arrested reportedly did not know how to direct underground activity. The women did their heroic and anonymous work in secret, which was never discovered because it would have conflicted with the established belief that women should not lead.

(Janion 2006, p. 99)

In the second half of the 19th century and the beginning of the 20th century, various female poets, such as Bronisława Ostrowska (1881–1928), Maria Komornicka (1876–1949), Maryla Wolska (1873–1930), Kazimiera Zawistowska (1870–1902), and Maria Konopnicka (1842–1910), write in accordance with the spirit and soul of the era, but they do not significantly influence either the social matrixes or the poetical schema of lyric in relation to the issue of religion. However, their presence in the literary scene is important because it allows the possibility of new developments in the 20th century.

The opinion regarding the impossibility of leadership by women throughout the 20th century is extant within the Polish Catholic Church, but is also often found in the opinions that Polish women have of themselves (Duch-Dyngosz et al. 2014).

5. Post-Romantic and Other Uses in the 20th Century

Throughout the 20th century (with some exceptions), the impact of Romanticism and its poetical and religious models remain alive, even if they undergo some modifications. The period of communism (1945–1989), as during the Period of Partitions and the Second World War, privileged the models of lyric where the images of women reproduce Romantic schema in poetics from the 19th-century canons. Even in a lyric of **Kazimiera Iłłakowiczówna (1892–1983)**, an independent woman fascinated by the beginning of feminism and at the same time fervently Catholic, one cannot find substantial changes in the models in her poems. As a secretary of Józef Piłsudski, but also due to her numerous talents and unpredictable character, she was famous in Warsaw during the 1920s and 1930s. In her poems from this period (Iłłakowiczówna 1926, 1927, 1928, 1930) and in her religious volume from the 1960s (Iłłakowiczówna 1967), no significant changes can be observed when comparing these works to the Baroque and Romantic periods, even if her style is recognisable, as is her particular traditional form and religious reflection (Głowiński 1960; Ratajczak 1998; Marzec 2018).

Such Romantic models continue to function in the lyric of most 20th-century female poets. Thus, over time, the "new" lyric monologue can be seen from poets such as Krzysztof Kamil Baczyński (1921–1944) *Elegia o chłopcu polskim* (Elegy about a Polish Boy) (Baczyński 1994). Here, the poet incorporates the role of a mother losing, in the sacrifice of a war battle, her son; this "iron" model of the mother is still alive throughout the 20th century, even if, as stated in the poem written by **Agnieszka Osiecka (1936–1997)** *Polska Madonna* (Polish Madonna), the nature of "motherhood" changes over time:

Madonno, polska madonno,	Madonna, Polish Madonna,
Panno z dzieckiem, mów,	Maiden with child, speak up,
Jak ty sobie radzisz nocą	How do you manage at night
Wśród złych snów.	Among bad dreams.
Madonno, polska madonno,	Madonna, Polish Madonna,
Skąd masz pieniądze na czynsz,	Where do you get the money for rent,
Czy cię nie przeraża pająk albo mysz?	Aren't you terrified of a spider or a mouse?
Damy ci kwiatek na święto kobiet,	We'll give you a little flower for Women's Day,
Czapkę na bakier i Polski obie.	A beanie on the back and both of Polands.
Damy ci wódki, bilet do kina,	We'll give you vodka, a ticket to the cinema,
kace i smutki, dwóje na szynach	flunks and sorrows, double digits on the rails.
(Osiecka 2009, p. 576)	

Osiecka contemporises the religious vision of the Madonna from the past, but still places it in the context of Polishness and as a kind of social and patriotic martyrology. Here, you are the woman who "has to" finally assume the ancestral role of the mother but also many other roles. In feminine poetry, there is no breaking with the repetitiveness of the model because in times of war and after the period of communism, it also serves as one of the identifiers of Polish society.

It was in the male lyrical sphere, which was traditionally more visible, where there were 20th-century metaphysical poetry and poetry regularly depicting themes of the sacred or being in dialogue with religion. For many readers of poetry, it is rather evident to mention well-known and popular names such as Tadeusz Miciński (1873–1918), Bolesław Leśmian (1877–1937), Jerzy Liebert (1904–1931), Roman Brandstaetter (1906–1987), Karol Wojtyła (1920–2005), and the immensely popular, Father Jan Twardowski (1915–2006), with his everyday topics. No such list for women is evident.

In the 20th century, the return of the other, shy model can be observed. This returns to the model of the Muse, the Poet herself, and to Sappho (mentioned previously by Kochanowski in the elegy for Ursula) after Poland regained independence in 1918, after which the poet could be finally free of martyrologic topics. The "motherhood" and "Madonnahood" seem to be of less interest here. We find in the 1920s and 1930s an interesting example in the poems written by **Maria Pawlikowska-Jasnorzewska (1891–1945),** a popular poet, who published a dozen volumes up to 1939. She breaks social rules and covenants with her three marriages and divorces; with the disputes with her uncles and ecclesiastics; with her sensual poetry, such as in her poems from cycle *Róże dla Safony* (Roses for Sappho), recalling the need for love rather than different religion; and with the still-existing ambiguous Sapphic themes:

Gdy świat Safonę odrzucał,	When the world to Sappho closed its ears,
Gdy jej dorobek palono,	When the fire consumed the work of years,
Buchnęły dymy różane	Rose-laden smoke rolled up to the skies;
Ociężała, chociaż szaloną	Gath'ring, heavy but delirious
Chmurą płynęły przez czas.	A cloud remaining through centuries.
Wciągnęłam ją z wiatrem w płuca,	I breathed it in on a windy day,
Poezja nie poszła w las …	Poetry is with us, here to stay …
(Pawlikowska-Jasnorzewska 1997, p. 491)	(Pawlikowska-Jasnorzewska 1978, p. 576)

Her poetic voice mostly appears in the form of frasca and miniatures freed from the eroticism of the parabolic and religious contexts so common in the Baroque period. She chose to avoid religious topics and replaced sacrifice with free love. Even if in an ironic half-humour, she criticises one of her uncles—a bishop. He is violet in his clothes but also violet in his dry and sad heart, always with the old women in the church, reading old books and unable to love and to laugh, and passing by the most important aspects of life, as described in the middle stanza of the poem *O fiołkowym biskupie* (About a Violet Bishop) (Pawlikowska-Jasnorzewska 1997, p. 67).

In the 20th century, other women follow Jasnorzewska in the replacing of matrixes, e.g., **Halina Poświatowska (1935–1967)**, who makes a kingdom of the feelings of love in her poems. With no interest for the traditional Catholic religion, and no interest for meanderings of reflections and political quarrels, she, like an antique Sappho, is the priestess of her own religion—love. Stanisław Grochowiak (1934–1976) writes about her (Grochowiak 1959), recalling her *Hymn bałwochwalczy* (The Idolatrous Hymn) (Poświatowska 1958) and her heart disease:

> Her love of life had something intrinsically religious about it, yet it was a forbidden religion, condemned in advance to ingratitude and lack of grace, an idolatrous religion. Here was a woman whose every living movement of the heart threatened death, worshipping unrestrained, sensual, carnal love.
>
> (Grochowiak 1967)

God, named, appears rarely in material, sensual love that is oriented firstly to another beloved human. When God is directly mentioned, it is in the context of suffering and questioning of existence:

Boże mój zmiłuj się nade mną	My God, have mercy on me
czemu stworzyłeś mnie na niepodobieństwo	why hast thou made me inimitable
twardych kamieni	of hard stones
pełna jestem twoich tajemnic	I am full of your mysteries
wodę zamieniam w wino pragnienia	I change water into wine of thirst
wino—zamieniam w płomień krwi	wine I change into the flame of blood
Boże mojego bólu	God of my pain
atłasowym oddechem wymość	with a satin breath sweep away
puste gniazdo mojego serca	the empty nest of my heart
Lekko—żeby nie pognieść skrzydeł	Lightly—so as not to crumple the wings
tchnij we mnie ptaka	breathe the bird into me
o głosie srebrnym z tkliwości	with a silver voice of tenderness
(Poświatowska 2000, p. 342)	

However, it was written in a study about Poświatowska's lyric that: "The last book which aroused her interest and which she read before her death was the Bible. And we will never know whether this had anything to do to the transformation of her sensualist worldview, which rejected religious consolation . . . " (Dumowska 1986). It is precisely the new expression of suffering that helps the poet refine the dialogue and interests manifested in the themes of love, self-affirmation, and interruption of the *pietà* model.

A new generation of female poets appears in the 1970s: the real internal aspects, inside a poem and in a poetic transformation, undergo a metamorphosis. An interesting example of this is **Anna Świrszczyńska (1909–1984)**. Her poems, especially those from the 1970s, contain a considerable amount of rebellion and a new distance from the canons of consciousness and an awareness of choice. One of her best-known volumes, *Jestem baba*, 1972 (I am the Old Woman), shows the frailty of imposed convention, explaining the vast range of perceptions—a kind of self-affirmation; it is pride in being a "baba" (in familiar speaking: a nasty broad). Zbigniew Bieńkowski writes about this volume that:

And suddenly there is such a confession, blatant, brutal, desperately brave. No woman has ever spoken out so emphatically, since poetry is poetry. I must make a great effort not to quote. Because here any quotation would only be an example of brutality, unbeauty, finally: unpoetry at last.

(Bieńkowski 1993, pp. 201–3)

Świrszczyńska poses a problem for critics who are unable to cope with the new, brutal, "meaty" writing of a poet previously known from artistic canons. Similarly, in the next volume, *Budowałam barykadę* (I Built the Barricade, 1974), the author refers to the thirty years after the war as the time of the Warsaw Uprising and demonstrates the need to create a new language of stories about difficult wartime experiences. Świrszczyńska points to the agency of the woman's "I", visible even in the title. A new feeling of experience, showing a separate, own perspective, which means more than a passive bearing of the consequences of wars, deaths of men, casemates and prisons, or even loneliness, needs to be expressed. Świrszczyńska's creative juxtaposition is worthy of attention because we also see from its example various readings of the poetry changing over time. Despite the poet regarding herself as a declared atheist and a person firmly rooted in the Earth, Czesław Miłosz (who, with a young team of researchers, translated her poetry into English) presents herself as a metaphysical poet. However, Magdalena Heydel follows the readings of the critics of her poetry (Miłosz, Bieńkowski, Balbus) and argues against this view by re-reading Świrszczyńska. Heydel does not agree with Miłosz and shows the poetry of Świrczyńska becoming *écriture féminine avant la lettre* (Heydel 2013, pp. 225–35); Heydel also looked for parallels between the poet's thrifty language and the concept of creativity presented in a theory by Hélène Cixous (Cixous 1975). However, the problems of re-reading and re-interpreting concern many other poems, not only these by Świrszczyńska, and belong to a new practice of the 21st century, when the question of the woman's "I" grows with the impact of religion in societal forms.

In the 1970s and 1980s, the work of two more poets, whose names are worth quoting in the context of our considerations, flourished: **Joanna Pollakówna (1939–2002)** and **Anna Kamieńska (1920–1986)**. The former often describes the dysfunctional and ageing body in a new form of lament linked to the image of Job. For the first time since the Romantic and blasphemous monologues of Mickiewicz and Słowacki, from the times of their Great Improvisations[4], and from Iłłakowiczówna's poetry, she transformed the monologue of a male poet into a monologue of a female poet. As in the famous monologues of the Polish Romantic hero Kordian at Mont Blanc or Konrad in a convent of Basilicans, the "I", the poet requests explanations from God, as in her *Kanikuła* (Canicule):

Czemu tak długo trwa ta próba bólu?	Why does this trial of pain take so long?
Już i zwątpiłam	I have already doubted
i wstałam w ufności;	and stood up in trust;
zawrzałam buntem	I have boiled by rebellion
i w pokorze ścichłam.	And in humility I have quieted down.
Doświadczasz nadal.	You are testing me still.
Puenta wciąż nie rychła	The puenta is not quick
i niepojęte sensów zawiłości.	And incomprehensible complexities of sens.
W niebieskim żarze kanikuły wiecznej	In the blue heat of the eternal canicule
miłość się Twoja nigdy w chłód nie skłania.	Your love never turns cold.
Nie ma schronienia i nie ma ucieczki.	There is no refuge and no escape.
Brak odpowiedzi	There is no answer
bo—nie ma pytania.	For—there is no question.

(Pollakówna 2012a, p. 423)

After a long period of seeking, we obtain a real, metaphysical question, which is not merely interest in attending to and confirming the ideal religious models (such as the mother and the Madonna) or oppositional or by-passing attitudes (Sappho, lover), but real, true dialogue: the will to dialogue with God:

W tę noc umarłą, bladą,	In this dead, pale night,
w ten wiatr.	in this wind.
Wypatruję się w Twoją ciemną jasność	I look out into your dark brightness
- badam.	- I explore.
Tym bolesnym i chromym duktem,	With this painful and lame track,
jakie mi prawdy czuło-okrutne	what tender and cruel truths
zawierzasz?	do You entrust to me ?
Czy wyrok mi wymierzasz,	Do You pass judgment on me,
czy przesłane tkliwe i unerwione	whether sent tender and innervated
jak płatki irysów, gdzie zarys	like the petals of an iris, where the outline
map nieczytelnych czyjejś marszruty plany	of illegible maps of someone's route plans
zarosłe z kwiatowym puchem.	overgrown with floral down.
Słów Twoich rytmy głuche,	Your words of deafening rhythms,
słów Twoich kształty zatarte	Your words of blurred shapes
- czarne szumy.	- black hums.

(Pollakówna 2012b)

This lyrical dialogue occurs in the intimate sphere far away from institutions, politics, and social engagement. Anna Kamieńska presents another attitude. Throughout the whole of the 20th century, Kamieńska is the only clear, decisive, widely read, and poetically recognisable example of religious engagement, whose works are accompanied by metaphysical questioning. An important breakthrough in her rich oeuvre occurred with the death of her beloved husband Jan Śpiewak (1908–1967) and then her meeting and friendship with a well-known personality and poet, Father Jan Twardowski. Kamieńska, a seeker, underwent a conversion and became a Dominican Tertiary involved in pastoral work. She participated in Christian Culture Weeks, signed protest letters against the communist authorities (Letter 59), and appealed in the defence of striking workers (Appeal 64). She also led and taught during Christian retreats. Her poems are engaged in a dialogue with the presence of God in his undoubted existence, as in *Przebudzenie* (Awakening):

Dzień dobry Panie Boże	Good morning my Lord, God
więc mamy zaczynać od nowa	so we have to start again
było już tak ciemno	anyway it was already so dark
a Ty obmyłeś wszystko	and You have washed everything
deszczem światła.	with a rain of light.
Dzień dobry Panie Boże	Good morning, my Lord, God
dałeś ptakom nowy głos	You gave a new voice to the birds
pszczoła przyszła do nowego miodu	the bee came to new honey
liść rozsunął pokrywy	The leaf has spread its lids
wszędzie lśnią kropelki łaski.	droplets of grace shine everywhere.
Nie znaj samego siebie	Do not know yourself
ale prędko otwieraj oczy i uszy	but quickly open your eyes and ears
krzew postrzępiony płonie	the frayed bush is burning
obłok zmartwychwstaje.	the cloud rises from the death.

(Kamieńska 1981a, p. 57)

The dark of the Polish 1980s was a period with no liberty of expression, no calm life in the society of the regime, and with threat of imprisonment and daily renunciations. Once again, the poetic romantic figure of the suffering mother returns, as in *Stabat Mater:*

Pod tym krzyżem, gdzie rozdarta umierała w nocy Polska
w lodach Wisły w śniegu grudnia stała Matka Częstochowska.

W tłumie matek popychana, osiwiała, ledwie żywa
pod tą bramą, pod tym krzyżem stała Matka Boleściwa.

Okutana szatą z lodu, poczerniała i milcząca
solidarnie z ludem swoim stała Matka Bolejąca

Zatrwożona i bezsenna jak dziś każda z polskich matek
przytuliła nas do serca, gorzko mocząc łzą opłatek.

Matko nasza, Matko Boża poorana, cała w troskach
większą wiarę i nadzieję daj nam Pani Częstochowska.
(Kamieńska 1981b, p. 61)

Under that cross, where Poland, torned, was dying at night
in the ice of the Vistula River, in the snow of December, Our Lady of Czestochowa was
[standing.

In the crowd of mothers—pushed, gray-haired, barely alive
Under that gate, under that cross, the Mother of Sorrows was standing.

Covered with a robe of ice, blackened and silent
the Sorrowful Mother was standing in solidarity with her folk.

Scared and sleepless like every Polish mother today
hugged us to her heart, bitterly wetting the hots-wafer with tears.

Our Mother, Mother of God, scarred and troubled
Give us greater faith and hope, Lady of Czestochowa.

One should not forget that in the 1980s—the most significant for Kamieńska's creations—
there was a strong presence of the Polish Catholic Church, which organised cultural and laic
events in the parishes in the period of censure when many scenes, theatres, and cultural organi-
sations were forbidden and closed. The Church also helped in caring for workers and strikers.
Censoring religious sentiments helped[5] strengthen the popularity of such religious poetry.

Furthermore, in the Polish poetic paysage, some poems abrogate the essence of the
old secular–religious debate. The female authors consider declarations of faith as a private
matter even if they are involved in its discussion. Wisława Szymborska (1923–2012),
Nobel Prize winner in 1996, is an example of such a voice. She shunned the model of
poetry, situating herself outside the issue of religion in poetry, flirting with the ideology of
socialism (but, of course, not outside human values common to religious universal codes
of values as: compassion, generosity and kindness towards the other (Tomasik 2009));
in interviews, she declared herself as an atheist. In her poem *Głos w sprawie pornografii*
(Voice on Pornography), with perspicacity, irony, and humour, she takes the floor, pointing
to the innocence of "pink buttocks" and nudity in the face of the influences of thoughts,
sometimes the most perverse (Szymborska 1996, pp. 130–31). In the social reality of the
new, liberal Poland of the late 1980s and in the 1990s after the fall of the Berlin Wall, sex
shops sprang up like mushrooms after the rain.

Yet, the new 21st century will bring interesting poetic phenomena by Polish female poets. The romantic topoi, e.g., of the ever-popular Polish mother, have returned, but in a renewed form, challenging the social and the literary stereotypes established over the centuries.

This time it is about the revindication of a different "I". One can notice an abandonment of "I"'s function in the poetry of irreligiosity and a clear return to the dialogue with the previous forms of religion and its contents functioning in Polish poetry.

For the first time in Poland, women and female poets have become present in this thematic climate in the mainstream literature, linked to their new roles in society and new debates concerning the entrenched role of the institution of the Catholic Church in Poland.

6. *HIC and NUNC*—The Burning Issue of Religion in the 21st Century

The female poets of the 21st century break with the hitherto existing ways of talking about religion in verse. Their presence is much more visible and linked to the social and cultural changes occurring at the turn of the 20th and 21st centuries in Poland, Europe, and the world. Models previously rare or non-existent appear as a worthy way of living: a businesswoman, a "singielka" (single, not in a couple), an LGBT+ activist, a feminist, a vegan, an ecologist, a militant woman, etc. Changes have also appeared in the Catholic Church and other Christian communities in Poland. Girls can now serve during mass near the altar, as the boys do. The first LGBT parades have appeared, as well as massive demonstrations connected to the tightening of the anti-abortion law (loudly broadcast and visible in the media, called the "black marches"). There have been responses to these changes in Poland, such as smaller but still apparent marches on the knees and men praying with rosaries, adepts of the traditional roles of women, and the interdiction of abortion. In 1990, religious education (only Catholic) was transferred from the parish to primary and secondary schools, sparking discussion on ministerial salaries of the priests already paid by Church in the years of runaway inflation.

In the second decade of 21st century, as a special and contesting answer to the new programme of governmental ideas after the 2015 elections, the first book on sexual education for young girls, *#sexedpl. Rozmowy Anji Rubik o dojrzewaniu, miłości i seksie* (#sexedpl. Anja Rubik talks about puberty, love and sex), written by the famous Anja Rubik, a top model, appeared and was immediately sold out (Rubik 2018). This is one of the rare European books on the subject of anticonception and sexuality—a subject still staunchly avoided during religion lessons in Polish schools. The older, pioneer Polish book by Michalina Wisłocka: *Sztuka Kochana* (Art of Loving) has been re-edited in large print (Wisłocka 2016), and the biography of Wisłocka has been transferred to the screen (Sadowska 2016). Interesting works on the Church and Catholic women were published in *Emancypantki. O kobietach, które zbudowały kościół* (Emancipantes. About Women who built the Church) (Radzik 2018) and in *Kościół kobiet* (The Church of Women) (Radzik 2020) by Zuzanna Radzik (born in 1983). As a theologian, journalist, and the president of the Forum for Dialogue (on Polish–Jewish dialogue), she constitutes an example of a possible and important input as a new, spiritual, and also feminist voice in the written history of women in the Church from the beginning of the Early Christian communities.

How do female poets—the vigilant recorders of reality—react to these huge changes? We observe a departure from linguistic shamefulness, a somewhat energetic, sometimes frenetic, new flow (Maliszewski 2020). In official poetical discourse, for the first time, strong, vulgar, and anatomic words appear, accompanied by theses, which belong to the "sacrum" sphere. For the first time, strongly expressed and distinguishable anti-canons have appeared. I indicate some of them here. They are written by contemporary female poets.

a. Genowefa Jakubowska-Fijałkowska (born in 1946) is one of the most expressive and interesting poets today, and a testimony that maturity and a late debut can be an asset. Her poetry, from the first poems printed on the pages of "Odra" and the volume *Dożywocie* (Life imprisonment) (Jakubowska-Fijałkowska 1994) until her final *Rośliny mię-*

sożerne (Carnivorous plants) (Jakubowska-Fijałkowska 2020a), is characterised by a sharply expressed observation of the world, perceived through the subject's painful experiences: the death of loved ones, experiences of illness, and the ageing of the body. The sharpness of her pen-razor is mixed here with humour; the analysed sense of one's own femininity is juxtaposed with the masculinity encountered in the necessary complementation, but also in incomprehension. Although parallels can be drawn with the world of Świrszczyńska's late poems (the disintegration of corporality, the discomforts of age, the contrast between enthusiasm and the stereotypical expectations of women of a certain age), in Jakubowska-Fijałkowska's poems, these contents are expressed with the force of fire and with ordinary communicative language.

Her works are a kind of older woman's manifesto, in the form of a dialogue with religion, which is mixed as a variable constant into the world of everyday human laic affairs. There is something Franciscan and tender in this, after all, very violent rebellion in defence of the weak (animals and plants) and those who cannot cope with the Polish version of capitalism: derailed, lost men and women who are tired of life and devoid of dreams. I quote what the critic and the reviewer, Wojciech Szot, wrote about this work, awarded the NIKE nomination in 2017:

> I call it, for my own use, 'postmenstrual poetry', where a mature, simply old poetess reveals to the person reading her world, which she experiences with a slight piss. It is beautiful that next to fatigue, a slight resignation, in Jakubowska-Fijałkowska's work there is a discord with old age, when "I dye my fucking grey hair/henna ruby flows down the whole tub with paint", which reminds the lyrical subject of menstruation and how "communist lignin rubbed my thighs every month". He writes:
>
> at seventy I'm fucking with it all again
>
> with hair dye like I did with blood then
>
> (. . .) If Jakubowska-Fijałkowska lived in the States, her poems would be recited by celebrities at political rallies like they now quote Maya Angelou. And so, we are a country of poets, where the so-called "Enzensberger constant" is a fact.
>
> (Szot 2017)

Although the poems of this female poet from Mikołów evoke common experiences in a communicative and seemingly ordinary, banal expression, we observe a shift in the canon. Here, in the poem: *ten diabeł ciągle mąci* (this devil is still messing about), Fijałkowska's alter ego, a lyrical subject, is not the child-bearing Madonna, but the tempted and bleeding Jesus:

jestem tam przez 40 dni	I've been there for 40 days
obnażona kuszona głodna spocona	naked tempted hungry sweaty
w tym głodzie szaleństwie opętaniu	in this hunger madness possessed
pocę się czerwoną krwią	sweating red blood
podobna do Jezusa	like Jesus
(Jakubowska-Fijałkowska 2016a, p. 8)	

A similar lyrical situation, according to the terminology of Edward Balcerzan (Balcerzan 1970, pp. 333–86), is repeated in a poem, *Wielkanoc 2018* (Easter 2018), where we read: "pocę się krwią przed ukrzyżowaniem/w noc wielkanocną w koszuli nocnej jęczę" (I sweat blood before the crucifixion/on Easter night in my nightgown I moan) (Jakubowska-Fijałkowska 2020b, p. 16).

It is not just a symbolic substitution, an exemplification of corporality, and strong carnal exhibitionism; in this poetry, the sensation—also brutal and ugly—of one's own body is the basis for asking important questions about the possibility of good religion, about the humanity of God, about the legitimacy of dogmas, and about one's own shape of faith, as in the poem *Bóg powrócił do miasta* (God has returned to the city):

<table>
<tr><td>

i mężczyzna
szukam go (boga i mężczyzny)

</td><td>

and a man
looking for him (god and man)

</td></tr>
<tr><td>

w Istambule na lotnisku
palę w miejscu wydzielonym

</td><td>

in Istanbul at the airport
I smoke in a designated area

</td></tr>
<tr><td>

przy barze wypijam brandy z lodem
za szybami samoloty kołują

</td><td>

at the bar I drink brandy on the rocks
behind the glass planes are taxiing

</td></tr>
<tr><td>

(ptaki daleko gdzie indziej)

</td><td>

(birds far away somewhere else)

</td></tr>
<tr><td>

i znowu palę w miejscu wydzielonym
i wypijam drugą brandy (przy barze)

</td><td>

and I smoke again in a designated area
and I drink another brandy (at the bar)

</td></tr>
<tr><td>

muezin nawołuje do modlitwy
jest czwarta nad ranem

</td><td>

the muezzin calls for prayer
it is four o'clock in the morning

</td></tr>
<tr><td>

minarety wbijają się we mnie wieżą wysoką
jak penis boga

</td><td>

the minarets are hitting me with their high
towers like a god's penis

</td></tr>
<tr><td>

gdy na lotnisku w toalecie spuszczam
dżinsy rajstopy majtki i sikam
(Jakubowska-Fijałkowska 2016b, p. 10).

</td><td>

when at the airport in the toilet I flush
jeans tights panties and pee

</td></tr>
</table>

God (this time written with a capital letter) can be found again in the juxtaposition of the pair, visible also in the title, *Mężczyzna i Bóg* (Man and God):

<table>
<tr><td>

Bóg jest moim mężczyzną w lipcu

</td><td>

God is my man in July

</td></tr>
<tr><td>

w innych miesiącach

</td><td>

in other months

</td></tr>
<tr><td>

odległością skalistą pustynią doliną śmierci
cudnym krajem astronomii

</td><td>

a distance a rocky desert a valley of death
the wondrous land of astronomy

</td></tr>
<tr><td>

szamanką w stringach

</td><td>

a shamaness in T-string

</td></tr>
<tr><td>

mam swoje obserwatorium Boga i mężczyzny
w średnicy Drogi Mlecznej

</td><td>

I have my observatory of God and man
in the diameter of the Milky Way

</td></tr>
<tr><td>

w paznokciu wrośniętym w duży palec stopy
w linii papilarnej do visy USA
(Jakubowska-Fijałkowska 2016c, p. 42).

</td><td>

in the nail embedded in the big toe
in the fingerprint to the USA visa

</td></tr>
</table>

The authors' poetry features many paraphernalia common to the age-old, established repository of religious and fatherland identifiers in the Polish space: a silver chain with a medallion of the Virgin Mary; a watch received at a celebration of communion, as in *Ucho van Gogha* (Van Gogh's ear) (Jakubowska-Fijałkowska 2016d, p.13); a holy picture with the pierced heart of Jesus, as in *Osiedlowy spleen* (Bloc quartier's Spleen) (Jakubowska-Fijałkowska 2016e, p. 19). Here, along the same lines, appear the presence of a James Bond-like handsome priest-steward in a white alba who brings the Body of the Lord in *W sobotę koło dziewiatej rano* (Saturday around Nine O'clock in the Morning) (Jakubowska-Fijałkowska 2016f, p. 22) and the biblical stories of Eve and Judith woven into the everyday as in *Ruletka z tobą,* (Roulette with You) (Jakubowska-Fijałkowska 2016g, p. 45).

When we carefully read her poems, we notice the presence of different fragments of rituals and in-church ceremonies, which include common agendas of religious and laic festivities. At different levels, the poetical "I" is involved in the community by questioning the sense of gestures and established models. "I" uses them as the signposts and time-sensitive signs, as in *Święcona woda* (Holy Water) (Jakubowska-Fijałkowska 2020c, p. 40),

Trzech Króli szóstego stycznia 2018 (Epiphany on 6 January 2018) (Jakubowska-Fijałkowska 2020d, p. 28). *Boże Narodzenie 1986 (wspomnienie)* (Christmas 1986 (souvenir)) (Jakubowska-Fijałkowska 2020e, p. 30), and *Wielkanoc 2017* (Easter 2017), (Jakubowska-Fijałkowska 2020f, p. 35).

In a simple way, Jakubowska-Fijałkowska shows the incoherence between dogmas, ecclesiastical laws, and common practices, as in *Jezus żonaty* (Jesus married) (Jakubowska-Fijałkowska 2020g, p. 37) or in *Niedziela* (Sunday):

kobiety w garsonkach z małymi torebkami Diora z lumpeksu	women in suits with little Dior handbags from the second-hand shop
modlą się na mszy	are praying at mass service
w nich (torebkach) komunijne skarbczyki różańce pamiątki chusteczki perfumy pomadki złamane [rzęsy	in them (bags) communion treasuries rosaries souvenirs handkerchiefs perfumes lipsticks broken [eyelashes
w każdą niedzielę ksiądz błogosławi wszystko w małej torebce prezerwatywę też	every Sunday the priest blesses everything in a small bag a condom too
to taki grzech seks przez gumkę (Jakubowska-Fijałkowska 2016h, p. 44)	it is such a sin to have sex through a condom

Finally, Genowefa Jakubowska-Fijałkowska points to ancient memories and stereotypes to comment on the recent black marches and protests against politics. Here, the experiences common across generations and the heavy and rich heritage of this Polish mother occur again in the poem *Rodzą się dziewczynki* (Girls are being born) (Jakubowska-Fijałkowska 2016i, p. 50). In many of her poems, we observe a militant tune and real anger. The result is the very reaction in the poem *Wszystko zawsze na matce wisi* (Everything always hangs on the mother) to the right-wing government:

Jarek płacze nad rozlaną jego spermą zakazuje in vitro	Jarek weeps over spilling his sperm bans in vitro
twoje ciało nie należy do ciebie PIS się wpierdala w twoją macicę jajniki	your body doesn't belong to you PIS fucks with your uterus, your ovaries
w pochwę wargi sromowe	into vagina into labia
w twoją pierś karmiącą w zapaleniu bolącą gdy odciągasz mleko	into your breast that feeds in inflammation that hurts when you pump milk
sutki bolą krwawią przeżyj to prezesie	nipples hurt, they bleed, live it up, president
sama sobie strzelę w krocze wtedy się odwalisz ode mnie (Jakubowska-Fijałkowska 2020h, p. 23)	I'll shoot myself in the crotch then you'll fuck away from me

With a strong lyrical subject, she discusses the body without taboos in all possible dimensions, and, above all, in its physiology and dysfunction, sexuality and pleasure in the ageing process. This voice of concern is, however, one of the rare poetic propositions that build a universe, a specific world where religion appears on different levels and "I" tries to cope with her entanglement, with its "now" evaluated, a distanced and ironic outlook, and the input of poems as well as with the emotional charge of histories, revealed in a language, symbols, designations, and names. Both coexist in a startling way.

b. The poetry of Justyna Bargielska (born in 1977) represents the second type of the poetical relation of a female poet with religion. Having received important awards—Rainer Maria Rilke Poetry Competition (2001), the Jacek Bierezin Special Award (2002), and Gdynia Literary Award (2010, 2011)—for her first volume, *Dating sessions* (Bargielska 2003), she explores the world from her own, subjective feminine perspective; at the same time, she breaks the conventional ways of writing. This portrays the freshness in her compositions, the smashing of existing hierarchies, her linguistic and conceptual collages with a kind of crazy selfness, and a simultaneous evident compositional and intellectual rigor. In a dialogue with religion—clearly visible in the last volumes in funny, sometimes falsely childish ways, and sometimes spoiled and ugly speaking—she examines the most important taboos concerning a woman's body to finally question the essential limits of life and death. In one poem, *Ona liczy na seks* (She Counts on Sex), the presence of Christ and his resurrection are parallel to a love-rebirth of "I", waiting for her lover. In this poem, in the very centre, we find a declaration of a belief, hidden in a word game—PIS (a main right-wing governmental party) off—piss off:

Kościół obstaje przy grzebaniu ciał,	The Church persists that corpses be buried
tłumacząc, że sam Chrystus chciał być	explaining that Christ himself wanted to be
pogrzebany. Pewnie, przecież miałem	buried. Sure, I was going to
zmartwychwstać, a nie odrodzić się	be resurrected, and not be reborn
z popiołów jak feniks, mówi Chrystus,	from ashes like a phoenix, says Christ,
ale to nie znaczy, że inni nie mogą	but it doesn't mean that others can't
się odradzać z popiołów jak feniks,	be reborn from ashes like a phoenix,
mówi Chrystus i jest naprawdę	says Christ and He's really
pissed off na ten głupi Kościół.	pissed off at this stupid Church.
Jak feniks. Za cztery godziny	Like a phoenix. In four hours
cię zobaczę, cokolwiek się	I'll see you, whatever
stanie, zobaczę cię	happens, I'll see you
za trzy godziny.	in three hours.
(Bargielska 2012a, p. 18)	(Bargielska 2012b)

Here, belief is much more important than the ultra-Catholic parties and bigger than the institution of the Catholic Church. The separated words "says Christ" and "He's really like a phoenix" are the most meaningful; but the reader has to join them to make sense and throw the "obstructing" words out: "pissed off at the stupid Church". In many of Bargielska's other poems, we see the presence of a childish idiolect. As Marta Koronkiewicz writes:

> The woman-child becomes at the same time *enfant terrible*, an expositor of games and conventions, of limitations and entanglements, whose method is to "overlook" certain boundaries. The figure of a child turns out to be a figure of shamelessness, although at the same time the term "shamelessness" is tempting. Bargielska's heroines are shameless, slutty maids from good homes. From the beginning of her work, the author has been exploring this specific register, such as reproof, the whole idiolect connected with the process of bringing up girls.
>
> (Koronkiewicz 2014)

Bargielska, as discussed in the critical paper, is problematic for critics because, inter alia, she is able to answer in an interview: "Why shouldn't I believe in God? I believe." (Nawarecki 2013, pp. 403–16), and in her poetry: "Why shouldn't I have a romance? I have." (Bargielska 2012c, p. 24). She is able, in the same disingenuous way, to build parallels between women, daughters, and mothers, as well as between herself and Jesus:

> Tak szczerze, to myślę, że nie wiesz, czym jest tęsknota.
>
> Czy twoja córka powiedziała ci kiedyś
>
> pięćdziesiąt trzy razy pod rząd, że pies ci je kapelusz?
>
> Niech je, powiedziałam. Nie wiem, czy choć raz. Twoje maile
>
> to nie maile, to pieszczoty, dzisiaj pójdę spać z tobą,
>
> mówi mi mail. Nie, mailu, dzisiaj pójdziesz spać z żoną,
>
> a ja pójdę spać z mężem. Niemniej nie dalej niż jutro
>
> planuję pozbyć się ze świata wszystkich naczyń,

do picia, do sikania, przechowywania prochów bliskich,
zbierania krwi naszego Zbawiciela, i będę ostatnim
naczyniem na świecie. I, umówmy się,
ja i krew Zbawiciela, tylko my dwie wiemy,
czym jest tęsknota.

(Bargielska 2012d, p. 59)

Has your daughter ever told you
fifty three times in a row that the dog's eating your hat?
Let it, I said. I don't know if it was even once. Your emails
aren't emails, they're caresses. Tonight I'll go to sleep with you,
the email says to me. No, email, tonight you'll go to sleep with your wife,
and I'll go to sleep with my husband. Nevertheless no later than tomorrow
I plan to rid the world of all vessels,
those for drinking, peeing in, storing loved ones' ashes,
for collecting our Saviour's blood, and I will be the last
vessel in the world. And, let's be clear,
our Saviour's blood and I, only we two know
what yearning is.

(Bargielska 2012e)

In *Jak to widzi sowa* (How the Owl Sees This), the poet's begging prepares the lyric "I" for maternity. She positions herself in a modern life, with the wise, considerate, and conscious (owl); but knowledge does not protect her. She will choose to believe, even if it cannot free her of hesitations and dilemmas, as shown in a fragment of the poem:

Sześć tygodni płaczu, że takie szczęście jest możliwe	Six weeks of crying that such happiness is possible
sześćdziesiąt kolejnych tygodni, że za nie dziękuję	sixty consecutive weeks that I thank for it
sześćset, że przepraszam, że w nie wierzyłam,	six hundred that I am sorry I did not believe in them,
sześć tysięcy, czy można je już zabrać ode mnie.	six thousand, can one take them from me away.
Więc wierzę, ale proszę, nie przychodźcie do mnie,	So I believe, but please do not come to me,
tym bardziej nie przysyłajcie jedni drugich nawzajem.	all the more reason not to send each other.
(Bargielska 2014a, p. 25)	

The palpable isolation seems to be doubled in the world of this poetry and concerns the life of a human being as well as a life according to the Church's code. The strong illustration of this solitude can be seen in *Cenne jest Jagniątko* (Baby Lamb is Precious):

Czy przygody Chrystusa nie były wymyślone?
I nagle idę łąką, a słyszę tunel. Ksiądz mówi: ciało Chrystusa,
a ja słyszę: proszę tę kartkę zabrać do domu
i ćwiczyć, ćwiczyć, ćwiczyć głoskę za głoską. To ćwiczę,
głoskę za głoską. Kto tu nie był, ten nie wie:
ani to limbo, ani plaża. Kto tu nie był, ten nie zrozumie.

Ale postaram się przybliżyć wam to miejsce,
to miejsce ma wiele ścian, a ty w nim siedzisz i myślisz:
Po co zabijać noworodki, przecież kupiłabym,
choćby do dalszej sprzedaży albo na narządy.

(Bargielska 2014b, p. 26)

Were the adventures of Christ not invented?

> And suddenly I am walking through a meadow, and I hear a tunnel. The priest
> says: the
>
> > [body of Christ,
>
> and I hear: please take this piece of paper home
> and practise, practise, practise vowel by vowel. So, I practise,
> vowel by vowel. Whoever has not been here does not know:
> this is neither a limbo nor a beach. Whoever has not been here will not under-
> stand.
>
> But I will try to bring this place closer to you,
> this place has many walls, and you sit in it and think:
> *Why kill newborn babies, I would buy them,*
> *if only to next sell or for organs.*

Since Rafał Wojaczek (1945–1971), an engaged and contested poet who proposed an innovating and individual tone, there has been no such voice in poetry. Since then, no poems have come to such a discussion with the Church. The position of a woman's voice, here, is different and new, based on strong antagonisms in a lyrical self-expression: very intimate and a kind of exhibitionism, with precision of announcement and a hidden sensum. I quote, one more time, her poem *Z głębi kontinuum* (From the deep continuum), from the volume *Bach for my baby*:

<table>
<tr>
<td>

Kotku, nie chcę Tego ciągnąć. Kotku,
może nie brnijmy w To dalej.
Kotku, o ile mi wiadomo, Nasze Szczęście
jest niemożliwe ze względów fizycznych
i mam na myśli tę naprawdę Dużą Fizykę,
tę, która mówi, że możesz być
w dwóch miejscach naraz, tylko jeśli jesteś
Bogiem lub Martwy. Będąc Martwym Bogiem,
o, wtedy to masz, możesz być Wszędzie.
A wiesz, kto miałby teraz w głowie
Martwego Boga, gdybym ja nim była?
Ale nie o tym chciałam. (...)
(Bargielska 2012f, p. 32)

</td>
<td>

Baby, I do not want to drag this out. Baby,
maybe we shouldn't go on.
Baby, as far as I know, Our Happiness
is physically impossible
and I mean the really Big Physics,
the one that says you can only be
in two places at once, only if you're
God or Dead. Being a Dead God,
oh, then you've got it, you can be Anywhere.
And you know who would have a
Dead God right now if I were him?
But that's not what I wanted to talk about it (...).

</td>
</tr>
</table>

The strength of Bargielska's poetry lies, as in the quoted excerpt, in her ability to deeply reflect upon the conflict between humans and God. Unlike the felt corporeality in Jakubowska-Fijałkowska's model of poetry, in Bargielska's poems, there is no possibility of self-identification with him as understood as the corporeal man, nor is he understood as a spiritual god. There can only be a body or spirit, closer or approximative rapprochement. There is no spiritual communion, and the choice is always dramatic.

c. In the poetry of Anna Augustyniak (born in 1976), we observe an interesting, new proposition and combinations of different unusual backgrounds. Here lies the questions of life and death, of redemption and religion, mixed with the problems of animal rights and the tendency of an ecological and vegetarian viewpoint. In the whole volume entitled *Między nami zwierzętami* (Among us animals) (Augustyniak 2020a), the poet deals with us: the barbaric, illogical human and our shaky opinions on the world of civilisation and the world of nature. The alter ego, named in the poems as Anna Q, asks eschatological questions in the logic of the Decalogue and in a reflection on physical (action–reaction, full–empty) words, as in the poem *o Annie Q i pytaniach* (about Anna Q and questions): "Czy pytanie o zasadność zabijania jednych dla drugich,/Nie jest najpilniejszą kwestią współczesnego świata? (Is the question of the legitimacy of killing some for others,/the most pressing issue of the modern world)" (Augustyniak 2020b, p. 42) or in the poem *o trawieniu* (about digestion) in more brutal and persuasive way:

Spożywani też czekają na zbawienie?
W teologii nadziei może chodzić o przyszłość trawionych.

Are the consumed also waiting for salvation?
The theology of hope may be about the future of

Kto liczy, ile trupich szczątków do ust włożył,

Who counts how many corpses remains he has put
[into his mouth,

Ile z odbytu jego wyszło
I spłynęło rurami kanalizacyjnymi
Gdzieś w głębiny szamb?
Ile przetrawionych istot czeka na zbawienie?
(Augustyniak 2020c, p. 11)

How many—came of his anus out
And flowed down the sewer pipes
Somewhere in the depths of septic tanks?
How many digested beings are waiting for salvation?

The poems of Augustyniak, in a new, empathic, but also documented, manner, investigate the established dogmas, important texts, and ways of their interpretation. The author annotates additional poems with footnotes, in which she clarifies the specific religious and ecclesiastical points of understanding for the possibly unfamiliar reader. The matter-of-fact and seemingly emotionless tone and the explanation of things apparently familiar to Christians, especially to Polish Catholics, result in the sharp and painful irony, as in the poem *o nieboszczyku* (about a dead man):

Ponoć wypieranie zwłok ze świadomości
To poważny problem społeczny.
Spójrzmy więc, jak anatom
Zwykł się przyglądać doczesnym szczątkom,
Na najsłynniejszy w dziejach organizm,
Na mięso i krew Chrystusa

Suppressing carcasses from consciousness
is a serious social problem.
So let's look at how an anatomist
Used to look at mortal remains,
On the most famous organism in history,
The flesh and blood of Christ

Jego mięso i krew zeszły się z nim najściślej,

His flesh and blood came together most intimately,

Bliżej nie mógłby być nawet z Bogiem Ojcem.

Closer he could not be even with God the Father.

W widowisku na krzyżu zmagał się
Z namacalnością w męczeństwie

In the spectacle on the cross He struggled
With palpability in martyrdom

I przetworzył agonię w najchodliwszy towar [4],

And transformed agony into the choicest commodity [4],

Lądujący teraz i na wieki na językach ludzi.

Landing now and for ever on the tongues of men.

[4] Bierzcie i jedzcie, to jest ciało moje—mówi Chrystus i mówi kapłan podczas liturgii mszy św., a wtedy hostia, którą ściska w palcach przeistacza się w ciało Chrystusa i można je konsumować. (Augustyniak 2020d, p. 8)

[4] Take and eat it, this is my body, says Christ, and the priest says it during the liturgy of the Mass, and then the host, which he squeezes in his fingers, is transformed into the body of Christ and can be consumed.

The volume's weight is heavy, especially when the reader follows, poem after poem, a kind of special "list of killing"; a "list of crimes" appears that obliges us to consider our human (or animal) motivation for acts. It can be read as a sort of answer, for example, to the volume by Rafał Gowin *Jem mięso* (I eat meat) (Gowin 2019) or to the world's statistics of excessive consumption of food in some parts of globe and poverty in others. Yet, this is also a sharp discussion within the framework of respecting fundamental social rules (Franciscan's brotherhood of being) and individual and institutional responsibility for words and acts:

O śmierci zamienionej w pokarm

Serce przestaje bić.
Mózg może jeszcze o czymś myśli

Za późno.

Ciało już jest mięsem
I przemienia się w potrawy
W ciemne ludzkie sprawy,
Z kręgu zabijania i jedzenia,
Wnika z każdym kęsem,
W każdą tkankę,
W węzły życia,
W więzy krwi,
Krew się z krwią zaczyna mieszać,
W żywe cmentarzysko konwertuje,
Przepoczwarza w monstrum.

Pijcie, bierzcie, jedzcie.
Na talerzu sen wieczysty
Krowy, wołu, owcy, kozy.
Grzech nie dotyczy zwierząt:
Św. Tomasz[5] to przesądził,
Dusze nierozumne,
Za to jakie smaczne mają ciała.

[5] Św. Tomasz z Akwinu przysposobił teorie Arystotelesa, odmawiając zwierzętom posiadania duszy i w katolickiej teologii do dzisiaj funkcjonuje taki model myślenia, a przecież Bóg, lepiąc z prochu ziemskiego ciała zwierząt, tak samo jak człowiekowi tchnął im w nozdrza dech życia. "Jednako tchną wszystkie, i nie ma człowiek nic więcej nad bydlę".

(Augustyniak 2020e, p. 9)

On death turned into food

The heart stops beating.
The brain maybe still thinking about something
Too late.

The body is already meat
And turns into food
Into dark human affairs,
From the circle of killing and eating,
It penetrates with every bite,
Into every tissue,
Into the knots of life,
Into the ties of blood,
Blood begins to mingle with blood,
It converts into a living graveyard,
It transforms into a monstrosity.

Drink, take, eat.

On a plate of eternal sleep

A cow, an ox, a sheep, a goat.
Sin does not concern animals:
St. Thomas has determined this,
Souls without understanding,
But what tasty bodies they have.

[5] St. Thomas Aquinas adopted Aristotle's theories, denying that animals have souls, and in Catholic theology, this model of thinking continues to operate to this day; yet God, in fashioning the bodies of animals from the dust of the earth, breathed into their nostrils the breath of life just as he did into human's. "They are all the same, and man has nothing more than a beast".

In this context, the poems of Augustyniak, with referees, deductions, and data, are like a new kind of manifesto, opposite to those of the futurists, and are joyful or a kind of new poetical treaty. Here, the hidden emotions of the poetical "I" are put into the shape of "reasonable". The personage, Anna Q, is a woman, but there is no distinguished gender role in the process of the devouring. When Augustyniak writes about the "holocaust of animals" in a shocking way (Augustyniak 2020c), the guilty laid on everyone, both men and women.

My next and the final example of recent poetry is yet another form of poetic proposal. Here, emotions are primordial and become a common shout.

d. **This poetry was written as protest songs, and the poems written by Malina Prześluga** (born in 1983) create a total composition with a performance (music, costumes, make-up, and movement). The model of the witch replaces the model of the Madonna here. For the first time since the solidarity strikes in the 1980s, we have a strong common subject in the text; the plural lyrical "I", called Chór Czarownic—The Witches' Choir. It was formed in Poznań, in one of its current districts (previously a neighbourhood) called Chwaliszewo. There, 500 years ago, for the first time in Poland, a woman was accused of being a witch and she was burned. Starting from then, women accused of witchcraft were burned at the stake all over Poland (but not as often in comparison with the other parts of Europe). However, the first performances by Chór Kobiet (Women's Choir, later renamed as The Witches' Choir) gained huge popularity. Women dressed in satin petticoats, barefoot, and with red lipstick, singing outdoors, playing the drums:

Twoja władza, twoja wiara	You are faith and government
Moja wina, moja kara	Mine are guilt and punishment
W twoich rękach jest mój świat	All my world is in your hands
Masz mnie w garści milion lat!	In your grasp for many years!
Popatrz na mnie w oczy prosto	Look at me now, in my eyes
Jestem twoją matką, siostrą	I'm your mother, I'm your wife,
Jestem twoją córką, żoną.	I'm your daughter, I'm your child.
Stoję z głową podniesioną.	I'm not coming here alone
Milion nas tak teraz stoi	Millions of us standing tall
Żadna z nas się już nie boi.	Not afraid of you no more
Milion nas tak teraz stoi	Millions of us standing tall
Żadna z nas się już nie boi.	Not afraid of you no more.

Stoję, krzyczę, stoję, krzyczę, Stand proud, stand shout
Stoję, krzyczę, stoję, krzyczę, Stand proud, stand shout

Milion nas tak teraz stoi Millions of us standing tall
Żadna z nas się już nie boi. Not afraid of you no more.
Potężniejsza niż myślałeś Stronger than I ever was
Oddaj wszystko, co zabrałeś Give back what you took from us!

Twoja wina, twoja wina Through your faults, through your faults
Twoja bardzo wielka wina! Through your faults, through your faults
Twoja wina, twoja wina, Through your faults, through your faults
Twoja wina, twoja wina! Through your faults, through your faults!
  (Prześluga [2016] 2017, performed by
  Chór Czarownic[6])

The "I", das Ewig-Weibliche (von Goethe 1982, p. 545), plays with cultural archetypes and stereotypes. For the first time in official streaming, one can hear a violent message with uses of vulgarisms. Why? The evoked primacy and the primitivism of the corporal attitude are here to underline the main problem—the desacralisation of woman—in a religious but also in a cultural meaning. It is not an accident; a similar message can be found in *Kąciki ust* (The Corners of the Mouth), where, in the refrain, we discover such a transformed woman talking in the singular first person, an affirmative "I":

Jestem pies obronny I'm a guard dog
Suka bez łańcucha a bitch unchained
Gryzę, warczę, wyję I bite, I growl
Na pieszczoty głucha. I howl, I'm deaf to your pet names.
  (Prześluga and Chór Czarownic 2021)

We are far from the secular and still-repeated model of the Polish mother, the Madonna, the sublime Muse, or the original and singular Sappho, and even further from the image of Mary Magdalene (rarely) appearing in Polish female poetry. Here, in the performances of the Witches' Choir, there is no place for penitence. I quote one more fragment from a poem in the repertoire of the choir, where not only a shout but also some amazing statistics can be found, as well as a reaction to the patriotico-religious "mixture" of recent years. The title, a fragment of the Catholic liturgy using the name of Poland, is significant *Polska—oto słowo pańskie* (Poland—this is the Word of the Lord) (Prześluga and Chór Czarownic 2012a):

Jesteśmy krajem tak bardzo katolickim We are such a Catholic country
Że nawet ateiści są tu w większości katolikami That even atheists here are mostly Catholics
Co prawda niewierzącymi, ale praktykującymi. Admittedly non-believers but practising.
Z badań wynika, że 70% bierze śluby kościelne mmm Surveys show that 70% get married in a church mmm
74% chrzci swoje dzieci mmmm 74% baptise their children mmmm
28% katolików wierzy w reinkarnację mmmm 28% of Catholics believe in reincarnation mmmm
(. . .) (. . .)
W kościele kobiety mogą robić rzeczy In church, women can do things
Podobne do ich obowiązków. Similar to their duties.
Tak jak opiekują się dziećmi Like they take care of children
Mogą się opiekować chorymi. They can take care of the sick.
Tak jak dbają o porządek Just as they take care of order
Powinny sprzątać kościół. They should clean the church.
Bo to taki zmysł trzeba mieć, For it is such a sense to have,
Żeby obrus był czysty To keep the tablecloth clean
Kobieta ma naturalne predyspozycje do tego. A woman has a natural aptitude for this.
  (Prześluga and Chór Czarownic 2012b)

This text, which is a fragment of the *Magnificat* performance, was presented during the Fourth Congress of Women in Poland in 2012. Through the language of messages and statistics mixed with fragments and collage statements of clergy and politicians, the liturgy mercilessly exposes the ongoing patterns in Polish space. Frustration is caused by the lack of new reforms and the lack of implementation of those already approved, for instance, about the possible and necessary laicisation of clothing for people in spiritual roles in Catholic clergy, as anounced by Second Vatican Council 1962–1965 in "Perfectae caritatis" document from 28 October 1965 (Pope Paul VI 1965). However, the poetics of rebellion and shouting cannot continue for long: the shock caused by the violation of taboos and the juicy novelty of expression—as the history of poetics shows—finally become banal. In response to the femininity and nudity, new Polish neoconservaitve trends were born—not in a protest song, but through business and affluent influences, as shown by the recent brand of Krzysztof Ziętarski. He is the founder and propagator of a new (and very lucrative) mode of dress for Catholic women (no trousers, only skirts and dresses)[7]. His "Marie Zélie" proposes polite clothes that cover most parts of the body. As he says:

> What I consider to be Catholic in fashion appears in the notes of St. Marie-Azélie Martin, who, while taking care of her daughter' dress, was also mindful of the danger of falling into vanity. For although clothing should follow trends, it should be treated with the attention it deserves, in the context of our other earthly concerns. In Catholic terms, the first place belongs to God.
>
> (Ziętarski 2020)

However, exposing the body in poetry, sometimes in a very exhibitionist way, aligns with the mechanism of taking the words off, finally using them as a form of sometimes brutal detachment and to uncover the bones—the essential part. This poetic treatment certainly links to the condition of our "useful and used body" (Agamben 2016) and the transformation we are living. One wonders then, if finally, this is not a revenge of the immaterial, stimulated by the concrete, now visible by the female poets who threw out in their creations not only their clothes but even the covering words.

7. Conclusions

In the considerations of poems by Polish women and the issue of religion, I have focused on Christianity, especially on the Catholic Church, as it is the most representative in Polish social and cultural space. There were no research results for Judaism, Protestantism, Orthodoxy, or Islam, for example, which were present in Poland of the Two Nations (16th–18th centuries), but are a small minority in today's Republic of Poland. Separate studies should be devoted to these issues. Additionally, it is the Catholic Church, with its impact on society, that seems, nowadays, to be the most important partner in the dialogue on religion in Polish society and literature. The poems by female poets quoted in this article are, of course, only several (deliberately) selected examples. Whereas they were relatively non-existent or small in number in earlier centuries, there are many more present from the end of the 19th century. However, in this paper, I have highlighted texts and authors that seem to me recognisable, as well as being recognised and noticed by readers and critics, but not often shown and compared together in such a study. These poems reflect historical and cultural social determinants and conditions, but also the influence, as we have seen, of important events in the heart of the country.

To draw some precise points, my conclusions from this poetic and fascinating journey are:

1. The accessible heritage of the canonical texts of Christianity, fundamental for European and world humanities as well as for Polish literary tradition, encourages neither a present nor a sympathetic view of the role of women in the sphere of religion and Church rituals. The educated strata of women, princesses and nuns, in the space of *civitas christiana*, which is broader than the state will, for centuries, work for change and the participation of women involved in and willing to speak out about issues of religion.

2. Poetry records social changes often invisible to the untrained eye, and sometimes anticipates them. The first text written in Polish is a poem. It is a hymn to the Virgin Mary, the anonymous poem *Bogurodzica*, written by a man or a woman, which conveys

to us the image of a woman—a model of the virtues of the Madonna, the Mother of Jesus. The Romantic period perpetuates the religious parallels between the image of Mary, the Mother of Christ, and the suffering Pietà, the Polish mother, whose image was transmitted to us by Mickiewicz in a poem under such a title. This concept-image is referred to by successive generations of women and poets, who either copy the established models or rebel against it. The rebellion is based on the change in roles; the lyrical "I" in a poem is no longer Mary, but a Christ-Saviour in the 21st century, for example, in the poetry of Jakubowska-Fijałkowska. The protest may also, in another dimension of a rebellious gesture, as in the poetic songs of Malina Prześluga and the Witches' Choir, strive for the annihilation of roles fixed in society and at the same time recall the existing models.

3. Until the beginning of the 20th century, Polish female poets, wishing to appear on the literary scene of the epoch (in which they lived and shared their experiences and religious knowledge) and to become part of the poetic mainstream, used well-known (by men) matrixes and models, such as the elegy after the loss of a dear person. The image of a suffering mother, modelled on the Madonna, and the poetic form of such an elegy, are so fused together that, again, to form a description and bring it closer to the reader, 20th-century poets reached for the lyricism of the role and suffering of the Polish mother, for example, in Baczyński's poem.

4. Poets not interested in reproducing this image and the established roles reach for the second type of lyrical expression, replacing the "mother" with the lover and the priestess of love. Distant in time and poorly present, the Sappho model is newly evoked (Maria Pawlikowska-Jasnorzewska). In the 20th century, the "religion" of love in women's work distances them from the problems of the poetry engaged in social and religious disputes, constituting a return to pagan rituals (*Hymn idolatrous* of Halina Poświatowska) or to carnality free from maturity, not necessarily in traditional, "polite" aesthetic (Świrszczyńska).

5. It is only since the 21st century that Polish poets have been speaking out on issues of state rights connected with the aspirations of the Catholic Church in Poland. Living in a Catholic society, female poets observe the incompatibility of rituals, catechesis, civil rights, and social behaviours (Jakubowska-Fijałkowska); they cultivate new poetic forms, mixing two separate models of faith and commentary on religion with carnality and sensuality, which does not always fit with the Decalogue or the Church's messages (Bargielska).

6. Religiously solidified poetry is the domain of few single female author-poets, and it appears at times when religion and faith become a form of solace and a choice of inner freedom, harmonising with a chosen life stance, as in the poetry of Kazimiera Iłłakowiczówna or Anna Kamieńska, or when it also becomes a form of therapy in a bodily illness, e.g., in Pollakówna. This, however, does not constitute an otherness or specificity of the "feminine" in relation to male models. In the 21st century, Polish female poets are able to declare themselves in their poems as engaged feminists and Catholic believers at the same time. As shown by Zuzanna Radzik, in her works quoted above, feminist and Catholic works can be fused to be a model for Polish social space without so deep contradictory frustrations for contemporary female poets. Even if – like in the case of Witches' Choir – the opposite, strong and rebel role model towards Catholic one is today also proposed.

Cesare Ripa, in his *Iconology*, represents "religion" as a woman with her face covered by a thin veil, holding a book and a cross in her right hand, and in her left hand, a flame. The woman is accompanied by an elephant (being a symbol of religion), more pious than the other animals (Ripa 2013, pp. 356–57). It seems to me a beautiful illustration to contemplate within the context of our considerations. Over the centuries, in Polish poetry, which is recognised internationally by prestigious juries of all kinds of awards and translations into many languages, there has not been a single female poet who has been truly spiritual and metaphysical, engaged in a real debate with God, never mind the political context. There is no single Polish Emily Dickinson. So, perhaps in this old iconological image evoked here lies the answer to the question: why?

Polish female poets, over the centuries, have tried to manage with the cross and the flame. They have certainly attempted to unveil the mystery of religion as well as uncovering-discovering their own face and body in their poems, through flames and books, and experiences across the long expanse of age—from a girl to an old woman—as if they become for themselves a new religion.

Funding: This research received no external funding.

Conflicts of Interest: The author declares no conflict of interest.

Notes

[1] I will base my discourse on the definition of religion and some of its components on various sources: definitions from *Encyklopedia Języka Polskiego* (2021), *Dictionnaire des religions* (Eliade and Couliano 1990) *Dictionnaire des faits religieux* (Levallois and Iogna-Prat 2019, pp. 1103–5), *Literatura a religia. Wyzwania epoki świeckiej* (Bielak and Tischner 2020), and *The Brill Dictionary of Religion* (von Stuckrad 2007).

[2] The year 966, the date of the beginning of Christianisation (baptism of Mesco I), is considered to be the beginning of Poland.

[3] All quotes and titles are translated by the author of this paper unless otherwise indicated.

[4] I mean here the most famous literary monologues in Polish Romantic poetry, which are in fact a verbal duel of the young men with God. One was pronounced by Konrad in *Dziady* by Adam Mickiewicz (Mickiewicz 1832, pp. 62–77), the other by Kordian, in *Kordian* by Juliusz Słowacki, (Słowacki 1834, pp. 70–74).

[5] Censorship in post-war Poland was officially in place from 1946 to 1990—firstly, through the Główny Urząd Kontroli Prasy, Publikacji i Widowisk (Main Office of Control of Press, of Publications and Performances); and after 1981, through the changed name Główny Urząd Kontroli Prasy, Publikacji i Widowisk (Main Office of Control of Publications and Performances).

[6] The translated text of the Prześluga and Witches' Choir in English is by Thimothy Williams; in the Project of Ewa Łowżył—see references below.

[7] This is not something exceptional in religious behaviours. Similar modes are found in contemporary Islam or Judaism.

References

Agamben, Giorgio. 2016. L'Usage des corps. In *Idem, Homo Sacer, L'integrale 1997–2015*. Translated by Marilène Raiola. Paris: Seuil.

Anonymous. 1984. O [Barbarze Giżance]. In *Patrząc na rozmaite świata tego sprawy. Antologia polskiej poezji renesansowej*. Edited by Jadwiga Sokołowska. Warszawa: Państwowy Instytut Wydawniczy, First written in 16th century.

Anonymous. 1998. Glińska, Anna. Przyczynek do dziejów poezji karmelitańskiej. Charakterystyka i wybrane wiersze z rękopisu o sygn. 3643 I ze zbiorów Biblioteki Jagiellońskiej. *Barok. Historia-Literatura-Sztuka* 10: 184–94, First written in 17th century.

Anonymous. 2004. [Zbiór kolęd i pastorałek], manuscript Archiwum Sióstr Karmelitanek Bosych w Krakowie na Wesołej, sygn. A 69, (mf.ABMK Lublin 2658), quotted after: Partyka, Joanna. In *"Żona wyćwiczona": kobieta pisząca w kulturze XVI i XVII wieku*. Warszawa: Instytut Badań Literackich PAN, First written in 17th century.

Aristotle. 1984. *The Politics*. Translated, Introduction, Notes and Glossary by Carnes Lord. Chicago: University of Chicago Press, pp. 42–80, First written in c.a. 335–323 BC.

Augustyniak, Anna. 2020a. *Między nami zwierzętami*. Gdańsk: Słowo/obraz terytoria.

Augustyniak, Anna. 2020b. o Annie Q i pytaniach. In *Eadem, Między nami zwierzętami*. Gdańsk: Słowo/obraz terytoria.

Augustyniak, Anna. 2020c. o trawieniu. In *Eadem, Między nami zwierzętami*. Gdańsk: Słowo/obraz terytoria.

Augustyniak, Anna. 2020d. o nieboszczyku. In *Eadem, Między nami zwierzętami*. Gdańsk: Słowo/obraz terytoria.

Augustyniak, Anna. 2020e. o śmierci zamienionej w pokarm. In *Eadem, Między nami zwierzętami*. Gdańsk: Słowo/obraz terytoria.

Baczyński, Krzysztof Kamil. 1994. Elegia o chłopcu polskim. In *Idem, Utwory zebrane*. v. 2. Edited by Aniela Kmita-Piorunowa and Kazimierz Wyka. Kraków: Wydawnictwo Literackie. First published in 1942.

Balcerzan, Edward. 1970. "Sytuacja liryczna"—Propozycja dla poetyki historycznej. In *Studia z teorii i historii poezji, Red*. Głowiński, Michał. seria II. Ossolineum: Wrocław-Warszawa-Kraków.

Bargielska, Justyna. 2003. *Dating sessions*. Kraków: Wydawnictwo Zielona Sowa.

Bargielska, Justyna. 2012a. Ona liczy na seks. In *Eadem, Bach for my baby*. Wrocław: Biuro Literackie.

Bargielska, Justyna. 2012b. *She Counts on Sex (Ona liczy na seks)*. Translated by Katarzyna Szuster. Available online: https://www.versopolis-poetry.com/poet/53/justyna-bargielska (accessed on 21 February 2021).

Bargielska, Justyna. 2012c. Suknia barwy pogody. In *Eadem, Bach for my baby*. Wrocław: Biuro Literackie.

Bargielska, Justyna. 2012d. Pies ci je kapelusz. In *Eadem, Bach for my baby*. Wrocław: Biuro Literackie.

Bargielska, Justyna. 2012e. *The Dog's Eating Your Hat (Pies ci je kapelusz)*. Translated by Maria Jastrzębska. Available online: https://www.versopolis-poetry.com/poet/53/justyna-bargielska (accessed on 21 February 2021).

Bargielska, Justyna. 2012f. Z głębi kontinuum. In *Eadem, Bach for my baby*. Wrocław: Biuro Literackie.

Bargielska, Justyna. 2014a. Jak to widzi sowa. In *Eadem, Nudelman*. Wrocław: Biuro Literackie.

Bargielska, Justyna. 2014b. Cenne jest Jagniątko. In *Eadem, Nudelman*. Wrocław: Biuro Literackie.

Beauvois, Daniel. 2004. *L'effacement de la carte européenne (1795–1918) in Idem*. La Pologne: Histoire, société, culture, Paris: La Martinière.

Bielak, Agnieszka. 2020. Alternatywy religii w świecie odczarowanym. Twórczość Wisławy Szymborskiej. In *Literatura a religia. Wyzwania epoki świeckiej. Literatura polska po 1945 roku—Kierunki, idiomy, paradygmaty*. Edited by Agnieszka Bielak and Łukasz Tischner. Kraków: Wydawnictwo Uniwersytetu Jagiellońskiego, vol. II, pp. 571–604.

Bielak, Agnieszka, and Łukasz Tischner, eds. 2020. *Literatura a religia. Wyzwania epoki świeckiej*. Kraków: Wydawnictwo Uniwersytetu Jagiellońskiego.

Bieńkowski, Zbigniew. 1993. *Ćwierć wieku intymności. Szkice o poezji i niepoezji*. Warszawa: Agawa.

Bogalecki, Piotr. 2020. "Boga zastąpiłam światem". Wiara i świeckość w poezji Krystyny Miłobędzkiej. In *Literatura a religia. Wyzwania epoki świeckiej. Literatura polska po 1945 Roku—kierunki, idiomy, paradygmaty*. Edited by Agnieszka Bielak and Łukasz Tischner. Kraków: Wydawnictwo Uniwersytetu Jagiellońskiego, vol. II, pp. 321–50.

Borkowska, Grażyna, Czermińska, Małgorzata, and Phillips Ursula. 2001. *Pisarki polskie od średniowiecza do współczesności. Przewodnik*. Gdańsk: Słowo/obraz terytoria.

Borowy, Wacław. 1978. *O poezji polskiej w XVIII wieku*. Warszawa: Państwowy Instytut Wydawniczy.

Cixous, Hélène. 1975. *Rire de la Méduse et autres ironies*. Paris: L'Arc.

Clark, Elisabeth A. 1998. The Lady Vanishes: Dilemmas of a Feminist. Historian after "Linguistic Turn". *Church History* 67: 1–31. Available online: www.jstor.org/stable/3170769 (accessed on 13 February 2021).

Drużbacka, Elżbieta. 1903. Opisanie czterech części roku. In *Stanisław Konarski, Elżbieta Drużbacka, Skarbiec Poezyj Polskiej. Antologia*. seria I. Edited by Antoni Lange. Warszawa: Wydawnictwo Alfreda Zonera, vol. 1. First published in 1752.

Duch-Dyngosz, Marta, Gawin Magdalena, and Radzik Zuazanna. 2014. Miejsce kobiety, "Znak" nr 714. Available online: http://www.miesiecznik.znak.com.pl/7142014z-magdalena-gawin-i-zuzanna-radzik-rozmawia-marta-duch-dyngoszmiejsce-kobiety/ (accessed on 2 February 2021).

Dumowska, Bogumiła. 1986. "Anegdota o istnieniu"—Wokół liryków Haliny Poświatowskiej. *Pamiętnik Literacki* 77: 139–57.

Eliade, Mircea, and Peter Couliano. 1990. *Dictionnaire des Religions*. Paris: Plon.

Encyklopedia Języka Polskiego. 2021. *Religia*. Warszawa: Wydawnictwo Naukowe PWN, Available online: https://encyklopedia.pwn.pl/haslo/religia;3966983.html (accessed on 10 January 2021).

Głowiński, Michał. 1960. W tonacji franciszkańskiej. *Twórczość* 7: 98–100.

Gowin, Rafał. 2019. *Jem mięso*. Warszawa: Convivo.

Graves, Robert. 1955. *Greek Myths*. London: Penguin Books.

Grądziel-Wójcik, Joanna. 2017. Poetyki i literatury momentalne. Między materią a metafizyką, czyli o chwycie paronomazji w poezji kobiet. *Poznańskie Studia Polonistyczne* 30. [CrossRef]

Grochowiak, Stanisław. 1959. Ciało. *Współczesność* 5: 8–9.

Grochowiak, Stanisław. 1967. Krzywda. *Kultura* 44: 3.

Hanusiewicz, Mirosława. 1998. *Święte i zmysłowe w poezji religijnej polskiego baroku*. Lublin: Wydawnictwo KUL.

Hernas, Czesław, and Mirosława Hanusiewicz, eds. 1995. *Religijność literatury polskiego baroku. Seria Tradycje Literatury Polskiej*. Lublin: Wydawnictwo KUL.

Hesiod VIII BC. 2021. *Works and Days*. Available online: https://chs.harvard.edu/primary-source/hesiod-works-and-days-sb/ (accessed on 19 January 2021).

Heydel, Magdalena. 2013. "Jeszcze żadna kobieta, jak poezja poezją…" Ku projektowi poetyckiemu Anny Świrszczyńskiej. In *Różne głosy: Prace Ofiarowane Stanisławowi Balbusowi na jubileusz siedemdziesięciolecia*. Edited by Dorota Wojda, Magdalena Heydel and Andrzej Hejmej. Kraków: Wydawnictwo Uniwersytetu Jagiellońskiego.

Iłłakowiczówna, Kazimiera. 1926. *Połów*. Warszawa: Wydawnictwo J. Mortkowicz.

Iłłakowiczówna, Kazimiera. 1927. *Płaczący ptak*. Warszawa: Wydawnictwo F. Hoesic.

Iłłakowiczówna, Kazimiera. 1928. *Z Głębi serca*. Warszawa: Wydawnictwo Gebether i Wolf.

Iłłakowiczówna, Kazimiera. 1930. *Popiół i perły*. Warszawa: Wydawnictwo F. Hoesic.

Iłłakowiczówna, Kazimiera. 1967. *Ta jedna nić. Wiersze religijne*. Poznań: Wydawnictwo Św. Wojciecha.

Jakubowska-Fijałkowska, Genowefa. 1994. *Dożywocie*. Kraków: Miniatura.

Jakubowska-Fijałkowska, Genowefa. 2016a. ten diabeł ciąge mąci. In *Eadem, Paraliż przysenny*. Mikołów: Instytut Mikołowski.

Jakubowska-Fijałkowska, Genowefa. 2016b. Bóg powrócił do miasta. In *Eadem, Paraliż przysenny*. Mikołów: Instytut Mikołowski.

Jakubowska-Fijałkowska, Genowefa. 2016c. Mężczyzna i Bóg. In *Eadem, Paraliż przysenny*. Mikołów: Instytut Mikołowski.

Jakubowska-Fijałkowska, Genowefa. 2016d. Ucho Van Gogha. In *Eadem, Paraliż przysenny*. Mikołów: Instytut Mikołowski.

Jakubowska-Fijałkowska, Genowefa. 2016e. Osiedlowy spleen. In *Eadem, Paraliż przysenny*. Mikołów: Instytut Mikołowski.

Jakubowska-Fijałkowska, Genowefa. 2016f. W sobotę koło dziewiątej rano. In *Eadem, Paraliż przysenny*. Mikołów: Instytut Mikołowski.

Jakubowska-Fijałkowska, Genowefa. 2016g. Ruletka z tobą. In *Eadem, Paraliż przysenny*. Mikołów: Instytut Mikołowski.

Jakubowska-Fijałkowska, Genowefa. 2016h. Niedziela. In *Eadem, Paraliż przysenny*. Mikołów: Instytut Mikołowski.

Jakubowska-Fijałkowska, Genowefa. 2016i. Rodzą się dziewczynki. In *Eadem, Paraliż przysenny*. Mikołów: Instytut Mikołowski.

Jakubowska-Fijałkowska, Genowefa. 2020a. *Rośliny mięsożerne*. Mikołów: Instytut Mikołowski.

Jakubowska-Fijałkowska, Genowefa. 2020b. Wielkanoc 2018. In *Eadem, Rośliny mięsożerne*. Mikołów: Instytut Mikołowski.
Jakubowska-Fijałkowska, Genowefa. 2020c. Święcona woda. In *Eadem, Rośliny mięsożerne*. Mikołów: Instytut Mikołowski.
Jakubowska-Fijałkowska, Genowefa. 2020d. Trzech Króli szóstego stycznia 2018. In *Eadem, Rośliny mięsożerne*. Mikołów: Instytut Mikołowski.
Jakubowska-Fijałkowska, Genowefa. 2020e. Boże Narodzenie 1986 (wspomnienie). In *Eadem, Rośliny mięsożerne*. Mikołów: Instytut Mikołowski.
Jakubowska-Fijałkowska, Genowefa. 2020f. Wielkanoc 2017. In *Eadem, Rośliny mięsożerne*. Mikołów: Instytut Mikołowski.
Jakubowska-Fijałkowska, Genowefa. 2020g. Jezus żonaty. In *Eadem, Rośliny mięsożerne*. Mikołów: Instytut Mikołowski.
Jakubowska-Fijałkowska, Genowefa. 2020h. Wszystko zawsze na matce wisi. In *Eadem, Rośliny mięsożerne*. Mikołów: Instytut Mikołowski.
Jamrożek, Wiesław, and Dorota Żołądź, eds. 1998. *Rola i miejsce kobiet w edukacji i kulturze polskiej*. Poznań: Instytut Historii Uniwersytetu Adama Mickiewicza, vol. 1.
Janion, Maria. 2006. *Kobiety i duch inności*. Warszawa: Sic!
Jay, Eldon Epp. 2005. *Junia. The First Woman Apostle*. Minneapolis: Fortress Press.
Judkowiak, Barbara. 1992. *Słowo Inscenizowane. O Franciszce Urszuli Radziwiłłowej—Poetce*. Poznań: Wydawnictwo WiS.
Kamieńska, Anna. 1981a. Przebudzenie. In *Eadem, Wiersze jednej nocy*. Warszawa: Ludowa Spółdzielnia Wydawnicza.
Kamieńska, Anna. 1981b. Stabat Mater. In *Eadem, Wiersze jednej nocy*. Warszawa: Ludowa Spółdzielnia Wydawnicza.
Karpiński, Adam. 2004. The Bogurodzica. In *Ten Centuries of Polish Literature*. Translated from Polish by Daniel Sax. Warszawa: Stowarzyszenie "Pro Cultura Litteraria"& Fundacja Akademia Humanistyczna & Instytut Badań Literackich PAN.
Kochanowski, Jan. 1986. *Treny, Red. Janusz Pelc. Ser. BN I*. Wrocław: Ossolineum. First published in 1580.
Koronkiewicz, Marta. 2014. Bargielska bez wymówek. *Śląskie Studia Polonistyczne* 5: 212.
Krzyżanowski, Julian. 1978. *History of Polish Literature*. Tansleted by Doris Ronowicz. Warsaw: PWB—Polish Scientific Publishers.
Levallois, Anne, and Dominique Iogna-Prat. 2019. Religion (approche: Historico-philologique). In *Dictionnaire des faits religieux*. Dir. Régine Azriz, Danièle Hervieu-Léger, Dominique Iogna-Prat. Paris: PUF.
Malinowski, Wiesław Mateusz, and Jerzy Styczyński. 2008. *La Pologne et les Polonais dans la littérature française*. Siècles XIV–XIX. Paris: L'Harmattan.
Maliszewski, Karol. 2020. *Bez zaszeregowania. O nowej poezji kobiet*. Kraków: Universitas.
Marzec, Lucyna. 2018. Kazimiera Iłłakowiczówna mniej więcej znana. *Poznańskie Studia Polonistyczne Seria Literacka* 32: 75–94. [CrossRef]
Michałowska, Teresa. 2000. Modlitewnik Gertrudy. In *Eadem, Średniowiecze*. Warszawa: Wydawnictwo Naukowe PWN.
Mickiewicz, Adam. 1832. Dziady part III. In *Idem, Poezye*. Paryż: A. Pinard.
Mickiewicz, Adam. 1993. Do matki Polki. In *Idem, Dzieła*. Wiersze, Red. Czesław Zgorzelski. Edited by Zbigniew Jerzy Nowak, Zofia Stefanowska and Czesław Zgorzelski. Warszawa: Spółdzielnia Wydawnicza "Czytelnik", vol. I. First published in 1830.
Milewska-Waźbińska, Barbara. 2018. The Literary Heritage of Jesuits of the Polish-Lithuanian Commonwealth. *Journal of Jesuit Studies Brill* 5: 421–40. [CrossRef]
Nawarecki, Aleksander. 2013. «Co mam nie wierzyć». Justyna Bargielska i polskie zawstydzenie wiarą. In *Więzi Wspólnoty. Literatura—Religia—Komparatystyka*. Edited by Piotr Bogalecki, Alina Mitek-Dziemba and Tadeusz Sławek. Katowice: Uniwersytet Śląski.
Osiecka, Agnieszka. 2009. Polska Madonna. In *Wiersze prawie wszystkie*. Warszawa: Wydawnictwo Prószyński i S-ka, vol. 1. First published in 1990.
Partyka, Joanna. 2004. *"Żona wyćwiczona": kobieta pisząca w kulturze XVI i XVII wieku*. Warszawa: Instytut Badań Literackich PAN.
Pawlikowska-Jasnorzewska, Maria. 1997. O biskupie fiołkowym. In *Eadem, Poezje zebrane, Red. Aleksander Madyda*, Preface: Krzysztof Ćwikliński. Toruń: Algo, vol. 1. First published in 1922.
Pawlikowska-Jasnorzewska, Maria. 1997. Róże dla Safony. In *Eadem, Poezje zebrane, Red. Aleksander Madyda*, Preface: Krzysztof Ćwikliński. Toruń: Algo, vol. 1. First published in 1937.
Pawlikowska-Jasnorzewska, Maria. 1978. *Roses for Sappho (Róże Dla Safony)*. History of Polish Literature. Translated by Doris Ronowicz. Warsaw: PWB—Polish Scientific Publishers. First published in 1937.
Perictyone 430 BC. 2021. (On Pythagoras). On the Duties of a Woman & On a Harmony of a Woman. In *The Complete Pythagoras*. Edited by Roussel Patrick. Available online: http://holybooks.lichtenbergpress.netdna-cdn.com/wp-content/uploads/The-Complete-Pythagoras.pdf?30f3d7 (accessed on 9 January 2021).
Pernoud, Régine. 1998. *Visages de femmes au Moyen Âge*. L'Abbaye Sainte-Marie de la Pierre-qui-Vire (Yonne): Éditions Zodiaque.
Pollakówna, Joanna. 2012a. Kanikuła. In *Eadem, Wiersze zebrane*. Edited by Jan Zieliński. Mikołów: Instytut Mikołowski. First published in 1979.
Pollakówna, Joanna. 2012b. Szumy. In *Eadem, Wiersze zebrane*. Edited by Jan Zieliński. Mikołów: Instytut Mikołowski. First published in 1979.
Pope Paul VI. 1965. 17. The Religious Habits. *Decree on the Adaptatio, and Renewal of Religious Life Perfectae Ca-ritatis*. Available online: https://www.vatican.va/archive/hist_councils/ii_vatican_council/documents/vat-ii_decree_19651028_perfectae-caritatis_en.html (accessed on 20 July 2021).
Poświatowska, Halina. 1958. *Hymn bałwochwalczy*. Kraków: Wydawnictwo Literackie.

Poświatowska, Halina. 2000. Inc. Boże zmiłuj się nade mną . . . . In *Eadem, Wszystkie wiersze*. Red. *Maria Smolarczyk*. Kraków: Wydawnictwo Literackie. First published in 1966.
Prześluga, Malina, and Chór Czarownic. 2012a. *Magnificat*. Project of Ewa Łowżył. Translated by Thimothy Williams. Available online: https://www.youtube.com/watch?v=lipdSIh1p_8&t=7s (accessed on 3 March 2021).
Prześluga, Malina, and Chór Czarownic. 2012b. *Oto słowo pańskie (Poland—This Is the Word of the Lord)*. Project of Ewa Łowżył. Translated by Thimothy Williams. Available online: https://www.youtube.com/watch?v=lipdSIh1p_8 (accessed on 19 January 2021).
Prześluga, Malina. 2017. *Twoja władza (Your Gouvernement)*. *Performed by Chór Czarownic*. Project of Ewa Łowżył. Translated by Thimothy Williams. First published 2016. Available online: https://www.youtube.com/watch?v=ls5u3uiawRA (accessed on 21 March 2021).
Prześluga, Malina, and Chór Czarownic. 2021. *Kąciki ust (The Corners of the Mouths)*. Project of Ewa Łowżył. Translated by Thimothy Williams. Available online: https://www.youtube.com/watch?v=sUOJFf315Dw (accessed on 26 February 2021).
Radzik, Zuzanna. 2018. *Emancypantki. O kobietach, które zbudowały Kościół*. Warszawa: Wydawnictwo WAM.
Radzik, Zuzanna. 2020. *Kościół kobiet*. Warszawa: Wydawnictwo WAM.
Ratajczak, Józef. 1998. Kazimiera Iłłakowiczówna—Homo religiosus. *Przegląd Powszechny* 2: 164–72.
Ripa, Cesare. 2013. Religia. In *Idem, Ikonologia*. Translated by Ireneusz Kania. Kraków: Universitas. First published in 1593.
Rubik, Anja. 2018. *#SEXEDPL. Rozmowy Anji Rubik o dojrzewaniu, miłości i sexie*. Warszawa: Wydawnictwo W.A.B.
Sadowska, Maria, dir. 2016. *Sztuka kochania. Historia Michaliny Wisłockiej*. Warsaw: Watchout Productions.
Słowacki, Juliusz. 1834. *Kordian*. Paryż: A. Pinard.
Stendahl, Krister. 1974. *The Bible and the Role of Women*. London: Facet Books.
Szot, Wojciech. 2017. Wojeciech Szot. Genowefa Jakubowska-Fijałkowska "Paraliż przysenny". April 19. Available online: https://zdaniemszota.pl/893-genowefa-jakubowska-fijalkowska-paraliz-przysenny (accessed on 22 February 2021).
Szymborska, Wisława. 1996. Głos w sprawie pornografii. In *Eadem, Ludzie na moście*. In *Edeam Widok z ziarenkiem piasku*. 102 wiersze. Poznań: Wydawnictwo a5. First published in 1986.
Tomasik, Wojciech. 2009. Pour la défense de «Tarcza» de Wisława Szymborska. Translated by Jeremy Lambert. *Slavica Bruxellensia* 2: 7–25. [CrossRef]
von Goethe, Johann Wolfgang. 1982. A. Faust I. In *Idem, Sämtliche Werke*. München: Hanser. First published in 1802.
von Stuckrad, Kocku. 2007. *The Brill Dictionary of Religion*. Amsterdam: Brill, vols. 1–4.
Weintraub, Wiktor. 1971. Tolerance and Intolerance in Old Poland. *Canadian Slavonic Papers/Revue Canadienne Des Slavistes* 13: 21–44. Available online: http://www.jstor.org/stable/40866318 (accessed on 14 July 2021).
Wisłocka, Michalina. 2016. *Sztuka Kochania*. Warszawa: Agora. First published in 1978.
Ziętarski, Krzysztof. 2020. In Kasia Czyżewska, "Marka dla katoliczek: Co powinno być ukryte". *Vogue*. Available online: https://www.vogue.pl/a/marka-dla-katoliczek-co-powinno-byc-ukryte (accessed on 23 April 2021).

Article

The Myth of Faust, "Titanism", and the Religious Topic of the Selling of the Soul in the Cultural Writings of Jan Patočka

Petra James

Department of Modern Languages and Literatures, Faculty of Letters, Translation and Communication, Université Libre de Bruxelles (ULB), 1050 Bruxelles, Belgium; petra.james@ulb.be; Tel.: +32-4-70-56-17-30

Abstract: The intensive and systematic scholarly interest in the relation of Patočka's phenomenology to religion and Christianity is recent and has only intensified over the last ten years. Thus far, the topic has mainly been studied from philosophical and theological perspectives, and the extensive body of Patočka's cultural writings has largely failed to attract the attention of scholars. Moreover, a culturological approach is virtually absent. Therefore, this article suggests focusing on the analysis of cultural archetypes in Patočka's cultural writings related to the topic of religion and Christianity from this perspective. The cultural archetypes of the Faustian figures of Patočka's cultural writings, whether Goethe's Faust, Goethe's Marguerite, or Mann's Adrian Leverkühn, are all Socratic-Christic avatars that personify Patočka's philosophical concept of "care for the soul" in the modern age. The legacy of Plato's Greek philosophy and that of Western Christianity as presented by Patočka insist on the universally shared existential experience of finitude that should be grasped as a positive challenge in the strife for meaning. Patočka's "titanism" and the archetypal titanic figures of his cultural writings are Patočkian manifestations of this universal effort. A culturological approach to Patočka's thinking on religion and Christianity might thus prove most relevant.

Keywords: Jan Patočka; titanism; Faust; Goethe; ethics; religion; Christianity; titanism; Socrates; myth; archetype

check for updates

Citation: James, Petra. 2021. The Myth of Faust, "Titanism", and the Religious Topic of the Selling of the Soul in the Cultural Writings of Jan Patočka. *Religions* 12: 528. https://doi.org/10.3390/rel12070528

Academic Editor: Dennis Ioffe

Received: 3 June 2021
Accepted: 6 July 2021
Published: 13 July 2021

Publisher's Note: MDPI stays neutral with regard to jurisdictional claims in published maps and institutional affiliations.

1. Introduction

Over the last thirty years, the philosophical writings of Jan Patočka (1907–1977), the most important representative of Czech phenomenology of the 20th century and one of the last disciples of Edmund Husserl, have been steadily acquiring their deserved place in the history of modern philosophy as well as in university programs in philosophy departments. During these decades, philosophers have been able to discover the rich and inspiring philosophical legacy of one of the leading Czech intellectuals of the 20th century, a legacy that might have been previously hidden behind the figure of Patočka as one of the key figures of the Czechoslovak dissident movement of the 1970s Charter 77. However, there are still many aspects of the rich oeuvre of Jan Patočka that deserve more in-depth scholarly investigation. As of 2011, the question of Jan Patočka's approach to religion was "virtually absent from the growing secondary literature on Patočka" (Hagedorn 2011, p. 245). Jindřich Veselý published an article in 2013 "Jan Patočka and Christianity", which is an effort to summarize the knowledge on this topic. In 2015, Eddo Evink wrote in his article that "only very recently start has been made in outlining the central status and importance of Christian ideas in his phenomenology" (Evink 2015). However, over the last decade, an important body of scholarly output was dedicated to the question, largely owing to the efforts and publication projects directed by Ludger Hagedorn. Indeed, an important milestone was the 2015 Special Issue of *The New Yearbook for Phenomenology and Phenomenological Philosophy XIV. Religion, War and the Crisis of Modernity*, edited by Ludger Hagedorn and James Dodd and dedicated to Jan Patočka (Hagedorn and Dodd 2015). The volume focuses on Patočka's understanding of myth and religion and is organized around

159

the English translation of two of Patočka's texts on these topics, namely "Time, Myth, Faith" (1952) and the last of Patočka's studies "On Masaryk's Philosophy of Religion" (1977).

An almost exhaustive bibliography of publications dealing with the relation of Patočka's philosophical system to religion is present in the article by Martin Kočí from 2019, entitled "Christianity after Christendom: Rethinking Jan Patočka's Heresy" (Kočí 2019). Kočí focuses systematically on the relation of Patočka's phenomenological thinking to theology, most recently in an article published in French "La phénoménologie est-elle une théologie?" (Kočí 2021). The interest in the dialogue of Patočka's phenomenology and theology certainly has its precursors in the figures of Erazim Kohák (Kohák 1989, pp. 16–22) and Henri Declève, who must both be given credit for advocating and introducing, through translations, a critical bibliography, and commentaries, the works of Jan Patočka in the English-speaking and the French-speaking realm in the 1980s. Patočka's personal ties to Belgium are not limited to the figure of Henri Declève but are also related to Husserl's legacy. Indeed, on the initiative of Herman Leo Van Breda, Professor at Catholic University in Leuven, Husserl's manuscripts were transferred to Leuven just before WWII and he worked with two of Husserl's assistants, the phenomenologists Egon Fink and Ludwig Landgrebe and later also with the philosopher Walter Biemel on the edition of Husserl's manuscripts. All three philosophers were friends of Patočka, a fact that contributed to the invitation of Patočka to Leuven in the 1960s. The lectures that he held there are edited under the title *Leuven lectures*.

The topic of religion and Christianity is central in the well-known commentary by Jacques (Derrida [1993] 1995), in his text *The Gift of Death* (*Donner la mort*). Thus, the topic returns with renewed topicality in critical literature after a break of almost two decades to be decisively present in Patočkian critical bibliography of the last ten years. The intensive interest in Patočka's work in international academia has also boosted English translations of Patočka's major texts, though still lagging behind the French ones, where the majority of Patočka's texts are available in the excellent translations of the indefatigable Erika Abrams. We can salute the coming edition of Patočka's texts by Bloomsbury, edited by Ivan Chvatík and Erin Plunkett, which should be available in 2021 or in 2022.

We can briefly summarize several biographical facts that are developed in detail by Erazim Kohák (Kohák 1989, pp. 16–22) and by Ludger Hagedorn. Patočka was born a Catholic but in 1927 (at the age of twenty) decided to leave the Church. However, he rejoined it a year later. The interpretation of these decisions is complex. It would either be out of respect for the family tradition or, on the contrary, as a rebellion against the paternal authority, with Patočka's father, a Classics scholar, having apparently been anti-clerical. Later in life, on several occasions, Patočka allegedly considered converting to Protestantism, largely due to contacts and friendship with the theologian and Dean of the Protestant Theological Faculty at Charles University in Prague Bohumil Souček, but this intention was never acted upon (Hagedorn 2011, p. 246). The private correspondence between Patočka and Souček (so far unpublished) shows Patočka's real interest in theology and Christian practice. Patočka read Souček's sermons and his theological works, notably, New Testament critical commentaries, and discussed with him current debates on the nature of myth in Protestant liberal theology.

As Veselý affirms in his article, it is certainly interesting to note that the first of Patočka's texts "Philosophy and theology" (1929) and "On Masaryk's Philosophy of Religion" (1977) both deal with the topic of religion and Christianity (Veselý 2013). Although, as all interpreters agree, Patočka dedicated texts to Christianity throughout his life (Veselý 2013), there is an important obstacle in dealing with the topic of religion in Patočka's work. As noted by Ludger Hagedorn, "(. . .) there is no text, no essay in which he deals explicitly, 'systematically', with the phenomenon of religion, or attempts a comprehensive commentary" (Hagedorn 2011, p. 245). The interpreters also agree that Patočka's concept of Christianity is in many ways "heretical". As he argues, Kočí Patočka favors the phenomenological attitude towards Christianity over the metaphysical one (Kočí 2017, p. 118). However, Patočka never dismisses the question of religion and spirituality as something totally irrelevant.

His interpretations of Christianity are multiple and are part of his argumentation in the texts dedicated to the philosophy of history and the historical and cultural meaning of Europe, but also in his cultural writings dedicated to the study of cultural archetypes. Thus far, the study of the topic of religion has been focused on Patočka's philosophical work and has been undertaken by philosophers and theologians. The present study thus aims to offer a culturological and aesthetic approach to Patočka's work with cultural archetypes related to religion and Christianity. The focus of this study will be the archetype of Faust and the concept of titanism. According to Jindřich Veselý, for Patočka, "Socrates is not a prefiguration of Christ, but Christ is an avatar of Socrates" (Veselý 2013, p. 79), and we add that we will argue in this study that Faust and Faustian avatars in Patočka's interpretations are emulations of the Socrates–Christ double model highlighted by Veselý. The evolution of these cultural archetypes throughout European history illustrates the changing form of the quintessentially European concept of "care for the soul". The Patočkian concept of "care for the soul" contains the classical and Christian legacy of European culture that Patočka strives to safeguard for the contemporary post-Christian and post-European society.

2. Patočka's Cultural Writings

As Patočka did not leave a systematic study of religion, we suggest turning our attention to an understudied corpus of Patočka's work, namely, a huge body of texts that we might label "cultural writings" that he had written for cultural and popular journals and that go back to the 1930s and 1940s and continue throughout Patočka's life in public debate platforms such as the magazines *Kritický měsíčník*, *Čin*, *Divadlo*, and *Tvář*. He is the author of influential interpretations of major figures of Czech literature, such as the key poet of Czech romanticism Karel Hynek Mácha (1810–1836), Josef Čapek (1887–1945), Jaroslav Durych (1886–1962), Ladislav Klíma (1878–1928), and Ivan Vyskočil (b. 1929). Jonathan Bolton claims in the conclusion of his recent article that "Patočka left behind a body of occasional writings that, even if he had not written a word of academic philosophy, even if he had not been a spokesman for Charter 77, would still mark him as a participant in Czech politics and culture of the twentieth century, and as a vital thinker who should not be overlooked" (Bolton 2020, p. 29).

This corpus of Patočka's cultural writings is the focus of the present article, with particular attention paid to the topic of titanism in connection to the Faust myth. The legend of Faust and its numerous variants all go back to one of the key religious and philosophical topics of Christian culture over the centuries, the "pact with the devil" and the "selling of one's immortal soul" (see James 2020; Svobodová 2021).

In his last study "On Masaryk's Philosophy of Religion" (1977)[1], a substantial text on religion and its crisis in modernity, Patočka focuses not only on Masaryk but also on the topic of religion in Dostoyevsky's *Brothers Karamazov*, in the philosophy of Immanuel Kant and in the writings of Friedrich Nietzsche (see the discussion of this essay in Hagedorn and Dodd 2015). In this text, Patočka quotes Masaryk's affirmation that he "was experiencing metaphysics through literature". We claim that Patočka shares this shift of metaphysics towards the realm of art with Masaryk and both are in this sense representative inheritors of the Central European legacy of Romanticism. In his study of 1968, "The Social Function of Literature", Patočka gives a glimpse of his approach to the study of literary archetypes:

> Literary work of art does not speak about a particular thing, but on the occasion of the demonstration of one possible relation it points out the constants of the world, which, contrary to all that is but unique, particular and arbitrary, last for ever and let shine the structure of the world as a whole. It is not about Antigone and Philoctetes, Faust and Mitya Karamazov, but about the relation of meaning, which articulates what is being, and which guides life, shapes destiny and decides about fulfillment and emptiness, about punishment and destruction. It is about the relation between nature and the supernatural, between human freedom and religious power, it is about man and God or gods, about man and

woman about together being in work and in fighting, about guilt, suffering and death. (Patočka [1968] 2006a, p. 179)

It is significant which cultural archetypes Patočka mentions in the demonstration of his interpretative method—two figures of classical mythology and characters of Sophocles' plays, Antigone and Philoctetes, Faust and one of the Karamazov brothers. They all represent figures through which subsequent historical European epochs renegotiate their relationship to religion and metaphysics. It also shows that we can trace certain constant topics that accompany Patočka throughout his life. Antigone is one of the key archetypes for Patočka (see Bolton 2020). In his text entitled "The Truth of Myth in Tragedies of the Labdacids by Sophocles" (1971), Patočka further develops his concept of myth and its use in literary interpretations:

> Against this proto-Enlightenment demand of the human being to take into his own hands his whole life and that of the polis, to let the law of the day rule everywhere, Antigone shows the supremacy of the myth, the supremacy of the whole, from which even the law of day is being nourished. The law of day, which is just a part and cancels itself, where it strives to become the whole. (Patočka [1971] 2004c, p. 466)

The emphasis on universal topics contained in literary archetypes is clearly visible. The figure of Antigone, parallel to the figure of Faust, questions human limits and thus offers a revealing mirror to the archetype of Faustian titanic figures.

3. "Titanism" in Jan Patočka's Thinking

Possibly the first mention of Faust in Patočka's writings dates to 1936 and his text "Titanism", which is a review of Václav Černý's (1905–1987) monograph *Essai sur le titanisme dans la poésie romantique occidentale entre 1815 et 1850* (Černý 1935).[2] The study, published in Prague in the publishing house Orbis in 1935, was Černý's habilitation thesis, which qualified him for the position of associate professor of Romance literatures at the Charles University of Prague. The discussion of Černý's book by Patočka was a way for the young and promising philosopher to engage in a dialogue with another bright mind of Czechoslovakia. Both Černý and Patočka (30 and 28 years old in 1935), each in their own way, will go on to become two major voices of intellectual dissent in Czechoslovakia of the 1970s. The expression "titanism" used by Černý in the title of his monograph is worth commenting on. *Dictionnaire historique de la langue française* indicates that the French word "titanisme" is rare in French and the occurrence dates back to the beginning of the 18th century in the text of Saint-Simon. Its meaning in French is "the spirit of revolt and usurpation". Indeed, the word is not a part of common French vocabulary of the 1930s and Georges Cirot (1870–1946), a leading French specialist of Spanish literature and author of an extensive review of Černý's monograph in the prestigious *Bulletin hispanique* that Cirot had co-founded, mentions this fact as one of the major criticisms of the study. Indeed, he writes that one of his major regrets is that the author "did not tell us anything about the history of the word "titanism", which he employs as if it was such a common expression in French (Cirot 1936, p. 239).[3] Černý was clearly unaware of this linguistic imbalance between Czech and French. This confusion might be partly explained by the fact that the word "titanismus" figures more importantly in Czech vocabulary and is also part of the standard dictionary of Czech language. Its Czech dictionary definition implies that it is often used in a pejorative way to denounce a certain pretentious grandeur and that it refers to characteristics of a titan. However, Černý uses it in the sense that was common among Czech symbolist art and literary critics. Indeed, the word titanism is frequently used by the most important Czech critic of the end of the 19th and the beginning of 20th century, F. X. Šalda (1867–1937), who happens to be the supervisor of Černý's habilitation thesis of 1935. The use of the word titanism in Czech is rooted in a specific Central-European aesthetic and philosophical tradition, largely based on the appropriation of German romanticism and its philosophical, social and political implications. Czech symbolist art is strongly anchored in this cultural legacy, as is the culture of independent Czechoslovakia after 1918. The aesthetic and philosophical preoccupations of both Černý and Patočka are striking testimonies of the strength of this Central-European cultural legacy. It constitutes a shared "cultural corpus" of all the Czech intellectuals that contribute to the debate on "titanism". Indeed, "titanism" is an important part not only of Šalda's vocabulary but also of that of Tomáš Garrigue Masaryk (1855–1937). In Masaryk's use, titanism is mostly linked to his interpretation of Goethe and his *Faust*. Thus, to engage in a discussion on the topic of titanism in the Czech context of the 1930s is a way of positioning oneself against the most important intellectuals of Czech cultural scene of that time. Indeed, in 1935, Masaryk is still the president of Czechoslovakia and Šalda is still considered an eminent authority in Czech literary criticism. Thus, both young men, Černý and Patočka, show their ambition of affirming their place among the greatest Czech minds of their generation.

Patočka's ambition to actively participate in a public intellectual debate is visible in the choice of the magazine for his review, *Čin*. Indeed, *Čin* was not a specialized philosophical revue but one of the major platforms for cultural debate of 1920s and 1930s with contributions by important Czech artists and intellectuals. The review shows Patočka's own personal interpretation of the concept of titanism. It is important for this study because it is related to Patočka's questioning of religion. Indeed, we can study the evolution of Patočka's thinking on this topic by comparing it with the last of his texts, "On Masaryk's Philosophy of Religion", written in 1977, forty years later. Indeed, in the second text, Patočka revisits the topic and, contrary to what the title might suggest, discusses the topic of the philosophy of religion not only in Masaryk's thinking but also in that of Kant, Nietzsche, and Dostoyevsky.

As Patočka affirms, "Masaryk condenses the entire problem of Western European metaphysics in the problem of titanism" (Patočka [1936] 1989, p. 140). He goes on and quotes Masaryk's definition of titanism as a form of modern subjectivism, whose exemplar for Masaryk is Goethe's *Faust*. Inspired by Černý and his polemics with Masaryk from 1934 (see Černý 1934) and 1935 (Černý 1935), Patočka then goes on to critically distance himself from this view and presents his own understanding of titanism.[4] We can thus claim with Veselý that the idea of titanism is key to Patočka's work on the topic of religion. Titanic figures of Patočka's cultural writings are various avatars of the cultural archetype of Faust. The concept of titanism is inspired by classical mythology and the myth of the clash of the Titans with the Gods. However, Patočka is mostly interested in the way European (and especially German) Romanticism incorporates and re-interprets the classical myth. Indeed, the Romantic appropriations of the Titans Prometheus and Hyperion, inherited from classical mythology, in the works of Shelley, Keats and Hölderlin are important both for Černý in his monograph and for Patočka in his cultural writings. In his interest in the concept of titanism, Patočka (the Classics professor's son) proves to be a typical inheritor of Germanic, Central European culture of his time, centered on the study of Classics and promulgated by classical high-school and university programs. As stated above, Central European cultures have been shaped by Romanticism, with its cult of classical and national heroes and ideals represented by Goethe and Schiller. Goethe's *Faust* is thus a commonly shared Central European intertext of Patočka's time, a text that Masaryk remembers learning by heart at school, a text whose author was revered by Nietzsche as a precursor of the Superman. We cannot state this fact enough as it constitutes a backdrop against which Patočka develops his own interpretation of titanism, closely connected to the Faust myth.[5]

Patočka's interpretations are the most inspiring when they go against the most well-known ones. Thus, contrary to Černý's and more general insistence on the element of revolt and rebellion (be it religious, metaphysical, social, or political), which is usually at the center of interpretations of Romantic titanism, Patočka insists on the element of moral, ethical responsibility that a world devoid of metaphysics casts upon man. Indeed, he writes in his review of 1936: "Titanism is a moralism, it is a moral viewpoint applied to both the world and God (. . .) a moral experiment in human freedom, a test whether the meaning of life can be found in life itself and under what conditions (…) it is the positive freedom of existence" (Patočka [1936] 1989, pp. 141, 143). Patočka thus does not share Masaryk's and Dostoyevsky's caution concerning titanism, which, for these thinkers, leads inevitably to subjectivism, nihilism, and moral decadence. Neither does he share Nietzsche's enthusiastic embrace of the vitalist forces unleashed by romantic titans. Contrary to Nietzsche, Patočka builds his concept of titanism around the key notions of personal freedom linked to a heightened responsibility for one's life in the face of the reinforced notion of one's own finitude: "Titanism no longer turns to a metaphysical power that guides the universe in resolving the question of the meaning of existence, but rather to an inner freedom that creates a personal world" (Patočka [1936] 1989, p. 141). In this sense, Patočka follows in the footsteps of Shelley (figuring importantly in Černý's monograph) when Shelley interprets the figure of the Titan Prometheus in the preface to his drama *Prometheus Unbound* in moral terms as "the type of the highest perfection of moral and intellectual nature, impelled by the purest and the truest motives to the best and noblest ends" (Shelley [1820] 2006, p. 776). Indeed, Patočka's understanding of titanism is not metaphysical ("a lack of an absolute faith"; a revolt against God or gods) but ethical, as a positive challenge in the strife for meaning. The review of Černý's monograph can be understood without much exaggeration as the basis for Patočka's meditations on the meaning of history and existence in his *Heretical Essays* of the 1970s.

Patočka's concept of titanism is thus an inheritor of Central European Romanticism, and although Patočka makes a point of distancing himself from Nietzsche, his titanism nevertheless shares with Nietzsche the fascination for the utopian ideal of the realization of human full potential. Patočka's titanism and Nietzsche's philosophy both presuppose a strong individual, capable of extraordinary (could we say superhuman?) existential courage and unflinching will. This exclusive definition of Patočka's titanism is also projected onto the various titanic figures of his interpretations, be it the idealized Socrates or his avatars evident in Patočka's take on the Faust myth. Patočka's original and counterintuitive (but also daunting) project of modern titanism points possibly to the inevitable limits related to all forms of utopian conceptualizations. Nevertheless, Patočka is unflinching in his titanic stance, and although his reading and understanding of the thinkers that he addressed in 1936 evolves, his understanding of titanism remains virtually unchanged over forty years. Thus, in his study of 1977, "On Masaryk's Philosophy of Religion", whose first version was interestingly entitled "Humanity and nihilism", he maintains the possibility and the moral value of the titanic stance that he defends against what he sees as a reductionist understanding by Masaryk and Dostoyevsky, who equal it with subjectivism and nihilism (Patočka [1977] 2006b, p. 402).

4. Selling One's Soul or Caring for It?

"On Masaryk's Philosophy of Religion" proves to be an essential text for the present study. Indeed, here Patočka criticizes Masaryk for "having misunderstood the real meaning of the Faustian titanism" (Patočka [1977] 2006b, p. 402). What Patočka means by that and his own understanding of the Faustian titanism are the subject of this section. Although our main focus is on the article "The Meaning of the pact with the devil-reflections on three phases of the Faust legend" (the first, unpublished version is entitled "Reflections on three phases of the Faustian Legend Today and Yesterday: On Thomas Mann's Novel *Doctor Faustus*") dating back to 1972, it is important to note that Patočka's interest in the Faust theme goes back further and can be linked to the topic of titanism that Patočka comments on in a review of 1936 and goes back to his study "On Masaryk's Philosophy of Religion" (1977). He reflects on the Faust myth in his interpretation of Ivan Vyskočil's works in 1963, where his reading of Faust is preeminently political (James 2020). Patočka writes the article in 1972 with the idea of commemorating the 25th anniversary of the publication of Thomas Mann's novel *Doctor Faustus* (1947). The importance of this text lies in the fact that it presents Patočka's vision of the evolution of the Platonic concept of care for the soul in European civilization over the ages. This notion of philosophy of history is expressed by the original title of the study, "Reflections on three phases of the Faust legend Faustian Legend Today and Yesterday". The way the archetype of Faust evolves through the ages reveals what "selling one's soul" means in any given historical moment and, thus, in a mirror image, what caring for one's soul looks like.

After the events of the Soviet and Warsaw Pact armies' invasion of Czechoslovakia in 1968, Patočka had to retire from his university position in 1972 and his publication opportunities in Czechoslovakia became limited. The Faust text was finally published in two revised and slightly different versions in Polish and in German. The text was clearly important for Patočka. Nevertheless, it took considerable time, effort, and intervention of Patočka's colleagues abroad before it could be published. The text was ready in 1972, and in a letter to his friend Walter Biemel (German phenomenologist of East-Central-European descent and editor of Husserl's and Heidegger's texts) from 20 September 1972, Patočka writes that, thus far, all the German journals that he had contacted rejected the article. According to Patočka, Thomas Mann did not seem like a topical theme either for philosophy or for literature or human sciences (Vojtěch and Chvatík 2004, p. 408). Patočka sums up the central theme of his article as the "reflection on the character of myth, on poetry as a means of working through basic mythical themes (such as the myth of the soul)" (Vojtěch and Chvatík 2004, p. 408). Biemel contacts Jean Améry and asks for his help. Améry likes the text and writes a recommendation. However, in the end, it is the phenomenologist Ludwig Landgrebe, another of Husserl's disciples and friend of Patočka's, who succeeds in securing the publication of the article in 1973 in the well-respected journal *Neue Zeitschrift for Systematische Theologie und Religionsphilosophie*, still published today by the renowned publishing house De Gruyter, and dedicated to "the exciting dialogue between Lutheran-Reformed theology and philosophy in the broadest sense". It was Landgrebe who suggested the title of the article, "The Meaning of the pact with the devil", so that the "religious-philosophical content" (Vojtěch and Chvatík 2004, p. 409) would be clearly evident. The Polish version was published owing to the efforts of Patočka's Polish translator (for details, see Vojtěch and Chvatík 2004, p. 409).

The text is significant for several reasons. Patočka writes in a private letter that this article is his attempt at a "philosophy of literature". The text is also an example of the centrality of philosophy of history in Patočka's thinking. The legendary topic of Faust and its variations by Goethe and Mann are Patočka's attempts at formulating the philosophy of German national destiny, with Faust being seen by Patočka as a typical German theme. Moreover, as he points out at the end of his essay, the gradual shifts in the representation of Faust illuminate at the same time European history and all of humanity at the threshold of what he calls the "post-European epoch" (Patočka [1972] 2004e, p. 119). Therefore, it is possible to read this text in parallel with the best-known text by Patočka, his *Heretical Essays in the Philosophy of History* (published in samizdat, in Ludvík Vaculík's Edice Petlice in 1975).

In the fifth of his *Heretical Essays*, Patočka also mentions Faust in his description of the evolution of the philosophical concept of responsibility (Patočka [1990] 1996, p. 105). Indeed, in Patočka's thinking, the interpretations of art and literature are inseparable from the evocation of religious, ethical, and civic categories. Among them, the concept of responsibility plays a central role. Thus, we can see this text as another effort by Patočka to illuminate his concept of "care for the soul". Indeed, the highly metaphorical theme of Faust is interpreted by Patočka as the motif of "selling one's immortal soul" (Patoč ka [1973] 2004f, p. 511).[6] However, the selling of one's soul is just the reverse side of the Platonic/Patočkian "care for the soul", a concept whose core theme is again individual "responsibility". The centrality of the concept of responsibility in Patočka's thinking is also highlighted by Jacques Derrida in his *Gift of Death (Donner la mort)*, based on the interpretation of the mentioned fifth *Heretical Essay* (Derrida [1993] 1995, p. 91). Thus, how does one care for his or her soul in Patočka's philosophical world? Patočka is Socratic and Platonic in the sense that his philosophy is a permanent effort to rise to the challenge to the fundamental Socratic question of "how should we live?" The Greek philosophical concept of "care for the soul" in its Socratic-Platonic variant is ever present in Patočka's writings in the 1970s, *Post-European Epoch and its Spiritual Problems* (1970), in the interpretations of Plato in the text *The beginnings of systematic psychology* (1971), *About soul by Plato* (1972), *Plato and Europe* (1973), and *Europe and Post-European Epoch* (1970–1977) (Josl 2018, pp. 23–24). In all of these texts, Patočka revisits the concept of "care for the soul" and explains in detail his understanding of it. The motif of immortality is understood and interpreted in ethical and philosophical terms. Indeed, the definition that Patočka gives of immortality is that of ethical integrity and responsibility: "The immortality in its right sense is achieved by those who prefer non-being to the destruction of one's soul" (Patočka [1972] 2004e, p. 511).[7] The figure of Socrates is clearly on Patočka's mind when he talks about true immortality as the courage to choose death as the price for preserving one's philosophical coherence and ethical integrity: "True immortality is for those, who overcame the horror of physical death by the horror of an absolutely negative existence, so that they could achieve what is achievable at the height of this life: one's finite absoluteness" (Patoč ka [1973] 2004f, p. 512).[8] Socrates is described in the text on Faust as a figure of "ethical being in the world". Socrates's choice is summed up by Patočka as the choice between "possible annihilation as a way of preserving one's authenticity" and "prolonging of one's life as a denial of one's true self" (Patoč ka [1973] 2004f, p. 511). The texts on Faust thus reflect Patočka's parallel work on the concept of soul by Plato and reflect the meditations on the sense of history. As Martin Kočí claims, "Patočka's whole project of caring for the soul is about caring for death, and thus searching for the meaning of life in spite of finitude" (Kočí 2019, p. 12).

In Patočka's texts on Plato, concern for ethics and for the concept of freedom are central, as in the interpretation of soul in *About Soul by Plato* (1972). As Jan Josl remarks, according to Patočka, Plato starts from "the human existence in this original crisis and problematic aspects, which is fundamentally ethical, that is such, that it is concerned with our own existence and non-existence, partially dependent on us, on our decision" (quoted according to Josl 2018, p. 25).[9] In his aforementioned interpretation of *Heretical Essays*, Jacques Derrida insists on Patočka's concept of responsibility inspired by Plato while quoting in the support of his argument the following passage from *Heretical Essays*: "Plato's doctrine of the immortality of the soul is the result of the confrontation of the orgiastic with responsibility" (Derrida [1993] 1995, p. 270).

5. Marguerite—A Socratic Faust?

It might come as a surprise that in a text on Goethe's Faust we learn more about Socrates than about Goethe or Faust. Patočka's cultural writings tend to be the most interesting, where they deliberately, although often tacitly, depart from previous, more well-known interpretations and challenge them with an original twist. Considering the fact that Goethe's Faust is one of the most frequently interpreted texts in the German language and of the key texts of European literature, Patočka cannot be accused of lacking in ambition. In the following central passage, Patočka gives his personal interpretation of Goethe's *Faust:* "and from here it is important to understand, how the mishearing of that proper inner motif arrives—that it is guilt that drives this movement—and why Faust is more and more understood as a man of energy, as a superman in the modern post-idealistic sense" (Patoč ka [1973] 2004f, p. 517). Patočka puts in the center of his interpretation moral categories and emphasis on the key role of Goethe's *Faust* existential experience of guilt ("it is guilt that drives this movement") as the true driving force behind his actions, not his pride or the thirst for knowledge. He goes on to claim that the previous interpretations are but a "mishearing of that proper inner motif", which basically amounts to claiming that the other interpretations did not understand the meaning of *Faust* correctly. And who might these authors, never directly mentioned in Patočka's text, be? The thinkers thus challenged are no minor intellectual figures: Hegel, Nietzsche, and Masaryk.

Hegel's reading of Goethe's *Faust* is probably the most well-known and the most influential. Hegel's philosophy and his thinking on aesthetics often function as Patočka's intertext, a philosophical work with which he is constantly in dialogue. Indeed, Patočka was intimately acquainted with Hegel's oeuvre, not only as a philosopher but also as a translator of two of his major works, *The Phenomenology of Spirit*, translated by Patočka in Czech and published in 1960, and his *Aesthetics*, published in Czech in 1966 with a large introduction by Patočka entitled *Hegel's Philosophical and Aesthetic Evolution*, which we Patočka's own interpretation of Hegel. Patočka, who literally knew both Hegel and Goethe by heart, refers to Hegel's interpretation of Goethe in the *Phenomenology of Spirit* in his study "The German Spirit in Beethoven's Era" (*Německá duchovnost Beethovenovy doby*) of 1971 (Patočka [1971] 2004d, p. 471). Patočka mentions Hegel twice (and only briefly) in the published version of his Faust text. Patočka was probably aware that the quote from Goethe's *Faust* in Hegel's *Phenomenology*, published in 1807, is that of the first version of *Faust, Part One*, the revised version being published in 1828–1829, and *Faust, Part Two* was published posthumously in 1831. It is also quite possible that Patočka was aware of a highly personal appropriation by Hegel of Goethe's text, in that he sometimes misquotes and uses it as a practical way of demonstrating his philosophical concept (see Champlin 2011). It is maybe this aspect of Hegel's use of Goethe's *Faust* that inspires Patočka to use the same text for the demonstration of his own philosophy. Indeed, an indirect polemics with Hegel might be seen in the central position that Patočka assigns to the character of Marguerite. Whereas in Hegel's reading all the victims of Faust's doings are dismissed as metaphorical "collateral damage" of Faust's gradual spiritual elevation, Patočka interprets Marguerite as the key character and the only real tragic figure of Goethe's *Faust*. Indeed, Marguerite and her destiny come to the forefront, and she is interpreted as a Platonic character: "what makes Faust in Goethe's conception into a real tragedy, what carries it, is the tragedy of Marguerite: the punishment is in a quintessentially platonic way experienced as the purification of soul and external help is refused as appearing opposed to being" (Patoč ka [1973] 2004f, p. 516). Patočka thus offers us a surprising and original interpretative angle with the reversal of roles. Marguerite is the female version of Hegel's Faust as a symbol of spiritual growth and elevation of the spirit. At the same, she is yet another avatar of Patočkian titanism of moral responsibility. Her choice is again interpreted by Patočka in Socratic terms as an ultimate ethical choice and a radical way of "care for the soul". The oblique dialogue with Hegel is a recurrent phenomenon in Patočka's literary interpretations. Indeed, Jonathan Bolton points out that one of the key aspects of Patočka's interpretation of Sophocles' tragedy *Antigone* is actually his refusal of Hegel's famous political reading of the play (Bolton 2020). The "spectral" presence of Hegel in Patočka's cultural writings would certainly deserve a more in-depth study.

Another crying absence in Patočka's Faust text is that of Masaryk. Masaryk as an intellectual and a symbolic figure is someone that every Czech thinker of the first half of the 20th century needed to come to terms with. Patočka's life-long engagement with Masaryk's thinking is thus coherent with Masaryk's status as one of the leading Czech intellectuals of his time. The title of Masaryk's most well-known take on Goethe and his Faust is self-explanatory: "Goethe's Faust: Superman", originally published in 1896 and later integrated into one of his major works, *Modern Man and Religion* (Masaryk [1896] 1938). As we already remarked in the section on titanism, Patočka is quite critical of Masaryk's reading of Goethe (whom Masaryk famously called "the Giant of Dilettantism") and the assimilation of Goethe's titanism with a negative notion of subjectivism, which for Masaryk leads to nihilism. Patočka proves to be a more sophisticated theoretician of literature than Masaryk when Patočka castigates Masaryk for confusing Goethe with his literary character of Faust. Patočka also refuses Masaryk's Nietzschean reading of Faust: "Faust is the real titan, he is the superman" (Masaryk [1896] 1990, p. 75). Nevertheless, and maybe surprisingly, we might find some parallels in Masaryk's and Patočka's interpretations of Faust. Indeed, Masaryk goes on to say about Faust that "He is a titan of reason but a coward of heart" (Masaryk [1896] 1990, p. 75). Indeed, Patočka does not shy away from reprimanding literary characters for their lack of moral courage and integrity. Thus, in a sketch comparing the characters of Ludvík Vaculík's novel *Axe* and Milan Kundera's novel *The Joke*, Patočka pitches the moral purity of the character of the father in Vaculík's novel (which Patočka calls "the law of the heart", *Gesetz des Herzens*) against Kundera's novel's "overall weak figures", whose main hero is incapable of *catharsis* (Patočka [1968] 2004b, p. 212). The classical aesthetic category of catharsis remains relevant for Patočka's work with cultural archetypes. Catharsis is central to the way Patočka describes both Goethe's characters, Faust, and Marguerite. According to Patočka, "Marguerite's grand will for self-sacrifice, which spiritualizes her and leads her to refuse earthly 'salvation', makes the Faust of Goethe's version (despite his seemingly cynical banality) believe in a higher meaning of life and take the path of purification" (Patoč ka [1973] 2004f, p. 525).[10] The English word "spiritualize" is the translation of the Czech verb "oduševňovat", which literally means "to infuse someone with soul". It is thus by acquiring soul, the strength of the heart, that Marguerite also acquires the moral and spiritual capacity to elevate Faust through her sacrifice. Masaryk's remark on the weakness of Faust's heart acquires a new resonance with Patočka's own interpretation of Goethe's *Faust*—it is also constructed around classical ethical categories inherited from Socrates and Plato, which put forward the value of personal responsibility. Indeed, this role given to Marguerite brings further distance from Hegel's interpretation. Instead of the elevation of the spirit coming from the outside through education, art, and philosophical training, it is driven by personal change; it comes first from the inside, from the "heart", from the "soul". This critical debate with Hegel and his concept of art continues in the article in Patočka's interpretation of the figure of Adrian Leverkühn.

6. Mann's Adrian Leverkühn: An *Übermensch* of Responsibility?

The second and larger part of Patočka's interpretation of the Faust myth is dedicated to Mann's novel. Patočka focuses on the question of how to stage the Faust motif of the pact with the devil and of the selling of one's immortal soul in an era and in a society where the very notion of a soul had lost all its relevance, the historical moment of Thomas Mann, that of post-WWII Germany. It is interesting to note that in this text, Patočka mentions Nietzsche only once and very briefly. This is of course highly significant, as Patočka must have been aware of the fact that Mann constructed the figure of Adrian Leverkühn around important biographical and philosophical details of Nietzsche and his work. Indeed, as Andrew Erwin remarks, "Since its inception in 1947, Thomas Mann's *Doktor Faustus* has been referred to as a Nietzsche-novel, and the parallels between Nietzsche and the novel's protagonist Adrian Leverkühn have been well documented" (Erwin 2010, p. 283). It is thus worth the effort to think about the reasons for this deliberate omission of the philosopher. In an article from 1966 dedicated to the history of Epos and Greek drama, Patočka also refers to Nietzsche indirectly, not mentioning his name but referring to him obliquely as a "spirited and dangerous thinker" (duchaplný a nebezpečný myslitel, Patočka [1966] 2004a, p. 348). We might even advance a provocative theory according to which Patočka sketches out a portrait of Nietzsche as he would have liked him, a responsible and ethical Nietzsche, an *Übermensch*, a titan of responsibility. By not directly mentioning the Nietzsche analogy of Leverkühn's character, Patočka projects into the figure his ideal version of Nietzsche, a Nietzsche who would have adhered to Patočka's concept of titanism. The fall of the main character of Mann's *Doctor Faustus* is therefore logically interpreted by Patočka as the taking on of universal responsibility, which means the loss of the titanic soul (and thus madness and subsequent death). However, by "losing his life", Adrian Leverkühn in Patočka's interpretation paradoxically acquires an immortal soul.

The destinies of individuals, art, and European civilization are put in a chain of cause and effect. In Patočka's view, where he continues in his tacit discussion with Hegel, Mann follows German romanticism in its faith in the central role of art as a form of the spiritual renewal of human beings and the tool of their education. As art unifies the spiritual and the sensual, it enables man to become whole again. Art is the "symbol and gate leading towards practical freedom" (*symbol a brána k praktické svobodě*, Patočka [1972] 2004e, p. 117). This was, according to Patočka, the project of spiritualization (*zduchovnění*, Patočka [1972] 2004e, p. 117), with which "the great men of their time stepped out against the total secularization of the modern era and against a partial Enlightenment" (Patočka [1972] 2004e, p. 117)[11]. After this personal paraphrase of Hegel and the focus of German Romantic philosophy, Patočka goes on to explain how the Hegelian project of German Romanticism of care for the soul gets modified in post-WWII Germany and how it is represented by the figure of Leverkühn. As Patočka affirms, Adrian (in accordance with his romantic forefathers) also insists on the intrinsic link between the renewal of art and of man. However, the relation is reversed. It is not art that causes spiritual renewal of man, but an ethically responsible, "soulful" man (the man who reacquired a soul, as Marguerite did) that will renew art of his era:

> [...] first, it is important to acquire the dimension of responsibility, therein should emerge 'the immortal soul', and then, as a result of this healing, a breakthrough towards a new art, which would thus be a suitable bearer of further renewal. The spiritual and artistic renewal is thus not divided into stages but is seen together. (Patočka [1972] 2004e, pp. 117–18)[12]

This peculiar quote creates a link between the acquisition of responsibility and the emergence (the possibility) of the "immortal soul". Only through this transformative experience can modern art also be transformed. It is thus the individual human being that is the starting point of the epochal spiritual renewal; a renewal does not happen in stages, as in Hegel's philosophy, but can occur all at once, through a transformative experience of personal *metanoia*. Indeed, Leverkühn is in Patočka's vision the titan of the new spirituality for contemporary Europe/the world, and, at the same time, he is the figure of a radical, "pure" sacrifice, which has for its aim the renewal of men and art. Indeed, Leverkühn is the post-WWII avatar of the Socratic–Christic archetype and a representation of a new type of "care for the soul" in an era that is both post-Christian and post-European. He is the modern titan of the Night, who, by his choice, refuses the survival of the bare life as the supreme value of existence (a theme that Patočka further develops in the sixth *Heretical Essay*) and chooses, as Socrates did, the sacrifice of life in the name of preserving a certain kind of transcendence and soulfulness. He thus, according to Patočka, needs to take on "not only the role of Faust but also that of Marguerite" (Patočka [1972] 2004e, p. 114), and their mutual salvation is combined in the final "salvation" of Leverkühn (Patočka [1972] 2004e, p. 118). The concept of responsibility comes again as a leitmotif of Patočka's work with cultural archetypes and is placed as a conclusion of the interpretation of the figure of Mann's Leverkühn: "This universal accountability and the will for it—the universal responsibility—is also the last legacy of Leverkühn as a man and that of his music, the music of that difficult era, which, as he knows, opens up during his lifetime" (Patoč ka [1973] 2004f, p. 523).

7. Jan Patočka and the Archetype of the Hero

Patočkian titanism is key to the understanding of Patočka's Socratic figures and their avatars (which also inspire his concept of political and civic dissent; see for example, Patočka's samizdat article of 1976 "The Heroes of our Time"). It is confirmed and revisited in many texts of the 1970s, for example, in his important lecture of 1975 "The Spiritual Person and the Intellectual" (*Duchovní člověk a intelektuál*), prepared at the same time as the *Heretical Essays*. In "The Spiritual Person and the Intellectual", Patočka makes a point of distinguishing between a pragmatic intellectual and a real "spiritual person" (*duchovní člověk*). Contrary to the pragmatic intellectual (we could say, the sophist of Socratic dialogues), he pitches the spiritual person, the person who can care for her/his soul. As a demonstration of proper "care for the soul", he surprisingly chooses the Homeric archetype of courage, the famous hero of the *Iliad*, Achilles: "The very figure of Achilles, the man who chooses that short life of glory! Something that goes against the direction of the ordinary life, something that belongs to the foundations of Greek political feeling" (Patočka [1975] 2002, p. 360).[13] The way he describes Achilles makes us think more of the Socratic type of courage and therefore is again a demonstration of the project of "care for the soul". Indeed, Patočka, well read in Socratic dialogues and well acquainted with Plato's reflections on courage and manliness (*andreia*), emulates Plato's renegotiation of the Homeric archetype of courage, adding to the traditional archetype of Achilles that of a philosophical courage, represented by the Socratic archetype (Hobbs [2000] 2006). Indeed, the courage of a philosopher (Patočka's spiritual person) is the central Platonic archetype of courage, the one capable of facing all adversities of destiny with calm and composure, even in the proximity of death, as Socrates did. Patočka's Achilles is less the famous Homeric fierce fighter and merciless adversary driven primarily by the desire for eternal glory gained on the battlefield and more a self-reflecting Socratic philosopher of Plato's dialogues dedicated to courage of the spiritual person that appears in the final passages of Patočka's sixth *Heretical Essay*, someone whose life's meaning has been shaken by the front experience (here, the analogy with Achilles is very appropriate). In the experience of the existential "shaking" (*otřesení*), Patočka's Achilles/Socrates/Faust/Marguerite/Adrian/Spiritual person chooses the maximal existence of life "at the peak", "the short and glorious life", in order to honor one's own inner integrity and achieve Patočkian immortality. The spiritual person experiences a real existential *metanoia*, which makes her/ him choose counter-intuitively against the rules of the ordinary life of the *polis*, against the rules and laws of the day, as Patočka so pointedly stresses in his interpretation of Sophocles' *Antigone*, the mirror image of Faust. The titanic aspect of this singular choice is evident here, as well as the parallels with the reflections on immortality developed by Patočka in his Faustian meditations. It is the titanism of heightened moral awareness and towering responsibility as Patočka redefined it in his text of 1936. The cultural archetypes of Faustian figures of Patočka's cultural writings, be it Goethe's Faust, Goethe's Marguerite or Mann's Adrian Leverkühn, are all Socratic-Christic avatars that personify Patočka's philosophical concept of "care for the soul" in the modern age.

8. Conclusions

Patočka's highly personal and unusual analyses of Faust, one of the key cultural archetypes of Central-European culture, also point at what Patočka considers to be the cultural and spiritual legacy of (Central) European civilization that "comes after (the end of) Christendom" (Koči 2019, p. 9). These ideas are further developed in *Heretical Essays* (1975) and in the study "On Masaryk's Philosophy of Religion" (1977). In this sense, Patočka is a quintessentially Central European philosopher. This "peripheral" perspective might be one of the reasons for the renewed interest in Patočka's work over the last two decades, which, paradoxically, and despite (or because) European integration, poses real questions about European identity and its place in the globalized world. Patočka offers Europeans the opportunity to think about their common and shared identity and their core historical values while focusing on cultural legacy rather than on politics or economics. In Patočka's Central-European perspective, "Europe (. . .) is the ideal of rational life directed toward truth based on insight. Europe is the culture of insight, that is, a scrupulous reflection on that which appears. Further, insight is the condition of the possibility of the 'spiritual life', 'life in truth', or alternatively, 'care for the soul'" (Koči 2019, p. 2). According to Patočka, it is care for the soul (*tes psychés epimeleia*) that gave rise to Europe (Heretical Essays 1975) and the self-reflecting soul is the core legacy of European culture, that is, the critical notion of the problematicity of one's own existence. In his final study, "On Masaryk's Philosophy of Religion", Patočka continues the dialogue with Nietzsche and Masaryk, this time openly and explicitly. It is also in this text that he returns to his initial reflections on the nature of titanism. Patočkian "care for the soul" is an original Central-European synthesis of the core cultural and spiritual legacy of classical Greece and Christianity. Patočkian "care for the soul" is built on the existential experience of finitude as the universally shared experience of being (Koči 2019, p. 12). According to Patočka, this experience must be accepted both by individuals and the community as a positive challenge in their strife for meaning. Patočkian titanism and archetypal titanic figures of his cultural writings are representations of this universal effort and manifestations of Patočkian "care for the soul".

Funding: This research received no external funding.

Institutional Review Board Statement: Not applicable.

Informed Consent Statement: Not applicable.

Data Availability Statement: Not applicable.

Acknowledgments: I would like to thank the director of Jan Patočka Archive in Prague, Jan Frei, for giving me the access to the private correspondence of the theologian Bohumil Souček and Jan Patočka.

Conflicts of Interest: The author declares no conflict of interest.

Notes

1 Tomáš Garrigue Masaryk (1850–1937) was one of the leading Czech intellectuals of the second half of the 19th and the first half of the 20th century. Philosopher, sociologist, and politician, he is credited with the foundation of Czechoslovakia. He was the country's president from its foundation in 1918 until 1935.

2 Václav Černý was one of the leading Czech intellectuals of the 20th century. Professor of Romance studies, his main scholarly interest was in the Baroque period and in Romanticism in Czech, Spanish and French literatures. Patočka does not mention in his review that in 1934 Černý published an article entitled "Quelques remarques sur la critique masarykienne du titanisme romantique" in *Revue de littérature comparée* (published by the renowned academic publisher Honoré Champion, vol. 14, 1 January 1934). The text becomes a chapter of the 1935 monograph. The article reacts to the publishing of Masaryk's book *Modern Man and Religion* in 1934 (*Moderní člověk a náboženství*, Prague, J. Laichter). Although the content of Masaryk's book had already been published in magazines in 1890s, the publication in a book form renewed the debate on the topic.

3 "(. . .) M. V. C. ne nous ait rien dit sur l'histoire du mot titanisme, qu'il emploie comme s'il était tellement courant chez nous".

4 Patočka's tacit dialogue with Černý's concept of tianism and its significance in Czech culture will be further explored in a separate study. For more detailed study of Černý's work, see the monograph of Júlis Vanovič (1999), *Osobnosť Václava Černého*.

5 Apart from the already mentioned culturological approach to the study of literature that Patočka adopts, there is another factor that shapes his study of the myth of Faust, and that is the intellectual discussion of the nature of myth, especially the one developed within the realm of German Protestant liberal theology of the 1940s and 1950s by theologians such as Karl Barth or Rudolf Bultman. An important testimony of the theoretical impulse of theology for Patočka's thinking on myth is represented by the private correspondence between Patočka and the Protestant theologian Bohumil Souček. A relationship that started as an intellectual confrontation developed into a lifelong friendship (see Veselý 2013). The debate on myth between the two thinkers was particularly lively in the 1940s and the first half of the 1950s. This unpublished correspondence is an under-researched source and should be explored in more depth by theologians.

6 "[. . .] problém prodeje nesmrtelné duše".

7 "Nesmrtelnost duše v pravém smyslu dosáhnou ti, kdo dají přednost nebytí před zkázou duše".

8 "Pravá nesmrtelnost je zde pro ty, kdo hrůzu z tělesné smrti překonali hrůzou z absolutně negativní existence, aby dosáhli toho, čeho je možno dosáhnout na vrcholku vezdejšího života: své konečné absolutnosti".

9 "[. . .] z *lidského bytí* v jeho základní krisis a problematičnosti, která je bytostně mravní, tj. taková, že v ní běží o naše vlastní bytí a nebytí v částečné závislosti na nás, na našem rozhodnutí [. . .]".

10 "Velkolepá vůle k sebeobětování, která oduševňuje Markétku a vede ji k pohrdnutí pozemskou 'záchranou', přivádí Fausta Goethovy verze k tomu, že přes zdání své cynické banálnosti uvěří ve vyšší smysl světa a vydá se na cestu očisty".

11 "[. . .] s níž velikáni tehdejší doby vystoupili proti úplnému zesvětštění moderní doby a jednostrannostem osvícenství." (Patočka [1972] 2004e, p. 117).

12 "Obnova umění a člověka patří k sobě, na tom trvá i Adrian, ale jejich vztah vidí spíše naopak: napřed je třeba získat dimenzi zodpovědnosti, v ní se má vynořit 'nesmrtelná duše', a potom, jako výsledek tohoto uzdravení, průlom k novému umění, které by tak bylo vhodným nositelem další obnovy. Duševní a umělecká obnova tak vlastně není rozložena do etap, nýbrž naopak viděna pohromadě". (Patočka [1972] 2004e, pp. 117–18).

13 "Samotná postava Achillea—člověk, který si vybere ten krátký a slavný život! Něco takového, co jde proti směru běžného života, něco takového, co patří k základu řeckého politického citu".

References

Bolton, Jonathan. 2020. A Guest from the Unknown: Antigone and Jan Patočka's Cultural Criticism. *Bohemica Litteraria* 23: 13–30. [CrossRef]

Černý, Václav. 1934. Quelques remarques sur la critique masarykienne du titanisme romantique. *Revue de Littérature Comparée* 12: 661–76. Available online: https://www.proquest.com/scholarly-journals/quelques-remqrques-sur-la-critique-masarykienne/docview/1293244331/se-2?accountid=201395 (accessed on 2 July 2021).

Černý, Václav. 1935. *Essai sur le Titanisme dans la poésie romantique occidentale entre 1815 et 1850*. Prague: Orbis.

Champlin, Jeffrey. 2011. Hegel's Faust. *Goethe Yearbook* 18: 115–25. [CrossRef]

Cirot, Georges. 1936. Compte rendu de Václav Černý. *Essai sur le Titanisme dans la poésie romantique occidentale entre 1815 et 1850. Bulletin Hispanique* 38: 235–41. Available online: https://www.persee.fr/doc/hispa_0007-4640_1936_num_38_2_2725_t1_0235_0000_3 (accessed on 2 July 2021).

Derrida, Jacques. 1995. *The Gift of Death*. Chicago: The University of Chicago Press. First published 1993.

Erwin, Andrew. 2010. Rethinking Nietzsche in Mann's Doktor Faustus: Crisis, Parody, Primitivism, and the Possibilities of Dyonisian Art in a Post-Nietzschean Era. *The Germanic Review: Literature, Culture, Theory* 78: 283–301. [CrossRef]

Evink, Eddo. 2015. *The Gift of Life. Jan Patočka and the Christian Heritage*. London and New York: Routeledge, Available online: https://bookshelf.vitalsource.com/#/books/9781317410010/cfi/6/32!/4/2/4@0:0 (accessed on 2 July 2021).

Hagedorn, Ludger. 2011. Beyond Myth and Enlightenment: On Religion in Patočka's Thouhgt. In *Jan Patočka and the Heritage of Phenomenology. Centenary Papers*. Edited by Ivan Chvatík and Erika Abrams. London and New York: Springer, pp. 245–61.

Hagedorn, Ludger, and James Dodd. 2015. *The New Yearbook for Phenomenology and Phenomenological Philosophy XIV. Religion, War and the Crisis of Modernity*. Edited by Ludger Hagedorn and James Dodd. London and New York: Routeledge.

Hobbs, Angela. 2006. *Plato and the Hero: Courage, Manliness and the Impersonal Good*. Cambridge: Cambridge University Press. First published 2000.

James, Petra. 2020. On Mastodons, Faust and Tight Hugs: Jan Patočka on Literature. *Bohemica Litteraria* 23: 51–68. Available online: https://www.phil.muni.cz/journals/index.php/bohemica-litteraria (accessed on 2 July 2021). [CrossRef]

Josl, Jan. 2018. *Umění a péče o duši*. Prague: Karolinum.

Kočí, Martin. 2017. The Sacrifice for Nothing: The Movement of Kenosis in Jan Patočka's Thought. *Modern Theology* 33: 594–617. [CrossRef]

Kočí, Martin. 2019. Christianity after Christendom. *The Heythrop Journal*, 1–14. Available online: https://d1wqtxts1xzle7.cloudfront.net/59548340/2019_Christianity_after_Christendom.pdf?1559824702=&response-content-disposition=inline%3B+filename%3DChristianity_after_Christendom_Rethinkin.pdf&Expires=1626079883&Signature=RPkELaAPjLO2fqc0bNorW7ZbdQ3E~{}NBqNjZBBKPEKZTfQ~{}-SkgEjec1Puj-8jOhMNgJ977vDujlnSMwBvJd0mdVZenJ8Wp9PCDiUfy~{}fptMt76GgVRpW9GpJ83JqXOrsZuCgQFByE-i03ZP86T13ML~{}y2FQiQ6jLCSAcQmiww3TjEylzdEot8uE-YOb4fh-jWk9bKpRzQU5ZAstJKwpayT4WLRo65PWnlFhWSNKMUBATsJgr196A37qpOBUzEgN7hCfWy1Bbyp5EJPivpq~{}XEOPz367swxVDOL0fGY4xl7NW~{}yhFG-4ydhF-VcyOFfGtpDVTyQZNWGXLUocF7kKVEg__&Key-Pair-Id=APKAJLOHF5GGSLRBV4ZA (accessed on 2 July 2021).

Kočí, Martin. 2021. *La phénoménologie est-elle une théologie? Jan Patočka entre le tournant théologique et la tâche de penser*. Paris: Institut Catholique de Paris, pp. 87–111.

Kohák, Erazim. 1989. *Jan Patočka: Philosophy and Selected Writings*. Chicago and London: The University of Chicago Press.

Masaryk, Tomáš Garrigue. 1938. Goethe's Faust: Superman. In *Modern Man and Religion*. English Translation by Ann Bibza and Václav Beneš. London: George Allen, pp. 256–86. First published 1896.

Masaryk, Tomáš Garrigue. 1990. *Masarykova Abeceda*. Edited by Jaroslav Dresler. Prague: Melantrich. First published 1896.

Patočka, Jan. 1989. Titanism. In *Jan Patočka: Philosophy and Selected Writings*. Edited by Erazim Kohák. Chicago and London: The University of Chicago Press, pp. 139–44. First published 1936.

Patočka, Jan. 1996. *Heretical Essays in the Philosophy of History*. Translated by Erazim Kohák. Edited by James Dodd. Chicago and La Salle: Open Court. First published 1990.

Patočka, Jan. 2002. Duchovní člověk a intelektuál. In *Péče o duši III. Sebrané spisy Jana Patočky*. Edited by Ivan Chvatík and Pavel Kouba. Prague: Oikoymenh Filosofia, vol. 3, pp. 355–71. First published 1975.

Patočka, Jan. 2004a. Epičnost a dramatičnost, epos a drama. In *Umění a čas I. Sebrané spisy Jana Patočky*. Edited by Daniel Vojtěch and Ivan Chvatík. Prague: Oikoymenh Filosofia, vol. 4, pp. 348–58. First published 1966.

Patočka, Jan. 2004b. Vaculík a Kundera. In *Umění a čas II. Sebrané spisy Jana Patočky*. Edited by Daniel Vojtěch and Ivan Chvatík. Prague: Oikoymenh Filosofia, vol. 5, pp. 211–13. First published 1968.

Patočka, Jan. 2004c. Pravda mýtu v Sofoklových dramatech o Labdakovcích. In *Umění a čas I. Sebrané spisy Jana Patočky*. Edited by Daniel Vojtěch and Ivan Chvatík. Prague: Oikoymenh Filosofia, vol. 4, pp. 461–67. First published 1971.

Patočka, Jan. 2004d. Německá duchovnost Beethovenovy doby. In *Umění a čas I. Sebrané spisy Jana Patočky*. Edited by Daniel Vojtěch and Ivan Chvatík. Prague: Oikoymenh Filosofia, vol. 4, pp. 468–88. First published 1971.

Patočka, Jan. 2004e. Faustovská legenda včera a dnes: Nad románem Thomase Manna Doktor Faustus. In *Umění a čas II. Sebrané spisy Jana Patočky*. Edited by Daniel Vojtěch and Ivan Chvatík. Prague: Oikoymenh Filosofia, vol. 5, pp. 105–19. First published 1972.

Patoč ka, Jan. 2004f. Smysl mýtu o paktu s d'áblem: Úvaha o variantách pověsti o Faustovi. In *Umění a čas I. Sebrané spisy Jana Patočky*. Edited by Daniel Vojtěch and Ivan Chvatík. Prague: Oikoymenh Filosofia, vol. 4, pp. 510–25. First published 1973.

Patočka, Jan. 2006a. Společenská funkce literatury. In *Češi I. Sebrané spisy Jana Patočky*. Edited by Karel Palek and Ivan Chvatík. Prague: Oikoymenh Filosofia, vol. 12, pp. 175–85. First published 1968.

Patočka, Jan. 2006b. Kolem Masarykovy filozofie náboženství. In *Češi I. Sebrané spisy Jana Patočky*. Edited by Karel Palek and Ivan Chvatík. Prague: Oikoymenh Filosofia, vol. 12, pp. 366–422. First published 1977.

Shelley, Percy Bysshe. 2006. Preface to Prometheus Unbound. In *The English Anthology of English Literature*, 8th ed. Edited by Stepehn Greenblatt. New York and London: W. W. Norton & Company, vol. 2, pp. 775–78. First published 1820.

Svobodová, Zuzana. 2021. Jak dnes může promlouvat pověst o Faustovi? *Studia Aloisiana* 12: 23–39.

Vanovič, Júlis. 1999. *Osobnosť Václava Černého*. Bratislava: Kalligram.
Veselý, Jindřich. 2013. Jan Patočka a křesťanství. *Studia Philosophica* 65: 63–84. Available online: https://digilib.phil.muni.cz/bitstream/handle/11222.digilib/128396/1_StudiaPhilosophica_60-2013-2_5.pdf?sequence=1 (accessed on 2 July 2021).
Vojtěch, Daniel, and Ivan Chvatík. 2004. Ediční komentář. In *Umění a čas II. Sebrané spisy Jana Patočky*. Edited by Daniel Vojtěch and Ivan Chvatík. Prague: Oikoymenh–Filosofia, vol. 5, pp. 367–428.

Article

Orthodox Christian Bulgarians Coping with Natural Disasters in the Pre-Modern Ottoman Balkans

Raymond Detrez

Department of Languages and Cultures, Ghent University, 9000 Ghent, Belgium; Raymond.Detrez@Ugent.be

Abstract: Premodern Ottoman society consisted of four major religious communities—Muslims, Orthodox Christians, Armenian Christians, and Jews; the Muslim and Christian communities also included various ethnic groups, as did Muslim Arabs and Turks, Orthodox Christian Bulgarians, Greeks, and Serbs who identified, in the first place, with their religious community and considered ethnic identity of secondary importance. Having lived together, albeit segregated within the borders of the Ottoman Empire, for centuries, Bulgarians and Turks to a large extent shared the same world view and moral value system and tended to react in a like manner to various events. The Bulgarian attitudes to natural disasters, on which this contribution focuses, apparently did not differ essentially from that of their Turkish neighbors. Both proceeded from the basic idea of God's providence lying behind these disasters. In spite of the (overwhelmingly Western) perception of Muslims being passive and fatalistic, the problem whether it was permitted to attempt to escape "God's wrath" was coped with in a similar way as well. However, in addition to a comparable religious mental make-up, social circumstances and administrative measures determining equally the life conditions of both religious communities seem to provide a more plausible explanation for these similarities than cross-cultural influences.

Keywords: Bulgarians; Orthodox Christianity; Islam; Ottoman Empire; natural disasters; plague

check for **updates**

Citation: Detrez, Raymond. 2021. Orthodox Christian Bulgarians Coping with Natural Disasters in the Pre-Modern Ottoman Balkans. *Religions* 12: 367. https://doi.org/10.3390/rel12050367

Academic Editor: Dennis Ioffe

Received: 30 April 2021
Accepted: 18 May 2021
Published: 20 May 2021

Publisher's Note: MDPI stays neutral with regard to jurisdictional claims in published maps and institutional affiliations.

In this contribution, I intend to shed some light on how in the pre-modern era, religious beliefs determined the way Orthodox Christians—more specifically, Bulgarians in the Ottoman Balkans—coped with natural disasters. Under "pre-modern era in the Ottoman Balkans", I understand the period from the fifteenth through the early nineteenth centuries. In the fifteenth century, a long series of disasters, caused predominantly by the violent Ottoman conquest of the Balkans, had come to an end; in the late eighteenth century and fully in the nineteenth century, as a result of the spread of Enlightenment ideas, attitudes towards natural disasters gradually changed, acquiring a more rational, scientific character. The focus on the Bulgarians is justified by the fact that among the South Slavs, they were—next to the Greeks—the most representative part of the Orthodox Christian flock, administered by the Patriarchate of Constantinople. (The Serbs under Ottoman rule had, from 1557 to 1766, a patriarchate of their own and were more directly exposed to Western influences, originating from Venetian Dalmatia and, after the 1699 Treaty of Passarowitz, from the Habsburg lands.) Bulgarians, Greeks, Serbs, Vlachs, and others all together constituted what in Ottoman Turkish was called the *Rum milleti*—the *millet* (community) of the *Rum* (from Greek *Rhomaios*, "Roman", "Byzantine", whence "Orthodox Christian"), which in the Balkans from the late eighteenth century onwards largely coincided with the jurisdiction of the Patriarchate of Constantinople. The position of the Bulgarians and the other members of the *Rum milleti* in the Ottoman Empire was that of *zimmis*—beneficiaries of the *zimma*, the covenant between Muslim rulers and Christian and Jewish subjects, which guaranteed, for the "people of the Book [the Old and New Testament]", "protection in exchange for submission", or, as Braude and Lewis (1982, p. 3) phrased it more adequately, they were "discriminated against without being persecuted".

In pre-modern times, in the Ottoman Balkans as elsewhere, ethnic belonging was far less important than it became after the rise of nationalism in the nineteenth century. The cultural environment in which the Balkan Orthodox Christians then lived and acted was not determined by ethnic but by religious identity, as the well-known terms referring to this environment—the "Byzantine commonwealth" (Obolensky 1971), "Slavia orthodoxa" (Picchio 1991)—suggest. Although, texts in Church Slavonic continued to be copied and occasionally new texts in that language were created. Greek, initially the liturgical and administrative language and the language of intellectual communication used by the Patriarchate of Constantinople, gradually developed into the common language of the entire Orthodox Christian community, replacing, especially in Bulgaria and Macedonia, Church Slavonic in all "high" functions excluded the merely liturgical. Only in the autonomous Serbian Patriarchate of Peć the Church Slavonic tradition was preserved.

Greek sources should be treated not necessarily as Greek sources (in an ethnic sense) but as "Orthodox Christian sources in Greek", pertaining to all Balkan Orthodox Christians of whatever ethnic appurtenance. One of the main sources, informing us about the Orthodox Christians' attitudes towards natural disasters, is the remarkable autobiographical chronicle, written in Greek by priest Synadinos (1600–1662), who lived and worked in the multiethnic and multiconfessional city of Serres in Northern Greece (Odorico and Asdrachas 1996, with French translation and copious comments). The "Christians" (*christianoi*) whom Synadinos systematically refers to are not only Greeks but also Slavs, Vlachs and others. Some of these Christians bear indisputably Slavic names like "Petkos" and "Asanis" (Bulgarian Petko and Asen or Asan). (Odorico and Asdrachas 1996, pp. 96, 102, 176).

In spite of the religious discrimination non-Muslims in the Ottoman Empire were subject to, everyday living conditions of ordinary Christians, Muslims and Jews did not essentially differ. Most of them were *raja*, "flock" guided and guarded by *askeri* or representatives of the sultan, as the Ottoman vision on the state had it. Inevitably, the shared life conditions in the Ottoman Empire and the age-long cultural interaction of the three creeds produced many instances of syncretism and similar ways of dealing with the fortunes of life.

Having sketched the societal framework in which they lived and acted, I now proceed to the examination of how Bulgarians—and by extension all Orthodox Christians—in the pre-modern Ottoman Empire dealt with natural disasters and how their attitudes were determined by religious beliefs. A discussion of the similarities and dissimilarities between the attitudes of Christian Bulgarians and Muslim Turks, from a genuinely "Balkanistic" standpoint, will shed an additional light on Bulgarian popular religiosity.

The events people perceived as "disasters" are summed up exhaustively by Synadinos. In addition to the disasters caused by humans (as wars, plunders, massive executions and suchlike), Synadinos points out "epidemics and arsons (. . .), earthquakes, thunder, lightning (. . .), starvation and hailstorm (. . .), wild animals pillaging what little we have" (Odorico and Asdrachas 1996, p. 225). Only floods and volcanic eruptions are missing from his list of calamities. Sources mentioning the occurrence of these natural disasters— marginalia (notes written in the margin of a text), chronicles, *sicils* (Ottoman legal or cadastral records), observations by Western travelers and diplomats to the Orient—are relatively numerous. Extremely scarce, however, are those that provide us with more than laconic and commonplace information about the people's reactions to these disasters. The marginalia mention the exact date and even the hour earthquakes occurred—unfortunately often omitting the location—but mostly in a stereotypical, "telegraphic" style—as in the following Bulgarian examples, borrowed from Gradeva (1999, p. 58; quoting Načev and Fermendžiev 1984, p. 97):

Be it known when the earth quaked in the year 7195 [=1678], [in] the month of October, 23rd day.

[Be it known] when the earth quaked in the year 1738, in May, at noon. There was a fair in Svishtov at that time.

Human emotions are paid attention to in a lapidarian way:

> 31 May 1738. At 3 a.m. there was a powerful earthquake. The frightened people ran back and forth but clinging to each other, so that nothing befell them. After three hours it occurred again, this time weaker.

While the toll of human lives is not always specified, the material damages are often listed up at length, with a focus on the cult buildings. Thus, priest Jovčo from Trjavna describes in his chronicle the sad results of an earthquake in Bucharest in 1802, where he spent some time as a student in a Greek school:

> In the same year, in the month of 14 October, the earth quaked severely, in villages and towns, many houses were demolished, also the mosques in the towns fell, and everywhere; in Bucharest the tower [Colțea, RD], the tallest belfry of the Church of the Three Hierarchs, fell too. And many other churches fell. The two domes of the Church of Saint George fell, and the belfry was also damaged, and there is a plane between the metropolitanate and Domna Valasha [the Church of Domnița Bălașa, RD] in front of the house of Brancoveanu, and a lot of boiling water sprang from the earth; and also in Moldavia many houses and churches were damaged. And the Monastery of the Three Hierarchs, where the relics of the Venerable Paraskevas [Paraskeva, RD] are laid, was also damaged and the belfry fell and was razed to its foundations (Gradeva 1999, p. 59; quoting Načev and Fermendžiev 1984, p. 291)

The lack of interest in human casualties might be explained by the triviality of death in a society with a high mortality and an understanding of human existence focused on afterlife rather than life. The special attention paid to the damaged churches and monasteries obviously resulted from their religious relevance, but there might also be a more prosaic explanation: the costly, complicated and long-running legal procedures which the Ottoman authorities imposed whenever non-Muslim cult buildings had to be restored. Applications for a building permit, decisions about possible tax reductions and other measures taken by the authorities were registered in *sicils*, records of rulings by local *kadis* (Islamic judges) (Gradeva 1999, pp. 61–63). Evidently, neither of these formalized administrative documents took into consideration the attitudes and feelings of the applicants.

In spite of the lack of detailed information, even a cursory glance at the available documents reveals that to the inhabitants of the Balkan Peninsula during the period in question, every accident—small or huge, individual or collective—was interpreted by all religious communities as a token of the God's providence. God's involvement may be active, when the disaster is understood as a punishment—*gněv božij* (God's wrath), *grěh radi naših* (because of our sins). A Bulgarian priest noticed:

> And be it known that there was a tremendous earthquake, the fear of God. And the earth and the buildings trembled from the fear of God. It was in the year 1781. The month of March, the fourth day, the sixth hour in the middle of the day. (Načev and Fermendžiev 1984, p. 111)

An other Bulgarian chronicler ascribes two locust plagues to God's wrath:

> And God sent His wrath onto His creatures from Zagore to the Danube, He sent locusts. It was in the year 1690. And God sent His wrath a second time—He sent locusts in the year 1711 (. . .). (Načev and Fermendžiev 1984, p. 90)

In 1713, a Serbian chronicler mentions that "at that time it happened that through God's will (*hotěnijem božije* [sic]) lightning struck a mosque and destroyed it completely" (Stojanović 1903, p. 87). However, God also intervenes in individual tribulations, as in the case of a father who opposed his daughter to become a nun, "but due to God's command (*s božie povelenie*) he was not able to do anything" (Načev and Fermendžiev 1984, p. 106). Some catastrophes seem to have occurred only by God passively consenting to them (*božijem popuštenijem*). In this case too, however, His role was considered to be decisive. As Hans-Georg Beck (1978, p. 263) remarks, "Physisches Übel lässt nach ihrer Lehre [the orthodox theology, RD] Gott nur zu, um damit höhere Zwecke zu erreichen,

Strafe für Sünde, Prüfung, Hinwendung weg von der Materie zu geistigen Werten usw". When a catastrophe has an evident, understandable and explainable cause and there is no clear reason for the necessity of divine intervention, it may also be ascribed to the devil. Synadinos thinks that both Manolis being hit by a stone at the head and becoming deaf, and Vasilis breaking his leg after he is knocked down by a horse, are due to "a temptation by the devil" (*diavolou peirasmos*), all the more so as in both cases the complications turned out to be fatal. (Odorico and Asdrachas 1996, pp. 71, 114) Although he is perfectly aware of the exact causes of the fire in the *čaršija* (business district) of Serres in 1630, Synadinos nevertheless blames it on the devil:

> In September, on Sunday 30 at daybreak, the workshops were burning and the fire started from the workshops of the cotton-workers. In the workshop of the shoemakers some people drank and smoked, and passing by, they beat out their pipes, the fire in them not being extinguished, while the workshop was filled with bales of cotton. And—temptation by the devil—the fire reached the cotton and the entire workshop was set ablaze. And all the cotton-workers' workshops were burning. (Odorico and Asdrachas 1996, p. 108)

Ultimately, the entire *čaršija* burned to ashes, a catastrophe the city hardly recovered from. A Serbian chronicle attributed the long and devastating Austro-Turkish war, which started in 1683, as a combination of "a devilish scheme" (*djavolsko navaždenije*) and "God's consent because of our sins" (*božije popuštenije greh radi naših*) (Stojanović 1902, p. 431).

The most devastating catastrophes—earthquakes and epidemics—occurred with such a high frequency that they must have been felt as a part of "daily life". Manolova-Nikolova (2004) counts 49 outbreaks of the plague (pp. 63–65) and 24 earthquakes (pp. 49–50) in Bulgaria in the eighteenth century alone. Although they were often preceded by omens as solar eclipses or the appearance of comets, earthquakes happened unexpectedly and momentarily and could not be prevented or remedied. During an earthquake, there was no time for deliberation, and running away was considered a natural and acceptable reaction. Epidemics were of a different nature; they could be observed and reflected upon during their appearance and the reaction to them was the result of ample consideration.

The usual words for "epidemic" were *mor* (in Bulgarian), *mora* in Serbian, *thanatiko* in Greek, which all mean "death" or "lethality". In most cases, the epidemic referred to was pestilence (*čuma*), which as a rule, though not always, was more or less clearly distinguished from other epidemics as smallpox or malaria. (Manolova-Nikolova 2004, pp. 170–71) Smallpox, however, which struck mainly children and was never that lethal, was not considered as a divine punishment, but rather as a "*mécanisme de selection naturelle divine*" (Kostis 1996, p. 584). The plague, on the other hand, spread rapidly and massively among all generations and made people die in painful agony. Synadinos's autobiography ends with an impressive description of the plague in his native Serres in 1641:

> The same year, from September onward and during the entire winter and the entire summer, the was a huge epidemic among all people in the entire world and it spread all over Egypt, Anatolia, Bursa, Constantinople, the islands, Rumelia, Thessaly, Thessalonica, Serbia, Bulgaria, Phillipopolis, Melnik, Siderokastron, Drama, Zich-nokhoria, and Serres, and it spared neither cities nor villages, not a single house, and no one among the elderly people could recall an epidemic that spread over such a large space. And wherever death arrived, it reaped everyone and even if one single person escaped, there was no single house that was spared. And the plague was so devastating that no one caught by it could escape, and it reaped people so quickly that they suffered no more than one or two days, sometimes, rarely, a week, and of a hundred of sick only one recovered. And how many Turks, Christians, Jews, and Gypsies in Serres died, about 12,000. At the same time as the plague also scabies and eye diseases spread. That year a great curse struck the people. (Odorico and Asdrachas 1996, pp. 169–71)

Like other writers of autobiographies, chronicles and marginalia at that time, Synadinos explicitly points out here that members of all religious communities fell victim to the plague. Synadinos's use of the overarching term "Christians" for people of various ethnic origins is typical of the Balkans in the pre-modern (pre-national) era. As we already mentioned, being an Orthodox Christian, a Catholic, an Armenian, a Jew, or a Muslim was far more important than belonging to some ethnic group. Religious belonging was of particular relevance also in relation to the natural disasters: punishing the sinners, God was believed to discriminate between the various kinds of believers. Synadinos points out that the plague killed more Turks (which means "Muslims") than Christians.

> And during winter many Turks died, but not so many Christians, and the Turks were amazed and all those that did not understand, said: "The Christians apply some sorcery in order not to die and they threw the evil upon the Turks and that is why they die." (. . .) And during winter the Christians did not die, but in May, June and July many of them died, but the Turks died all year long in great numbers. (Odorico and Asdrachas 1996, p. 175)

In 1725, an anonymous Bulgarian author wrote that "[i]n that good winter began a bad disease, its name is the plague. From October onward, it killed only a few Christians, but Turks all the time" (Načev and Fermendžiev 1984, p. 83). Although these authors do not explicitly claim that God punishes non-Orthodox Christians more severely than Orthodox Christians, it is clearly implied. Throughout the ages following the Great Schism in 1054, Orthodox Christians believed God to be particularly resentful of the Catholics. Partenij Pavlovič (ca. 1700–1760), a highly educated Bulgarian cleric who spent his life in Serbia, did not hesitate to ascribe three major disasters that happened during his lifetime to Catholic outrages against Orthodox Christians:

> I think that the Gracious God in our times too, as He destroyed the ancient city of Nineveh, so did He destroy now Lisbon, the glorious city in Portugal, because they chased away our faith and tortured our Orthodox Christians, as I heard from some that immigrants and foreigners are cruelly tortured. And this city was destroyed in 1755, on December 1, just before noon.

Some years ago, in the Hungarian city of Buda, they wanted to create a Roman altar in one of our Orthodox churches. Through the wrath of God, the gunpowder beneath the city exploded and provoked an earthquake, and all people started fleeing far away out of fear for the fiery ignitions and the thundering of the canons.

In the glorious Venetian city of Corfu in 1719, the same happened. The entire palace of the general was destroyed by thunder and fire because he wanted to make a Latin altar in a church of the Orthodox Christians, in which the holy relics of our Holy Father Spyridon of Trimythous the Wonderworker rest (Angelov 1964, p. 206).

The eagerness with which the writers of the marginalia sum up the number of collapsed mosques may also be inspired by the idea that God preferably punishes non-Orthodox Christians. However, the fact that in all Ottoman cities, mosques were the highest and therefore the most vulnerable constructions is probably a more trustworthy explanation.

Although God's wrath apparently strikes entire communities, "because of their sins", nevertheless each case of disease or death is considered to be individual. The Byzantines made a distinction between God's providence "in general" and individual "fate" (*tychē*), a concept inherited from antiquity. *Tychē* corresponds to Bulgarian "*sădba*" or *kăsmet* (from Turkish *kısmet*) and Serbian "*sudbina*". In times of plague, some people do not fall ill, others fall ill but recover, and still others die—each case a question of *sădba*. Byzantine theologians had not been able to completely remove the contradiction that obviously exists between "fate" and "divine providence" (Beck 1978, pp. 271–73). Synadinos, for his part, strongly keeps to a relation between (individual) "sin" and "divine punishment" to explain the seeming arbitrariness of God's providence: "[God} does not punish us because of the sins of others, but when a man commits a sin, it is him whom God punishes." (Odorico and

Asdrachas 1996, p. 180). For some victims of the plague, Synadinos suggests some rather far-fetched reasons for their death. Priest Georgis, who died after being infected when burying the dead, allegedly had the unpleasant habit to take revenge for insults during the liturgy, and "maybe that is the reason why God deprived him of his life" (Odorico and Asdrachas 1996, p. 178). About a deceased female roe vendor, he points out that "as we have heard everyone telling, she was not that innocent. (...) Maybe God took her life for that reason" (Odorico and Asdrachas 1996, p. 180).

Conversely, those who escaped death also did so as a result of divine intervention. Western travelers in the Balkans love to comment on the—in their opinion—radically different attitudes of Christians and Muslims towards natural disasters—a distinction that, curiously enough, is seldom commented on in the Slavic or Greek sources. Muslims are alleged to fatalistically accept the plague as sent by God, finding comfort in the idea of individual predestination: they believed that one's lifetime was completely determined by God's providence or "written on the forehead". "I had two notable examples", writes Sir Henry Blount (1602–1682), an English landowner and traveler, in 1636,

> one was at *Rhodes*, where just as we entered the Port, a *French* Lacquey of our company died with a great plague for which he had taken from the *Gunners Mate*, who with one running upon him, conversed and slept among us: The rest was so far from fear, at his death, as they sate presently eating, and drinking, by him, and within half an hour, after his removal, slept on his Blanquet, with his clothes instead of a Pillow, which when I advised them not to do, they pointed upon their foreheads, telling me it was written there at their *birth* when they should die; they escaped, yet divers of the passengers thereof before we got to *Egypt*: The other was at my passage to *Adrianople* in *Thrace*, myself, the *Ianizary*, and one other being in a *Coach*, we passed by a man in good quality, and a Soldier, who lying along, with his Horse by, could hardly speak so much, as to intreat us to take him into *Coach*; the *Ianizary* made our companion ride his Horse, taking the man in; whose breast being open, and full of plague tokens, I would not have him received; but he in like manner, pointing to his own forehead, and mine, and told me we could not take hurt, unless it were written there, and that then we could not avoid it; the fellow died in the night (. . .) (Blount 1636, pp. 85–86)

The Greek historian Kostas Kostis (1996, p. 585) remarks that "[l]es textes du XVIIIe et des débuts du XIXe siècle, abondent en commentaires moqueurs des voyageurs occidentaux, qui, du haut de leur supériorité, accusent les Musulmans d'attitude fataliste envers la peste et critiquent leur immobilité dans les lieux touchés par l'épidemie". Particularly revealing is the observation made by the French traveler (of Hungarian origin) Baron François de Tott (1733–1793) in his *Mémoires sur les Turcs et les Tartares* in 1785. He notices that "les Turcs trouvent encore dans une aveugle prédestination une plus grande sécurité". "Exempts de l'excès du meme préjugé, les Grecs [the Orthodox Christians, RD], les Arméniens, les Juifs ont étudié une sorte de remède dont ils paraissent user avec une espèce de success". However, "[l]es Européens sont les seuls qui prennent quelques précautions contre la contagion" (Tott 1785, vol. I, pp. 30–31). Tott's observation perfectly fits into Maria Todorova's concept of "Balkanism", a set of biased perceptions of the Balkan peoples attributing them a particular mental make-up that culturally locates them permanently "in-between" the Orient, "les Turcs", and the West, "les Européens", and makes them "semi-European", "semi-Asian". (Todorova [1997] 2009) Ironically, the fatalistic Ottomans were familiar with vaccination already in the early eighteenth century, before it found its way to Western Europe. The English Lady Mary Wortley Montagu described the procedure in 1718 in one of her letters from Istanbul: "The smallpox, so fatal and so general amongst us, is here entirely harmless by the invention of engrafting [which is the term they give it]. (. . .) Every year thousands undergo this operation (. . .) There is no example of anyone who has died in it (. . .)" (Montagu 1988, p. 121).

A closer look reveals that Christians and Muslims facing the plague actually behaved in a rather similar way. In some cases, the ill were isolated, visitors were put into quarantine and massive religious gatherings were prohibited, albeit mostly to little avail (Panzac 1985, pp. 298–99). Both religious communities, however, considered collective and continuous praying as the most effective way to cope with the epidemic. In the eighteenth century, after being introduced in the Peloponnesus and the Aegean Islands, Saint Haralampij enjoyed an increasing popularity in Bulgaria and Serbia as a protector against the plague, as the mushrooming *vitae*, icons and special liturgies, dedicated to the saint, indicate (Manolova-Nikolova 2004, pp. 121–67). His feast day on February 10 was called *Čuminden*, "day of the plague", in Bulgaria. A curious phenomenon, probably occurring rarely but typical of Ottoman multiconfessional society, was the shared praying sessions between Christians and Muslims. Manolova-Nikolova (2004, pp. 159, 182) quotes a Catholic priest in Trănčovica in North Bulgaria, where in 1806 a kind of "prophet" appealed to "all priests, to whatever sect they belonged" to bring a sacrifice to "the goddess Plague". "Catholics, schismatics [Orthodox Christians, to the Catholic author, RD] and Turks [Muslims, RD] came in droves to hear the old inspirer". As with many other cases of religious syncretism in the Balkans, this one too seems to have pagan roots: the personification of the plague as an evil spirit—a *veštica* (witch) in Bulgarian, a *cin* (demon) in Turkish. (Panzac 1985, p. 291)

Both Muslims and Orthodox Christians struggled with the problem to what extent it was permitted to oppose the Lord's will by attempting to escape His wrath. Was it allowed to flee the location where the plague had spread in order to avoid contagion? Within the Islamic community, this problem had sporadically been discussed since the time of Mohammed already. (Dols 1977, pp. 22–25). It became topical in the middle of the fourteenth century, after the Black Death had made millions of victims in the Middle East and Europe. Ibn Hajar al-Asqalani (1372–1449), referring to the Hadith, recommended not to flee from the plague, but also not to enter regions struck by it. Other Islamic theologians, the most influential among them being Jalal al-Din al-Suyuti (1445–1505), denied that the disease was transmitted from one person to another (the role of rats and fleas was not yet known), but that it spread through miasmas (poisonous bad air). The fact that everyone came in contact with the miasmas but not everyone was infected and not everyone who was infected died was explained by God's will. Al-Suyuti added that not only to die from the plague but also to recover from it was of equal worth as dying as a *şehit*, a martyr fighting for the faith (Ayalon 2015, p. 26; Manolova-Nikolova 2004, pp. 179–80; Panzac 1985, pp. 282, 290).

The Ottomans adopted the same principles but tended nevertheless to be more tolerant towards flight. The learned Taşköprüzade Ahmed (1495–1561) and *şeyḫ ül-islām* Ebussuud Efendi (1490–1574) were of the opinion that flight in certain circumstances was allowed (Ayalon 2015, p. 76; Varlık 2008, pp. 172–76, 202–4). Elias Abesci, an English servant at the sultan's court, wrote in his description of the Ottoman Empire (published in French):

> Tant que ce précepte du Koran est resté sans commentaire, les Turcs y ont adhéré constamment à la lettre; mais un jour un Mufti moins superstitieux que ses prédécesseurs, a trouvé, dans ce meme livre sacré, un passage, qui, à l'aide de quelque petit supplément, interprète ainsi le premier. Quoiqu'il soit indubitable qu'un homme ne doit tenter de contrarier la volonté de Dieu, il peut cependant en cas de peste ou dans toute autre maladie contagieuse dans la ville, en sortir, pourvu qu'il ne s'en éloigne pas à une plus grande distance de six lieues. (Abesci 1792, vol. II, p. 223)

Anyhow, not fleeing was not a question of mere fatalism. It was also considered a holy duty to take care of ill relatives and to bury the deceased according to the religious prescriptions. In the patriarchal Muslim community, these duties were strong imperatives not to flee. (Panzac 1985, pp. 283–84)

At first sight, Orthodox Christians seem to have fled for the plague without any qualms, although to them too the duty to look after the ill and the dead was a serious reason to stay (Manolova-Nikolova 2004, p. 171). However, this permissive attitude towards fleeing for the plague was relatively new. Just like the Muslims, the Christians thought that the epidemic spread through miasmas with which all people got in touch, but not all died. "Donc, les causes de la mort ne se trouvent pas dans le phénomène physique, c'est-à-dire la maladie pestilentielle, mais bien ailleurs, dans 'le terme de la vie qui est atteint'"(Kostis 1996, p. 589). Because the end of one's life was predestined, running away had no sense. Theophanes of Medeia (1400–1474) had advised that "il ne faut pas partir vers des lieux déserts ni meme chercher des vents, mais se réfugier vers Dieu, celui qui peut sauver tous ceux qui s'adressent à Lui" (Kostis 1996, p. 590). Christian tolerance towards those who flee for the plague, combined with an outspoken preference for abiding by God's providence by staying and hoping for divine protection, clearly transpire from the following passage from the Chronicle of the Monastery of Trojan in Bulgaria:

> And again in the that year 1837 there was a great lethal disease among the people, which is called the plague; it was all over the universe and many people and children, and countless women died, not only Christians, but also infidel Turks and Jews; and people ran away from the towns and the villages hiding from the plague in the vineyards, but there they died just the same. Others fled to the monastery, and came to God's monastery, to hide under its roof. People from Loveč and Sopot fled [to the monastery, RD].

Additionally, when the feast day of the Dormition of the Mother of God had come, the notables that were in the monastery gathered and consulted with the abbot, and they promised to give him money if he did not allow visiting pilgrims on the feast day [to enter the monastery, RD] and to infect the healthy, and if he closed the gates, so that ill and infected people could not enter.

"But the abbot", Mister Partenij replied, "I cannot do that, my children, I cannot close the gates, because these our brothers and Christians run away to the monastery and to us out of need as well—some in order to confess, others to partake in the holy sacraments. How can I send them away without letting them participate in the communion. But do not be afraid, put your hope on the Mother of God, pray to Her incessantly and do not be afraid. Among them some retired to the cabins [in the neighborhood of the monastery, RD], others remained in the monastery and the abbot gathered the monks and convinced them to stay and to make preparations for the feast day and the vigils as usual. Those among the brothers that were disposed to obedience, remained, others did not obey and ran away to the woods. And those who were obedient and remained in the monastery and put their hope on the Most Holy Mother of God were saved by Her from the lethal ulcer. And those who ran away from the monastery perished. The Mother of God accomplished a great miracle, to the astonishment of all. The brothers that remained in the monastery welcomed the pilgrims and served them. There were pilgrims who were infected and fallen ill from that cruel ulcer, but no one died. This happened because the abbot, Mister Partenij, and the other brothers in the holy monastery devoted themselves to prayer and vigils and the Mother of God protected them and gave health and life to all those that were under Her canopy and She saved them from the cruel ulcer" (Načev and Fermendžiev 1984, pp. 281–82).

Although advised against by Christians and disapproved of by Muslims, flight was resorted to by both communities. However, the decision to flee or not depended not only on religious considerations. Administrative measures and social circumstances seem to have been important factors as well. According to Sam White (2011, p. 89; quoted by Ayalon 2015, p. 79), most Ottoman subjects remained where they were "not because flight violated divine law, but because it violated imperial law". Flight created chaos and allowed people to avoid paying taxes. Before leaving, they had to fulfill their financial and fiscal obligations (Manolova-Nikolova 2004, p. 176). In addition, one's social position played a significant role too. Well-to-do citizens, who could isolate themselves in huge mansions enclosed by gardens and walls usually preferred to remain, also to prevent their properties from being robbed. On the other hand, if they possessed a house in the countryside, which was often the case, they could find shelter there. As a rule, poor city-dwellers for financial reasons preferred not to leave their working places and few belongings (Ayalon 2015, pp. 156–57). In general, peasants were more inclined to flee and, doing so, displayed a greater solidarity and mutual support than the more individualistically minded city-dwellers (Manolova-Nikolova 2004, pp. 176–77). M. Mičev (1964) describes the case of the inhabitants of 42 Bulgarian villages in the region of Vraca, north of Sofia, who in 1762–1763 fled for the plague. The inhabitants of four of these villages found shelter in a cave; others built temporary accommodations in the neighborhood of their native village and returned to it after the plague had come to an end; still others transformed their temporary accommodations into new, lasting settlements. According to the Bulgarian historian Hristo Gandev (1976, pp. 539–32), the behavior of Christians and Muslims was so divergent that it resulted not only in the foundation of new villages but also in a demographic shift, whereby Christian Bulgarians in the cities became a majority at the expense of the Turkish Muslims, with dramatic political consequences during the national revival period. However, such resettlements occurred also among the Turkish population. Manolova-Nikolova (2004, p. 181) mentions the names of about a dozen of villages in Bulgaria that were left by Turks and/or Pomaks (Bulgarian speaking Muslims) after an outbreak of the plague. According to Manolova-Nikolova, they did so under the influence of the Christian practice.

One should not think of flight too much as a rational solution, made in disregard of religious prescriptions. Lady Montagu, describing the opening of the veins during "engrafting" (vaccination), relates that "[t]he Grecians have commonly the superstition of opening one in the middle of the forehead, in each arm, and on the breast to mark to sign of the cross" (Montagu 1988, p. 121). Obviously, even remedies based on empirical evidence needed some religious substantiation. Fleeing should not be merely tolerated or permitted by God but somehow required by God Himself. This idea is expressed in the margin of a Bulgarian liturgical book from 1725, saying that "whoever says 'I do not fear the plague' be accursed by God". (Načev and Fermendžiev 1984, p. 93) An elaborate "theological" justification offers Synadinos:

> Because when you stay and do not try to escape, you display in front of God the attitude of one who did not fail and does not fear Him (…) It is fair and fitting to escape from his just wrath, as the prophet Isaiah has ordered: "My child, enter your chambers, and shut your doors behind you; hide yourselves for a little while until the fury has passed by." The words of the prophet make clear that you should escape without hesitation, not at your place, but in a hidden place where the wrath of God cannot find you. (Odorico and Asdrachas 1996, p. 233)

Muslims in Hejaz came up—"non sans humour", according to Panzac, quoting a letter by Johann Ludwig Burckhardt—with an other excuse to justify their fleeing from the plague:

> Je vis alors qu'un grand nombre d'habitants fuyaient vers la montagne. Lorsqu'on leur demandait pourquoi ils avaient peur, puisque si leur destin était de mourir, la mort irait aussi les chercher sur la montagne, ils répliquaient: 'La peste est une grâce, qu'Allah a envoyé sur la terre, afin d'appeler rapidement au ciel les hommes vertueux. Nous sentons que nous ne sommes pas encore digne de cette grâce et nous nous en écartons jusqu'à d'autres temps. (Panzac 1985, p. 286)

To twenty-first century readers, these justifications probably sound rather casuistic. However, as Lucien Febvre ([1942] 1947) demonstrated in relation to Rabelais and his contemporaries in sixteenth-century France, people in the pre-modern world were incapable of reasoning The authors declare no conflict of interest outside of religion. To the Bulgarians in the Ottoman Balkans too, life was permeated by religion, all phenomena had a spiritual meaning and all actions needed a religious justification. They could think but in religious categories, according to an unshakeable religious logic.

Funding: This research received no external funding.

Institutional Review Board Statement: Not applicable.

Informed Consent Statement: Not applicable.

Data Availability Statement: The study does not report any data.

Conflicts of Interest: The author declares no conflict of interest.

References

Abesci, Elias. 1792. *État present de l'Empire Ottoman. I–II*. Paris: Villette.

Angelov, Bonju. 1964. *Săvremennici na Paisij II [Contemporaries of Paisij II]*. Sofia: BAN.

Ayalon, Yaron. 2015. *Natural Disasters in the Ottoman Empire. Plague, Famine, and Other Misfortunes*. New York: Cambridge University Press.

Beck, Hans-Georg. 1978. *Das byzantische Jahrtausend*. München: Beck.

Blount, Henry. 1636. *A Voyage into the Levant. A Breife (sic) Relation of a Journey, Lately Performed by Master H. B.(…)*. London: Andrew Crooke.

Braude, Benjamin, and Bernard Lewis. 1982. *Christians and Jews in the Ottoman Empire: The Functioning of a Plural Society*. London and New York: Holmes & Meier.

Dols, Michael. 1977. *The Black Death in the Middle East*. Princeton: Princeton University Press.

Febvre, Lucien. 1947. *Le problème de l'incroyance au 16e siècle: la religion de Rabelais*. Paris: Albin Michel. First published 1942.

Gandev, Hristo. 1976. *Problemi na bălgarskoto văzraždane [Problems of the Bulgarian National Revival]*. Sofia: Nauka i izkustvo.

Gradeva, Rossitsa. 1999. Ottoman and Bulgarian Sources on Earthquakes in Central Balkan Lands (17th–18th Centuries). In *Natural Disasters in the Ottoman Empire*. Edited by Elizabeth Zachariadou. Rethymnon: Crete University Press, pp. 55–66.

Kostis, Kostas. 1996. La peste n'est pas une mort naturelle, mais un châtiment de Dieu. In *Conseils et mémoires de Synadinos, prêtre de Serrès en Macédoine (XVIIe siécle)*. Edited by Paolo Odorico and Synadinos. Paris: Association "Pierre Belon", pp. 581–96.

Manolova-Nikolova, Nadja. 2004. *Čumavite vremena 1700–1850 [Times of Pestilence 1700–1850]*. Sofia: IF-94.

Mičev, M. 1964. Vremenni razselvanija ili premestvane na njakoi selišta văv Vračansko prez XVIII v. poradi čumna epidemija [Temporary resettlement or relocation of some villages in the region of Vraca during the eighteenth century because of the plague]. *Izvestija na etnografskija institut i muzej* 8: 25–44.

Montagu, Lady Mary Wortley. 1988. *Embassy to Constantinople. The Travels of Lady Wortley Monagu*. London: Century.

Načev, Venceslav, and Nikola Fermendžiev, eds. 1984. *Pisahme da se znae. Pripiski i le-topisi [We Wrote So That It Be Known. Marginalia and Chronicles]*. Sofia: Otečestven front.

Obolensky, Dimitri. 1971. *The Byzantine Commonwealth: Eastern Europe, 500–1453*. London: Weidenfeld & Nicolson.

Odorico, Paolo, and Spyros I. Asdrachas, eds. 1996. *Conseils et mémoires de Synadinos, prêtre de Serrès en Macédoine (XVIIe siècle)*. Paris: Association "Pierre Belon".

Panzac, Daniel. 1985. *La Peste dans l'Empire Ottoman 1700–1860*. Leuven: Peeters.

Picchio, Riccardo. 1991. *Letteratura della Slavia ortodossa: IX–XVIII sec*. Bari: Dedalo.

Stojanović, Ljubomir. 1902. *Stari srpski zapisi i nadpisi I [Old Serbian Marginalia and Inscriptions I]*. Beograd: Državna štamparija.

Stojanović, Ljubomir. 1903. *Stari srpski zapisi i nadpisi II [Old Serbian Marginalia and Inscriptions II]*. Beograd: Državna štamparija.

Todorova, Maria. 2009. *Imagining the Balkans*. New York: Oxford University Press. First published 1997.

Tott, François de. 1785. *Mémoires de Baron de Tott sur les Turcs et les Tartares. I*. Amsterdam.

Varlık, Nükhet. 2008. Disease and Empire: A History of Plague Epidemics in the Early Modern Ottoman Empire (1453–1600). Ph.D. dissertation, University of Chicago, Chicago, IL, USA.

White, Sam. 2011. *The Climate of Rebellion in the Early Modern Ottoman Empire*. New York: Cambridge University Press.

Article

To Be or Not to Be God—The Issue of Authorial Power in Dostoevsky [†]

Alexander Zholkovsky

College of Letters, Arts and Sciences, University of Southern California, Los Angeles, CA 90007, USA; alik@usc.edu

† This is a revised version of an article originally published in Russian in a collected volume *Paradoksy russkoi literatury*. Edited by Wolf Shmid and Vladimir Markovich. St-Petersburg: Inapress, 2001, pp. 234–47.

Abstract: This paper problematizes the now widely accepted concept of Dostoevsky's dialogism, which alleges the 'Author's' equal empowerment of all his characters. Using examples from *Crime and Punishment* and *The Brothers Karamazov*, Zholkovsky focuses on instances of 'scene-staging' based on the 'scripts' devised and enacted by some characters, that are 'read,' with varying success, by their targets. He documents the resulting 'discursive combat' among the characters, with special attention paid to those 'playing god' and thus, the more 'authorial' among them. In several cases, the would-be 'divine' manipulation is shown to be consistently subverted by the Dostoevskian narrative. However, in one instance, where Aliosha Karamazov charitably scripts Captain Snegirev's behavior, the ensuing discussion of this episode, in Aliosha's conversations with Lise Khokhlakova, upholds Aliosha's right to play god with the Other—"for the Other's own good", of course (not unlike the Grand Inquisitor).

Keywords: authorial power; scripting others; reading each other; discursive combats; playing god; Raskolnikov; Luzhin; Aliosha Karamazov; the Grand Inquisitor

check for
updates

Citation: Zholkovsky, Alexander. 2021. To Be or Not to Be God—The Issue of Authorial Power in Dostoevsky. *Religions* 12: 506. https://doi.org/10.3390/rel12070506

Academic Editor: Dennis Ioffe

Received: 3 June 2021
Accepted: 30 June 2021
Published: 7 July 2021

Publisher's Note: MDPI stays neutral with regard to jurisdictional claims in published maps and institutional affiliations.

'God is dead, long live God, that is, Man'. This slogan by various voices resounded in Russian even before Friedrich Nietzsche.

Ivan Turgenev's Bazarov claims that the nihilistic 'cause' needs such slaves of its discourse as Sitnikov, as noted to Arkady:

-You, brother, are still stupid, I see. We need Sitnikovs. I, understand this, I need such dummies. It's not the Gods' job to fire pots … !

-"Wow," Arkadii thought to himself, and only then did he discover the bottomless abyss of Bazarov's ego.—"You and I are gods, aren't we? That is to say, you are a god, while I am just a dummy?"

-Yes, Bazarov repeated sullenly, You are still stupid. (Fathers and Sons, chp. 19, vol. 3, pp. 202–3; Turgenev 1961–1962)

(-Ты, брат, глуп еще я вижу. Ситниковы нам необходимы. Мне, пойми ты это, мне нужны подобные олухи. Не богам же всамом деле горшкиобжигать! …

Эге, ге! … —подумал про себя Аркадий, и тут только открылась ему … вся бездонная пропасть базаровского самолюбия.—Мы, стало быть, с тобой боги? то есть—ты бог, а олух уж не я ли?

-Да—повторил угрюмо Базаров—тыеще глуп.)

Bazarov's impertinent discursive move consists of reversing the perspective of the Russian proverb, "It doesn't take Gods to fire pots", which is, in fact, quite reverent towards gods—in the spirit of rendering to potters the things that are potters' and to God the things that are God's. Incidentally, Turgenev's choice of this proverb has rich theological overtones, since according to one of the biblical versions ("And, God Yahweh created Man from the dust of the ground . . . "; Genesis 2:7), "God's creation of Man [*wayyisar*] is like a potter's work [*yosar*]" (Shifman 1993, vol. 58, p. 270).

In an opposite, kenotic, gesture, Leo Tolstoy, sated with his absolute power over the text, would wonder: "Who Should Learn to Write from Whom: Peasant Kids from Us or We from Peasant Kids?" and even signed some of the Iasnaia Poliana peasant schoolkid's essays—as containing the ultimate truth.

These two moves—up and down the ladder of discursive authority—merge in the figure of the Author. The Author both acknowledges his inevitable human limitations, akin to those of his characters, and acts as the creator of an artistic world where these characters walk under him as under God. Thus, Nikolai Gogol confessed that his comic characters were his "own trash . . . demoted from generals to privates" (*Selected Passages from Correspondence with Friends*, chp. XVIII, p. 3; Gogol' 1976–1979, vol. 6, pp. 259–60), and yet he was accused of authorial despotism by a literary descendant of one of these "privates": the main protagonist of Fedor Dostoevsky's *Poor Folk* (Bocharov 1985, pp. 161–209).

Abolishing the distance between the divine top and the human bottom takes different forms. By projecting himself into Platon Karataev or a peasant kid, the Author sheds his social and institutional prerogatives, strips off his epaulettes, goes slumming. Often such reversal of the power hierarchy is accompanied by a Rousseauist emphasis on the soulful, 'soilist', intuitive, and other advantages of identifying with the 'social bottom'. Another variant is the author's co-opting of the 'morally low,' that is, an attempt to place representatives of 'evil' in the service of the 'good' and its narrative discourse—in the spirit of John Milton's use of Satan and Johann Wolfgang von Goethe's utilization of Mephistopheles. Such, for instance, was the intention of the author/narrator of *Dead Souls*, in his own words, to gainfully "harness" a true "scoundrel", i.e., Chichikov (*Dead Souls*, I, chp. 11; Gogol' 1976–1979, vol. 5, p. 214).

This harnessing was twofold. In the distant Dantesque perspective of Gogol's unrealized project, the "scoundrel" was to become the engine of Russia's providential transfiguration; while in the immediate 'infernal' reality of the book's first volume, the "scoundrel" was trusted with the better part of plot-building. Co-authorship with a trickster-protagonist has a venerable literary pedigree; and ethical problems do arise when purely narratorial services rendered to the author by the protagonist's deceitful artistry lead to an ambiguous moral symbiosis.

According to Mikhail Bakhtin, Dostoevsky's radical contribution to literary discourse consisted in the carnivalesque overthrow of the author's 'divine' power over his characters: conferring on "privates" equal dialogical rights with the "generals", and thus empowering them as narrators and ideologically authoritative personae. One of Dostoevsky's favorite plot devices was to have a rationalistic manipulator fail as a result of arrogantly ignoring the human unpredictability of the 'Other', usually a modestly 'ordinary' character. (Dostoevsky might have learned this from the way Gogol had Chichikov exposed by none other than Korobochka, the most obtuse of his gentlefolk partners).

An important aspect of this motif is its focus on discourse. The manipulator 'prescribes' to his 'naïve' partner a certain 'script' and does so in an 'authorial' manner that schematizes, objectifies, 'automatizes' him. However, the latter, being in touch with 'reality,' 'reads' and 'rewrites' the script in his own way, thereby refuting his 'automaticity' and thus beats the 'author' of the script at his own game of script-writing. Examples abound, e.g.:

-Raskolnikov's skillful, but eventually failed, attempt to pretend to laugh nonchalantly as he enters, along with Razumikhin, Porfirii's apartment;

-Smerdiakov's surprising Ivan and others by suddenly speaking up—like the "Balaam's donkey";

-and the title character of "A Gentle Creature" similarly surprised her husband, who had planned their entire life together in advance and on seeing her dead body thought he was "only five minutes late".

This Dostoevskian invariant was eloquently formulated by Fed'ka Katorzhnyi, an archetypal 'low-life' character, in his conversation with Stavrogin about Peter Verkhovenskii:

"[He] has … learned all God's pre-destinations but he is also subject to criticism … If he has said a man is a scoundrel, he knows nothing about him but that he is a scoundrel. Or if he has said one is a fool, he has no other title for that man but a fool. While I may be only a fool on Tuesdays and Wednesdays, but on Thursday I am smarter than him … Now he knows about me that I am in a bad need of a passport … so he thinks he's got my number. He … has it very easy living in the world, 'cause he just imagines a person and then lives with one as such" ('The Demons', II, 2; Dostoevsky 1972–1990, vol. 10, p. 205).

("Петр Степаныч < … > все божии планиды узнал, а и он критике подвержен… У [него] коли сказано про человека: подлец, так уж кроме подлеца он про него ничего и не ведает. Али сказано—дурак, так уж кроме дурака у него тому человеку и звания нет. А я, может, по вторникам да по средам только дурак, а в четверг и умнее его … Вот он знает теперь про меня, что я оченно паспортом скучаю … так уж думает, что он мою душу заполонил. Петру Степановичу … оченно легко жить на свете, потому он человека сам представит себе, да с таким и живет.")

'The war of the scripts' is by definition: strategic, discursive, and hierarchical. The question is always, who will outdo whom in 'reading' and 'rewriting' the other. And the carnivalesque mode of Dostoevsky's narrative inevitably generates paradoxes. The protagonist-narrator of "Notes from Underground," whose philosophical kinship with the implied (and, indeed, the real) author poses a serious problem, is referred to as a "paradoxalist." And, of course, Raskolnikov, true to his telling surname, is a walking paradox: a moral immoralist, a clever fool, an under-Overman.

Not all of Dostoyevsky's paradoxes are equally elaborate and well-thought-through. Some look like involuntary contradictions, dictated by narrative conventions and the *Zeitgeist*; others, fraught with meaning, are due to the author's own inconsistency. The former case is exemplified by the Victorian purity of Dostoevsky's fallen women: Sonechka Marmeladova's profession, rhetorically so important in *Crime and Punishment,* is never actually plied by her in the plot; likewise, a close reading of *The Brothers Karamazov* leaves one with the impression of the *fatale* Grushenka's immaculate chastity. In what follows, we will focus on the latter case: the unplanned dead ends of the Author's innovative attempt to renounce his divine powers.

Let us turn to Part V of *Crime and Punishment*, Chapters 1 and 3, where Luzhin accuses Sonya, in dramatic fashion, of having stolen money from him. The episode unfolds as a series of discursive duels, with unexpected outcomes that end up undermining the power claims of the arrogant 'script-writers'.

Chapter 1 (Dostoevsky 1972–1990, vol. 6, pp. 276–90) begins with Luzhin's regretting he had miscalculated in trying to get Dunia cheaply ("why the hell did I get so yid-like greedy?"; p. 277) and being angry at Raskolnikov for upsetting those plans. The news that Raskolnikov is coming to the wake gives Luzhin "an idea" (p. 278). This is followed by a description of his relationship with Lebeziatnikov, with whom he is staying: he begins by flattering him as a bearer of a new authoritative discourse, but soon recognizes his harmless adherence to an ideological vogue. As Lebeziatnikov spouts his ridiculous utopian ideas about marriage, Luzhin keeps counting money while venomously teasing him. Then, at Luzhin's request, Lebeziatnikov brings in Sonia, and Luzhin gives her a ten-ruble bill. Delighted by his kindness, Lebeziatnikov (who "heard and saw everything"; p. 288) forgives him his mocking tone and launches into utopian rants again. Luzhin chuckles again, but "even Lebeziatnikov finally . . . noticed" that he was "thinking of something else... and was rubbing his hands" (p. 290).

The discursive dynamics here is approximately as follows:

(1) Luzhin has lost to Dunia and Raskolnikov.

The author is on their side[1], despite Raskolnikov's sharing some features with Luzhin.

(2) Luzhin plots against Raskol'nikov.

The author prefers Raskol'nikov.

(3) Luzhin mocks Lebeziatnikov, who looks foolish.

The author uses one negative character to defeat another, a deliberate paradox by Dostoevsky (in the words of another of Dostoevsky's 'authorial' characters, Ivan Karamazov, "One viper eats another viper"; IV, 5; Dostoevsky 1972–1990, vol. 14, p. 170; incidentally, Lebeziatnikov is, in essence, a Dostoevskian version of Turgenev's Sitnikov, while Luzhin and Raskol'nikov in tandem form an approximate counterpart of Bazarov).

(4) Luzhin presents himself in a positive light in front of Sonia and Lebeziatnikov.

The author makes it clear that there may be something behind this.

In Chapter 3 (vol. 6, pp. 300–11), Luzhin publicly accuses Sonia of stealing a 100-ruble bill from him, calls on Lebeziatnikov as a witness, and insists that Sonia be taught a lesson. Katerina Ivanovna then offers to search Sonia, turns her pockets inside out, and a bill falls to the floor. Luzhin triumphs over Sonia and Raskolnikov and declares that the lesson has been taught, so he is willing to generously close the case. However, then Lebeziatnikov accuses Luzhin of meanness, stating that he saw Luzhin surreptitiously put the money in Sonia's pocket, which he, Lebezitnikov had then took to be an act of secret beneficence. However, Luzhin denies this, pointing out that he has no motive for such actions. Lebezyatnikov launches into ridiculous ethical assumptions about the motives of Luzhin, who, in turn, accuses him of personal ideological ill-will towards him. Lebeziatnikov is at a loss as to what to think, when Raskolnikov speaks up to unravel Luzhin's scheme: by disgracing Sonia he tried to undermine Raskolnikov's authority in the eyes of his sister.

In terms of the war of the scripts, the following is happening:

(1) Luzhin tries, according to a script he has devised, to teach Sonia and Raskolnikov a lesson while parading his own generosity.

The author keeps ambiguously silent.

(2) Luzhin's script stumbles upon Lebezyatnikov's unforeseen abandoning of his prescribed role, unexpected even for Lebeziatnikov himself, who thus takes the upper hand in his personal tug-of-war with Luzhin.

The author sides with the good and honest fool Lebeziatnikov (in an invariant Dostoevskian paradox).

(3) Luzhin, however, once again outplays Lebeziatnkov, who is unable to 'read' his motives, and imputes to Lebeziatniukov base intentions.

The author implicitly prefers Lebeziatnikov.

(4) Raskolnikov defeats Luzhin by correctly reading his script.

The author sides with Raskolnikov.

Raskolnikov's triumph, both practical and moral, crowns the pyramid of rival scripts but does not abolish the ambiguity of his entire discourse strategy. In the next chapter, he will try to justify his immoralism in the eyes of Sonia by invoking her own defenselessness in the world of Luzhins, while confessing to her his crime and asking her for support. His, a criminal's, successful advocacy of Sonia and indictment of Luzhin form one of the major paradoxes of *Crime and Punishment*. And, in the larger scheme of things, this episode is a negative foreshadowing of the subsequent speeches of Sonia, Porfirii and Svidrigailov, who, each in their own way, will 'read' and refute his script (first published as an article and then enacted in real life).

In this discursive respect, the relationship between Raskolnikov and Svidrigailov is of particular significance. Despite his general immoralism and self-destructive end,

Svidrigailov is, just like Sonia and Porfirii, complicit in the novel's authorial agency. Sonia, of course, carries the main load as the book's principal righteous and sermonizing Christian; but her potential moral authoritarianism is softened by her sinful profession, her general humility, and, most importantly, by the lack of scriptural and directorial manipulation on her part: she persuades Raskolnikov by straightforward arguments and her personal moral integrity.

Porfirii embodies a rationalist stance presented in a positive light, which is rare in Dostoevsky's world. He arrives at an adequate reading of Raskolnikov on his own—in contrast to Sonia, who relies on faith in God and gets a reading of Raskolnikov from Raskolnikov himself and only then intuitively makes a deeper moral sense of it.

Svidrigailov, on the other hand, has much in common with Raskolnikov and so he owes his correct reading of him not only to the coincidental eavesdropping but also to an empathy based on their moral kinship (inherent in Dostoevsky's polyphonic portrayal of characters). Svidrigailov's penetration into the unconscious paradoxes of Raskolnikov's mentality is emblematized by his remark about Raskolnikov's apparent conviction that "eavesdropping is all bad, while cracking old women's skulls is OK" (VI, 5; 6: 373).

Here, Svidrigailov achieves the kind of discursive victory over Raskolnikov that the author had denied to Luzhin (this is yet another round in the ongoing war of the scripts). The fact is that Luzhin's soulless and mechanistic bourgeois mentality is a greater anathema to Dostoyevsky than Svidrigailov's demonic and, therefore, somewhat romantic, i.e., 'spiritual,' cynicism. Besides, as has been noted (albeit without decisive proof), Svidrigailov's predilection for underage girls (Nabokov's future "nymphets") may have been a reflection of his author's similar tastes.

Last but not least, the character of Svidrigailov boasts yet another 'authorial' function: his role in arranging the novel's denouement and its protagonists' future destinies. To begin with, by his suicide, he frees Dunia for marrying Razumikhin (and Raskolnikov, from having to protect her from him). What is more, he takes upon himself the financial settlement not only of his own newly found bride, but also of Sonia and her siblings and even of Raskolnikov ("When I give you, it is all the same as if I give him"; VI, 6; 6: 385). Thereby he fulfils a typically authorial function, and Dostoevsky, so to speak, does not shy away from soiling his hands with Svidrigailov's money. This—after having presented money in a consistently negative light throughout the plot: in the hands of Alena Ivanovna the old moneylender, Raskolnikov the robber–murderer, Luzhin the would be bridegroom of Dunia and accuser of Sonia, Marfa Petrovna as Svidrigailov's "landlady-owner" ("khoziaika"), and of Svidrigailov himself as Dunia's potential seducer!

Overall, however, the 'authorial' functions in the novel are distributed among the several quasi-authorial characters, who are 'disempowered' in one way or another, so that no sense of unquestionable, authoritarian control emerges. Even in the Epilogue, which has been frowned upon by critics for having been "fixed" by the author, there is no authorial manipulation of Raskolnikov by other characters. In the figure of Sonia one can see a number of stereotypes ('a strong Russian woman', 'a hooker with a heart of gold', 'a saintly sinner'), but none of the many types of Dostoevsky's 'manipulative scripters'.

For a more challenging case, let us turn to *The Brothers Karamazov*—the end of Book IV and the beginning of Book V (Dostoevsky 1972–1990, vol. 14, pp. 178–202). We will focus on Aliosha, the author's favorite positive hero, a disciple of the perfectly righteous (although kenotically "stinking" once a corpse) elder Zosima.

Confronted by schoolboys and Iliushechka, who had bravely stood up for his father, staff-captain Snegirev humiliated by Dmitry Karamazov, and, among other things, bit Aliosha on the finger, Aliosha visits Snegirev at home and meets him and his variously sickly family (chp. 6, "A Laceration in a Hut"). Snegirev meets him with defiance, seeing in his visit a continuation of Dmitrii's aggression, and Iliusha shouts that Aliosha has, of course, come to complain about him. Aliosha tells how Iliusha threw a stone at him and bit him, and Snegirev, sneering in the spirit of mock modesty, offers to have Iliusha flogged right there and then. But then, he promptly scoffs at the very idea of flogging his son—"for your complete satisfaction"—and probably even cutting off "four of my own fingers . . . with this very knife" ("four fingers . . . will be enough to quench your thirst for vengeance won't they, so you won't demand the fifth, sir?'; vol. 14, p. 183). Aliosha says that he has not come to complain, but rather to convey that Dmitri has repented and wants to beg forgiveness and is ready to publicly kneel before Snegirev. "Pierced to the heart"—replies Snegirev (p. 183). Nevertheless, in spite of Iliusha's objections, Snegirev goes out with Aliosha to talk "outside these walls".

The war of discourses here is so far of the usual Dostoevskian sort:

(1) The 'little man' (Snegirev), humiliated by a 'significant person' ("his lordship' Dmitri; p. 183), contrary to expectations, turns out to be morally, intellectually and verbally sophisticated (especially in comparison with the script of "The Overcoat"—as seen in *Poor Folk*), indeed, uncompromisingly bold, as he plays out a scene of consummate mockery at the expense of the script he attributes to his opponent.

The author presents this from the point of view of Aliosha, whose unquestionably good intentions are known to the reader.

(2) Aliosha neutralizes Snegirev's game, showing real humility—on behalf of his brother, and most importantly, on his own behalf; Snegirev hesitates, but Iliusha does not give up his confrontational stance.

The author sympathizes with both sides and retains hope for their reconciliation.

In the following chapter ("And in the Open Air"; IV, 8; 14: 185–93), Snegirev frankly tells Aliosha about himself, about the impossibility for him to sue Dmitrii, about the wounded pride of Iliusha, who "alone against all rebelled for his father" (p. 187) and about Iliusha's plans to grow up and take revenge on Dmitri ("I will strike him down and tell him: I could have killed you now, but I forgive you"; p. 189). Aliosha repeats that his brother has repented and is ready to ask for forgiveness, while his bride, Katerina Ivanovna, also insulted by Dmitrii, asks Snegirev to accept money ("from her alone, not from Dmitrii... And not from me... This means that the sister goes to her brother with help... She specifically instructed me to persuade you to accept these two hundred rubles from her as from a sister. No one will know about this... You have a noble soul... you must understand this...!"; p. 190). Snegirev at first agrees to accept the money, although asking, "I will not then be a scoundrel, will I?" (p. 191); he makes plans of using the money for the benefit of the family and Iliushechkas', and worries whether "the money will suffice" (to which Aliosha promises to add as much as needed of his own; p. 192). Yet, Snegirev ends up refusing, with perverse pride ("And what would I tell my boy, if I took money from you for our shame?"; p. 193), and, "throwing both crumpled bills on the sand—did you see?" (p. 193), Snegirev asks Aliosha to tell those who sent him that he is "not selling his honor!" and runs off (p. 193). Aliosha picks up and smooths out the banknotes.

The discourse dynamic here is as follows:

(1) Aliosha's Christian, 'brotherly-sisterly' script begins to materialize, as mutual understanding emerges in the course of a heart-to-heart talk "in the open air".

However, the author makes it clear that not everything is going smoothly. Firstly, Aliosha acts not only on his own initiative, but according to the script of Katerina Ivanovna, whose motives are not simple and who in no way embodies Christian humility. Secondly, Snegirev keeps slipping into unhealthy 'alien' discourses: of an exaggeratedly heroic scenario (in the spirit of Alexander Pushkin's Sylvio), of Iliusha's anticipated revenge, and of his own exaggeratedly sentimental utopian plans about using the money.

(2) Snegirev thwarts the fulfillment of Aliosha's script (in accordance with Dosto-evsky's favorite paradox of a 'modest' character evading the line of conduct prescribed for him by a 'cleverer' partner). He does so by following yet another ready-made script, once again borrowed from Pushkin: the station master's throwing away the money received "for Dunia" from Minsky. As Aliosha picks up the money, he varies this subtext in an interesting way: in Pushkin's short story, the money is also picked up—not by Minski, but by some well-dressed accidental passer-by.

The author sides with Aliosha, but his position is not fully clear yet.

Book V opens with a chapter in which Aliosha tells Lise Khokhlakova about what happened.

-"Well … You gave that money, and now how is this unfortunate person?"

-"That is just it, I did not give it, and there's a whole story here", Aliosha answered … He spoke … well and thoroughly. He used to … when Lise was a child, he liked … to tell what had just happened to him then, or from what he had read, or to remember from his childhood. Sometimes they even dreamed together and wrote whole novellas together … Now it was as if both of them were suddenly transported back in time … Lise was extremely touched by his story. Aliosha with fervent emotion managed to draw before her the image of Iliushechka, and then explained in detail why Snegirev refused the money: he "hated" Aliosha, before whom "he was ashamed that he had bared the depth of his soul" (p. 196) and who had committed the mistake of injuring his pride with the offer "to give[s] from his own as much as needed." But, Aliosha continued, this mistake was "for the best", for if Snegirev had taken the money by overstepping his pride, he would have returned it the next day; while now that his pride had been won back, he would, on the contrary, be ready to accept it with a clear conscience.

-"And that's when I am going to show up, meaning, yes, you're a proud man, you've proven it, well, now take it and forgive us. And then he will take it!"

In a kind of rapture, Aliosha exclaimed: "And then he'll take it!" Lise clapped her hands.

-" … Oh, Aliosha, how do you know all this? You're so young, and you already know what's in everyone's heart. I would never have thought this up."

-"The main thing is to convince him now that he's on an equal footing with all of us, despite the fact that he takes money from us... and not only on an equal footing, but even on a higher footing..."

-"'On the higher ground' is lovely, Alexei Fedorovich... But... Isn't there in all this reasoning of ours, that is, of yours... better ours... some contempt for him—in the fact that we are now parsing his soul in this way, as if looking down on him, eh? That we've decided for sure now that he'll accept the money, eh?"

-"No, Lise, no disdain," said Aliosha firmly, as if he had prepared for the question.—"Because we, too, are same as he, no better... You know, Lise, my elder once said: people should always look after people like after children, and after some, like after patients in hospitals". (p. 197)

(-Что же … отдали вы эти деньги, и как же теперь этот несчастный? …

-То-то иесть, что не отдал, и тут целая история—ответил Алеша … Он говорил … хорошо и обстоятельно. Он и прежде … еще в детстве Lise, любил … рассказывать то из случившегося с ним сейчас, то из прочитанного, то вспоминать из прожитого им детства. Иногда даже оба мечтали вместе и сочиняли целые повести вдвоем … Теперь они оба как бы вдруг перенеслись в прежнее … время … Lise была чрезвычайно растрогана его рассказом. Алеша с горячим чувством сумел нарисовать перед ней образ 'Илюшечки', а затем подробно объяснил, почему Снегирев отказался от денег: он «возненавидел» Алешу, перед которым «стыдно стало за то, что он так всю душу мне показал» (p. 196), и который совершил ошибку, уязвив его гордость предложением «да[ть] из своих сколько угодно». Но, продолжил Алеша, эта ошибка «к лучшему», ибо если бы Снегирев взял деньги, переступив через свою гордость, то на другой день вернул бы их, теперь же, когда его гордость уже отыграна, он, наоборот, будет готов принять их с чистой совестью.

-А я-то вот тут и явлюсь: Вот, дескать, вы гордый человек, вы доказали, ну теперь возьмите, простите нас'. Вот тут-то он и возьмет!

-Алеша с каким-то упоением произнес: "Вот тут-то он и возьмет!" Lise захлопала в ладошки.

- ... Ах, Алеша, как вы все это знаете? Такой молодой и уж знает, что в душе. Я бы никогда этого не выдумала..

-Его, главное, надо теперь убедить в том, что он со всеми нами на равной ноге, несмотря на то, что он у нас деньги берет... и не только на равной, но даже на высшей ноге...

-"На высшей ноге"—прелестно, Алексей Федорович ... но ... нет ли тут во всем этом рассуждении нашем, то есть вашем ... нет, уж лучше нашем.. нет ли тут презрения к нему, к этому несчастному ... в том, что мы так его душу теперь разбираем, свысока точно, а? В том, что так наверно решили теперь, что он деньги примет, а?

-Нет, Lise, нет презрения—твердо ответил Алеша, как будто уже приготовленный к этому вопросу ... —Потому что ведь и мы такие же, не лучше ... Знаете, Lise, мой старец сказал он раз: за людьми сплошь надо как за детьми ходить, а за иными, как за больными в больницах.)

In Chapter 5, Book X, the main narrator informs us that Snegirev accepted those two hundred rubles from Katerina Ivanovna exactly as Aliosha had predicted. "And then Katerina Ivanovna ... visited their flat herself ... After that her hand hasn't grown scarce, and the staff-captain himself ... forgot his former haughtiness and humbly accepted alms" (vol. 14, pp. 486–87).

In contrast to all the previous cases, this episode is not so much an unfolding combat of scripts but rather an authorial/critical meta-discussion of a discourse combat that had already taken place (between Aliosha and Snegirev) and of the forecast (by Aliosha) of its expected denouement.

(1) The meta-discourse mode is set at once, not only by the very format of the retelling/discussion, but also by a direct reference to the usual practice of such 'literary' meetings and joint writing of "novellas", as well as by the use (in the author's narrative) of literary-critical terms, such as "he managed to draw her an image of Iliushechka".

The author, by transferring part of the authorship to his favorite hero, complicates the narrative hierarchy, maintaining a certain distance from Aliosha 'the author', for instance, by noting his "rapture" ("upoenie") with his own script.

(2) The relationship between the two interlocutors is not limited to Lise's idyllic agreement with Aliosha (who is older and has a certain moral mandate based on his closeness to Zosima), emphasized by her respectfully phrased 'co-authorial' statements ("so young, and you already know what is in a heart"; "in all this reasoning, that is, ours, no, yours, no, better ours"). Lise, the "little devil," plays (here and elsewhere) the role of the devil's advocate, teasing and provoking Aliosha in every way about his "monastic holiness" and revealing in his speech a characteristic 'hierarchical' clause (about the "equal/higher footing"). In the end, she poses the ultimately troubling question, namely: Does not Aliosha's 'omniscient parsing from above' (a few pages later, he will return again to the "martyr's" question of "anatomizing" another person's soul; 199), in fact, calculating and directing the behavior of another person—does not this mean "contempt" for Snegirev (that is, in today's terms, objectifying a live human being and his/her unfinished discourse).

The author, sensing, perhaps, that this is the mother of all questions for a truly dialogic poetics, lets Lise develop her position quite eloquently and even with caustic irony.

(3) Aliosha, however, is not taken by surprise by Lise's question (unlike the usual case, where the 'strong' partner is caught unawares by the 'weak'). He "himself had thought about this on his way here," and now gives her a prepared (!) answer: yes, it is possible and even necessary to manipulate people for their own good, provided we remember that we are "just like them, no better." This argument is supported by nothing else but a reference to the authority of the elder (Zosima), which is enough for Lise: "Ah... sweetie, let us look after people as if they were ill!" (p. 197; Lise omits the second component of the elder's

instructions (central to the novel that is about the relationship between fathers and sons): "we should look after them like after children"; in fact, in the person of Snegirev, Aliosha adopts—as a son—one of the novel's distinctly paternal figures.

The author implicitly but clearly endorses this position: Zosima's moral authority in the novel is never in doubt, and Aliosha's remark is accompanied by the narrator's words: "Aliosha answered firmly."

This is the deadend paradox of Dostoevsky alleged dialogicity, which this article set out to demonstrate.[2] Aliosha, acting in an authorial role, 'reads' in the heart of another, 'little,' man, assumes a parental, authoritative, practically 'divine' role towards him and successfully scripts for him his 'mechanical' behavior (implementing one after another the two opposite versions of Dostoyevsky's typical 'action in reverse': in Book IV, first, humility, then pride, in Book X, first pride, then humility) and "firmly" bolsters his directorialism with a "prepared" reference to 'authority.' It is possible, of course, to invoke the 'openness' of the text, the questionable traits in Aliosha's character ("I may not believe in God"; p. 201), his subsequent failure to prevent the murder of his father and in the conviction of Dimitri in court, but this does not eliminate the problem. The author lets Aliosha play God and get away with it.

In the context of the novel as a whole and of Dostoevsky's other narratives, this episode looks all the more paradoxical as it echoes them as an unexpected variation on familiar themes. Note that, like in the cases of Luzhin-Sonia and Svidrigailov-Sonia, this episode, too, involves the transfer of money. Luzhin's vile provocation fails, Svidrigailov's charitable—quasi-authorial—action succeeds, although it does so close to the novel's epilogue, thus not allowing for any further dramatic turns, whereas Aliosha's two-step combination not only does succeed in the heat of events, but also receives approval from above (the God and the Author) in the course of a special meta-discursive dialogue.

Meanwhile, Aliosha's script is structurally no less manipulative than, say, that of Kolia Krasotkin, who delays bringing Iliusha the retrieved pet Zhuchka—solely for the effect he considers beneficial (Book X; cf. the "Only five minutes late!" line in "A Gentle Spirit"). In a broader perspective, one cannot help drawing a comparison with the discourse of the Grand Inquisitor (which, incidentally, appears only four chapters later, in the same Book V, and also within the framework of multi-level 'authorship'), whose problem is precisely manipulatively assuming responsibility for "these little ones"—allegedly in their own interest and with reliance on the Church and God.[3] Bakhtinian metaphysical hermeneutics[4] was to a certain degree motivated by the invisible and ineffable *hidden image of God,*[5] being generally related to the common issue of religion par excellence. Mikhail Bakhtin himself, as we may gather from multiple memoirist sources, was deeply religious (no less than say, someone like Aleksey Losev). Many typical Bakhtinian concepts (including those of *polyphony* and *dialogue*) betray a concealed religious subtext—to one extent or another.

The 'God,' it is true, is provided with a *dialogical* opportunity to speak, but, as the Inquisitor observes correctly, "You have no right to add anything to what you have already said before" (V, 5; 14: 228). Despite this silence, however, the Author's subversion of the Inquisitor's script is evident. Aliosha, on the other hand, simply refers to the elder, on whose side, according to the novel's overall script, is God himself, thus cutting off the possibility of any further discussion. The polyphonic principle: "Trust in God, but do the right thing", is suddenly suspended. The 'firing' not only of pots, but also of men, turns out to be entrusted, with divine approval, to a kind of overman. This, contrary to Dostoevsky's and Bakhtin's refusal to stop the dialogue and be 'mechanistically final,' produces the categorical discursive effect of a *deus ex machina.*

But then, this deadend paradox, apparently so un-Dostoevskian, does stem from the very essence of the writer's favorite Scripture. Indeed, if 'fishers of men' (Matthew 4:19) are OK, why not those who fire clay pots, i.e., 'potters of men' as well?!

Funding: This research received no external funding.

Acknowledgments: The author is happy to thank Dennis Ioffe and Frederick White for their kind assistance with the preparation of this text for print.

Conflicts of Interest: The author declares no conflict of interest.

Notes

[1] Here, and below, for the sake of simplicity, the "Author" is perceived as the one who produces the "horizons of perception" (which are constantly in the proces of changing in the course of the narrative development). We employ this term in the sense of Wolfgang Iser (1978).

[2] Bakhtin, naturally does not miss this passage and refers to it as an example of the fact that "[t]ruth turns out to be unjust if it touches upon some inner depths of someone else's personality," and places it in a class with two similar ones: Aglaya's remark to Myshkin in *Idiot* (III, 8; 8: 354) and Stavrogin's to Tikhon in *Demons* (III, 9; 11: 11). See (Bakhtin 1972, pp. 101–2).

[3] On the transformation of Aliosha from "Son" into "Father", dealing with him as with the "author" of the Life of Zosimus and about the parallel between Zosimus and the Grand Inquisitor, see (Holquist 1977, pp. 177–97) ("[Zosima] has precisely the kind of power sought by Ivan's Grand Inquisitor"; p. 189). On the authority/compromise of the dicourses of characters in "The Brothers Karamazov" and on the 'vita-sanctorum'-type image-construction of Aliosha based on the hagiographic figure of *Alexei the Man of God*, in particular see (Vetlovskaia 1977).

[4] On the suggestive issue of Bakhtinian hermeneutics and religion one may consult, see (Bagshaw 2016; Felch and Contino 2001; Mihailovich 1997; Coates 1999).

[5] If to use this tellng expression borrowed from Bakhtin's younger Parisian contemporary Lucien Goldmann which was used in his seminal Marxist-religious celebrated monograph entitled *Le dieu caché* (Goldmann 1955).

References

Bagshaw, Hilary. 2016. *Religion in the Thought of Mikhail Bakhtin. Reason and Faith.* London: Routledge.

Bakhtin, Mikhail. 1972. *Problemy Poetiki Dostoevskogo.* Moscow: Khudozhestvennaia Literatura.

Bocharov, S. G. 1985. Perekhod ot Gogolia k Dostoevskomu. In *O Khudozhestvennykh Mirakh.* Moscow: Sovetskaia Rossiia.

Coates, Ruth. 1999. *Christianity in Bakhtin: God and the Exiled Author.* Cambridge: Cambridge University Press.

Dostoevsky, Fyodor M. 1972–1990. *Polnoe Sobranie Sochinenii v Tridtsati Tomakh.* Leningrad: Nauka.

Felch, Susan, and Paul Contino. 2001. *Bakhtin and Religion: A Feeling for Faith (Rethinking Theory).* Evanston: Northwestern University Press.

Gogol', Nikolai V. 1976–1979. *Sobranie Sochinenii v Semi Tomakh.* Moscow: Khudozhestvennaia Literatura.

Goldmann, Lucien. 1955. *Le Dieu Caché: Etude sur la Vision Tragique dans les Pensées de Pascal et dans le Théâtre de Racine.* Paris: Gallimard.

Holquist, Michael. 1977. *Dostoevsky and the Novel.* Princeton: Princeton UP.

Iser, Wolfgang. 1978. *The Act of Reading. A Theory of Aesthetic Response.* Baltimore and London: The Johns Hopkins University Press.

Mihailovich, Alexandar. 1997. *Corporeal Words: Mikhail Bakhtin's Theology of Discourse (Studies in Russian Literature and Theory).* Evanston: Northwestern University Press.

Shifman, I. S., ed. 1993. *Uchenie: Piatiknizhie Moiseevo.* Moscow: Respublika.

Turgenev, Ivan S. 1961–1962. *Sobranie Sochinenii v Desiati Tomakh.* Moscow: Khudozhestvennaia Literatura.

Vetlovskaia, V. E. 1977. *Poetica Romana Bratia Karamazovy.* Leningrad: Nauka.

 *religions*

Article

Gogol's "The Nose": Between Linguistic Indecency and Religious Blasphemy

Igor Pilshchikov [1,2]

[1] Department of Slavic, East European and Eurasian Languages and Cultures, University of California, Los Angeles, CA 90095, USA; pilshchikov@ucla.edu

[2] School of Humanities, Tallinn University, 10120 Tallinn, Estonia

Abstract: Focused on Nikolai Gogol's absurdist tale, "The Nose" (1835), this article is an investigation into the concealed representation of suppressed and marginalized libertine and anti-religious discourses in nineteenth-century Russian literature. The author identifies overlooked idiomatic phraseology, forgotten specificities of the Imperial hierarchy (the Table of Ranks), and allusions to religious customs and Christian rituals that would have been apparent to Gogol's readers and shows how some were camouflaged to escape censorship in successive drafts of the work. The research builds on the approaches to Gogol's language, imagery and plot developed earlier by the Russian Formalists, Tartu-Moscow semioticians, and a few other scholars, who revealed the latent obscenity of Gogol's "rhinology" and the sacrilegious meaning of the tale's very specific chronotope. The previous scholars' observations are substantially supplemented by original findings. An integrated analysis of these aspects in their mutual relationship is required to understand what the telling details of the story reveal about Gogol's religious and psychological crisis of the mid-1830s and to demonstrate how he aggregated indecent Shandyism, social satire, and religious blasphemy into a single quasi-oneiric narrative.

Keywords: The Nose (tale); Gogol; religious crisis; sacrilege; blasphemy; linguistic indecency

Citation: Pilshchikov, Igor. 2021. Gogol's "The Nose": Between Linguistic Indecency and Religious Blasphemy. *Religions* 12: 571. https://doi.org/10.3390/rel12080571

Academic Editor: Liam Francis Gearon

Received: 4 June 2021
Accepted: 15 July 2021
Published: 24 July 2021

Publisher's Note: MDPI stays neutral with regard to jurisdictional claims in published maps and institutional affiliations.

1. Introduction

Encyclopedia Britannica presents Nikolai Gogol as "one of the finest comic authors of world literature, and perhaps its most accomplished nonsense writer" (Morson 1998, p. 1007). Gogol's reputation of a preacher of moral asceticism and a religious author is less felicitous. His final book, *Meditations on the Divine Liturgy*, attracts more attention now than it used to in the Soviet period, when it was not even included in the fourteen-volume edition of his complete works (for obvious ideological reasons). Nevertheless, *The Journal of the Moscow Patriarchy* expresses regret that "even today Gogol as a religious writer and spiritual thinker is little known to a broad readership" (JMP 2010, p. 87). Gogol always wanted to describe the best of the human soul but excelled in describing the worst. Although questions of morality and religion were always central to his thinking, his comic and satirical genius overshadowed his dubious achievements in practical theology (*Meditations*) and notorious failure as a moral teacher (*Selected Passages from Correspondence with Friends*). It is generally (and justly) believed that his worldview found a more adequate expression in his fiction rather than in his critical and instructional prose, and this explains why I selected Gogol's most grotesque and absurdist work to represent him in a volume on East Slavic religion(s), using *Meditations* only as a point of comparison in the last section of my analysis.

2. God and Devil in "The Nose"

Gogol's peculiar position between Russian and Ukrainian languages and mentalities (today somewhat anachronistically referred to as his "hybrid identity") simultaneously represents two regions of the East Slavic world—"Great Russia" and "Little Russia", i.e., Imperial Russia and East Ukraine (Koropeckyj and Romanchuk 2003; Bojanowska 2007; Ilchuk 2021; Lounsbery 2021, pp. 492–94). Post-Petrine Russia was a powerful but relatively recent and artificial formation (Saint Petersburg's "artificiality" was vividly experienced by both Gogol and his "Great Russian" contemporaries). "Little Russia", a Ukrainian and partly Russified area, was a liminal and borderline "contact zone" between the East Slavic and West Slavic worlds, as well as between the Eastern Orthodox and Roman Catholic churches.[1]

In 1831–1832 Gogol published his first collection of prose tales, *Evenings on a Farm Near Dikanka* (in two volumes). This groundbreaking work, a combination of comic and supernatural elements drawing on Ukrainian folklore, resounded with the Romantic interest in local color and ethnic roots. It was an instant success. In *Evenings*, supernatural characters from folkloric demonology (devils and the like) are put on stage together with Ukrainian peasants (Holquist 1967; Putney 1999; Butler 2017). Gogol's vision of Saint Petersburg in his so-called "Petersburg Tales" (to which "The Nose" belongs) is different from how the city appears in works by Petersburgians or Muscovites—it reflects some distinctive aspects of his Ukrainian origins. For Gogol's narrator, the Russian capital remains "vaguely familiar but ultimately incomprehensible", just as are Ukrainian words and habits for his Russian readers (Koropeckyj and Romanchuk 2003, p. 543).

On 10 December 1834, Gogol obtained the censor's permission to print his new book, *Arabesques*, a two-volume collection of tales and essays published soon thereafter, in January 1835. It features the first three "Petersburg Tales"—"Nevsky Prospect", "The Portrait", and "The Diary of a Madman." All three of these texts were substantially revised for the third volume of Gogol's works published in 1842. This volume, entitled *Povesti* (*Tales*), opens with "Nevsky Prospect", which is followed by "The Nose", so that the former could be read as a sui generis introduction to the latter.

"The Nose" came down to us in three successive redactions (this count does not include the earliest one-page draft of 1832): the first complete manuscript redaction (1833–1834), the first published version (1836), and the censored definitive redaction (1842). The uncensored definitive version never existed materially in Gogol's lifetime, being a later critical reconstruction. The text we are using today is a critical text published in the third volume (1938) of the 14-volume academic edition (Gogol 1937–1952). It mainly reproduces the 1842 redaction, in which, however, the censored passages are restored on the basis of the manuscript redaction and some editorial interventions of the 1842 text are eliminated and replaced by the more authentic 1836 version. Other critical editions mostly follow the 1842 publication and make less use of the manuscript variants.

Like *Evenings,* the "Petersburg Tales" also combine descriptions of local communities, landscape, and cityscape ("estranged" by the gaze of a provincial newcomer) with elements of the comic absurd and surrealistic grotesque, but they do not feature bearers of a folkloric worldview whose mythological vision might motivate the routine presence of the "uncanny" in everyday life. Yuri Mann showed that in the tales of the Petersburg cycle, various latent or implicit fantastic elements evident in details of the characters' everyday life and behavior—or even in the language they or the narrator use—are a substitute for the folk demonology of Gogol's early stories. Mann referred to this discursive strategy as "the omission of the exponent of the fantastic [*sniatie nositelia fantastiki*]" (Mann 1973, 1996, pp. 76–91; Cornwell 1990, pp. 19–20; Bocharov 1992, p. 23; Meyer 1999).

To make this point clear, consider a brief summary of "The Nose."

The tale consists of three chapters. In Chapter I, the barber Ivan Yakovlevich, a terrible drunkard, wakes up in the morning and finds a human nose in a loaf of freshly baked bread. When he tries to get rid of the nose by throwing it in the river, he is caught by a policeman. "But here the event is completely covered by a fog, and absolutely nothing is known about what happened next" (III: 52/201). In Chapter II, a certain Major Kovalyov, whom Ivan Yakovlevich had shaved a day or two before, wakes up in the morning to find his nose missing. Later, he unexpectedly encounters his own nose, dressed as a high-ranking official, on Saint Petersburg's Nevsky Prospect. Kovalyov follows the nose to the city's main cathedral, where he sees it/him "praying with an expression of the greatest piety" (III: 55/204). The nose disappears again, and Kovalyov searches for it/him around Saint Petersburg—but in vain. A policeman later appears unexpectedly and restores Kovalyov's nose to its rightful owner, explaining that it was apprehended at the border while attempting to flee abroad. Kovalyov tries to reaffix his nose, but it does not stick. "Following this... but here again the whole event is hidden by a fog, and absolutely nothing is known about what happened then" (III: 72/223). In Chapter III, Kovalyov wakes up in the morning to find his nose back in its proper place. The narrator is astonished: "No, I can't understand this at all, I absolutely do not understand!" But "no matter what you say, such events do happen in the world—they happen rarely, but they do happen" (III: 75/227).

The personified supernatural forces no longer interfere with the action. They are, as it were, present in the background and break through from there into everyday phraseology. When Kovalyov tries to explain the tale's events to himself, he stops, embarrassed: "By what means, by what fates did this occur? Only the devil can figure it out!" (III: 71/222).[2]

The word used for the devil is *chort*. This word denotes a minor devil or demon and is never applied to Satan (*satana*), who can also be called Diabolus (*diavol*, from the Greek διάβολος)[3] but not *chort*.[4] Unlike the English translations of Gogol, the original text is always quite explicit as to who is mentioned: Devil or the devil. The former is a rare guest in everyday life, whereas the latter is always nearby and ready to interfere.

"The Nose" is notable for its characters' instantaneous and seemingly unmotivated transitions from swearing by God (*bozhba*) to swearing by the devil (*chertykhanie*). The basic verbs for these verbal nouns are *bozhit'sia*, from *bog* 'god,' and *chertykhat'sia*, from *chort*. The permanent verbal mixture of the sacred and the demonic (Lachmann 1999, p. 22; Zyrianov 2009, pp. 132–33; Kondakova 2009, p. 139) becomes an iconic depiction of the struggle between the two driving forces of the human world. In the uncensored critical edition of the definitive version of the tale (1842), *chort* is mentioned eleven times, whereas *Bog* is mentioned ten times (this includes the archaic vocative case *Bozhe* but excludes the adjective *nabozhnyi* ("pious") and the noun *nabozhnost'* ("piety") of the same root, which are used twice, i.e., once each.) The first complete manuscript redaction (1833–1834) featured five instances of *chort* and eight instances of *Bog/Bozhe* (plus *nabozhnyi* and *nabozhnost'*). Therefore, in the definitive version, the balance of forces is symmetrical because both sides are equally represented (pace Meyer 1999, p. 201).[5]

Moreover, *bozhba* and *chertykhanie* succeed each other on a regular basis. "*The devil knows* how this happened. [...] I can't understand it at all!" the barber says to himself (III: 50/199), and soon he swears to the policeman: "*Honest to God*, sir, [...] I was just looking to see whether the river was running fast" (III: 52/201).[6] Kovalyov is talking to himself in front of the mirror:

> "Well, *thank God* nobody's here." he said. "Now I can take a look." He went timidly up to the mirror and took a look: "*The devil only knows!* What rubbish!" he said, and spat. "If only there were something instead of my nose, but there's nothing!" (III: 54/203)

"Oh, *the devil take it*!" says Kovalyov and decides to drive to the chief of police (III: 57/207). However, he is afraid that "all searching would be in vain or could take a whole month, *God forbid*. Finally, it seemed that *heaven itself* made him see the light" and gave him an idea to go to the newspaper office instead (III: 58/208). It smells bad there, but Kovalyov cannot experience the scent "because his nose itself was in *God knows* what locality" (III: 60/209).[7] However, he is scared to reveal his name because the rumor might reach his female acquaintances: "What if they find out, *God forbid*!" (III: 60/210). After that, he explains the situation to the clerk: "*The devil* wanted to play a trick on me!" (III: 60/210). The clerk does not believe him but Kovalyov exclaims: "I swear to you, *as God is my witness*!" (III: 62/211; literally: "*as God is holy*", Gogol 1998, p. 313). At the end of the visit, the clerk offers him snuff; Kovalyov indignantly replies: "Do you really not see that the thing I would need for sniffing is precisely what I don't have? *The devil take* your snuff!" (III: 63/213). These are just "words, words, words", but phraseology plays a salient role in Gogol's plot-construction. As we shall see, the entire plot of "The Nose" can be understood as a "realization" of several idiomatic metaphors.[8]

The characters and the narrator explain the antilogism of what is happening through the intervention of the devil. However, the author does not want to simply present all the events as "devil's shenanigans", a device he previously used in the folklore-inflected Ukrainian stories. Quasi-explanations do not help, and the devil slips away: *"the devil knows"* is a dysphemism for *"God knows"*, and, furthermore, both expressions mean that no one in the human world knows or understands what is going on.

Yuri Mann thus comments on Kovalyov's phrase *"The devil wanted to play a trick on me!"* (III: 60/210): "Because of its routine, 'mundane' coloring, the phrase balances on the verge of the verbal-image form of the fantastic and of ordinary everyday speech" (Mann 1996, p. 77). Tellingly, in the early manuscript redaction, the infernality, which in the definitive version would be veiled by the conventional colloquialism, was emphasized and highlighted: "It was Satan-Diabolus himself [*sam satana-diavol*] who wanted to play a trick on me!" (III: 393). When the nose returns to its place, Kovalyov exclaims twice: "Good, the devil take it!" (III: 73, 74/224, 226), as if invoking the point when the devil—perhaps—took his nose indeed.

In the fourth letter apropos *Dead Souls* (1846) included in *Selected Passages*, Gogol explicitly described *samonadeiannost'* (arrogance, conceit, or presumption) as self-estrangement or a split in man, when one of his parts escapes from the other (just like it happens in "The Nose"). There is no question where this escape leads: "he runs away from himself straight into the hands of the devil, the father of presumption" (VIII: 298; Gogol 2009, p. 109; Bocharov 1992, p. 32). In this way, the realized metaphor of the Nose's running off indicates not so much absurdity as sin.

3. Nosology as Rhinology: Noses in Russian Idioms and in Gogol's Tale

3.1. The Problem: Why Is "The Nose" "Filthy"?

Devilry, either metaphorical or real, is accompanied by other abnormalities.

Initially, Gogol prepared "The Nose" for publication in the just-launched magazine *Moskovsky Nabliudatel'* (*The Moscow Observer*) where his friends and admirers Stepan Shevyryov and Mikhail Pogodin played leading roles. Pogodin received a manuscript from the author in March 1835. However, according to the critic Vissarion Belinsky, who became an unofficial editor of this magazine a few years later, the editorial board rejected the tale "because of its banality and triviality" (Belinskii 1955, p. 504). The word "triviality" (*trivial'nost'*) could have a more specific meaning at that time and refer to frivolous vulgarity.[9] In another account of why "The Nose" was not published in *The Moscow Observer*, Belinsky writes that the editorial board "found it filthy [*griaznaia*, literally: 'dirty']" (ibid., pp. 406–07). Apparently, the *Moscow Observer* Slavophiles rejected the tale because of the indecent associations it exploited. Another prominent Slavophile, Konstantin Aksakov, wrote of "The Nose" in 1836: "This jest has its merit but it is indeed somewhat bawdy [*nemnozhko sal'na*, literally: 'a little bit greasy']" (Lanskoi 1952, p. 550). Unlike Aksakov and Pogodin, Pushkin, who had no issue with ambiguities and profanities, published "The Nose" in his magazine *Sovremennik* (*The Contemporary*) in September 1836, preceded by an editor's note:

> N. V. Gogol has objected to the printing of this jest for a long time; but we found in it so much that was unexpected, fantastic, merry, and original that we persuaded him to allow us to share with the public the enjoyment afforded us by his manuscript. (Gogol 1836, p. 54; quoted in Setchkarev 1965, p. 155)

What are then the associations that so displeased Pogodin and Aksakov, but did not at all disturb Pushkin? To find an answer, we need to take a closer look at the phraseology of the noun "nose" (*nos*) in the Russian language and specifically in the language of Gogol.

3.2. "The Nose" as a "Linguistic Metaphor"

In a book-length essay on Gogol, Vladimir Nabokov devoted a few eloquent pages both to Gogol's nose and to the nasal motifs in Gogol. "His big sharp nose was of such length and mobility that in the days of his youth he had been able [...] to bring its tip and his underlip in ghoulish contact; this nose was his keenest and most essential outer part" (Nabokov [1944] 1961, p. 3). Gogol's nose was remembered by his contemporaries: Mikhail Longinov described as "thin and crooked", Ivan Turgenev—as "long and pointed", and Ivan Panaev—as "gaunt, long and sharp, like the beak of a bird of prey" (Mashinskii 1952, pp. 70, 532; Veresaev 1933, p. 216). Gogol also called his nose "decisively avian [or 'bird-like': *ptichii*], pointed and long" (IX: 25). The quotation is from an entry in Elizabeth Chertkova's album (May 1836), which did not escape the attention of the "nosologists" from Viktor Vinogradov ([1921] 1929, pp. 36–37) to Vladimir Nabokov ([1944] 1961, p. 3). Gogol used the same prose in his letter to Maria Balabina of 15 March 1838, written in Italian (a language Gogol knew well): "il mio naso, lungo e simile a quello degli uccelli" ["this nose of mine, long and similar to a bird's"] (XI: 127). The reference to a bird is, above all, a pun: the Ukrainian and Russian *gogol'* designates *Bucephala clangula*, a bird of the genus of diving ducks. As a hint or a devil-like cameo, Gogol's Kovalyov insists that "without a nose a person is the devil knows what: *not a bird*, not a citizen" (III: 64/214; see also Gregg 1981, pp. 375–76).

Nabokov noted the presence of "the nasal leitmotiv" throughout Gogol's fiction: "Noses drip, noses twitch, noses are lovingly or roughly handled" (Nabokov [1944] 1961, p. 3). Below I cite a few verbal portraits from *Evenings on a Farm Near Dikanka* and the "Petersburg Tales", which anticipate their combination and culmination in "The Nose".

A "barber who tyrannically holds his victim by the nose" (I: 215/105)—a figure that suddenly appears as part of an extended simile in "The Night Before Christmas" (1831)—portends the character of Ivan Yakovlevich, the barber from "The Nose", who was holding Kovalyov by his nose while shaving him and then found the nose in a loaf of bread.

The motif of the nose's "perversity" and "insolence" (II: 253/194–95) is found in "The Tale of How Ivan Ivanovich Quarreled with Ivan Nikiforovich" (1833): "The judge's lips were close under his nose, and the nose could sniff his upper lip to his [or *its*? The original is ambiguous.—*IP*] heart's content. This upper lip served him [or *it*? The original is ambiguous again.—*IP*] instead of a snuffbox" (II: 245/188, translation slightly modified). At one moment,

> [the judge's] nose unconsciously sniffed his upper lip, which it commonly did only from great satisfaction. Such perversity on the part of his nose caused the judge even more vexation: he took out his handkerchief and swept from his upper lip all the snuff, to punish its insolence. (II: 253/194–95)

In "Nevsky Prospect" (1834), the drunken shoemaker Hoffmann is going to cut the nose of the drunken tinsmith Schiller, who exclaims: "I do not want it, I do not need a nose!" and insists: "Cut my nose off! Here's my nose!" (III: 37–38/143–44).[10] In "The Diary of a Madman" (1834), the protagonist believes that the moon "is such a tender orb that people cannot live there, and only noses live now. And for that very same reason we cannot see our noses, for they are all located on the moon" (III: 212/178).[11] Osip Senkovsky (Józef Julian Sękowski), a Polish-born Russian critic, journalist and writer, in his derisive review of *Dead Souls* mocked at Gogol's predilection for noses:

> "Tell me, by grace, [...] why does the nose play such an enduring role here? The whole poem pivots on nothing but noses!"—"Because," I reply as the poem's insightful commentator, "the nose is perhaps the foremost source of 'sublime, excited, lyrical laughter'." —"I, however, can't see anything funny in it," you protest.—"You can't but we can," I argue back. "You must agree that this tri-angular piece of flesh which stands prominently in the center of man's face is surprisingly, excitedly, lyrically funny. And it has already been proved that you

can't create anything truly amusing without a nose. (Senkovskii 1842, p. 37; also quoted in Vinogradov 1926, p. 151)

In the manuscript version of the review, these interlocutors are an ass and a camel (Senkovskii 1936, p. 240); therefore, their speculations about the disproportions of a human face are "estranged" by their animal nature. Whatever Senkovsky's intentions were when he was writing this (see Debreczeny 1966), he was right: Gogol's works abound in various idiomatic expressions with the word *nose* as well as related images, which often are comic and hyperbolic. Compare: "In fact, for a villager like me to poke his nose [*vysunut' nos*] out of his hole into the great world [. . .]" (I: 103/3); "If you, messieurs colonels, do not know your own rights, then let the devil lead you around by the nose [*chort vodit za nos*]" (II: 284); "To think that I'd let any old hussy hoity-toit around before me with her turned-up snub nose in the air! [*podymat' . . . nos svoi*]" (V: 125); "He'll spread so much nonsense around town I won't be able to show my face [literally, 'nose': *nosa pokazat'*] anywhere" (V: 134; more examples in Pilshchikov 2019).

Yuri Tynianov considered the Nose in "The Nose" as a "linguistic metaphor" "realized as a mask", a device found not only in Gogol's literary texts, but also in his letters (Tynianov [1921] 2019, p. 32). In April 1838, Gogol wrote to Maria Balabina (this time in Russian):

> Do you believe that I often have a fierce desire to turn into nothing but a nose—
> so that there would be nothing else—neither eyes, nor arms, nor legs, just one
> super-tremendous nose the nostrils of which would be the size of big buckets so
> that I could inhale as much of the fragrance and spring as possible? (XI: 144;
> Gogol 1967, pp. 74–75; Shukman 1989, pp. 81–82 fn. 43)

There are many such examples in Gogol's correspondence, although not always as detailed and extensive. I cite only the most eloquent ones, focusing primarily on potential parallels to the idioms relevant to "The Nose" (for the full list of these idioms in the Russian language, see Lubensky 2013, pp. 361–65, idioms from H-216 to H-249).

> *Showing one's nose*, that is "going out", as in "X doesn't show his face/nose
> anywhere": "But now I've gotten so terribly afflicted that I cannot show my nose
> anywhere [*nikuda ne mogu nosa pokazat'*]". (X: 317)

Compare Kovalyov, who cannot go out because he literally cannot show his nose anywhere.

> *Not seeing farther than one's nose*, that is "being narrow-minded": " . . . who cannot
> see anything [*nichego ne vidit*] farther than his German nose [*dalee svoego . . . nosa*]
> and his merchantry". (X: 341)

Compare Kovalyov, who is always preoccupied with his nose and his rank and who is not interested in anything but his nose when it escaped; also compare his Nose, which/who embodies all his career dreams (see Section 4 below).

> *Sticking one's nose in the air*, or *turning up one's nose*, that is "behaving arrogantly",
> as in "X turns up his nose at everyone": "Believe me, every human is a wader-
> bird [*kulik*], and if he sticks his nose up then his tail necessarily gets dunked. And
> so it must be, so that he not turn up his nose [at everyone] too much [*ne slishkom
> podymal svoego nosa*] and always bear in mind that he is mere rubbish". (XII: 326)

Kovalyov is indeed a kind of person who acts high and mighty, as discussed in Section 3 below.

> *Leading someone by one's nose*, that is "deceiving": " . . . [to find out] who leads
> others by the nose [*vodit za nos*], who is just making fools of them [*durachit*], and
> who has or can have influence on them". (XII: 409)

Perhaps the devil does this with Kovalyov, as Kovalyov thinks.

Gogol's letters contain examples of a synecdoche in which the nose is detached from the rest of the body: " . . . only my nose [*odin tol'ko nos moi*, literally: 'my nose alone'] reached Frankfurt" (XII: 462), as well as a synecdoche in which the nose substitutes for human beings: " . . . my own nose [*moi sobstvennyi nos*] might stick out [*vysunut'sia*] in *Dead Souls*, instead of people" (XIII: 280; quoted in Gippius [1924] 1989, p. 164). More examples are easy to find (see Pilshchikov 2019).

3.3. A Nose as a Phallus

Nabokov notes that Gogol's "nose-consciousness resulted at last in the writing of a story, *The Nose*, which is verily a hymn to that organ. A Freudian might suggest that in Gogol's topsy-turvy world human beings are turned upside down [. . .], so that the part of the nose is played by some other organ and vice-versa" (Nabokov [1944] 1961, p. 4). Such a Freudian really existed—it was Ivan Yermakov (also transliterated as Ermakov), the author of psychoanalytic analyses of Pushkin, Gogol, and Dostoevsky. His foreword to a separate edition of "The Nose" was published in 1921, and his *Essays in the Analysis of Gogol's Oeuvre*, which contain a chapter on "The Nose", came out in 1924.[12]

According to Yermakov, the nose in Gogol's texts is a phallic symbol, so the fear of losing the nose symbolizes the fear of castration. In addition, "the polysemantic meaning of the nose-symbol" "indicat[es]—in witticisms, jokes and ambiguities—both the sex organ and the act of defecation" (Yermakov [1924] 1974, pp. 194, 166). For a later Freudian critic, "The Nose" is "a satirical comic fantasy born of an impotence complex" (Spycher 1963, p. 361). On the contrary, Simon Karlinsky did not find any sexual and psychological implications in "The Nose", except Kovalyov's active unwillingness to marry, which is quite common among Gogol's characters (Karlinsky 1976, pp. 129–30). For Daniel Rancour-Laferriere, "the nose of 'The Nose' is merely phallic" and "fail[s] to partake of a *double psychoanalytic symbolism*" (Rancour-Laferriere 1982, p. 192; original emphasis).

I do not tackle here the complex issue of Gogol's sexuality and its reflection in his works. This theme requires serious research, and most of the earlier works on this topic are methodologically outdated and factually inaccurate. Perhaps Nabokov was right when he warned against excessive psychologism and urged "to treat Gogol's olfactivism—and even his own nose—as a literary trick allied to the broad humor of carnivals in general and to Russian nose-humor in particular" (Nabokov [1944] 1961, p. 4). Two decades earlier Vinogradov explained, as if in passing, that this sort of humor was based on "the semantics of the 'rhinological' calembour innuendos implied the ancient omen that remained popular in the eighteenth and nineteenth centuries", i.e., the omen that links "the size of the nose to men's dignity" (Vinogradov [1921] 1929, pp. 25 fn. 1, 34 fn. 2).

Elsewhere Vinogradov wrote: "Under the banner of 'Shandyism' on the canvas of the Romantic fantastic, Gogol marked out 'dirty' patterns, using [. . .] popular anecdotes surrounded by less than modest associations" (Vinogradov [1925] 1987, p. 79). *Shandyism* refers to Lawrence Sterne's novel *The Life and Opinions of Tristram Shandy, Gentleman*, the third volume of which contains three chapters (XXXI–XXXIII in the first edition) that play on the obscene connotations associated with the nose. They are written as an explanation of the protagonist's great-grandmother's statement: *"you have little or no nose, Sir"* (Sterne 1761, pp. 145, 149). The narrator warns the readers with comic gravity:

> For by the word *Nose*, throughout all this long chapter of noses, and in every
> other part of my work, where the word *Nose* occurs,—I declare, by that word I
> mean a Nose, and nothing more, or less. (Sterne 1761, p. 149; original emphasis)

"The gentle obscenity of this passage was quite likely to amuse both Pushkin and Gogol (both rather uninhibited in their letters)," Vsevolod Setchkarev commented (Setchkarev 1965, p. 156).

In his celebrated book on Rabelais, Mikhail Bakhtin found a correlation between the "supposition prevalent in those days even among physicians that the size of the phallus could be surmised by the size of the nose" and the fact that "in both classical and medieval grotesque the nose had usually this link with the phallus", and, moreover, "that it always symbolizes phallus" (Bakhtin [1965] 1968, pp. 86–87, 316; translation slightly modified). He also mentioned that the phallus is "predominantly subject to positive exaggeration, to hyperbolization", so that it can "even detach [itself] from the body and lead an independent life", and "the nose can also in a way detach itself from the body" (ibid., p. 317). However, when Bakhtin contrasted the elements of "folk humor" in Rabelais and Gogol, he refused to recognize any hint of the nose–phallus inversion in Gogol's tale, even after he linked "the most grotesque nose, which strove for an independent existence", with both Sterne's tradition and the tradition of "Ukrainian folk-festival" forms of art (Bakhtin [1975] 1976, pp. 285, 288). Meanwhile, in Russian popular prints (*lubok*), the nose also acts both as a metaphor for the phallus and as an independent character. Apparently, Gogol was not unaware of these metaphors and associations (Dilaktorskaia 1984, pp. 162–63; Pletnyova 2003). An allusion to a phallus is particularly obvious in the first, one-page draft of the tale. When the barber finds a nose in the loaf of bread, the finding is described as having phallic qualities: "Ivan Fedorovich [a character to be renamed into Ivan Yakovlevich in the next redaction—*IP*] broke the bread and what a surprise it was for him when he saw a nose there. It was a male nose, quite strong and thick . . . [*Nos muzhskoi, dovol'no krepkoi i tolstyi*]" (III: 380).

The idioms built on the metaphorical and metonymic use of the word *nose*—either those actually used in the text or even just implied by the context—play a pivotal role in creating an atmosphere of frivolous ambiguity. These idioms include: *drat'/podnimat' nos* ("to stick one's nose in the air"), *ne videt' dal'she svoego nosa* ("to be unable to see beyond one's nose"), *pokazat' nos* ("to show one's nose [face] somewhere"), *vodit' za nos* ("to lead someone by the nose"), *ostavit' s nosom* (literally "to leave someone with a nose", i.e., "to leave someone empty-handed"), *byt'/ostat'sia s nosom* ("to be/to be left with a nose", i.e., "to be left empty-handed"), *ostat'sia bez nosa* ("to be left without a nose", i.e., "to lose the nose as a consequence of syphilis").[13] Pushkin played on the parallelism of the two last idioms in an epigram of the early 1820s aptly noted by Vinogradov ([1921] 1929, p. 25):

> Lechis'— il' byt' tebe Panglosom,
>
> Ty zhertva vrednoi krasoty —
>
> I to-to, bratets, budesh' s nosom,
>
> Kogda bez nosa budesh' ty. (Pushkin 1937–1949, vol. 2.1, p. 206)

> Literally: "Get treatment, otherwise you will become a Pangloss,/you are a victim of harmful beauty,/and thus you surely will be 'with the nose'/when you find yourself without one" (the translation of the two last lines is Erlich (1969, p. 86); Pangloss, a philosopher in Voltaire's *Candide*, suffered from syphilis). [14]

Mikhail Lermontov used the same device in 1837 in a brief epigram on a deceived syphilitic:

> Ostat'sia bez nosu—nash Makkavei boialsia,
>
> Priekhal na vody—i s nosom on ostalsia.
>
> (Lermontov 2014, p. 285)

> Literally: "Our Maccabee[15] feared finding himself without a nose./He came to the mineral waters for treatment and was left 'with a nose'". (see Okhotin's commentary in Lermontov 2014, pp. 605–6).

Nikita Okhotin has shared a few other similar examples from Russian literature with me (private communication). The first text to introduce this phraseological parallelism in Russia may have been an epigram that Viazemsky attributed to Denis Davydov, a poet-hussar and violator of all decorums:

> Vozvratu tvoemu s pokhoda vsiak divitsia:
>
> Kak bez nosu poiti, a s nosom vozvratit'sia?
>
> (Viazemskii 1893, p. 492)

> Literally: "Everyone is surprised by your return from the military campaign:/How could you leave without a nose and return with one?"

The memoirist Aleksandr Galakhov reports a pun by the famous vaudeville writer and actor Dmitry Lensky (dated the mid-1840s): "in married life, it is better to be 'with a nose' than without one [*v brachnoi zhizni luchshe byt' s nosom, chem bez nosa*]" (Galakhov 1886, p. 187).

Last, but definitely not least, in *The Brothers Karamazov*, Dostoevsky devotes almost two pages to the pun under discussion (Vinogradov [1921] 1929, p. 71; Martinsen 2004, pp. 59–60). The passage that crowns the literary tradition of Russian rhinological puns is found in Chapter 9 of Book 11, entitled "The Devil [*Chort*]. Ivan Fyodorovich's Nightmare." The existing translations render the wordplay only approximately—the idioms *"to get one's nose pulled"* (Constance Garnett) or *"to put someone's nose out of joint"* (David Magarshack; Julius Katzer; Pevear and Volokhonsky) are used instead of *ostat'sia s nosom*:[16]

> "My friend," the visitor observed sententiously, "it's sometimes better to have your nose put out of joint than to have no nose at all [*s nosom vse zhe luchshe otoiti, chem inogda sovsem bez nosa*], as one afflicted marquis (he must have been treated by a specialist) uttered not long ago in confession to his Jesuit spiritual director. I was present—it was just lovely. "Give me back my nose!" he said, beating his breast. "My son," the priest hedged, "[…] If a harsh fate has deprived you of your nose, your profit is that now for the rest of your life no one will dare tell you that you have had your nose put out of joint [*chto vy ostalis' s nosom*]." "Holy father, that's no consolation!" the desperate man exclaimed. "On the contrary, I'd be delighted to have my nose put out of joint [*ostavat'sia s nosom*] every day of my life, if only it were where it belonged [*na nadlezhashchem meste*, literally, 'on its proper place']!" "My son," the priest sighed, "[…] if you cry, as you have just cried, that you would gladly have your nose put out of joint [*ostavat'sia s nosom*] for the rest of your life, in this your desire has already been fulfilled indirectly; for, having lost your nose, you have thereby, as it were, had your nose put out of joint all the same [*poteriav nos, vy tem samym vse zhe kak by ostalis' s nosom*] … "
> (Dostoevskii 1976, pp. 80–81; Dostoevsky 1991, p. 646)

In light of these persistent associations, the opening paragraph of the second chapter of "The Nose" sounds far from innocuous: "Collegiate Assessor Kovalyov jumped up from the bed and shook himself: There was no nose!" (III: 53/201). Kovalyov tried to explain the problem to the Nose when he met it/him:

> You must agree that it is improper for me to walk around without a nose. […] with plans to obtain … and moreover being acquainted with ladies in many homes: Mrs. Chekhtaryova, the wife of a state councillor, and others … […] Forgive me … if one looks at this in conformity with the rules of duty and honor … you yourself can understand … (III: 56/205)

However, the Nose refuses to understand the allusions in the best tradition of Sternean literature: "'I understand absolutely nothing,' the Nose answered. 'Express yourself in a more satisfactory manner'" (III: 56/205). A similar ambiguity arises when Kovalyov attempts to explain himself in the newspaper office: "I have many acquaintances: Chekhtaryova, the wife of a state councillor, Palageya Grigoryevna Podtochina, the wife of a staff officer … What if they find out, God forbid!" (III, 60/210). Both the narrator and the characters constantly talk about the indecency of not having a nose but never directly refer to the reason. Advertising the absence of a nose in a newspaper is, for instance, "unseemly, it's awkward, it's not good!" (III, 75/226); or, in another translation, "it is indecent, inept, injudicious!" (Gogol 1998, p. 325). The police commissioner is absolutely sure "that a respectable man would not have had his nose torn off, and that there are all kinds of majors in the world who don't even keep their underwear in a seemly condition and who hang around in indecent places" (III: 63–64/213–14). The narrator fully agrees with the commissioner: "That hit him right between the eyes!" (III: 64/214; the Russian phrase *"ne v brov', a (priamo) v glaz"*—literally: "not in the brow but (right) in the eye"—means "exactly, precisely, absolutely right").

Finally, Kovalyov writes an indignant letter to Podtochina, suspecting her of having stolen his nose in order to make him marry her daughter (*sic!*). In her reply, Podtochina changes the topic by referring to a more innocent idiom *"ostavit' vas s nosom"*, literally: "to leave you with a nose", i.e., "to leave you empty-handed" (Slonimskii [1923] 1974, pp. 359–60; Erlich 1969, p. 86; Peace 1981, p. 139):

> You also mention a nose. If you mean by this that I wished to lead you around by the nose [The idiom in the original reads *"ostavit' vas s nosom."*—*IP*] that is, to give you a formal refusal, I am amazed that you yourself are saying that, when I, as you well know, was of the exact opposite opinion, and if you are now asking in a legitimate fashion for my daughter's hand in marriage, I am prepared this very minute to give you satisfaction, for this has always been the object of my most keen desire. (III: 71/221–22)

Thus, Gogol plays on the same parallelism of idiomatic phrases as Pushkin, Lermontov, and others (compare Budagov 1953, pp. 26–27).

The first redaction of the tale finished with explicit erotic overtones:

> "Hey, Ivan!"—"What would you like, sir?"—" What, wasn't there a girl, a very pretty one, asking for Major Kovalyov?"—"No, sir!"—"Hmm!" said Major Kovalyov and looked in the mirror, smiling. (III: 399)

Noses are in one way or another associated with the erotic and matrimonial theme both in the definitive redaction of "The Nose" (Shukman 1989, pp. 76–77) and in Gogol's other works. In *The Marriage*, Agafia Tikhonovna's main flaw is her long nose (this theme is particularly emphasized in the first redaction of the comedy). The ladies of the town of N. discover the same flaw in Chichikov's appearance (*Dead Souls*, chap. IX): "People have been spreading rumors that he's handsome, but he's not handsome at all, not handsome at all, and his nose … a most unpleasant nose … » (VI: 182; Gogol 2004, p. 206). The extended metaphorical pun in the tale of the two Ivans is especially noteworthy:

> I must admit I do not understand why it has been ordained that women should take us by the nose as easily as they take hold of the handle of a teapot: either their hands are so created or our noses are fit for nothing better. (II: 241/184)

Last but not least, the parallel between the nose and the phallus in "The Nose" is supported by the latent obscene phraseology. The climax of the story of how Kovalyov discovered the absence of the most important part in his face is his exclamation: "If only there were something instead of my nose, but there's nothing! [*a to nichego!*]" (III: 54/203). The main dysphemism for *nichegó* ("nothing") in Russian is *ni-khuyá* ("nothing"; literally, "not a prick") (Dahl and Baudouin de Courtenay 1903–1909, vol. 4, p. 1244). This kind of dysphemization started to flourish in the twentieth century, but *ni-khuyá* and *né-khuy(a)*, which stand for *nichegó* and *néchego* (both meaning "nothing"), belong to some of the earliest documented examples of obscene dysphemisms. They are found in obscene folklore proverbs collected by the lexicographer Vladimir Dahl (also transliterated as Dal') and folklorist Alexander Afanasyev in the middle of the nineteenth century (Afanas'ev 1997, pp. 498, 505, 525, 585; Carey 1972, pp. 59, 97; see (Pilshchikov forthcoming) for details and analysis). *Khuya* is the genitive case of *khui*, the main Russian obscene term for the phallus. Kovalyov's *"nothing"* sounds like a euphemism of this dysphemism, so that the phallus (*khui*) *almost* sticks out on the surface of the tale's text—but eventually it does not, and its narrative non-appearance iconically depicts its absence.

4. How the Table of Ranks Brought the Nose to the Kazan Cathedral

4.1. "The Nose" as a Cornerstone of the "Petersburg Text"

The setting of the tale is not so much the anthropologically authentic description of capital-city life as the mythopoetic space of the "Petersburg Text" of Russian literature, which Vladimir Toporov discovered and described in a series of studies (from the inaugural programmatic manifesto Toporov 1984 to the posthumous compendium Toporov 2009). Toporov credits Pushkin and Gogol as the "initiators" of this "synthetic hypertext" and Dostoevsky as its "ingenious designer"; Gogol's "Petersburg Tales" are named as a foundational oeuvre, one that strongly influenced other writers, including Dostoevsky (Toporov 1984, pp. 14–15; compare Lotman 1990, p. 195). Despite its specific temporal situatedness, the urban text becomes, at a deeper layer of analysis, ahistorical, like a myth (Kalinin 2010). Toporov compared this to Propp's vision of the magic folktale: every tale is a different tale, but each realizes a different variant of the same invariable plot or its part; similarly, every "Petersburg tale" (Pushkin's, Gogol's, etc.) can be read as a variant of the "Petersburg Text" (Toporov 1984, p. 17 fn. 17).[17]

The intrinsic features of the reinvented Saint Petersburg as represented in the "Petersburg Text" are its duality, ambiguity, artificiality, fictitiousness, and illusoriness (Toporov 1984, pp. 4, 6, 12, 25–26, 2009, pp. 644, 657, 665–66, 676–77, 682, 711; Lotman 1990, p. 196). Saint Petersburg is "artificial" because it is not an old city formed that took shape over centuries in a "natural" way, but a new one, which the Tsar created by decree in 1703 (Toporov 1984, p. 12 fn. 12; Lotman 1990, p. 194). The duality of Saint Petersburg was determined from the very beginning: it was envisaged as the city of Saint Peter but realized as a willful and arbitrary creation of Peter the Great (Bocharov 2009, p. 13). This fundamental ambiguity underlies Saint Petersburg's "foundational myth" and is thus at the core of the "Petersburg Text." The dualism of the sacral and the profane is even reflected in the name of the city (ibid.): in colloquial Russian, the city is never called by the full name "Saint Petersburg" but always shortened to "Petersburg" or the even more unceremonious "Piter." In such a city, the secular is constantly substituted for the sacred, and the secular itself is ambiguous because the city is simultaneously treated as terrestrial "paradise", "the embodiment of Reason", and as "the terrible masquerade of the Antichrist" (Lotman 1990, p. 194). This ambiguity is a leitmotif in all the "Petersburg Tales", including "The Nose." Moreover, Gogol always emphasizes the evil and infernal aspect of the city:

> In "The Nose", just as in the other "Petersburg Tales", the main object of description is the satanic city taken in the confluence of both of its symbolic functions—secular (the brainchild of Peter the Great) and ecclesiastical (the city of the Apostle Peter), the latter being embodied in the Kazan Cathedral (which is, as is known, stylized as Saint Peter's Basilica). (Weiskopf 1993, pp. 236–37)

The Kazan Cathedral, the ex officio church of the Imperial family and Saint Petersburg's de facto Mother Church in 1811–1858, with its colonnade—a replica of Lorenzo Bernini's Colonnade of Vatican's San Pietro—is the point where the Nose's earthly career reaches its zenith. Like all of Saint Petersburg, its main cathedral also embodies both the sacred and the secular hierarchy, and the Nose putting in an appearance there makes a travesty of both. The secular hierarchy was then represented by the Table of Ranks, a system of correspondence between the military, civilian, and court ranks that was introduced by Peter I in 1722. In Imperial Russia, not only the military hierarchy but also the court and administrative hierarchies were visually reinforced by the corresponding dress uniforms worn by everyone serving in these hierarchies. This layer of representation was so important that the monarchs themselves personally reviewed the designs for every institution's uniforms (see Shepelyov 1991). The Table of Ranks, the dress uniforms, and the Kazan Cathedral all play a conspicuous role in "The Nose."

4.2. A Collegiate Assessor or a Major?

For the tale's texture, the details of everyday social life are as important as the idioms of everyday speech:

> Kovalyov was a collegiate assessor of the Caucasus. He had only been at that rank for two years and therefore could not forget about it for a single moment, and so as to lend himself nobility and weight, he never called himself "Collegiate Assessor", but always "Major". (III: 53/202)

In the newspaper office, Kovalyov suggests: "You can simply write: a collegiate assessor, or even better, a person occupying the rank of major" (III: 60/210). Why was this substitution possible and, if possible, why did it then give Kovalyov more "nobility and weight"?

Equating one rank to another was possible thanks to the Table of Ranks, which affirmed the meritocratic principle of social organization, placing it above the principles of nobility and wealth (a person's status is determined by his service to the Empire). However, the bureaucratized Russian state gradually eroded the essence of Peter's reform, rendering the system of ranks and positions the only regulator of relations between people in all spheres of life, including outside the service.

All ranks were divided in fourteen classes (the first was the highest; the fourteenth was the lowest). Collegiate assessor was a civilian rank of the eighth class, which corresponded to the military rank of major. This rank was, so to say, the lowest of the higher ranks. Beginning with the eighth class, the official was honored with a special mention in the newspapers when entering and leaving the city or town. Thus, every issue of the newspaper *Moskovskie Vedomosti* (*Moscow Gazette*) published "News about persons of the first eight classes who have arrived in and departed from this Capital." Before 1845, the eighth class afforded civilians hereditary nobility. It remains unclear if Kovalyov was born noble or earned nobility through his service (Reyfman 2016, p. 108). To compare: the main characters of "The Diary of a Madman" (1834) and "The Overcoat" (1841), Aksenty Ivanovich Poprishchin and Akaky Akakievich Bashmachkin, are only one class below Kovalyov. They are titular councillors (ninth class) and have been granted only "personal" nobility (not to be passed on to their children); their fate is generally believed to symbolize the "tragedy of the little man."[18]

Why, then, did Kovalyov prefer to be addressed as major, rather than collegiate assessor? This was not simply "a more impressive title" (Seifrid 1993, p. 384). The answer is prompted by the full name of the Petrine Table of Ranks: *The Table of Ranks of all Grades, Military, Civil, and Court, which grades are in which class; and those which are in one class have seniority among themselves according to the time of entry into the grade; however, military ranks are higher than civilian ones, even when [civilians] had been promoted into that class earlier* (Complete Collection of the Laws of the Russian Empire, First Series, No. 3890; Dukes 1978, p. 10). In other words, within the same grade, the military rank was considered higher than the civilian rank. By identifying himself as "major", Kovalyov cheated, tacking on an extra half-rank in comparison with other collegiate assessors. He used the same stratagem for purposes of flattery when addressing "a court councillor of his acquaintance" (a civil rank of the seventh class) as "lieutenant colonel, especially if there were other people around" (III: 57/206). It should be pointed out that this practice was expressly prohibited by the law (Shepelyov 1977, p. 53, 1991, p. 115; Dilaktorskaia 1984, p. 154): "Civilian officials are not allowed to be addressed using military ranks" (Statutes of Civil Service 1833, p. 119). Therefore, Kovalyov is a Khlestakov-type dodger who likes to take advantage of privileges, both legal and illegal, or simply enjoys being taken for a person of a higher rank.[19] Another telling detail: Kovalyov was once seen "buying the ribbon for an Order of some sort, no one knows for what reason, because he himself was not a knight of any Order at all" (III: 75/226). In light of these details, the Nose's brazen imposture is a grotesque hyperbole of Kovalyov's petty imposture.

Yet another significant detail has already been commented on in Gogol scholarship: "Kovalyov was a collegiate assessor of the Caucasus" (III: 53/202), or, literally: "Kovalyov was a Caucasian [*kavkazskii*] collegiate assessor" (Gogol 1998, p. 305). According to the government's decision, "to prevent a shortage of capable and worthy officials", the promotion to collegiate assessor in the Caucasus, "because of the remoteness of those places", was simplified in relation to the established order (Statutes of Civil Service 1833, p. 105). Namely, officials were promoted without taking a special exam and without obtaining a university education (Dilaktorskaia 1984, p. 154). According to the Statutes of Civil Service, "officials in the Caucasus region, Georgia, the provinces of Transcaucasia [and a few other remote regions] [...] are promoted to the 8th and 5th classes [...]

without the examination and certificates required of other civil servants" (Statutes of Civil Service 1834, pp. 850–51). This is what the narrator implicitly refers to:

> The collegiate assessors who receive that rank with the help of learned diplomas cannot at all be compared with those collegiate assessors who are created in the Caucasus. These are two quite particular types. (III: 53/201–2)

4.3. In Search of a Plumed Hat

> *Colonel Skalozub.* The uniform's the way to tell, ma'am.
>
> The braid, the shoulder-tabs, and gorget-tabs as well ma'am.
>
> (Griboedov 2006, p. 98)

And what about the Nose? Gogol's contemporaries understood immediately that the Nose was "wearing the uniform of the Ministry of Education" when Kovalyov met him (the words of Prince Pyotr Vyazemsky in his letter to Alexander Turgenev of 9 April 1836; Viazemskii 1899, p. 314). This is how Gogol depicts the Nose's appearance:

> He was in a uniform with gold embroidery, with a large stand-up collar; he was wearing suede trousers and had a sword at his side. Judging by his plumed hat he bore the rank of state councillor. (III: 55/203–4)

In the administrative semiotics of Imperial Russia—or, indeed, in the framework of Russian "semiotic totalitarianism" (Morson 1992)—"uniform signifies the place of service, as well as a degree of rank and position" (Statute on Civil Uniforms 1834, § 1; Complete Collection of the Laws of the Russian Empire, Second Series, No. 6860, in Complete Collection of the Laws 1835, vol. 9.1, p. 169). State councillor is a rank of the fifth class. The corresponding military rank of brigadier was abolished in 1799 (Raskin 1989, p. 663). The next-lower rank was that of colonel (sixth class), and the next-higher rank was that of major-general (fourth class). Beginning with the fifth class, the corresponding rank was conferred only "with the Supreme [=His Imperial Majesty's] permission" (Statutes of Civil Service 1833, p. 106). The fifth class was the lowest of the five highly privileged classes.

The plumed hat (*"shliapa s pliumazhem"*) deserves special attention. Irina Reyfman in her otherwise illuminating book on the Imperial Table of Ranks in Russian literature insists that Kovalyov and the Nose allegedly "misread" each other's uniforms. In particular, she argues that "the uniform of state councillors did not include plumes on the hat; plumes could signal either a high military rank [...] or a court title. [...] The Nose's declaration that he is [...] an educator [...] undermines this assumption and leaves the reader as confused as Kovalyov" (Reyfman 2016, p. 108). Olga Dilaktorskaya, who was the first scholar to specifically investigate the meanings of official titles and forms of their semiotic expression in "The Nose" (Dilaktorskaia 1984), made the same arguments earlier when commenting on the tale in the *Literaturnye pamiatniki* (*Literary Monuments*) edition of Gogol's "Petersburg Tales": "By all this we can assume that the Nose is not just a state councillor, but a person who, in the author's opinion, is close to the Imperial Court, a VIP" (Dilaktorskaia 1995, p. 273).

The real historical reception of this passage and other available documents do not confirm these claims. Vyazemsky, who was present at Gogol's reading of the uncensored version of "The Nose" to the élite of Petersburg litterateurs at the poet Vasily Zhukovsky's apartment on 4 April 1836, found it amusing that Kovalyov met the Nose in the church (more on that below), but he did not find anything special in the Nose's dress:

> Zhukovsky's Saturdays are flourishing [...]. Only Gogol [...] livens them up with his stories. Last Saturday he read us a story about a nose which all of a sudden disappeared from the face of some Collegiate Assessor and turned up later in the Kazan Cathedral wearing the uniform of the Ministry of Education. Killingly funny! A lot of real humour.[20] After having met his nose, the Collegiate Assessor tells him: "I am surprised at seeing you here; it seems to me you should know your place. (Viazemskii 1899, pp. 313–14; partly quoted in Gogol 1967, p. 7)

Characteristically, the text also provoked Vyazemsky (as a real "ideal reader" of the tale) to make puns on the idioms with the nouns *nose* and *place*: "And in order for my letter not to be lost, but to get to its place [*k svoemu mestu*], that is, under your nose [*tebe pod nos*], and for me not to be 'left with a nose' [*s nosom*], I will finish writing and send my letter to the Ministry of Foreign Affairs" (Viazemskii 1899, p. 314; the addressee, Alexander Turgenev, was travelling around Europe).

The uniform of the Ministry of Education was introduced on 28 August 1810; the uniform of the Petersburg educational district was introduced earlier, on 20 January 1809. The uniforms of the ministry, educational districts, and universities were similar although differentiated by minor details (Shepelyov 1999, pp. 304–13). The law describes the ministerial uniform briefly:

> A dress of dark blue cloth with a stand-up collar and velvet cuffs of the same color; blue lining; the camisole and undershirt are white cloth; gilded plain buttons […] Gold embroidery in accordance with the attached pattern". (quoted in Shepelyov 1999, pp. 311–12)

A later statute that was enacted as part of the larger reform of civil uniforms on 27 February 1834 introduced only minor changes (Statute on Civil Uniforms 1834, § 84; Shepelyov 1999, pp. 226–31, 312–13). This is precisely the period when Gogol was working on the first redaction of "The Nose."

The form of the hat was not defined by those laws (we only know it should have been triangular), but the existing practices are attested in at least one legal document. On 17 October 1834, a Supremely Ratified Decree of the Committee of Ministers Declared by the Senate was published entitled *On granting permission for officials of the first five classes, who retired before the Decree on the Civil uniforms was made effective, to wear a plumed hat as they did before* (Complete Collection of the Laws of the Russian Empire, Second Series, No. 7437). The permission was founded on the Emperor's decision of 5 October:

> HIS IMPERIAL MAJESTY, upon the provision of this [issue] by the Committee, on 5th of this October, did MOST EMINENTLY deign […] to decree […] to the Civil Service that officials of the first five classes, who retired before 27 February 1834, be given the right to wear a plumed hat as before [*nosit' po-prezhnemu shliapu s pliumazhem*] […] if they retired with the uniform approved for their last place of service. (Complete Collection of the Laws 1835, vol. 9.2, pp. 54–55)

Furthermore, a plumed hat as a sign of a high rank in the civil service (state councillor and above) is mentioned in various memoirs (Kirsanova 1989, pp. 182–83). The following is an excerpt from Count Mikhail V. Tolstoy's memoirs about what the Doctor of Medicine, Professor of Moscow University, and State Councillor Vasily Kotelnitsky looked like in the early 1830s (between 1831 and 1834):

> [When Kotelnitsky] took a cab he would tell the cabman: "Look out, drive carefully—you are driving a state councillor!" The passion for ranks was then almost ubiquitous, and the good old boy Kotelnitsky was fascinated by the grandeur of his rank! […] [He] would go down after a full-dress function from the grand porch of the old university building wearing a uniform and triangular plumed hat [*v treugol'noi shliape s pliumazhem*] (at that time, officials of the 5th class and above wore plumes on their hats [*pliumazh na shliapakh*]) […]. (Tolstoy 1881, p. 50)

All the persons mentioned above were "within one handshake" from Gogol: when Mikhail Tolstoy entered the university, one of his examiners was Mikhail Pogodin, who would reject "The Nose" from the *Moscow Observer* four years later. Tellingly enough, Pogodin was also among those who did not find anything unusual in the Nose's hat. Meanwhile, this detail was important for Gogol, and he repeated it again ("[Kovalyov] remembered very well that his hat was plumed and that his uniform had gold embroidery", III: 57/206).

4.4. The Two Thresholds

The Nose says to Kovalyov: "Judging by the buttons on your uniform, you must be employed in the Senate or at least in the Ministry of Justice. As for me, I am in the scholarly line" (III: 56; Struve 1961, p. 49). Reyfman argues that, although "the buttons of the Senate and the Ministry of Justice uniforms were indeed similar", Kovalyov could not have served in either institution "during his time in the Caucasus […] since both were located in Saint Petersburg" (Reyfman 2016, p. 109). The structure of civil administration in the Caucasus was complicated and constantly changing, so that we cannot say for sure what Kovalyov's position was when he served there. We only know that his functions were similar to those in the institutions subordinate to the Ministry of Justice, since he "had been sent on investigations several times when he was still in the Caucasus region" (III: 71/222).

Furthermore, Kovalyov was not even wearing a full dress uniform (*mundir*): he was in "vice-uniform" (*vits-mundir*). This difference is rarely commented on and never translated. The *vits-mundir* was a dress uniform for ordinary (not festive) days; for civil officials who did not belong to the three highest classes, the function of the *vits-mundir* was performed by the *mundirnyi frak*—the "uniform tail-coat" (Shepelyov 1999, pp. 229–30). The "uniform tail-coat" of the Senate Chancellery was "the same as that of the Ministry of Justice" (Statute on Civil Uniforms 1834, § 64; Complete Collection of the Laws of the Russian Empire, Second Series, No. 6860, in Complete Collection of the Laws 1835, vol. 9.1, p. 174). Therefore, the Nose was unable to precisely identify Kovalyov's affiliation.[21]

The Senate and ministerial positions that a collegiate assessor could occupy in a standard career were junior secretary in the Senate or junior "desk chief" [*stolonachal'nik*] in ministerial departments (Raskin 1989, p. 662). A state councillor would be appointed a vice-director in a Department of the Ministry of Justice or an official at the Chief Procurator's desk in the Senate. However, the Nose served "in the scholarly line", i.e., in science and/or education; therefore, it/he could have been a university rector or a deputy trustee of the educational district (ibid., p. 661). Kovalyov's Nose—being part of the same body—must have been awarded its (considerably higher) rank of the fifth class in the Caucasus as well. For the promotion of Caucasian officials to the fifth class, the same exemptions were made as for the promotion to the eighth class. In other words, the Nose may not have had a proper education either, which did not hinder promotion "in the scholarly line."

There were two "thresholds" in the Table of Ranks: Class 8 and Class 5. Kovalyov only mastered the first threshold, whereas the Nose somehow managed to get over the second one as well. As Grigory Gukovsky put it,

> … this is the meaning of the story's conflict: the struggle is between a *man* with the rank of collegiate assessor and a *nose* with the rank of state councillor. The *nose* is three ranks higher than the *man*. This conditions its victory, its invulnerability, and its superiority over the man. (Gukovskii 1959, p. 283)

4.5. To Know One's Place

The interaction between Kovalyov and the Nose is framed in the social-psychological field that Gogol himself dubbed "the electricity of rank" (V: 142): "We must note that Kovalyov was a person who was extremely quick to take offense. He could forgive everything that was said about himself but could never excuse anything that related to his office or rank" (III: 64/214). In Gogol's world, the rank makes people arrogant, and Kovalyov is not an exception. Various unpleasant characters in Gogol's texts have their noses in the air just because they are higher in rank than their neighbors. In the preface to the second volume of *Evenings on a Farm near Dikanka* (1832), the beekeeper Rudy Panko complains:

> I tell you what, dear readers, there is nothing in the world worse than these high-class people. Because his uncle was a commissar once, he *turns up his nose at everyone* [*nos nesyot vverkh*]. As though there were no rank in the world higher than a commissar! Thank God, there are people greater than commissars. No, I don't like these high-class people. (I: 195–96/90–91).

A similar character is featured in "A May Night; or The Drowned Maiden" (1830–1831): "He thinks [...] that because he is the mayor, [...] he can *turn up his nose at everyone* [*nos podnial*]!" (I: 160/56). In the early draft version of *The Marriage*, the bride's aunt laments (act 1, scene 12): "The nobleman just sticks his nose in the air [*deryot ... nos*] when commoners are around, but with his own sort that's just a bit higher in rank, he bows and scrapes 'nough to break [his neck]" (V: 295). In an early manuscript redaction of the first volume of *Dead Souls*, "a new-made statesman sits there like a peacock with his nose and bald head craning up in the air [*podniavshi kverkhu nos* ...]" (VI: 413).

"It seems to me ... you should know your place [*znat' svoe mesto*]," Kovalyov says to his Nose (III: 55/204). He had held his nose so high that it *forgot its place*. Perhaps the odious Vladimir Yermilov, an "orthodox" Marxist critic of the Stalin era, was not too far off the mark when he suggested the apparently simplistic formula: "Kovalyov turned up his nose too high—and it flew away from him!" (Yermilov 1952, p. 157).[22] If the Nose is a hypertrophied projection of Kovalyov (compare Bem [1928] 1979, p. 244), then the words about "knowing one's place" could be addressed to Kovalyov himself. Gogol makes a pun on this when he writes that Kovalyov came to Saint Petersburg "to find *a position* becoming to his rank [the Russian word for 'position' being *mesto*, i.e., 'place']: if he could manage it, a position as vice-governor, and if not, then as administrator in some prominent department" (III: 54/202). A position of vice-governor required a rank of the fifth class, which the Nose had already been awarded but Kovalyov had not.[23]

"It's strange to me, my dear sir ... it seems to me ... you should know your place And suddenly I find you, and where? In a church" (III: 55/204)—this is the full text of the major's remark quoted in the previous paragraph. The Nose's attending the church service is indeed unusual, but doing so in a uniform makes a little more sense at least, because he was there not as a private person but as a high-ranking official: it was allegedly his duty. However, Dilaktorskaya's explanation (Dilaktorskaia 1984, p. 156) accepted by Yuri Mann (1996, p. 90)—that officials were legally required "to be at the Divine Service in festive uniform" on all the twelve major church holidays (or "great feasts") of the Orthodox calendar, including the Day of the Annunciation when Kovalyov meets the Nose—seems to be an anachronism. The scholars refer to a later law of 8 March 1856 (No. 30247) enacted in the second year of the reign of Alexander II (Description of Changes 1856, p. 9; Complete Collection of the Laws 1857, vol. 31.2, p. 55). It has nothing to do with the age of Gogol unless it reaffirms the status quo. Furthermore, in earlier versions of the tale, the day when Kovalyov meets the Nose in Kazan Cathedral is not a great feast day. Therefore, the legal reasons for the Nose's presence in the cathedral in full dress uniform require further investigation.

The uniforms immediately reveal the hierarchical inequality between Kovalyov and his double the Nose and turn physical distance into social distance that Imperial semiotics further transforms into existential distance: "The Table of Ranks was a direct product of the imperial representation policy, where the 'natural' (divine) hierarchy, according to which existence is organized, is replaced by an artificial imperial Table" (Iampolski 2007, p. 16). For Theophanus Prokopovich, one of the most ardent ideologists of the Petrine reforms, "if all rank is derived from God [...] then the thing most necessary to us and pleasing to God that the rank demands is that I should have mine, you yours and likewise for everyone" (Feofan Prokopovich 1961, p. 98; quoted in Lotman 1990, p. 262; translation modified). However, for Gogol, a hierarchy that pretends to be divine and replaces the divine could only be diabolical.

5. Liturgy, Sacrilege and the Calendar

5.1. The Problem: Why the Annunciation?

Yermakov insightfully observes that the Nose disappeared from Kovalyov's face on the day of the Annunciation according to the Julian calendar (25 March) and returned on 7 April, that is, on the day following the Annunciation according to the Gregorian calendar (6 April), "so the whole episode took place in one day—or one night" (Yermakov [1924] 1974,

p. 185 fn. 48). Yermakov noticed this fact but did not actually explain it (Seifrid 1993). An explanation was found by Boris Uspensky: according to popular beliefs characteristic of "interfaith frontier areas" (a category which perfectly fits Gogol's native Ukraine), the entire period separating the Orthodox and Catholic holidays of the same name is considered an "unclean" time, when various evil things usually happen (Uspenskii 2004, pp. 335–37). What is more, the absurd events of the tale take place over thirteen days—an unlucky number, "the devil's dozen", according to a widespread superstition.

In the uncensored definitive version, Kovalyov meets the Nose in the Kazan Cathedral, where it/he was praying on the occasion of the Feast of the Annunciation. In earlier redactions, the story was not yet timed to coincide with the Annunciation, but Gogol nevertheless insisted that these characters should meet in the church: "In case your stupid censorship starts to fuss about having a nose in Kazan Church, it can be transferred to a Catholic church," Gogol wrote to Pogodin on 18 March 1835 (X: 355; quoted in Oulianoff [1959] 1967, p. 162; see also Glyantz 2013, p. 100). Predictably enough, the censors excluded the mentions of the liturgy from both printed versions (1836 and 1842), and the entire scene was transferred to the city's main shopping center, known as Gostiny Dvor, which is situated 0.3 miles to the east of the Kazan Cathedral along Nevsky Prospect (Gillel'son et al. 1961, pp. 183–84).

The dialog between Kovalyov and the Nose is remarkable in all respects. Kovalyov: " ... you should know your place. And suddenly I find you, and where? In a church" (III: 55/204). The Nose pretends he does not understand anything: "Pardon me, I cannot make any sense of what you wish to say ... " (III: 56/205). The blasphemous innuendo is followed by an obscene one ("You must agree that it is improper for me to walk around without a nose"), to which Nose again responds with feigned incomprehension (III: 56/205). "One could see from the nose's own replies that *nothing was sacred* for this person," Kovalyov concludes (III: 58/207).

5.2. The Proskomedia in "The Nose" and in Meditations on the Divine Liturgy

Having found Kovalyov's nose in the loaf of bread, Ivan Yakovlevich exclaimed in surprise "It's solid! [...] What could it be?" (III: 50/198) and conjectured that "bread is a baked thing, and a nose is something else entirely" (III: 50/199). Boris Uspensky comments that the Annunciation usually falls on the days of Lent, when meat is forbidden to eat; therefore, the nose found in bread was especially improper, being quite literally a piece of flesh (Uspenskii 2004, p. 335). The Russian word for "solid" that Ivan Yakovlevich used is *plotnoe*, literally, "fleshy." However, as the scholar himself notes, these phrases were added to the text as early as in the 1836 redaction (or even earlier) when the action had not yet been synchronized with the Annunciation or Lent (ibid., p. 343 fn. 30). Perhaps, Gogol initially meant Friday fasting (compare Weiskopf 1993, pp. 234–35).

In the context of the religious symbolism of the tale, the bread motif that was present in the text from the very first draft may have a different meaning, which is not connected with Lent but refers to the Divine Liturgy. According to Mikhail Weiskopf, the barber's manipulations with bread and the nose can be interpreted as a disguised parody of the preparation of bread and wine for the Eucharist—Proskomedia (Greek: Προσκομηδή), or the Liturgy of Preparation, which Gogol later described in detail in his *Meditations on the Divine Liturgy* (Weiskopf 1987, pp. 28–29; 1993, pp. 231–33).

Gogol thus explains the meaning of the Proskomedia:

> All this part of the service consists in the preparation of what is required for the celebration, i.e., in the separation from the gifts, or little loaves of bread, of those sections which at first represent and are later to become the Body of Christ. (Gogol 1960, p. 5)

The parallels between "The Nose" and Gogol's explication of the Liturgy of Preparation found by Weiskopf are striking. "For the sake of propriety, Ivan Yakovlevich put his tailcoat on over his shirt" (III: 49/198), similarly to the priest who is "proceeding to don the sacred vestments" (Gogol 1960, p. 1; translation slightly modified). The barber's spouse took "some freshly baked loaves of bread out of the oven" and "threw a loaf on the table" (III: 49/197–98).[24] In the service, "the priest takes one of the little loaves of altar bread, called *prosphori*, in order to separate from it the section which will become the Body of Christ—the middle portion of the upper part, on which the seal bearing the name of Jesus Christ" (Gogol 1960, p. 5). This portion is called the Lamb of God and symbolizes He "Who offered Himself in sacrifice for the whole world" (ibid., p. 6). After that, the priest takes a knife, "which has the form of a spear in commemoration of the spear with which pierced the Body of the Savior upon the Cross" and cuts the middle portion from the *prosphora* (the Host, Eucharistic bread). "Then with the spear he lifts out the middle section of the bread which has been cut round" (ibid.). Ivan Yakovlevich did almost the same: he "took a knife in his hands, and assuming a dignified air, started cutting the bread. Having cut the loaf into two halves, he looked into the middle [. . .] and pulled out—a nose" (III: 49–50/198). To these parallels we can add a no less important difference. "The priest who is to celebrate the Liturgy should from the evening before hold himself sober [*trezvit'sia*] in body and mind" (Gogol 1857, p. 4; 1960, p. 1), whereas the barber was so drunk the evening before that in the morning he does not even remember how drunk he was: "Whether [. . .] I came home drunk last night or not, I can't tell for sure" (III: 50/199).

In the first draft, where Ivan Yakovlevich was called Ivan Fedorovich, not only the phallic connotations of the nose were more explicit (see Section 3 above), but also a travesty of the Liturgy was manifested from the very beginning. However, initially it started as a travesty of the Eucharist, rather than a travesty of its preparation. The barber did not *cut* the bread but *broke* it: "Ivan Fedorovich perelomil khleb [. . .]" (III: 380). The verb *perelomil*—a past tense form meaning "broke"—is a pleophonic vernacular Russian equivalent of the non-pleophonic Church Slavonic *prelomi* and ecclesiastic Russian *prelomil* with the same meaning (in Russian, *prelomil* sounds very much like the past tense form *brake* in KJV instead of *broke*). This verb is unambiguously associated with the rite of breaking of the bread (Latin: *fractio panis*; Church Slavonic and ecclesiastic Russian: *prelomlenie khlěba*), which is the central part of the Eucharist. According to Mark (14:22), as the Apostles were eating at the Last Supper, "Jesus took bread, blessed and broke it [KJV: *brake*; Synodal Slavonic Bible, 1751: *prelomi*; Russian Synodal Translation, 1819: *prelomil*], and gave it to them and said, Take, eat; this is My body" (NKJV). Luke (22:19) reports Jesus also said: "do this in remembrance of Me." This is the meaning of the Eucharist. After Resurrection, the Apostles recognized Jesus when "He sat at the table with them, [. . .] took bread, blessed and broke it, and gave it to them" (Luke 24:30–31); "He was known to them in the breaking of bread" (Luke 24:35; compare Weiskopf 1987, p. 30).

In the next redaction of Gogol's tale, the rite of the breaking of bread was substituted for by the rite of the cutting of bread, which, according to Gogol, "is merely a preparation for the Liturgy" (Gogol 1857, p. 14; cf. Gogol 1960, p. 5), i.e., a less important ritual. One of the reasons for this change may have been another pun—this time not obscene but blasphemous.

On the Host of the Lamb is found what Gogol refers to as "the seal bearing the name of Jesus Christ." This is an abbreviated form of the Greek Christogram ΙΗΣΟΥΣ ΧΡΙΣΤΟΣ ΝΙΚΑ (Jesus Christ Conquers) spelled in Greek or Cyrillic letters (which in this case look similar or identical):

IC	XC
NI	KA

The priest takes the Host and starts cutting out the seal from the right part (with the letters "IC NI"). In view of Gogol's parody, the result is overwhelming. This part as seen from the other participants' side reads as "NI IC", that is, the Ukrainian word *nis* "nose", spelled in Cyrillic as НИС(Ъ) or НІС(Ъ).[25] The identity of the Church Slavonic N and the secular Cyrillic Н is obvious for churchgoers (i.e., everyone in the nineteenth century), whereas a straightforward correspondence between the Ukrainian *nis* "nose" and the Russian *nos* "nose", spelled in Cyrillic as НОС(Ъ) (the title of Gogol's tale) is equally obvious for the speakers of both languages, to say nothing of Ukrainian–Russian bilinguals such as Gogol.[26] Then, the priest cuts the seal from other sides (see Figure 1) and eventually "cuts it crosswise", as the shortened and censored text of Gogol's *Meditations* reads (Gogol 1857, p. 17; 1913, p. 14; the same wording is used in Dmitrevskii [1803] 1807, Explanation of the Divine Liturgy, part 1: On the Proskomedia, chap. 4, p. 12). Gogol's original formulation was more ample: " . . . in token of His death upon the Cross, traces the sign of sacrifice on it, along which it will then be divided during the coming rite [i.e., the Eucharist]" (Gogol 1889, p. 419; cf. Gogol 1960, p. 6).[27]

Figure 1. The *Prosphora* of the Lamb. Adapted from Slobodskoi (1967, p. 645).

Taking into consideration Gogol's exploration of the same theme in his *Meditations*, his insistence on a church as the meeting place for Kovalyov and his nose and the choice of the Annunciation as the day when the story begins in the final version, it becomes possible to read the tale as permeated with leitmotifs of blasphemy and sacrilege (compare Uspenskii 2004, pp. 335, 342–43 fn. 30). Moreover, its entire plot can be considered a travesty of the Gospel story from the Annunciation to the Ascension,[28] which the rites of the cutting and breaking of bread represent "in complete semiotic entirety" (Weiskopf 1987, p. 28; 1993, p. 231).[29] At the same time, the parodies in "The Nose" of the Life of Christ and the rites of the Divine Liturgy that symbolize its various events are not carried out consistently: "The allusions to the main events of Christ's life are in disorder; they do not follow the evangelic sequence" (Glyantz 2013, p. 107). All in all, it is not a coherent travesty of the Gospel plot but rather "a humorous midrash on it", as John E. Cornell put it on another occasion (Cornell 2002, p. 271).

5.3. The Ascension and the Chronotope of "The Nose"

The Ascension is explicitly although inconspicuously alluded to by the address of "the barber Ivan Yakovlevich, who lives on Voznesensky Avenue/Prospect" (III: 49; Gogol 1985, vol. 2, p. 216; 1998, p. 301). This is the second sentence of the tale, and the name of the street/avenue is repeated again when the police officer brings Kovalyov his nose: "And it's strange that the chief participant in this affair is that crook of a barber on Voznesenskaya Street," he says (III: 66; Gogol 1998, p. 317; cf. Gogol 1985, vol. 2, p. 232). The English translations do not usually inform the readers that *Voznesensky* and *Voznesenskaya* are adjectival forms of *Voznesenie*, "Ascension." Only three translations make it "Ascension Prospect/Avenue" (Gogol 1916, pp. 67, 93; 1984, pp. 46, 59; 2020, pp. 197, 217). The Russian readers do not usually pay any attention to this detail either: the names sound too familiar and therefore inconspicuous.

The Ascension of Christ was preceded by His Resurrection. Both events are considered a prefiguration of the future Resurrection of the Dead. In Gogol's travesty, the Resurrection is alluded to in the mention of the *Voskresensky* (Resurrection) Bridge (III: 56/205; Glyantz 2013, p. 106).[30] The Resurrection is closely associated with baptism; only those baptized will be resurrected (John 3:3–5). According to Saint Paul's epistle to the Romans, those who "were baptized into Christ Jesus were baptized into His death":

> Therefore we were buried with Him through baptism into death, that just as Christ was raised from the dead by the glory of the Father, even so we also should walk in newness of life. (Rom 6:3–4)

Gogol undoubtedly remembered Paul's further explanation:

> Why then are they baptized for the dead? […] what you sow is not made alive unless it dies [KJV: "that which thou sowest, is not quickened except it die"]. […] So also is the resurrection of the dead. The body is sown in corruption, it is raised in incorruption. […] It is sown a natural body, it is raised a spiritual body. (1 Cor 15:29, 36, 42, 44).

In the fourth letter on *Dead Souls* (1846) already quoted in Section 2 above, Gogol recalls 1 Corinthians 15:36:

> "It is not quickened except it die," says the Apostle.[31] It is first necessary to die in order to be resurrected. (VIII: 297; cf. Gogol 2009, p. 108)

Linked directly to the idea of rebirth, baptism is symbolized by immersion in water. This is exactly what happens with the nose (Weiskopf 1993, p. 236). The barber decided "to throw it into the Neva", went to St. Isaac's Bridge, and "quietly threw in the rag with the nose on it" into the water [river] (III: 50–51/199–200). After that, it/he is blasphemously reborn as a high official, whose social—rather than spiritual—rank has been upgraded. The Nose as "a *spiritual* body" (Church Slavonic and Russian: *dukhovnoe*) implies another sacrilegious pun: the carnal olfactory organ represents the Spirit (Church Slavonic and Russian: *dukh*, also meaning "smell") (Virolainen 2004, pp. 106–7).[32]

The chronotope of "The Nose" is characterized not only by elapsed time but also by distorted space. Yuri Lotman described it as ambiguous "phantasmagoric space", in which two quasi-spatial forms can be discerned: physical space and social, bureaucratic space. According to Lotman (1968, pp. 40–41), in the material world of "The Nose", the several intertwined plots do not fit together, forming gaps in discontinuous physical space. However, the social space of the authoritarian State preserves its continuity, although "within this space, continuity exists only at the level of forms, but not that of meanings" (ibid.). The Table of Ranks continues to operate, giving the Nose legitimacy as long as he is properly classified according to his rank. However, as soon as it turns out that the Nose has a fake passport, he instantly ceases to be a "Sir" (*gospodin*) and becomes just a "nose" again. To the physical (material) and social (semiotic) spaces, we can add a third (symbolic) space—sacral (Toporov 1984, p. 23), or, given Gogol's blasphemous travesty, sacrilegious. It exploits the real urban sacral toponymy—names such as the Kazan Cathedral dedicated to Our Lady of Kazan and mentioned in connection with the Feast of the Annunciation, Ascension Prospect, and the Resurrection Bridge. The symbolic meaning of other toponyms, for example Saint Isaac's Bridge near Saint Isaac's Cathedral, requires more research. Particularly important is the fact that the Cathedral is dedicated to Saint Isaac of Dalmatia, a patron saint of Peter I (Weiskopf 1987, p. 36; 1993, p. 538 note 348).

5.4. The Psychological and Cultural Background of Gogol's Religious Blasphemy

"The Nose" is on par with the demonological works of the period of Gogol's religious and psychological crisis (tales such as "Nevsky Prospect", 1834, and "Viy", 1835) and his sacrilegious mockery of the Eucharist in a letter to Alexander Danilevsky of 28 September 1838:

> Imagine that all along the way, in all cities, the temples are poor, the worship too, the priests are ignorant and unkempt [...], to say nothing about the taste and fragrance of the sacrifices [...].Thus I must confess that, against my will, I am visited by free-thinking and apostatical thoughts, and I feel my religious rules and faith in the true religion are weakening every minute, so that if only another religion could be found with skilled priests and especially sacrifices, like tea or chocolate, then farewell to the last piety [*nabozhnost'*]. (XI: 173; Weiskopf 1993, p. 538 note 348)

The ill-fated Kovalyov suffered the same hesitation: in the church he "felt so upset that he had no strength to pray", whereas his Nose "was praying with an expression of the greatest piety [*nabozhnost'*]" (III: 55/204). "The monster clearly is parodying prayer and those who pray. This is a caustic mockery of Christian piety" (Oulianoff [1959] 1967, p. 169).

Viktor Zhivov considered the ambiguous and problematic nature of the separation of the religious and the secular in post-Petrine Russian culture to be the origin of blasphemy in Russian literature of the eighteenth and early-nineteenth century. The sacralization of the poet's role in the neoclassical and romantic period leads authors to blasphemous parodies of church rites as well as biblical and liturgical texts in the works of various genres (Zhivov 1981).[33] Robert Maguire demonstrated the implicit equation between priest and artist in Gogol's *Meditations on the Divine Liturgy* (Maguire 1994, pp. 308–10). This can be taken as the cultural and semiotic background of Gogol's making a comedy of Proskomedia. Apparently, it also had psychological grounds.

At the same time that Gogol was working on "The Nose", he was writing *The Government Inspector*. As the author of this play, Gogol thought of himself not only as a comedic playwright but also as a teacher of life (a role to which Gogol would infamously return later). His expectations that his great comedy (it is great indeed) would change the world overnight did not materialize (Mochul'skii 1934, p. 42). A moral and religious crisis began, and Gogol left Russia. However, his sense of fear and despair began right when he was planning to take on the comedy and other projects (Mann 2012a, pp. 353–59). He wrote to Pogodin on 1 February 1833:

> I don't want to produce anything minor! I can't seem to think up anything great! In a word, mental constipation. Send me sympathy and wish me the best! Let your word be more effective than an enema. (X: 257)

This looks like a typical neurotic reaction: Gogol wants to perform a great deed, but he is afraid of a possible failure. "By the end of 1833, Gogol's letters begin to carry notes of terrible anguish, almost despair" (Mann 2012a, p. 353; cf. Oulianoff [1959] 1967, pp. 159–60). On 28 September 1833, he wrote to Pogodin:

> What a terrible year 1833 has been for me! My God, how many crises! Will there be a benign restoration for me after these destructive revolutions? How much I started, how much I burned, how much I gave up! Do you understand the terrible feeling of not being content with oneself? [...] The person who has been possessed by this hell of a feeling [*eto ad-chuvstvo*] is turned all to anger, he alone forms the opposition to everything, he makes a terrible mockery of his own ineptitude. My God, may it all be to the good! Say this prayer for me, too. (X: 277)

However, the problem was much broader: Gogol "was seeking a literary equivalent to the incarnation of the Word, but he was unable to go beyond a parodic consciousness" (Iampolski 2007, p. 20). An example is *Dead Souls*. It is a mock epic—a comic poem in prose (the author himself subtitled his work a "poem") and a travesty of Dante's *Divine Comedy*.

Instead of three volumes, Gogol wrote only his version of "Inferno"—*Dead Souls*, Volume One. Only scattered fragments remained from his "Purgatory" (in February 1852, Gogol burned the manuscripts of the second volume and starved himself to death), whereas "Paradise" was not even started. Volume One is a laugh riot, but it is laughter through tears, "laughter visible to the world and tears invisible and unknown to it" (Gogol 2004, p. 150).

Gogol's ambivalent carnivalesque laughter and *parodia sacra* are intimately linked not only with tears but also with fear and horror[34]: "The theatricalized world […] is at the same time a diabolical, illusory world, in which 'the demon himself lights the lamps for the sole purpose of showing everything not as it really is'" (Lotman [1974] 1976, p. 299; the quotation is from "Nevsky Prospect", III: 46/153). Lotman linked this ambivalence with the Great Russian, rather than Ukrainian, aspect of Gogol's personality, and explained it through "the tradition of medieval Eastern Orthodoxy" that "separated the divine principle, as real, from the apparent, the imaginary, the 'dreamy,' and the diabolical. The former is constant, the latter is many-faced and variable" (ibid., pp. 297–98).

5.5. The Calendar of "The Nose"

The latent blasphemy of "The Nose" explains yet another detail: in the definitive version, Kovalyov lost his nose on Friday. We know this because "the barber Ivan Yakovlevich had shaved him on Wednesday, and all day Wednesday and even all day Thursday his nose had been intact" (III: 65/215). Friday is an unlucky day because it is the day of the events that make up the Passion of Christ (Mk 15:42). In this context, Kovalyov's exclamation when he realized the horror of his situation sounds (un)ambiguous: "My God! my God! Why such a misfortune?" (III: 64/214). It sounds like a travesty of the Crucifixion (noted by Weiskopf 1987, p. 28; also, Weiskopf 1993, p. 231):

> And about the ninth hour Jesus cried out with a loud voice, saying, "Eli, Eli, lama sabachthani?" that is, "My God, My God, why have You forsaken Me?" (Mt 27:46)

Again—Kovalyov ends his phrase with devilry: "but without a nose a person is the devil knows what" (III: 64/214).

In the early redactions of the tale, there was no symbolism of dates, and the dates themselves varied. In the initial draft (1832), the story begins "on the 23rd of the year 1832 [*23 chisla 1832-go goda*]" (III: 380), month not specified. This omission immediately reminds us of "The Diary of a Madman", where the dates are confused ("January of the same year, which happened after February"), improbable and incomplete dates are introduced ("The Year 2000, April 43", "Marchtober 86", "The first [Chislo 1]", "The 25th [Chislo 25]"), and—eventually—time totally disappears ("No date of any sort. The day had no date"; "I do not remember the date. There was also no month. It was the devil knows what."—III: 207–13/173–79).

In the first complete manuscript redaction of "The Nose" (1833–1834), the story begins "on the 23rd of this February [*sego fevralia 23 chisla*]" (III: 484), i.e., "February 23 of this year." In 1834, February the 23rd according to the Julian calendar fell on Friday, but there is no indication of a Friday in the text yet—one might rather think that the nose disappeared the day after shaving, i.e., on a Thursday. This could mean, then, that the events took place not in 1834, but in 1833, when February 23 was a Thursday. *Terminus post quem* for the first complete manuscript redaction is 21–23 September 1833, and *terminus ante quem* is 27 August 1834 (III: 651). Therefore, the words "this February" could very well refer to 1833.

In the first published version (the text of The Contemporary, 1836) Kovalyov's nose "had been intact" "all day Wednesday and even all day Thursday" (Gogol 1836, p. 78; the same phrase in the definitive version, III: 65/215). The nose was lost "on the 25th of this April [*sego aprelia 25 chisla*]" and was regained "at the beginning of May—on the 5th or 6th" (Gogol 1836, pp. 54, 88; III: 399, 484). It is unlikely that Gogol was describing future events in this version because it was finished much earlier than 25 April 1836.[35] In any case, it is not clear in what year the story happened: 25 April 1836 was a Saturday; in 1835, April 25 was a Thursday; and in 1834, it was a Wednesday. In other words, in these years there was no Friday, 25 April.[36] Does it mean that the events started on a non-existent day?

The importance of days of the week for the tale's chronotope is indirectly evidenced by another seemingly random detail: "Major Kovalyov wore a multitude of carnelian seals, some with coats of arms and some engraved with 'Wednesday,' 'Thursday,' 'Monday,' etc." (III: 54/202). With this in mind, what significance could the choice of a particular day of the month have in each of these redactions?

Nicholas Oulianoff pointed out that 23 February according to the Julian calendar is "the feast day of the holy martyr St. Polycarp and Sts. John and Alexander", when "the first chapter of Isaiah is read at the sixth hour, and this would be quite enough to disturb Gogol if he were in a susceptible state" (Oulianoff [1959] 1967, p. 164). This is, however, just a conjecture and does not explain much in the text. The next date Oulianoff attempted to explain was "April 23 of this year", as it allegedly stands in *The Contemporary*: "April 23 is the feast day of St. George the Conqueror, a saint who battled with a serpent, with a monster. Letting loose a monstrous creature into the world on this very day meant imparting a special point to the story" (ibid., p. 169). Alas, this is a mistake or, in fact, an illusion: the first published version reads "April 25" and not "April 23" (*The Contemporary* used a font-face in which *5* looks similar to *3*). Although the academic edition correctly reproduces the 1836 text as "on the 25th of this April" (III: 484), Oulianoff's mistake and his "explanation" of the phantom date continues to reappear in later scholarship (Woodward 1981, p. 84; Glyantz 2013, p. 101).

Presumably, the selection of dates proceeded consecutively, by way of approximation. In the first published version, the nose disappears on a Friday, and the time of action is about twelve days (this is the difference between the Julian and Gregorian calendars in the nineteenth century). In the definitive version, the Nose's escape was timed to coincide with a major Christian holiday, and a dozen turned into a "devil's dozen" of the interfaith temporal collapse (12 + 1). Kovalyov's meeting with the Nose was designed to happen at church from the very beginning, but Gogol wanted more:

> ... he needed a Christian holiday to serve as the background against which the appearance of the Nose would be the ultimate expression of the idea of the devil's triumphant onslaught. For such a holiday, he chose the Annunciation, March 25, the day which brings the tidings of the forthcoming advent of the Savior. (Oulianoff [1959] 1967, p. 169)

Boris Uspensky points out that the final change of dates occurs during the years of Gogol's long stay abroad. When Gogol witnessed the Annunciation in Rome in 1837, 1838, 1839, and 1841, he first watched the Catholic celebration in Rome's cathedrals and then participated in the Eastern Orthodox celebration at the Russian embassy church (Uspenskii 2004, p. 339). If Gogol checked the calendar (and apparently, he did), then the events in the final version of the tale take place either in 1832 or 1838—only in these two years the day of the Annunciation (25 March according to the Julian calendar) was a Friday (Pilshchikov 2019, pp. 234–35). Most likely, the events in the definitive version happen in 1832 as the Nose walks around wearing a plumed hat abrogated in 1834.

6. Nosology as Hypnology: HOC as COH in Documents and Legends

More than any other of his works, this tale which Gogol's contemporaries failed to appreciate (Pushkin and Vyazemsky excepted) resists a realistic interpretation. It displays in the highest degree the bizarre combination of fantasy and naturalism that Vasily Rozanov noted in Gogol's work generally. In a series of articles written between 1891 and 1909, Rozanov performed a complete deconstruction, as we would say today, of Gogol's poetics. He declared the established view of Gogol as a realist to be a misunderstanding, and Gogol's social criticism to be a slander of reality (Mondry 2003). In Gogol's works, details are realistic, but his overall picture of the world is phantasmagoric.

The pinnacle of Rozanov's "deconstruction" comes in the article "The Genius of Form (On the Centenary of Gogol's Birth)", most of which is devoted to "The Nose." According to Rozanov, Gogol "finds for incarnation the least important details, vulgarity, monstrosity, distortion, disease, insanity, or a dream that resembles insanity. After all, 'The Nose' is—literally—a chapter from 'The Diary of a Madman', [whereas] 'The Diary of a Madman' is a thread of intertwined 'Noses'" (Rozanov [1909] 1995, p. 348). "The fabula outline of Gogol's 'The Nose' strongly resembles the ravings of a madman", Tynianov agreed (Tynianov [1926] 2019, p. 235). However, in "The Diary of a Madman", the absurdity of the world is "depicted in the light of the hero's insane consciousness", whereas in "The Nose", it is manifested "without any motivation, by the author's arbitrary will" (Gippius 1966, p. 83). This is why "The Nose" is "radically inexplicable" (Morson 1992, p. 227).

Gogol combines disparate devices and approaches in one text: idiomatic puns, grotesque social satire (even—caricature), and blasphemous travesty. Granted, these strategies of construction and interpretation do not even need to be semantically coherent and can contradict each other to produce a stronger absurdist effect: "*Absolutely* meaningless makes sense; sense is therefore fictitious" (Pumpianskii 2000, p. 326; see Slonimskii [1923] 1974, p. 369; Todorov [1970] 1975, p. 73; Fanger 1979, pp. 120–22; Shukman 1989, pp. 64, 78; Morson 1992, pp. 207, 230–33; Maguire 1994, pp. 338–39). That said, how does Gogol manage to do this formally? Is there a single unifying device—"the dominant" (to use the Formalist term)?

The finale of the first complete redaction of the tale explains all the events as a nightmare: "However, everything here described was seen by the major in a dream" (III: 399), but the definitive version is ambiguous. On the one hand, the described events are presented as a real occurrence, and Kovalyov even pinches himself to make sure that he is not dreaming or drunk (III: 65/215). It seems that he is awake and sober, but who knows for sure? On the other hand, his awakening could have been false, and the nightmare could have extended into a daydream (Erlich 1969, pp. 88–90; Rowe 1976, pp. 100–4; Shukman 1989, p. 72). Furthermore, if "The Nose" is a bad dream, it is not individual and personal but collective: the barber seems to be having the same dream as Kovalyov (Pumpianskii 2000, pp. 328–29).

Gogol's contemporaries compared the poetics of the "Petersburg Tales" with those of E. T. A. Hoffmann, who created an ironic narrative open to both supernatural and realistic explanations of the fantastic events, with the narrator's point of view vacillating between the two mutually exclusive interpretations. A supernatural interpretation would involve mystical worlds, occult practices, or magic; a realistic-psychological explanation would favor phenomena from everyday life: falling asleep and dreaming, excessive consumption of alcohol, or an unbridled imagination. However, even against this background, "The Nose" stands out for its undisguised parody that sublimates irony to the grotesque—a fantastic combination of incompatible elements (Vinogradov [1921] 1929; Slonimskii [1923] 1974; Mann 1966, 1973; Günther 1968, pp. 131–47; Erlich 1969; Karlinsky 1976, pp. 123–30). In Hoffmann, the events can be explained as *either* unreal *or* real, whereas in Gogol, the reader's "fantastic hesitation" over how to understand a nose in a plumed hat, a carriage on Nevsky Prospect, or a post chaise planning to go to Riga with a fake passport is sustained throughout, never resolving into either a rational, materialist explanation (it was a dream), or the acceptance of supernatural events at face value (a nose can abscond if it pleases) (Todorov [1970] 1975, pp. 72–73). In the "Petersburg Tales" cycle, absurdity can be motivated by demonism, but this motivation (in the Russian Formalists' sense of the term) can shift from explanation to comparison, as in the celebrated passage from "Nevsky Prospect": "It seemed [. . .] that a demon had crumbled the whole world into a multitude of various pieces and had mixed them all together with no meaning or sense" (III: 24/129). Victor Erlich noted that this passage could serve as a motto for "The Nose" (Erlich 1969, p. 82). The world of Gogol's Saint Petersburg is a world turned upside down and ruled by the devil, "the main hero of almost all of Gogol's works" (Čiževsky 1938, p. 193; cf. Čiževsky [1951] 1952).[37]

All that notwithstanding, there is a commonly held notion that that the "nose" points to a pun on "sleeping/dreaming" because the Russian *nos* ("nose") is a palindrome of *son* ("sleep, dream"). This observation belongs to Yermakov, who even claimed that "at first [the tale] bore another title: 'The Dream'", but—allegedly, later—"Gogol discarded the original title and renamed the story 'The Nose'" (Yermakov 1921, p. 99; Yermakov [1924] 1974, pp. 173–74). Unfortunately, this tempting hypothesis is not supported by documentation. Its apocryphal nature was already noted by Vinogradov ([1921] 1929, pp. 43–44 fn. 1). Nonetheless, this legend, popularized in Janko Lavrin's biography of Gogol in "The Republic of Letters" book series (Lavrin 1926, p. 116),[38] is still alive and references to the non-existent draft reappear in Anglophone Gogol scholarship at least once or twice every decade (see Bowman 1952, p. 210; Erlich 1969, p. 89; Karlinsky 1976, p. 123; Rancour-Laferriere 1982, p. 84; Shukman 1989, p. 64; Sobel 1998, p. 335; Kutik 2005, p. 67). In later redactions of "The Nose", the initial motivation of the absurdity by the oneiric illogic was removed or at least veiled—or, in Vinogradov's perfect formulation, "dismantled like the scaffold of an artistic construction" (Vinogradov [1921] 1929, p. 46). The dream, like the devil, was left in the subtext.

7. Conclusions

However, the devil, as is well known, is in the details, and "The Nose" is "a fantasy rich in enigmatic detail" (Gregg 1981, p. 370). In order to understand the many-pronged functions of this richly-detailed masterpiece, we have had to recall or reconstruct many features of the language and everyday life of Gogol's time. In particular, there are three groups of linguistic and social-cultural phenomena with which we need to be familiar to competently read the tale: idiomatic phraseology, forgotten specificities of the Imperial hierarchy (the Table of Ranks), and allusions to religious customs and Christian rituals that would have been obvious to Gogol's readers. An integrated analysis of these aspects in their mutual relationship both reveals how the telling details of the story are associated with Gogol's religious and psychological crisis of the mid-1830s and demonstrates how he aggregated indecent Shandyism, social satire, and religious blasphemy into a single oneiric narrative that mischievously rejects its dreamlike nature (compare Bocharov 1985, pp. 204–5). Correspondingly, the quasi-oneiric chronotope of the tale combines three spaces—physical, social, and sacral—each of which is presented as *mutatis mutandis*, mockingly indecent, grotesquely satirical, and latently sacrilegious. While creating one of the first fragments of the "Petersburg Text", Gogol makes a parody of what would become its main idea and main plot: finding "the way to moral salvation, to spiritual rebirth under conditions where life is being conquered by death, and lies and evil triumph over truth and good" (Toporov 1984, pp. 16–17; quoted in Shapovalov 1988, p. 43). Viewed from this perspective, "The Nose" is not only the most absurdist but also the most subversive and nullifidian work of Gogol.

Funding: This research was supported by Estonian Research Council grant PRG319.

Acknowledgments: I warmly thank Ainsley Morse, Joe Peschio, Dennis Ioffe, and the anonymous reviewers for their constructive criticism and valuable suggestions.

Conflicts of Interest: The author declares no conflict of interest.

Notes

[1] "Contact zone" is Mary Louise Pratt's term for "social spaces where cultures meet, clash and grapple with each other, often in contexts of highly asymmetrical relations of power" (Pratt 1991, p. 34).

[2] References are provided to Gogol's fourteen-volume complete collected works (Gogol 1937–1952) cited by volume: page (volumes are given in Roman numerals, and pages in Arabic numbers). The translations of Gogol's Petersburg tales used here are those by Susanne Fusso (Gogol 2020). Earlier stories are quoted in Constance Garnett's translation revised by Leonard J. Kent (Gogol 1985, vol. 1). The translation pages are cited after the original pages and divided by a slash. Other translations from Russian are by Igor Pilshchikov and Ainsley Morse if not indicated otherwise.

[3] Etymologically, "slanderer." From διαβάλλω [δια- ("across") + βάλλω ("throw")], "throw over or across, bring into discredit, mislead, calumniate, slander" (LSJ).

[4] An East and West Slavic word with uncertain etymology; not used in South Slavic languages, including Church Slavonic. The Proto-Slavic *čьrtъ ("demon") may be cognate with either the Proto-Slavic *čьrta ("line, boundary") and its descendants or—less

likely—with *čьrъ/*čьra* ("spell, magic") and its descendants (Trubachev 1977, pp. 164–66), or with both, as in the Czech *čára*, which means "a borderline up to which something is permitted or magically prohibited, e.g., the line which marks the so-called magic circle (where the evil demons retain or lose their power)" (Jakobson 1959, p. 276). Compare Andrei Bely's intuition that "Gogol's 'devil' [*chort*] derives from a line [*cherta*], and a line is a boundary" (Bely [1934] 2009, p. 182, see also p. 227; one needs to consult the original (Bely 1934, pp. 148, 186) to get the point). The *chort*'s synonym *běsъ* < *běd-s-*, also meaning "demon" (Church Slavonic: *běsъ*; Russian: *bes*; Ukrainian: *bis*) is cognate with the Proto-Slavic *běda* ("trouble, calamity") and *běditi* ("compel, persuade") (Trubachev 1975, pp. 88–91, 54–57). The Russian word *chort* is more colloquial and folkloric than *bes*. The Old Believers in the town of Mogilev shared a folk-belief that the devil rejoices when he is called by the Russian word *chort* (pronouncing this word is itself a sin), but does not like to be called by the Church Slavonic and ecclesiastic Russian word *běsъ/bes* because he hates everything related to church (Zelenin 1930, p. 89; Uspenskii 2012, p. 26).

5   The word *bes* is not used in "The Nose" (there is only one occurrence of the verb *vzbesit'* of the same root, meaning "enrage"). In Gogol's works and letters, the word *bes* is used 23 times (and the Ukrainian *bis* once), whereas *chort* occurs almost ten times as frequently. Pushkin, whose style sets the norm for this period, used *bes* 59 times and *chort* 91 times (a 2:3 ratio). On Pushkin's usage of these words, see Vinogradov (1956–1961, vol. 1, pp. 96–97, vol. 4, pp. 926–27).

6   Emphasis added here and below if not attributed to the author.

7   The olfactory aspects of "The Nose" has recently been studied by Klymentiev (2009).

8   Compare Tynianov ([1921] 2019, p. 32); Čiževsky (1952, pp. 270–73).

9   For example, Emperor Nicholas I suggested that Alexander Pushkin should change "too trivial passages" in the first version of his *Boris Godunov*, where phrases such as *"bliadiny deti"* were used (the latter, meaning "sons of whores," was later changed for *sukiny deti*, "sons of bitches", and, eventually, *postrely*, "scamps") (Vinokur 1935, pp. 423, 426; Pushkin 1937–1949, vol. 14, p. 59).

10 The names of these characters mockingly refer to the German literary tradition, whose actuality for Gogol was discussed by his readers and critics (see Meyer 2000, pp. 69–70; Kutik 2005, pp. 59–60). Ironically, Lieutenant Pirogov, another character in "Nevsky Prospect" who saw this scene, "could not understand what was going on"—not for mystical reasons, however, but simply because they spoke German, whereas his "knowledge of German did not go beyond *'guten Morgen'*" (III: 37/143).

11 See Kutik (2005, pp. 69–76), for a discussion of the "moon" topic and, in particular its connection with Ariosto's *Orlando furioso*, canto 34.

12 On Yermakov see Young (1979).

13 Etymologically, *nos* in the idiom *byt'/ostat'sia s nosom* has perhaps nothing to do with the nose, being supposedly a suffixless substantive of the verb *(pri)nosit'* "to bring" (compare the pairs like *beg//begat'* and the like). If so, *nos* meant "what is brought (as a gift or bribe)," and the meaning of the whole idiom was "to be left with what is brought, i.e., to be rejected a gift or bribe (such as a ritual bribe in wedding mediation)." However, in modern usage the original meaning of the idiom has been lost, and the speakers interpret *nos* in this phrase as "nose," just like in all other idioms of this series (Shanskii 1960, p. 77, 1963, p. 67 fn. 1). The situation was not different in Gogol's time (Dahl and Baudouin de Courtenay 1903–1909, vol. 2, p. 1443). Compare Bierich et al. (1994, p. 180).

14 See also Pushkin (1984, p. 123).

15 On the name of Maccabee in this context see Zhatkin (2002).

16 Roman Jakobson cited this Dostoevsky passage as an example of "the abrogation of the boundary between real and figurative meanings" of idioms in poetic language (Jakobson [1921] 1973, pp. 67–68). On the same issue in Spanish translations of Dostoevsky see Obolenskaia (1980, pp. 58–59).

17 "A Petersburg Tale" is the subtitle of Pushkin's narrative poem "The Bronze Horseman" (1833). Gogol's Tales (1842) started to be routinely called "Petersburg Tales" as early as the 1880s. This non-authorial title was codified in academic parlance by the Gogol scholar Vladimir Shenrok (1893, p. 78).

18 On the Table of Ranks in Gogol's fiction see Reyfman (2016, pp. 102–16).

19 Khlestakov, an irresponsible, frivolous young man, who is mistakenly identified as an inspector by the corrupt officials of a small Russian town in Gogol's comedy *The Government Inspector* (1836), is a liar *par excellence*. According to Yuri Lotman: "Lying intoxicates Khlestakov because, in his imaginary world, he can cease to be himself, escape from himself, become someone else. [. . .] The split personality which was to become the central focus of Dostoevsky's *Double* [. . .] was already present in Khlestakov" (Lotman [1975] 1985, p. 162). Lotman compares Khlestakov with the protagonist of "The Diary of a Madman": "The deliverance from self and flight to life's summits that Khlestakov receives from 'an uncommon addle-headedness in thinking' and a potbellied bottle of provincial Madeira, Poprishchin experiences as the price of insanity" (ibid., p. 163 fn. 29). Kovalyov belongs to the same type; if "'Diary of a Madman' is a tragic parallel to *The Inspector General*" (ibid.), then "The Nose" is a surrealist parallel to both.

20 *Humour*: English word in the original.

21 It should be added here that the censor excluded the references to specific ministries in the Nose's conversation with Kovalyov from the printed text, so that the phrase under discussion was reduced to "Judging by the buttons on your uniform [*vits-mundir*], you must serve in a different department" (III: 486; Gogol 1836, p. 65, 1842, p. 97). This censored variant, reproduced in some

twentieth-century editions, is translated in all English versions from 1916–2020, except the *en regard* translation by Gleb and Mary Struve.

This statement was disputed on the both sides of the Iron Curtain: the émigré scholar Nicholas Oulianoff (Oulianoff [1959] 1967, p. 160) and the Soviet scholar Yuri Mann (Mann 1966, pp. 37–38) objected to Yermilov's interpretation.

See Peace (1981, pp. 136–37) on *mesto* ("place, position, job") as a key word throughout "The Nose."

"The barber's wife baking bread can be likened to one of the minor characters behind the scenes who participate in the process of the Mass: the special woman who bakes the liturgic bread. There were specific rules fixed by the ecclesiastics which obliged them to 'choose as bakers of liturgic bread' either 'widows living in purity' or virgins no younger than 50 years of age" (Glyantz 2013, p. 99). She obviously was neither virgin nor widow—nor, in fact, did she live in purity, judging by "complaints of an erotic nature, addressed to her husband" (ibid.).

In Gogol's times, the word *nis* could have spelled as нисъ (the spelling of the printed editions of Ivan Kotlyarevsky's foundational mock-epic *Eneyida* [the Aeneid]), нісъ (the spelling of Oleksiy Pavlovsky's 1818 Ukrainian grammar) or ніс (the "spelling of The Mermaid of the Dniester" introduced by Markiyan Shashkevych in 1837). See (Ohiyenko 1927, pp. 5–8; Nimchuk 2004, pp. 6–8).

As an example, in the Ukrainian spelling that Mykhaylo Maksymovych proposed in 1827, the word *nis* was spelled etymologically, as нôсъ. See Ohiyenko (1927, pp. 6–7).

On the differences between the manuscript and the first published version see Gogol (1889, pp. 593–94) (Nikolai Tikhonravov's commentary) and Voropaev (2000, pp. 189–90).

It has been proposed that the phrase "it seemed that *heaven itself* made him see the light" and sent him "directly to the advertising department of the newspaper" (III: 58/208) to place an *announcement* (III: 58/208) can also be read as a travesty of the Annunciation (Cornell 2002, p. 276). This is a questionable claim. First, in Russian, there is no straightforward correspondence between the words for "announcement" (*ob"yavlénie*) and "Annunciation" (*Blagovéshchenie*). In the latter, the first part, *blago-*, reflects the Greek εὐ- in Εὐαγγελισμός. The other part, *-veshchenie*, is used with a different stress in the words like *izveshchénie* ("notice") but not in the word for "announcement." Second, this episode is already contained without any allusion to the Annunciation in the 1836 redaction of "The Nose," in which the strange events begin on 25 April and not on 25 March (see below).

On the dogmatic and anagogical meanings of the Proskomedia and the Eucharist accepted by the Russian Orthodox Church see Ivan Dmitrevsky's *Historical, Dogmatic and Mysterious Explication of the Liturgy* (Dmitrevskii [1803] 1807) and Archbishop Benjamin's *The New Stone Tablet* (Veniamin [1803] 1823)—the two main sources for Gogol's *Meditations on the Divine Liturgy* (Frank 1999, p. 87 fn. 3; Voropaev 2000, p. 186).

Mistranslated as "the Ascension bridge" in Gogol 1916, p. 78.

Gogol quotes the Slavonic version: *"ne ozhivet, ashche ne umret."*

Compare Iampolski (2007, pp. 560–61), on excessive materiality and imaginary illusion as exaggerated extremes in Gogol's travesties of the Transubstantiation. In his mockery, Gogol comes close to mis/reinterpretations of the Eucharist as symbolic cannibalism (see Kitson 2000; compare Utz and Baatz 1998).

For a broader context see (Levy 1981, 1993; Lawton 1993; Nash 1999, 2007; Whickman 2020).

On horror in "The Nose" see Bely ([1934] 2009, pp. 20–21, 227–28).

We know that on 4 April 1836, Gogol read his story at one of the "Saturdays" at Zhukovsky's, as Vyazemsky reported to Alexander Turgenev (discussed above; see also Mann 2012b, p. 41).

The nearest April 25 that fell on Friday occurred in 1830.

Čiževsky's observation goes back to Merezhkovsky ([1906] 1974, esp. pp. 57–58). On Merezhkovsky's *Gogol and the Devil* see Hashemi (2017, pp. 157–62).

"The most outdated and least reliable of [Gogol's] biographies" (Karlinsky 1976, p. 320).

References

Afanas'ev, Aleksandr. 1997. *Narodnye russkie skazki ne dlia pechati, zavetnye poslovitsy i pogovorki, sobrannye i obrabotannye A. N. Afanas'evym, 1857–1862*. Edited by Ol'ga Alekseeva, Valeriia Eremina, Evgenii Kostiukhin and Leonid Bessmertnykh. Moscow: Ladomir.

Bakhtin, Mikhail. 1968. *Rabelais and His World*. Translated by Helene Iswolsky. Cambridge: MIT Press. First published 1965.

Bakhtin, Mikhail. 1976. The Art of the Word and the Culture of Folk Humor (Rabelais and Gogol'). In *Semiotics and Structuralism: Readings from the Soviet Union*. Edited with an Introduction by Henryk Baran. New York: International Arts and Sciences Press, pp. 284–96. First published 1975.

Belinskii, Vissarion. 1955. *Polnoe sobranie sochinenii*. Moscow and Leningrad: Izdatel'stvo AN SSSR, vol. 6.

Bely, Andrei. 1934. *Masterstvo Gogolia: Issledovanie*. Moscow and Leningrad: GIKhL.

Bely, Andrei. 2009. *Gogol's Artistry*. Translated from the Russian and with an Introduction by Christopher Colbath. Foreword by Vyacheslav Ivanov. Evanston: Northwestern University Press. First published 1934.

Bem, Alfred. 1979. "The Nose" and "The Double". In *Dostoevsky & Gogol: Texts and Criticism*. Edited and Translated by Priscilla Meyer & Stephen Rudy. Ann Arbor: Ardis, pp. 229–48. First published 1928.

Bierich, Alexander, Valerii Mokienko, and Liudmila Stepanova. 1994. *Istoriia i etimologiia russkikh frazeologizmov (Bibliograficheskii ukazatel') (1825–1994) = Bibliographie zur Geschichte und Etymologie der russischen Phraseme (1825–1994)*. Munich: Otto Sagner.

Bocharov, Sergei. 1985. Zagadka "Nosa" i taina litsa. In *Gogol': Istoriia i sovremennost': K 175-letiiu so dnia rozhdeniia*. Edited by Vadim Kozhinov. Moscow: Sovetskaia Rossiia, pp. 180–212.

Bocharov, Sergei. 1992. Around "The Nose". In *Essays on Gogol: Logos and the Russian Word*. Translated by Susanne Fusso. Edited by Susanne Fusso and Priscilla Meyer. Evanston: Northwestern University Press, pp. 19–39.

Bocharov, Sergei. 2009. Peterburgskii tekst Vladimira Nikolaevicha Toporova. In *Peterburgskii Tekst*. Essays by Vladimir Toporov. Pamiatniki Otechestvennoi Nauki. XX vek. Moscow: Nauka, pp. 5–18.

Bojanowska, Edyta. 2007. *Nikolai Gogol: Between Ukrainian and Russian Nationalism*. Cambridge: Harvard University Press.

Bowman, Herbert E. 1952. The Nose. *Slavonic and East European Review* 31: 204–11.

Budagov, Ruben. 1953. *Ocherki po yazykoznaniiu*. Moscow: Izdatel'stvo AN SSSR.

Butler, Peter. 2017. Up the Dnepr: Discarding Icons and Debunking Ukrainian Cossack Myths in Gogol's "Strashnaia mest'". *Russian Literature* 93: 199–249. [CrossRef]

Carey, Claude. 1972. *Les Proverbes Érotiques Russes: Études de Proverbes Recueillis et Non-Publiés par Dal' et Simoni*. Slavistic Printings and Reprintings 88. The Hague and Paris: Mouton.

Čiževsky, Dmitry. 1938. O "Shineli" Gogolia. *Sovremennye Zapiski* 67: 172–95.

Čiževsky, Dmitry. 1952. Gogol: Artist and Thinker. *The Annals of the Ukrainian Academy of Arts and Sciences in the U.S.* 2: 261–78.

Čiževsky, Dmitry. 1952. The Unknown Gogol'. *Slavonic and East European Review* 30: 476–93. First published 1951.

Complete Collection of the Laws. 1835. *Polnoe Sobranie Zakonov Rossiiskoi Imperii. Seriia Vtoraia*. Vols. 9.1 and 9.2: (1834). Saint Petersburg: The Typography of the Second Section of His Imperial Majesty's Own Chancellery.

Complete Collection of the Laws. 1857. *Polnoe Sobranie Zakonov Rossiiskoi Imperii. Seriia Vtoraia*. Vol. 31.2: (1856). Appendices. Saint Petersburg: The Typography of the Second Section of His Imperial Majesty's Own Chancellery.

Cornell, John F. 2002. Anatomy of Scandal: Self-Dismemberment in the Gospel of Matthew and in Gogol's "The Nose". *Literature and Theology* 16: 270–90. [CrossRef]

Cornwell, Neil. 1990. *The Literary Fantastic: From Gothic to Postmodernism*. New York and London: Harvester Wheatsheaf.

Dahl, Vladimir, and Ivan (Jan) Baudouin de Courtenay. 1903–1909. *Tolkovyi slovar' zhivogo velikorusskogo yazyka Vladimira Dalia*, 3rd ed. Revised, and Substantially Enlarged. Edited by I. A. Baudouin de Courtenay. Saint Petersburg and Moscow: M. O. Wolf, 4 vols.

Debreczeny, Paul. 1966. *Nikolay Gogol and His Contemporary Critics*. Transactions of the American Philosophical Society. New Series 56.3. Philadelphia: American Philosophical Society.

Description of Changes. 1856. *Opisanie izmenenii v forme odezhdy chinam grazhdanskogo vedomstva i pravila nosheniia sei formy*. Saint Petersburg: The Typography of the Second Section of His Imperial Majesty's Own Chancellery.

Dilaktorskaia, Olga. 1984. Fantasticheskoe v povesti N. V. Gogolia "Nos.". *Russkaia Literatura* 1: 153–66.

Dilaktorskaia, Olga. 1995. Primechaniia. In *Peterburgskie povesti*. Tales by Nikolai Gogol'. Saint Petersburg: Nauka, pp. 258–95.

Dmitrevskii, Ivan. 1807. *Istoricheskoe, dogmaticheskoe i tainstvennoe iz"iasnenie na Liturgiiu*, 4th ed. Moscow: The Synodal Typography. First published 1803.

Dostoevskii, Fyodor. 1976. *Polnoe sobranie sochinenii v 30 tomakh*. T. 15. Leningrad: Nauka.

Dostoevsky, Fyodor. 1991. *The Brothers Karamazov*. Translated from the Russian by Richard Pevear and Larissa Volokhonsky. New York: Vintage Classics.

Dukes, Paul. 1978. *Russia under Catherine the Great*. Edited, Translated and Introduced by Paul Dukes. Select Documents on Government and Society. Newtonville: Oriental Research Partner, vol. 1.

Erlich, Victor. 1969. *Gogol*. New Haven and London: Yale University Press.

Fanger, Donald. 1979. *The Creation of Nikolai Gogol*. Cambridge and London: Belknap Press of Harvard University Press.

Feofan Prokopovich. 1961. *Sochineniia*. Moscow and Leningrad: Izdatel'stvo AN SSSR.

Frank, Susi. 1999. Negativity Turns Positive: "Meditations Upon the Divine Liturgy". In *Gøgøl: Exploring Absence: Negativity in 19th-Century Russian Literature*. Edited by Sven Spieker. Blsoomington: Slavica, pp. 85–98.

Galakhov, Aleksandr. 1886. Literaturnaia kofeinia v Moskve v 1830–1840 gg.: Vospominaniia. [Part 1]. *Russkaia Starina* 50: 181–92.

Gillel'son, Maksim, Viktor Manuilov, and Anatolii Stepanov. 1961. *Gogol' v Peterburge*. Leningrad: Lenizdat.

Gippius, Vasilii. 1966. Tvorcheskii put' Gogolia. Written in 1942. In *Ot Pushkina do Bloka*. Moscow and Leningrad: Nauka, pp. 46–200.

Gippius, Vasilii. 1989. *Gogol*. Edited and Translated by Robert A. Maguire. Durham and London: Duke University Press. First published 1924.

Glyantz, Vladimir. 2013. The Sacrament of End: The Theme of Apocalypse in Three Works by Gogol'. In *Shapes of Apocalypse: Arts and Philosophy in Slavic Thought*. Edited by Andrea Oppo. Boston: Academic Studies Press.

Gogol, Nikolai. 1913. *Meditations on the Divine Liturgy*. Translated by L. Alexéieff. London and Oxford: A. R. Mowbray & Co.

Gogol, Nicholas. 1916. *The Mantle and Other Stories*. Translated by Claud Field, and with an Introduction on Gogol by Prosper Merimée. New York: Frederick A. Stokes.

Gogol, Nikolai. 1836. Nos. *Sovremennik* 3: 54–90.

Gogol, Nikolai. 1842. *Sochineniia*. Saint Petersburg: A. Borodin i K°, vol. 3.

Gogol, Nikolai. 1857. *Razmyshleniia o Bozhestvennoi Liturgii*. Saint Petersburg: P. A. Kulish.

Gogol, Nikolai. 1889. *Sochineniia*, 10th ed. Edited by Nikolai Tikhonravov. Moscow: Knizhnyi magazin V. Dumnova, pod firmoiu "Nasledniki br. Salaevykh", vol. 4.

Gogol, Nikolai. 1937–1952. *Polnoe sobranie sochinenii*. Moscow and Leningrad: Izdatel'stvo AN SSSR, 14 vols.

Gogol, Nikolai. 1960. *The Divine Liturgy of the Eastern Orthodox Church*. Translated by Rosemary Edmonds. London: Darton, Longman & Todd.

Gogol, Nikolai. 1967. *Letters*. Selected and Edited by Carl R. Proffer. Translated by Carl R. Proffer in Collaboration with Vera Krivoshein. Ann Arbor: University of Michigan Press.

Gogol, Nikolai. 1984. The Nose. Translated by Robert Daglish. *Soviet Literature* 433: 46–65.

Gogol, Nikolai. 1985. *The Complete Tales*. Edited, with an Introduction and Notes, by Leonard J. Kent. The Constance Garnett Translation Revised by the Editor. Chicago: University of Chicago Press, 2 vols.

Gogol, Nikolai. 1998. *The Collected Tales*. Translated and Annotated by Richard Pevear and Larissa Volokhonsky. New York: Pantheon Books.

Gogol, Nikolai. 2004. *Dead Souls: A Poem*. Translated with an Introduction and Notes by Robert A. Maguire. London: Penguin.

Gogol, Nikolai. 2009. *Selected Passages from Correspondence with Friends*. Translated by Jesse Zeldin. Nashville: Vanderbilt University Press.

Gogol, Nikolai. 2020. *The Nose and Other Stories*. Translated by Susanne Fusso. New York: Columbia University Press.

Gregg, Richard. 1981. À la Recherche du Nez Perdu: An Inquiry into the Genealogical and Onomastic Origins of "The Nose". *Russian Review* 40: 365–77. [CrossRef]

Griboedov, Aleksandr. 2006. *Woe from Wit*. A Commentary and Translation by Mary Hobson. Studies in Slavic Languages and Literature 25. Written in Russian in 1824. Lewiston, Queenston and Lampeter: The Edwin Mellen Press.

Gukovskii, Grigorii. 1959. *Realizm Gogolia*. Moscow and Leningrad: GIKhL.

Günther, Hans. 1968. *Das Groteske bei N. V. Gogol': Formen und Funktionen*. Slavistische Beiträge 34. Minich: Otto Sagner.

Hashemi, Oshank. 2017. Simvolistskie mify o N. V. Gogole (po materialam literaturnoi kritiki rubezha XIX–XX vv.). *Russian Literature* 93/94: 153–98.

Holquist, James M. 1967. The Devil in Mufti: The *Märchenwelt* in Gogol's Short Stories. *PMLA* 82: 352–62. [CrossRef]

Iampolski, Mikhail. 2007. *Tkach i vizioner: Ocherki istorii reprezentatsii, ili O material'nom i ideal'nom v kul'ture*. Moscow: Novoe Literaturnoe Obozrenie.

Ilchuk, Yuliya. 2021. *Nikolai Gogol: Performing Hybrid Identity*. Toronto: University of Toronto Press.

Jakobson, Roman. 1959. Marginalia to Vasmer's Russian Etymological Dictionary (R–Ya). *International Journal of Slavic Linguistics and Poetics* 1/2: 265–78.

Jakobson, Roman. 1973. Modern Russian Poetry: Velimir Khlebnikov [Excerpts]. In *Major Soviet Writers: Essays in Criticism*. Edited by Edward J. Brown. Oxford and New York: Oxford University Press, pp. 58–82. First published 1921.

JMP. 2010. Izdatel'stvo Moskovskoi Patriarkhii provelo prezentatsiiu sobraniia sochinenii Nikolaia Gogolia v Kieve. *Zhurnal Moskovskoi Patriarkhii* 5: 86–87.

Kalinin, Il'ya. 2010. "Peterburgskii tekst" moskovskoi filologii. *Neprikosnovennyi Zapas* 70: 319–26.

Karlinsky, Simon. 1976. *The Sexual Labyrinth of Nikolai Gogol*. Cambridge and London: Harvard University Press.

Kirsanova, Raisa. 1989. *Rozovaia ksandreika i dradedamovyi platok: Kostium—veshch' i obraz v russkoi literature XIX veka*. Moscow: Kniga.

Kitson, Peter J. 2000. "The Eucharist of Hell"; or, Eating People is Right: Romantic Representations of Cannibalism. *Romanticism on the Net* 17. [CrossRef]

Klymentiev, Maksym. 2009. The Dark Side of "The Nose": The Paradigms of Olfactory Perception in Gogol's "The Nose". *Canadian Slavonic Papers* 51: 223–41. [CrossRef]

Kondakova, Yuliya. 2009. Tekhnika "zavualirovannoi fantastiki" v proizvedeniiakh Gogolia. In *N. V. Gogol' kak germenevticheskaia problema*. Edited by Oleg Zyrianov. Ekaterinburg: Ural University Press, pp. 137–56.

Koropeckyj, Roman, and Robert Romanchuk. 2003. Ukraine in Blackface: Performance and Representation in Gogol's *Dikan'ka Tales*, Book 1. *Slavic Review* 62: 525–47. [CrossRef]

Kutik, Ilya. 2005. *Writing as Exorcism: The Personal Codes of Pushkin, Lermontov, and Gogol*. Evanston: Northwestern University Press.

Lachmann, Renate. 1999. The Semantic Construction of the Void. In *Gøgøl: Exploring Absence: Negativity in 19th-Century Russian Literature*. Edited by Sven Spieker. Bloomington: Slavica, pp. 17–33.

Lanskoi, L., ed. 1952. Gogol' v neizdannoi perepiske sovremennikov (1833–1853). In *Pushkin. Lermontov. Gogol'*. Literaturnoe Nasledstvo 58. Moscow: Izdatel'stvo AN SSSR, pp. 533–772.

Lavrin, Janko. 1926. *Gogol*. London: George Routledge & Sons, New York: E. P. Dutton & Co.

Lawton, David. 1993. *Blasphemy*. Philadelphia: University of Pennsylvania Press.

Lermontov, Mikhail. 2014. *Sobranie sochinenii v 4 tomakh*. Volume Editor: Nikita Okhotin. Saint Petersburg: Izdatel'stvo Pushkinskogo Doma, vol. 1.

Levy, Leonard W. 1981. *Treason against God: A History of the Offense of Blasphemy*. New York: Schocken Books.

Levy, Leonard W. 1993. *Blasphemy: Verbal Offense against the Sacred, from Moses to Salman Rushdie*. New York: Alfred A. Knopf.

Lotman, Yuri. 1968. Problema khudozhestvennogo prostranstva v proze Gogolia. *Uchenye zapiski Tartuskogo universiteta* 209: 5–50.

Lotman, Yuri. 1976. Gogol' and the Correlation of "The Culture of Humor" with the Comic and Serious in the Russian National Tradition. In *Semiotics and Structuralism: Readings from the Soviet Union*. Edited with an Introduction by Henryk Baran. New York: International Arts and Sciences Press, pp. 297–300. First published 1974.

Lotman, Yuri. 1985. Concerning Khlestakov. In *The Semiotics of Russian Cultural History*. Essays by Iurii M. Lotman, Lidiia Ia. Ginsburg, Boris A. Uspenskii. Translated from the Russian. Edited by Alexander D. Nakhimovsky and Alice Stone Nakhimovsky. Ithaca and London: Cornell University Press, pp. 150–87. First published 1975.

Lotman, Yuri. 1990. *Universe of the Mind: A Semiotic Theory of Culture*. Translated by Ann Shukman. London and New York: I. B. Tauris.

Lounsbery, Anne. 2021. Nikolai Gogol, Symbolic Geography, and the Invention of the Russian Provinces. In *The Palgrave Handbook of Russian Thought*. Edited by Marina F. Bykova, Michael N. Forster and Lina Steiner. Cham: Palgrave Macmillan, pp. 491–506.

Lubensky, Sophia. 2013. *Russian-English Dictionary of Idioms*, Revised Edition. New Haven and London: Yale University Press.

Maguire, Robert A. 1994. *Exploring Gogol*. Stanford: Stanford University Press.

Mann, Yuri. 1966. *O groteske v literature*. Moscow: Sovetskii Pisatel'.

Mann, Yuri. 1973. Evoliutsiia gogolevskoi fantastiki. In *K istorii russkogo romantizma*. Edited by Yuri Mann, Irina Neupokoeva and Ulrich Vogt. Moscow: Nauka, pp. 219–58.

Mann, Yuri. 1996. *Poetika Gogolia: Variatsii k teme*. Moscow: Coda.

Mann, Yuri. 2012a. *Gogol'. Kniga 1: Nachalo, 1809–1835*. Moscow: RGGU.

Mann, Yuri. 2012b. *Gogol'. Kniga 2: Na vershine, 1835–1845*. Moscow: RGGU.

Martinsen, Deborah A. 2004. Shame's Rhetoric, or Ivan's Devil, Karamazov Soul. In *A New Word on The Brothers Karamazov*. Edited by Robert Louis Jackson. Evanston: Northwestern University Press, pp. 53–67.

Mashinskii, Semën, ed. 1952. *N. V. Gogol' v vospominaniiakh sovremennikov*. Moscow: GIKhL.

Merezhkovsky, Dmitry. 1974. Gogol and the Devil. In *Gogol from the Twentieth Century: Eleven Essays*. Selected, Edited, Translated, and Introduced by Robert A. Maguire. Princeton: Princeton University Press, pp. 55–102. First published 1906.

Meyer, Priscilla. 1999. Supernatural Doubles: Vii and The Nose. In *The Gothic-Fantastic in Nineteenth-Century Russian Literature*. Edited by Neil Cornwell. Amsterdam and Atlanta: Rodopi, pp. 189–209.

Meyer, Priscilla. 2000. The Fantastic in the Everyday: Gogol's "Nevsky Prospect" and Hoffmann's "A New Year's Eve Adventure". In *Cold Fusion: Aspects of the German Cultural Presence in Russia*. Edited by Gennady Barabtarlo. New York and Oxford: Berghahn Books, pp. 62–73.

Mochul'skii, Konstantin. 1934. *Dukhovnyi put' Gogolia*. Paris: YMCA Press.

Mondry, Henrietta. 2003. Gogol's Body, Rozanov's Nose. In *Gogol 2002: Gogol Special Issues in Two Volumes*. Edited by Joe Andrew and Robert Reid. Essays in Poetics Publication 8. Keele: Keele Students Union Press, vol. 1, pp. 72–87.

Morson, Gary Saul. 1992. Gogol's Parables of Explanation: Nonsense and Prosaics. In *Essays on Gogol: Logos and the Russian Word*. Edited by Susanne Fusso and Priscilla Meyer. Evanston: Northwestern University Press, pp. 200–39.

Morson, Gary Saul. 1998. Russian literature. In *The New Encyclopædia Britannica*, 15th ed. Chicago: Encyclopædia Britannica, Inc., vol. 26, pp. 1003–11.

Nabokov, Vladimir. 1961. *Nikolai Gogol*, Corrected edition. New Directions Paperbooks 339. New York: New Directions. First published 1944.

Nash, David. 1999. *Blasphemy in Modern Britain: 1789 to the Present*. Aldershot and Brookfield: Ashgate.

Nash, David. 2007. *Blasphemy in the Christian World: A History*. Oxford and New York: Oxford University Press.

Nimchuk, Vasil'. 2004. Perednie slovo. In *Istoriia ukraïns'koho pravopysu XVI–XX stolittia: Khrestomatiia*. Edited by Vasil' Nimchuk and Nataliia Puriaieva. Kyiv: Naukova Dumka, pp. 5–26.

Obolenskaia, Yuliya. 1980. Kalambury v proizvedeniiakh F. M. Dostoevskogo i ikh perevod na ispanskii yazyk. In *Tetradi perevodchika 17*. Mosow: Mezhdunarodnye Otnosheniia, pp. 48–60.

Ohiyenko, Ivan. 1927. *Narysy z istoriï ukraïns'koï movy: Systema ukraïns'koho pravopysu. Populiarno-naukovyi kurs z istorychnym osvitlenniam*. Warsaw: Zakłady Graficzne E. i dr K. Koziańskich.

Oulianoff, Nicholas. 1967. Arabesque or Apocalypse? On the Fundamental Idea of Gogol's Story "The Nose". *Canadian Slavic Studies* 1: 158–71. First published 1959.

Peace, Richard. 1981. *The Enigma of Gogol: An Examination of the Writings of N. V. Gogol and Their Place in the Russian Literary Tradition*. Cambridge: Cambridge University Press.

Pilshchikov, Igor. 2019. K poetike i semantike gogolevskogo "Nosa", ili Chto skryvaiut govoriashchie detali. In *Literaturoman(n)iia: K 90-letiiu Yuriia Vladimirovicha Manna*. Moscow: RGGU, pp. 218–37.

Pilshchikov, Igor. Forthcoming. Russkii mat: Chto my o nem znaem? (O proiskhozhdenii i funktsiiakh russkoi obstsennoi idiomatiki). *Matica Srpska Journal of Slavic Studies*, 100.

Pletnyova, Aleksandra. 2003. Povest' N. V. Gogolia "Nos" i lubochnaia traditsiia. *Novoe Literaturnoe Obozrenie* 61: 152–63.

Pratt, Mary Louise. 1991. Arts of the Contact Zone. *Profession* 91: 33–40.

Pumpianskii, Lev. 2000. Gogol'. Written in 1922–1923. In *His Klassicheskaia traditsiia: Sobranie trudov po istorii russkoi literatury*. Moscow: Yazyki russkoi kul'tury, pp. 257–379.

Pushkin, Aleksandr. 1937–1949. *Polnoe sobranie sochinenii*. Moscow and Leningrad: Izdatel'stvo AN SSSR, 16 vols.

Pushkin, Aleksandr. 1984. *Epigrams and Satirical Verse*. Edited and Translated by Cynthia Whittaker. Ann Arbor: Ardis.

Putney, Christopher. 1999. *Russian Devils and Diabolic Conditionality in Nikolai Gogol's "Evenings on a Farm near Dikanka"*. New York: Peter Lang.

Rancour-Laferriere, Daniel. 1982. *Out from under Gogol's Overcoat: A Pychoanalytic Study*. Ann Arbor: Ardis.

Raskin, David. 1989. Chiny i gosudarstvennaia sluzhba v Rossii v XIX–nach. XX vv. In *Russkie pisateli: Biograficheskii slovar', 1800–1917*. Moscow: Sovetskaia Entsiklopediia, pp. 661–63.

Reyfman, Irina. 2016. *How Russia Learned to Write: Literature and the Imperial Table of Ranks*. Madison: University of Wisconsin Press.

Rowe, William Woodin. 1976. *Through Gogol's Looking Glass: Reverse Vision, False Focus, and Precarious Logic*. New York: New York University Press.

Rozanov, Vasilii. 1995. Genii formy (K 100-letiiu so dnia rozhdeniia Gogolia). In *O pisatel'stve i pisateliakh*. Moscow: Respublika, pp. 345–52. First published 1909.

Seifrid, Thomas. 1993. Suspicion toward Narrative: The Nose and the Problem of Autonomy in Gogol's "Nos". *Russian Review* 52: 382–96. [CrossRef]

Senkovskii, Osip. 1842. Review of: Pokhozhdeniia Chichikova, ili Mertvye Dushi: Poema N. Gogolia. *Biblioteka dlia Chteniia* 53: 24–54.

Senkovskii, Osip. 1936. Rukopisnaia redaktsiia stat'i o "Mertvykh dushakh." Written in 1842. Annotated by Nikolai Mordovchenko. In *N. V. Gogol': Materialy i issledovaniia*. Edited by Vasilii Gippius. Moscow and Leningrad: Izdatel'stvo AN SSSR, pp. 226–49.

Setchkarev, Vsevolod. 1965. *Gogol: His Life and Works*. Translated by Robert Kramer. New York: New York University Press.

Shanskii, Nikolai. 1960. Iz russkoi frazeologii: Frazeologicheskie oboroty s tochki zreniia ikh leksicheskogo sostava. *russkii yazyk v natsional'noi shkole* 1: 74–77.

Shanskii, Nikolai. 1963. *Frazeologiia sovremennogo russkogo yazyka*. Moscow: Vysshaia shkola.

Shapovalov, Veronica. 1988. A. S. Pushkin and the St. Petersburg Text. In *The Contexts of Aleksandr Sergeevich Pushkin*. Edited by Peter I. Barta and Ulrich Goebel. Lewiston, Queenston and Lampeter: Edwin Mellen Press, pp. 43–54.

Shenrok, Vladimir. 1893. *Materialy dlia biografii Gogolia*. Moscow: Tipografiia A. I. Mamontova.

Shepelyov, Leonid. 1977. *Otmenennye istoriei: Chiny, zvaniia i tituly v Rossiiskoi Imperii*. Leningrad: Nauka.

Shepelyov, Leonid. 1991. *Tituly, mundiry, ordena v Rossiiskoi Imperii*. Leningrad: Nauka.

Shepelyov, Leonid. 1999. *Chinovnyi mir Rossii: XVIII—nachalo XX v*. Saint Petersburg: Iskusstvo-SPB.

Shukman, Ann. 1989. Gogol's The Nose or the Devil in the Works. In *Nikolay Gogol: Text and Context*. Edited by Jane Grayson and Faith Wigzell. Basingstoke and London: Macmillan, pp. 64–82.

Slobodskoi, Serafim. 1967. *Zakon Bozhii: Dlia sem'i i shkoly so mnogimi illiustratsiiami*, 2nd ed. Jordanville: Holy Trinity Monastery.

Slonimskii, Aleksandr. 1974. The Technique of the Comic in Gogol. In *Gogol from the Twentieth Century: Eleven Essays*. Selected, Edited, Translated, and Introduced by Robert A. Maguire. Princeton: Princeton University Press, pp. 323–74. First published 1923.

Sobel, Ruth. 1998. The Nose: Nos: Short story. In *Reference Guide to Russian Literature*. Edited by Neil Cornwell. London and Chicago: Fitzroy Dearborn, pp. 334–35.

Spycher, Peter. 1963. N. V. Gogol's "The Nose": A Satirical Comic Fantasy Born of an Impotence Complex. *Slavic and East European Journal* 7: 361–74. [CrossRef]

Statute on Civil Uniforms. 1834. *Polozhenie o grazhdanskikh mundirakh*. Saint Petersburg: The Typography of the Second Section of His Imperial Majesty's Own Chancellery.

Statutes of Civil Service. 1833. *Svod Zakonov Rossiiskoi Imperii, poveleniem Gosudaria Imperatora Nikolaia Pavlovicha sostavlennyi: Uchrezhdeniia: Svod uchrezhdenii gosudarstvennykh i gubernskikh. Chast' 3. Ustavy o sluzhbe grazhdanskoi*. Saint Petersburg: The Typography of the Second Section of His Imperial Majesty's Own Chancellery.

Statutes of Civil Service. 1834. *Prodolzhenie Svoda Zakonov Rossiiskoi Imperii. 1832 i 1833 gody*. Saint Petersburg: The Typography of the Second Section of His Imperial Majesty's Own Chancellery.

Sterne, Lawrence. 1761. *The Life and Opinions of Tristram Shandy, Gentleman*. London: R. and J. Dodsley in Pall-Mall, vol. 3.

Struve, Gleb. 1961. *Russian Stories = Russkie Rasskazy: Stories in the Original Russian*. Edited by Gleb Struve. With Translations, Critical Introductions, Notes and Vocabulary by the Editor. A Bantam Dual-Language Book. New York: Bantam Books.

Todorov, Tzvetan. 1975. *The Fantastic: A Structural Approach to a Literary Genre*. Translated from the French by Richard Howard. With a Foreword by Robert Scholes. Ithaca: Cornell University Press. First published 1970.

Tolstoy, Mikhail. 1881. Vospominaniia. Chapters IV and V. *Russkii Arkhiv* II.1: 42–131.

Toporov, Vladimir. 1984. Peterburg i peterburgskii tekst russkoi literatury. *Uchenye zapiski Tartuskogo universiteta* 664: 4–29.

Toporov, Vladimir. 2009. *Peterburgskii Tekst*. Pamiatniki Otechestvennoi Nauki. XX vek. Moscow: Nauka.

Trubachev, Oleg, ed. 1975. *Etimologicheskii slovar' slavianskikh yazykov: Praslavianskii leksicheskii fond*. Moscow: Nauka, vol. 2.

Trubachev, Oleg, ed. 1977. *Etimologicheskii slovar' slavianskikh yazykov: Praslavianskii leksicheskii fond*. Moscow: Nauka, vol. 4.

Tynianov, Yuri. 2019. Dostoevsky and Gogol (Toward a Theory of Parody). In *his Permanent Evolution: Selected Essays on Literature, Theory and Film*. Translated and Edited by Ainsley Morse and Philip Redko. Boston: Academic Studies Press, pp. 25–61. First published 1921.

Tynianov, Yuri. 2019. On Plot and Fabula in Film. In *His Permanent Evolution: Selected Essays on Literature, Theory and Film*. Translated and Edited by Ainsley Morse and Philip Redko. Boston: Academic Studies Press, pp. 235–36. First published 1926.

Uspenskii, Boris. 2004. Vremia v gogolevskom "Nose" ("Nos" glazami etnografa). *Welt der Slaven* 49: 335–46.

Uspenskii, Boris. 2012. Oblik cherta i ego rechevoe povedenie. In *Umbra: Demonologiia kak semioticheskaia sistema 1*. Moscow: Indrik, pp. 17–65.

Utz, Richard J., and Christine Baatz. 1998. Transubstantiation in Medieval and Early Modern Culture and Literature: An Introductory Bibliography of Critical Studies. In *Translation, Transformation and Transubstantiation in the Late Middle Ages*. Edited by Carol Poster and Richard J. Utz. Evanston: Northwestern University Press, pp. 223–56.

Veniamin. 1823. *Novaia Skrizhal' ili Dopolnenie k prezhdeizdannoi Skrizhali s tainstvennymi ob"iasneniiami o tserkvi, o razdelenii eia, o utvariakh i o vsekh sluzhbakh v nei sovershaemykh*. Saint Petersburg: The Typography of the Imperial Orphanage. First published 1803.

Veresaev, Vikentii, ed. 1933. *Gogol' v zhizni: Sistematicheskii svod podlinnykh svidetel'stv sovremennikov*. Moscow and Leningrad: Academia.

Viazemskii, Pëtr. 1893. *Polnoe sobranie sochinenii*. Saint Petersburg: M. M. Stasiulevich, vol. 8.

Viazemskii, Pëtr. 1899. *Ostaf'evskii arkhiv kniazei Viazemskikh*. Edited by Vladimir Saitov. Saint Petersburg: M. M. Stasiulevich, vol. 3.

Vinogradov, Viktor, ed. 1956–1961. *Slovar' yazyka Pushkina*. Moscow and Leningrad: Gosudarstvennoe Izdatel'stvo Inostrannykh i Natsional'nykh Slovarei, vols. 1–4.

Vinogradov, Viktor. 1926. *Etiudy o stile Gogolia*. Voprosy Poetiki 7. Leningrad: Academia.

Vinogradov, Viktor. 1929. Naturalisticheskii grotesk: Siuzhet i kompozitsiia povesti Gogolia "Nos.". In *Evoliutsiia russkogo naturalizma Gogol' i Dostoevskii*. Leningrad: Academia, pp. 7–88. First published 1921.

Vinogradov, Viktor. 1987. *Gogol and the Natural School*. Translated by Debra K. Erickson, and Ray Parrott. Introduction by Debra K. Erickson. Ann Arbor: Ardis. First published 1925.

Vinokur, Grigorii. 1935. "Boris Godunov": [A Commentary]. In *Polnoe sobranie sochinenii*. Edited by Aleksandr Pushkin. Leningrad: Izdatel'stvo AN SSSR, vol. 7, pp. 481–96.

Virolainen, Maria. 2004. Gorod-mir i sakral'nyi siuzhet u Gogolia. In *Gogol' i Italiia*. Edited by Rita Giuliani and Mikhail Weiskopf. Moscow: RGGU, pp. 102–12.

Voropaev, Vladimir. 2000. Posledniaia kniga Gogolia (K istorii sozdaniia i publikatsii "Razmyshlenii o Bozhestvennoi Liturgii") *Russkaia Literatura* 2: 184–94.

Weiskopf, Mikhail. 1987. Nos v Kazanskom sobore: O genezise religioznoi temy. *Wiener Slawistischer Almanach* 19: 25–46.

Weiskopf, Mikhail. 1993. *Siuzhet Gogolia: Morfologiia. Ideologiia. Kontekst*. Moscow: Radiks.

Whickman, Paul. 2020. *Blasphemy and Politics in Romantic Literature: Creativity in the Writing of Percy Bysshe Shelley*. London and New York: Palgrave Macmillan.

Woodward, James B. 1981. *The Symbolic Art of Gogol: Essays on his Short Fiction*. Columbus: Slavica.

Yermakov, Ivan. 1921. Posleslovie. In *N. V. Gogol'. Nos: Povest'*. Moscow: Svetlana, pp. 83–129.

Yermakov, Ivan. 1974. "The Nose". In *Gogol from the Twentieth Century: Eleven Essays*. Selected, Edited, Translated, and Introduced by Robert A. Maguire. Princeton: Princeton University Press, pp. 156–98. First published 1924.

Yermilov, Vladimir. 1952. *N. V. Gogol'*. Moscow: Sovetskii Pisatel'.

Young, Donald. 1979. Ermakov and Psychoanalytic Criticism in Russia. *Slavic and East European Journal* 23: 72–86. [CrossRef]

Zelenin, Dmitrii. 1930. Tabu slov u narodov vostochnoi Evropy i severnoi Azii: II. Zaprety v domashnei zhizni. *Sbornik Muzeia Antropologii i Etnografii* 8: 1–166.

Zhatkin, Dmitrii. 2002. Makkavei: Opyt osmysleniia bibleiskogo obraza. *Tarkhanskii Vestnik* 15: 50–53.

Zhivov, Viktor. 1981. Koshchunstvennaia poeziia v sisteme russkoi kul'tury kontsa XVIII—nachala XIX veka. *Uchenye zapiski Tartuskogo universiteta* 546: 56–91.

Zyrianov, Oleg. 2009. Traditsiia "fantasticheskogo realizma". In *N. V. Gogol' kak germenevticheskaia problema*. Edited by Oleg Zyrianov. Ekaterinburg: Ural University Press, pp. 120–37.

Article

The "Christology" of Bely the Anthroposophist: Andrei Bely, Rudolf Steiner, and the Apostle Paul

Monika Spivak [1,2]

1 Department of "Literary Heritage", IWL RAS, 121069 Moskow, Russia; monika_spivak@mail.ru
2 Department of "Memorial Apartment of Andrei Bely", The Pushkin State Museum of Fine Arts, 119019 Moscow, Russia

Abstract: The article focuses on R. Steiner's perception of the Gospels and the impact of that view on Bely's works. The latter had always valued Steiner's lectures on Christ and the Fifth Gospel, the "Anthroposophic" (relating to the philosophy of human genesis, existence, and outcome) Gospel, the knowledge of which had been received in a visionary way. In addition, Bely was an esoteric follower of Steiner and often quoted from Apostle Paul's 2 Corinthians, "Ye are our epistle written in our hearts, known and read of all men". The citation occurs in Bely's philosophical works (*The History of the Formation of the Self-Conscious Soul*, "Crisis of Consciousness"), autobiographic prose (*Reminiscences of Steiner*), the essay "Why I Became a Symbolist … ", and letters (to Ivanov-Razumnik and Fedor Gladkov). Bely's own anthroposophic and esoteric ideas relating to the gospel sayings are also examined. The aim of the research is to show through the example of one quotation the specifics of Bely the Anthroposophist's perception of Christian texts in general. This provides a methodological meaning for understanding other Biblical quotations and images in the works of Bely because anthroposophical Christology is also the key to their deciphering.

Keywords: Andrei Bely; Rudolf Steiner; Apostle Paul; gospels; anthroposophy; *The History of the Formation of the Self-Conscious Soul*; Crisis of Consciousness

Citation: Spivak, Monika. 2021. The "Christology" of Bely the Anthroposophist: Andrei Bely, Rudolf Steiner, and the Apostle Paul. *Religions* 12: 519. https://doi.org/10.3390/rel12070519

Academic Editor: Dennis Ioffe

Received: 4 June 2021
Accepted: 6 July 2021
Published: 10 July 2021

Ye are our epistle written in our hearts,
known and read of all men: Forasmuch as
ye are manifestly declared to be the epistle
of Christ ministered by us, written not with
ink, but with the Spirit of the living God; not
in tables of stone, but in fleshy tables of
the heart. And such trust have we through
Christ to God-ward. (2 Cor. 3: 2–4)

One of the eccentric peculiarities of the unique culture of the literature at the end of the turn of the 19th–20th centuries is well expressed by the term "the Occult Renaissance", suggested by English historian James Webb (see the "Classical Monography", Webb 1971). The scientific–technological revolution, the social program of the radical social transformation, the crisis of positivism, the success of ethnography, and so on, created in the second half of the 19th century the impression of a global change in the worldview paradigm. In these conditions the demand for spiritualism and similar "psychic phenomena" traditionally classified as superstitions was created, and mystically minded intellectuals were at that time trying to explain it with the aid of the classical sciences combined with occult knowledge. In addition to the usual Freemasonry, theosophy, the teaching of George Gurdjieff, and so forth also flourished. Examples of the famous West European writers involved in the Occult Renaissance include Edward Bulwer-Lytton, Joris-Karl Huysmans, Gustav Meyrink, René Daumal, William Butler Yeats, and Christian Morgenstern (Hanegraaff 2013).

In Russia, the Occult Renaissance (Bogomolov 1999; Szilard 2002) coincided with the Religious Renaissance (Zernov 1963), appearing in the heyday of religious and philosophical thought (V. Solovyov, P. Florensky, V. Sventsitsky, V. Ern, S. Frank, N. Berdyaev, S. Bulgakov, and others), in the activity of religious and philosophical societies in Moscow and St. Petersburg, and in the interest expressed towards symbolist writers in the religious foundations of life (D. Merezhkovsky, Z. Gippius, V. Ivanov, and others).

Andrei Bely (whose real name was Boris Nikolayevich Bugaev; 1880–1934), the poet, prosaic, linguist, thinker, symbolist theorist, and mystic, was in that sense a bright voice of this epoch.

The religious exaltation peculiar to him since childhood appeared not owing to his family tradition and education, but, quite the opposite, contrary to them. The future writer's father, the famous mathematician Professor Nikolai Vasilyevich Bugaev "was a determined denier of the church, dogmas, traditions and he hated 'mysticism'. However he didn't oppose rites, i.e., he received a priest with a cross out of social politeness". Nevertheless he actively promoted "the basic principles of science", "the slogans of Darwinism", and others. "All his aunts and uncles from the side of his father were either outright atheists or indifferent; the brothers and sisters from the side of his mother expressed the same unconcern … ". His mother, Alexandra Dmitrievna Bugaeva was also indifferent to the questions of religion, and his "only traditionally believing grandmother was constantly mocked by his father and mother". Bely remembered " … tradition was mocked in our home; < … > I was mechanically taught two or three prayers and they didn't demand any religious signs … " (Bely 1994, pp. 423–24).

However, it was typical of Bely to perceive with great enthusiasm the images of the Old and New Testaments coming from the depths of his soul, as he thought. At the age of seven or eight, he devises "games of New Testament" hiding it from his parents, he lives through "the Descent of the Holy Spirit on two or three tiles of his parquet floor" in his parents flat (Bely 1994, p. 423). Having entered the scientific faculty of Moscow University under the influence of his father, he was obsessed with the idea of the "private" apocalypse (happening in one country or even in one city—Moscow) and in his first book, the *Second Dramatic Symphony* (1902), he describes how the "private" resurrection from the dead was happening at the cemetery of Novodevichy Convent (Lavrov 1978; Lavrov 1995).

His great interest in Christianity was combined with his interest in other religions and mysticism. From 1896 he was "getting interested in the problems of Hypnosis, Spiritualism and Occultism; he is under the huge impact of reading *Quotes from the Upanishads, The Tao* by Lao Tzu" (Bely 2016, p. 41). He read with great interest the books by Elena Blavatskaya and theosophic literature (see *Touching to Theosophy* in Bely 2016, pp. 751–54 and On Theosophy in Russia, Carlson 1993). He must solve a dilemma—how to join the Christian esoterics with oriental religions and mysticism. He was helped in solving this dilemma by Rudolf Steiner (1861–1925), an Austrian philosopher and visionary, and the author of the anthroposophy that began to spread all over Europe and was popular in Russia in 1910s. Bely regarded Steiner's teaching as a revelation: it was Steiner who introduced the Christian element in theosophy. This allowed Bely to remain a true Steinarian and Anthroposophist (Spivak 2006) and to consider himself at the same time a true Christian (although from the point of view of the Orthodox Church this was certainly rather dubious).

The aim of this article is to show the acceptance by Bely the Anthroposophist of the Christian texts. Reading them in the occult key with regard to Steiner's statements and relying on his own esoteric experience, the writer radically rethinks them. The research is based on the example of one quotation from the Epistle of Apostle Paul which was many times quoted by Bely. This has a methodological meaning for the understanding of other Biblical quotations and images in the works of Bely because anthroposophic Christology is the key to their deciphering.

Andrei Bely met Rudolf Steiner in May 1912 in Cologne, where rather unexpectedly even for himself, he had come with A.A. Turgeneva from Brussels. He had been driven there by vague intuitions and mystical omens. In Cologne, Steiner welcomed his Moscow guests and invited them to his lecture "Christ and the XX century" (March 8th) (Bely 2014, pp. 737–88).[1]

Soon Bely became a regular participant at Steiner's lectures and he began practicing esotericism under Steiner's guidance. Many of his friends who had remained in Russia were less than enthusiastic that Bely had readily accepted the anthroposophical discipleship. "All Christians suspect Steiner to be a follower of Luciferianism and to have a biased interpretation of Christ"[2] (Bely 2006, p. 199)—thus did Bely explain the reasons for the negative and unfair (in his view) attitudes of his previously likeminded acquaintances to his new idol, and he argued with them. Among the arguments in defense of his idol, the main one, according to Bely, was that Steiner, who at the time was the leader of the Theosophic Society in Germany, was more than just a regular theosophist, but instead a true Christian (Bely 2006, p. 199). What followed was that accepting anthroposophy was not an abandonment of the former spiritual path of Bely the Symbolist, but rather its rightful and logical continuation.

On May 7th (20) 1912 he wrote to E. K. Medtner (Bely and Metner 2017, vol. 2, p. 302; here and further highlighted by Bely):

"< . . . > I was, I am and I will be one who professes the name of Christ and truly experiences *His Forthcoming*. < . . . > And that is why I now going to Steiner: Christ and Russia!"[3].

Bely spoke enthusiastically about his impression shortly after he had heard that first lecture of Steiner, "while the trail is still hot". The very title of the lecture seemed to have already played a decisive role in the choice of Steiner as teacher. Bely proclaimed to N.P. Kiselev on May 7th (20) 1912:

"I must declare that I listened to the lecture of Steiner *'Christ and the Twentieth Century'*. It was as if this lecture was read specifically for me. All my doubts of his understanding of Christ were *dispelled* by this lecture. His understanding does not impinge upon the *Creed*, nor does it hamper the Orthodox teaching *revealed in reason*, rather it extends it, and speaks about what has not yet been revealed in history < . . . >". (Bely and Metner 2017, vol. 2, p. 303)[4]

He wrote about this to his mother in May 1912:

"We lived for 3 days in Cologne and we heard three lectures. We had a half an hour talk with the Doctor. < . . . > You simply cannot imagine what kind of man he is: you can see his aura (the light around him) with your own eyes. He was giving a lecture about the coming of Christ being close at hand. I have never heard such a thunderous, powerful speech in my life. It was as if his face was being torn apart, and from his face there shone another face, etc. We were absolutely amazed < . . . >". (Bely 2013, p. 153)[5]

In his letter to A.A. Blok from May 1st (14) 1912 he covered the themes touched upon at the lecture in greater detail (Bely and Blok 2001, pp. 459–60), but still concluded his narrative by giving a portrait of the lecturer as preacher.

"Toward the middle of the lecture his voice strengthens and he seems to be cutting himself off the crowd with his palms drawing some radiant line between the crowd and himself. After drawing each line he seems to loom larger, then leaps at the crowd—with his palms: and Asya and they hear these slaps in the face again. Beams of light fill the hall, and among these beams I see his transparent face shouting terribly enormous things. < . . . > He finishes by shouting at fourfold volume: *'Those who have understood what is the superhistorical Christ cannot help knowing that the Jesus of history was real. And He is coming'*. That is how his lecture on 'Christ and the XX Century' concludes. When he finished I was so shocked I couldn't help shouting from amazement: *'What is this*?!'". (Bely and Blok 2001, p. 460)[6]

Bely would refer to his impressions from this lecture in later life. In his work, *Notes of an Eccentric*, Bely wrote that "in Cologne, at the lecture announced on billboards as 'Christ and Our Century' he had heard the inner 'Voice' that had subsequently directed him in his life" (Bely 1997, p. 302). In his later reminiscences Bely emphasized that his first impression of Steiner the lecturer and Steiner the teacher had been the most faithful one:

> "< . . . > the first moment of our meeting < . . . > remained in me to the last: *"Steiner speaks in hearts at the very moment* when all words have been exhausted'".
> (Bely 2000, p. 260)[7]

Andrei Bely joined the anthroposophical movement at the very time when the themes of the New Testament and Christ in the life of a man and humanity ("The Christ Impulse") dominated Steiner's lectures. Steiner's "Christology" from the middle of the 1910s can be regarded as the basis of Bely's autobiographical works, as well as of his works of creative writing and nonfiction. Bely developed these ideas of Steiner most fully in *The History of the Formation of the Self-Conscious Soul* ("Историястановлениясамосознающей души", 1926–1931). This two-volume treatise, only recently published (Bely 2020, vols. 1–2), is Bely's most fundamental philosophical, historiosophical and anthroposophical work[8] (Odessky 2011; Spivak 2011; Stahl 2011).

In it the writer proceeds from the fact that "Christianity is the moment of changing the ideas about man, God, the universe, the spirit, the flesh of history". "The Christ Impulse", as Bely demonstrated, changed the very course of humanity's history, the result of its influence is the coiling of the straight line of history into a helix (Bely 2020, vol. 1, p. 184).

The theme in the first volume of *The History of the Formation of the Self-Conscious Soul* is the process of the birth of Christian Gnosis. Bely analyzes in detail flashes of Christian intuitions in antiquity, parses the teachings of gnostic sects, then passes on to the Gospels, regarding them originally as "historical documents". In the chapter devoted to this question, the writer refers to the opinions of the Church historians on the questions of textology and dating of the Gospels: the authorship and the time of writing, the sources of the Biblical texts, their most ancient layers, and later insertions. He literally stuffs the chapter with the names of authoritative scholars and a retelling of their sensational concepts. Bely, however, does not do that to align with any one of these viewpoints. His aim is just the opposite: to stigmatize the criticism of the Gospels as being flagrantly inadequate to the essence of the subject. According to Bely's scathing characterization, to write like the luminary scholars "they must be either absolute cranks as far as a living perception is concerned < . . . > or respectable grave-diggers without eyes, or ears, or bel-esprits, their wit derived from logical abstractions, or half maniacs, no matter how convincing their arguments sound < . . . >" (Bely 2020, vol. 1, p. 127):

> "The formal exegetes of the Gospels such as German scholars, or those claiming that Christ never existed . . . do not see the details that a more mature eye, developed by constant exercise can see. They do not see the style that is unique in the Gospels, no matter which parts we might divide them into; they do not hear the sound that is also unique and which the Apostle recommends that we listen to: 'Discern the spirits'[9]. Those who consider that Christianity appeared a few centuries earlier or later do not have the talent of telling the *times* which the Apostle refers to: *'one must have the ear, the eye, the spirit, the rhythm of the time'*[10]—this is the Leitmotif running through the early Christian records < . . . >".
> (Bely 2020, vol. 1, p. 125)[11]

Bely opposes those who interpret the Gospel into parts not on the basis of the facts or scholarly doctrines, but on arguments of a different sort. According to his view, to perceive the Gospels correctly one needs to "appeal to the style, the eye, the ear, the time, the colors of the images that had never been encountered either before, or after (not in poetry, nor in *narrative poems* of the Gnostics, nor in systems of thought)" (Bely 2020, vol. 1, p. 126). The Gospels, broken apart into quotations, die, in the sight of the "learned men", but in the sight of those who have mastered "the art of *hearing* and *seeing*"—they "like the dying mustard seed, begin to yield fruit". As Bely categorically states, "it has been proven item by item that they [the Gospels] do not exist on paper, they are being restored < … > in us and they become the writings in our *hearts*, rather than in *letters*".

According to Bely, the Gospels are a living tradition of not just words, but of gestures, rhythm, intonation: "from mouth to mouth, from ear to ear, from the sparkle in the eye to the sparkle in the eye" (Bely 2020, vol. 1, p. 134).

True Christianity, according to Andrei Bely, is by no means "what the grave-diggers consider Christianity (dogma, cult, custom, ritual, etc.)", but something else:

> "< … > Christianity in Christianity is what penetrates the images of Christianity given in the Gospels; and if the Gospels are not Gospels, i.e., the Gospels of Gospels: *'the epistle in the hearts'*, that Paul refers to: and it is the style, the spirit, the rhythm that the Gospels are imbued with < … >". (Bely 2020, vol. 1, p. 128)[12]

Bely's long-winded elaborations lead to an inescapable conclusion: "the Gospels appeared like an "epistle written in our hearts" (Paul), but the canonical Gospels are just reflections of those living Gospels" (Bely 2020, vol. 1, p. 1280).

It appears that Bely, on the one hand, refers to the well-known, even popular quotation from the Second Epistle of Paul to Corinthians:

> "Ye are our epistle written in our hearts, known and read by all men: For as much as ye are manifestly declared to be the epistle of Christ ministered by us, written not with ink, but with the Spirit of the living God; not in tables of stone, but in fleshy tables of the heart. And such trust have we through Christ to God-ward < … >". (2 Cor. 3: 2–4)

On the other hand, Bely associates the "epistle written in the hearts" that Paul mentions with a source not included in the Biblical canon, and nonexistent in the material world, not "existing in writing": that is the "living Gospels", "the Gospel of Gospels", "an inner Gospel connected with the four written Gospels" that sprang from "the depth of the personal experience of the Pentecost" (Bely 2020, vol. 1, p. 164).

> "Such is the amazing, in my personal view, explanation of the event of the Descent of the Holy Spirit provided by Rudolph Steiner as the key to the tonality for the theme of the Gospels. There is one key to them: the *Gospel* from the Holy Ghost as the *Fifth* of the four Gospels; the key to the fifth Gospel is the experience of the life in Christ by those who have developed in themselves this life as witness of the experience that the life of the *'I'* in Christ is the inner veracity". (Bely 2020, pp. 178–79)[13]

Steiner began speaking about the Fifth Gospel in the course of the lectures he gave in Christiania (Oslo) on October 1st, 2nd, 3rd, 5th and 6th, 1913[14] (Steiner GA 148 2009[15]). Bely emphasized that "he was one of very few people present at the *revelation*" who were honored "to see the doctor at the moment when he first discovered the crown of all his words about Jesus Christ" (Bely 2000, p. 507). From the first lecture, Bely perceived Steiner's performance as a special symbolic "gesture" as a "step" to "establishing another connection with us (our 'New Testament' with him)" (Bely 2000, p. 515).

In these lectures in Christiania, Steiner claimed that the Fifth Gospel, or the Anthroposophical Gospel, is as ancient as the other four Gospels, but it does not exist in the written form, rather it can only be revealed to the eyes of a clairvoyant. It was exactly like a visionary's story that the writer perceived the whole course of lectures. Bely remembered Steiner as "pale, excited, agitated, as a person who just that previous instant had seen a Vision" (Bely 2000, c. 508) and "for the first time giving an account of what had been seen" (Bely 2000, p. 510): "< … > he arranged the facts of the course < … > as they had been seen in the Astral world: the facts were placed in the order opposite to the natural perception" (Bely 2000, p. 510).

The starting point of the narration for the lecturer in the astral world in the so-called "Akasha Chronicle" was the moment when the Apostles awoke from their sleep. According to Steiner that sleep lasted much longer than described in the canonical Gospels: it started when Christ was praying over the chalice and it ended when the Holy Ghost descended on the fiftieth day—Pentecost (Steiner GA 148 2009). However, this was not a dream in the usual sense. It did not, as Steiner described it, "prevent the Apostles from performing their daily routine activities, going out or coming back ... So those who lived with them did not seem to notice the state of mind they were in" (Steiner GA 148 2009).

Nevertheless the consciousness of the Apostles was clouded and they lived as sleep-walkers, not perceiving adequately anything that passed before their eyes. It was as if they had slept through everything that happened at Golgotha and thereafter: the death on the cross, the laying in the tomb, the resurrection, the meetings with the resurrected Christ, the ascension ... They saw all these events in a dreamlike manner. "But the time came when the Apostles had the sensation that they had awoken after a long sleep. The Pentecost commemorates that awakening. < ... > They were awoken by the primeval force of love that fills and warms the universe, as if that primeval force had sunk into the soul of every one of them", and "little by little, like dreams, rising from the depth of our subconsciousness, of our soul, the memories of days they had lived rose in their consciousness, in the souls of the Apostles. < ... > They lived through the entire time again, day by day". However "now they recognized everything they had seen before in a normal way". The Apostles could remember and appreciate the events that they had witnessed only because on the Day of the Descent of the Holy Spirit they were "impregnated by universal love" (Steiner GA 148 2009), or the Christ Impulse, that descended onto them, piercing and enlightening their hearts.

In the *Reminiscences of Steiner* Bely noted the main points of Steiner's course:

"Lecture One: we are in the Impulse; that is why we observe the history of the impulse in reverse: from ourselves to the Apostles, i.e., seeing ... beyond Jesus Christ and the hearts of the Apostles: the heart is the Round Table at which all the 12 Apostles are sitting with Christ among them < ... >.

Lecture Two—the foundation of such an opportunity: the descent of the Holy Spirit, the source of the Impulse; 12 Apostles in the Holy Spirit and the 13th Apostle—Paul in Damascus (and now each of us is a 'Saul', who can become Paul); the connection of the '12th' with '13th' is the connection of '12' in the Impulse with each of us. < ... > That is the source of the 4 Gospels: earthly memories through the prism of one who has slept through and later had revealed to—in the regions where the 13th, the villain-persecutor, from Damascus, already sees the same light of the event; nowadays everyone can be a 'rememberer', a participant in Golgotha, like a thief-persecutor; to whom it is said: 'Today shalt thou be with me in paradise!'. < ... >

The Biography of Jesus—these are the final lectures summoning us to hear the Christ Impulse in ourselves. < ... > It is to the *Beginning*, prior to the baptism, before history, before Christianity, that this course leads us; but we and the XXth century are the 'end' < ... >

It is we who are revealed in the imminent Second Coming of Christ—that is the very impact of the course!". (Bely 2000, p. 511)[16]

For Bely, the most important thesis for understanding Gospel textology was that the Fifth Gospel was "the reality of the witness of the Apostles taken, not at the moment of writing, but at the moment of awareness informed with the Descent of the Holy Spirit," (Bely 2000, p. 510). In *The History of the Formation of the Self-Conscious Soul* it is specified that the Gospels would appear absolutely contradictory if one perceived them as external evidential testimony, and their value cannot be understood without "a profound spiritual experience", without that "gnostic experience that Paul was teaching about; the Gospels are revealed in it, and not only in the Gospels of the four evangelists read rationally" (Bely 2020, vol. 1, p. 164).

Bely juxtaposes Paul, "the Apostle of self-awareness" (Mischke 2011) and the ide-ologist of the "epistle in our hearts", to the Apostle Peter who symbolizes the past (the traditional church) and the Apostle John who symbolizes the future. Paul's testimony of Christ proves to be the most important and meaningful for the modern world ("Paul teaches us the approach to the Gospels"—Bely 2020, vol. 1, p. 163), because unlike the other Apostles, and like a person of the XXth century, "he was not acquainted with Christ personally", but he knew "Christ's light that had flashed on him; he then opened Christ within his 'self'; and recognized Him as already leading humanity < . . . >" (Bely 2020, vol. 1, p. 163). Bely writes about this in "Crisis of Consciousness"[17]: "Paul did not see Christ, but he knew: he had seen the coming in his heart, humanity had become free < . . . >" (Bely 1996, p. 31). He explores this more extensively in *The History of the Formation of the Self-Conscious Soul*:

> "Let us emphasize that we rely mostly on the Apostle Paul's experience for our knowledge about Christ; this was an internal experience: Christ came to Paul from the depth of his heart < . . . >; Paul understood very well that Jesus Christ is the light of the world, the bread of life, the key unlocking the door of the heart, the very door of your exit or the way, the resurrection of life, the truth and the true vine; from the wise light cast into his heart open like a door, he came to an understanding of the personality of Christ < . . . >". (Bely 2020, vol. 1, p. 177)[18]

The factor of "heart" in the description of Paul's mission is constantly emphasized by Bely with repeated references to his Epistles: "Paul's heart is enlarged like the sun: 'our heart is enlarged' (2 Cor. 6:11)" (Bely 1996, p. 69). Or: "The Church of Paul is the connection through the heart; or—the correspondence of hearts: 'Ye are our epistle written in our hearts' < . . . >" (Bely 1996, p. 34).

The well-known quotation acquires new meaning in Bely's interpretation. Just as "the other Apostles were chosen in a human way", i.e., they came to the Church directly through Jesus, "but Paul came through the Spirit", so that "Paul was the chosen one" (Bely 1996, p. 40). Steiner emphasized several times that "Apostle Paul had passed through the Hebrew prophetic school of his time" (Steiner GA 148 2009). The Apostle Paul received the "initiation given to him by grace": "For he came to Him not by proper study through the ancient mysteries, but through grace on the way to Damascus when the Risen Christ appeared, so I call that an initiation given through grace < . . . >. He recognized the Risen Christ. And since then he has proclaimed Him . . . " (Steiner GA 142 2009).

Bely goes further, claiming that the reading from the heart of the Gospels by Paul is explained by the fact that he is "esoteric" or, in fact, an "Anthroposophist".

> "Paul here is someone esoteric; < . . . > the key to the Wisdom *'of wisdoms'* of this world, from *anthropism* (paganism) and *sophism* (Judaic law) selected by Paul; he is a true *Anthroposophist* < . . . >". (Bely 1996, p. 59)[19]

In this connection, it is precisely as an esoteric experience that we should understand the words from *The History of the Formation of the Self-Conscious Soul* noted above about "Paul's experience < . . . > being an "internal experience" and that one cannot understand the Gospels without a profound spiritual experience < . . . >—the very gnostic experience that Paul had taught . . . " (Bely 2020, vol. 1, p. 164). The Commandments of Paul, the esoteric, reverberates in Bely's recital absolutely anthroposophically, including some occult work on different parts of the human body and primarily on the heart:

> "< . . . > the Apostle advises: 'be ye transformed by the renewing of your mind!'[20]. The renewing of one's mind is the path of meditation, of yoga of perception; thought, as it is growing stronger, is introduced into the body via the heart < . . . >". (Bely 1996, p. 63)[21]

This provides the esoteric and anthroposophical context to the semantics of the "heart" in Bely's beloved quotation from the Second Letter to Corinthians:

> "Paul naturally shows us that we are a fiery *'correspondence of the heart'* with Christ: 'You are . . . Christ's epistle, written . . . by the Spirit of the Living God

... on the tables of thine hearts'. 'The Tables' of thine heart < ... >". (Bely 1996, p. 68)[22]

In *Reminiscences of Steiner* Bely described his teacher as a disciple of the "Esoteric Apostle" and his successor: "Doctor Steiner loved and understood the "Unjust" ardent Paul burning in his heart with all his heart". "He spoke like Paul; he was silent like John" (Bely 2000, pp. 272, 497). Undoubtedly Steiner is likened in the memoirs to other evangelical personages (Lagutina 2015), but the continuity concerning Paul goes by way of the "heart". Bely emphasizes "heartedness" as his most important feature: "I would rather point to the heartedness in the Doctor, that could never be answered equally < ... > (Bely 2000, p. 296). We can hardly speak about the heartedness as an exclusively psychological feature. Steiner, according to Bely, was one who "reads in hearts" (Bely 2000, p. 388), "listens to hearts" (Bely 2000, p. 343), "speaks with all the strength of his thought and fire of his heart: from *heart to heart*" (Bely 2000, p. 338).

The words "from heart to heart" sounded extremely clearly in Steiner's lectures about Christ:

> "< ... > at other moments he spoke to *hearts*; the expression: 'from heart to heart'—he would say this with such a clear, loving smile when he spoke about the 'infant' Jesus < ... >; he was himself a heart; or to be more precise: his mind was in the place of his heart; and the *mindful heart* blossomed; the 'heart', not the 'mind of the heart'". (Bely 2000, p. 496; also: Bely 2000, p. 343)[23]

In a number of descriptions, Bely "shines light" on the complex initiation methods that Steiner practiced himself and which he taught his esoteric students (Kazachkov 2015; Stahl 2015):

> "Meditation over the Name is the way < ... >. It summoned one to something greater: to the ability to praise the Name by inner breathing extinguishing the outer verbal sound: toward the birth of the *word* in the heart". (Bely 2020, p. 497)[24]

Sometimes it is stated directly that the source of Steiner's knowledge of the "heart" about Christ was "spiritual-scientific research", i.e., clairvoyance:

> "He was an inspiration: not merely imagination! The words about *Christ* are inspirations: they are thoughts of the heart < ... >. The Doctor kept silent about Christ with his head; but he spoke with his *sun-heart*; the words of his courses about Christ are expirations: not of oxygen, but merely of carbon dioxide, hinting at the process of the secret of life. < ... > *he* stood not at *these* doors—*he* was standing by other doors < ... >,—consciousness grew dim. There was another door—the *heart*! He was summoning us to *that* door ... < ... >". (Bely 2000, p. 496)[25]

In his course of lectures in 1923, "Contemporary Spiritual Life and Education", Steiner said that the "new initiation", already available to the contemporary individual familiar with anthroposophy, "will introduce with a clear light into the human heart that which leads to the awakening of the spirit in the heart and soul of a person, to his religious perception" (Steiner GA 307 2018). This process will require from anthroposophists the creation of a new communicative medium—a "hearted" one.

The language for communication between people needs an airy, sensual medium. If people can understand each other through some deeper elements of our soul, through the thoughts that carry with them feelings and the warmth of the heart, they find a means of communication besides language. But for this international means of understanding of each other we need a heart (Steiner GA 307 2018).

"In this case we speak, as it were, about the language of those dedicated that will function < ... > in the pure element of light passing from soul to soul, from heart to heart" (Steiner GA 307 2018).

Symbolically, Bely placed these very words of Steiner about the language "from heart to heart", "to which Anthroposophy aspires", as the epigraph to the chapter "Rudolph

Steiner in the Theme of Christ" in his *Reminiscences of Steiner* (Bely 2000, p. 493). This, obviously, also explains the meaning of the "language of the heart" in which Steiner spoke about Christ and the words of Apostle Paul:

> "There was another door—*the heart*! He summoned us *to that* door … < … >. Outside the language of the *heart* ('*ye* are an epistle *written in our hearts*'—says the Apostle to us) there is silence". (Bely 2000, p. 496)[26]

The anthroposophical esoteric praxis described in relation to Steiner and Paul, the "Apostle of self-awareness", was also well known to Bely who had been accepted in 1913 into the "'*Esoterische Stunde*' (meetings for the disciples of Steiner using the methods of spiritual science for themselves < … >" (Bely 2016, p. 137).

In his letter to P.A. Florensky from Dornach of 17th February 1914, Bely compares the "school of experience" in Orthodoxy to the "experience" of anthroposophy, pointing to the fact that "both schools, while accepting the heart as the spiritual Sun and life center, differ in the way of '*the immersion of the mind into the heart*'" (Bely 2004, p. 479). Naturally, Bely proves the advantage of the anthroposophical way as the method "< … > not the training of the mind juxtaposed to the heart, but rather a free immersion of the mind realizing itself in the heart < … >; this is the rule of the school that drew close to me < … >" (Bely 2004, p. 479), as well as the aim-setting guidelines:

> "The Heart is the Sun; < … > inside the heart you perceive the gleaming of the sun; it becomes Christ's heart. But *Christ* came not only for the Earth, < … > but for the entire Cosmos: the Church has failed to stress the cosmological meaning of Christ < … >. We must < … > draw a *starry line through the sun* into our earthly heart". (Bely 2004, p. 481)[27]

In the *Material for a Biography* it is mentioned that after Steiner's course of lectures in Christiania about the Fifth Gospel, Bely "< … > became acquainted with the "Christ Impulse" (Bely 2016, p. 140), in the *Reminiscences of Steiner*"—that "in Christiania the moment of the Descent of the Holy Spirit was demonstrated" (Bely 2000, p. 514). At that time, as the writer admits, he was convinced that "the Holy Spirit would soon descend" upon him as it had on the Apostles and that "God's voice" would emanate from him as well (Bely 2016, p. 141). Bely perceived Steiner's words about the mystical role of the Pentecost and the "Christ Impulse" piercing the Apostles if not as offering guidelines to action, then as a guideline to experience the world, as the basis for comparison of himself with Paul: "< … > in the 'Fifth Gospel' I'm 'an Apostle' among 'the Apostles' < … >" (Bely 2000, p. 514).

Bely recounts his work on the meditation about Christ assigned by Steiner (Kazachkov 2015) and he dropped the "Word" into "the heart", following the advice of experienced occultists: "< … > one must be able to pronounce the words known to you not with the lips, nor the tongue, nor the larynx; then the words sink into the heart; and acquire enormous power!" (Bely 2016, p. 144).

These experiments resulted in visions in which the events occurring were perceived as the way of initiation[28]:

> "< … > somebody (it seems like the Doctor) < … > slit the sign of the cross on my forehead < … >, and so a drop of either blood from my forehead, or a drop of balm, or my own '*I*' dripped into the Chalice, into the Holy Grail; but this chalice was no longer a chalice, but my own heart, and the drop was my consciousness having plunged into my heart: < … > and when the drop touched the Chalice, Christ united with me: and streams of indescribable love and the Christ Impulse surged from me, in me, through me; and at that moment I awoke < … >. It became clear to me: no, this was not a dream, it was a real initiation". (Bely 2016, p. 145)[29]

Soon, however, the way of initiation was interrupted. Not only did the regular meditations cease providing the desired result, but, in fact, they led to the opposite, a serious illness: "Cardiac neurosis is the name of that bizarre illness" (Bely 1997, p. 363).

Bely had to find another meaning and way in anthroposophy, other than via the occult. However, the mystical experience that he had acquired proved to be enough to remember the esoteric and anthroposophical meaning of the Apostle's words about the "epistle of our hearts", the best of which was derived from works such as the "Crisis of Consciousness", *The History of the Formation of a Self-Conscious Soul*, and the *Reminiscences of Steiner*.

It should be noted that the quotation from the Second Letter to the Corinthians, that was so dear to his heart, was used by Andrei Bely in another context as an epistolary formula. For example, in the letter to Ivanov-Razumnik from 18th March 1926:

> "< ... > Our correspondence is always in my heart, i.e., I always write to you in the heart, in the words of the Apostle Paul, 'You are an epistle written in our hearts'. (I might be misquoting, because I'm writing this off the top of my head); I walk constantly with an epistle to You in my heart; < ... > there is an enormous need to transform *this correspondence of the heart* into *a conversation from the heart* < ... >". (Bely and Razumnik 1998, p. 346)[30]

In the letter to Fyodor Gladkov from 17 June 1933:

> < ... > я был радостно взволнован Вашим письмом; но эта радость, радость отклика (со-вестия: «сердце сердцу весть подает», «вы—письмо, написанное в сердцах», ап<остол> Павел) тут же стала переходить в горечь < ... >.

> "< ... > I was joyfully excited by Your letter; but this joy, the joy of response (co-news 'one heart sending news to another heart', 'You are an epistle written in our hearts', Ap<ostle> Paul)[31] has begun to turn to bitterness < ... >". (Bely 1988, p. 765)

In both cases the words of the Apostle Paul, it seems, do not contain any esoteric "measure". They are in the realm of psychology: they express a warm feeling towards a correspondent who is far removed from the anthroposophical discourse and thus incapable of recognizing it. Quite another case is the use of that quotation in an autobiographical essay "Why I Became a Symbolist ... " (1928). The essay was addressed to anthroposophists and contained scathing criticism of the entire institution of the Anthroposophical Society. Understanding the polemical ardor of his own work and obviously foreseeing a possible negative reaction from "the people of his own circles", Bely concluded it with a defiant attack against those who would not want to accept the "epistle of the heart":

> "The time of writing has passed; now comes the time of reading of what is written in the heart; for nothing is secret, that shall not be made manifest. But whoever does not have the *writings* in the heart and who refuses to understand the words of the Apostle ('*Ye are the epistle written in our hearts*'), will not understand me.

> I know that very well". (Bely 1994, p. 493)[32]

It is possible, however, that here Bely was attempting to follow the dictates of Steiner who, in his course "Modern Spiritual Life and Education", established the necessity of a new language "coming from soul to soul, from heart to heart". "Contemporary civilization needs such a means of communication", Steiner proposed, emphasizing that "it will be used not only for the higher order, but for everyday life" (Steiner GA 307). Bely used the image of a "epistle of our hearts" in this way. If in the "Crisis of Consciousness", *The History of the Formation of the Self-Conscious Soul*, and the *Reminiscences of Steiner* the Apostle Paul's words serve to clarify questions of the "higher order", in contrast, the letters to his friends and the essay "Why I Became a Symbolist ... " express feelings and thoughts of "everyday life"[33].

Funding: This research received no external funding.

Institutional Review Board Statement: Not applicable.

Informed Consent Statement: Informed consent was obtained from all subjects involved in the study.

Data Availability Statement: Not applicable.

Conflicts of Interest: The author declare no conflict of interest.

Notes

[1] The reference is to Steiner's lecture "Vorverkündigung und Heroldtum des Christus-Impulses der Christus-Geist und Seine Hüllen: Eine Pfingstbotschaft".

[2] The letter of Bely to N.P. Kiselev, E.K. Medtner, A.S. Petrovsky, M.I. Sizov dated May 7(20), 1912 from Brussels (Bely and Metner 2017, vol. 2, p. 303).

[3] "< … > я был, есмь и буду исповедующим имя Христово и реально чувствующим *Его Приближение*. < … > И потому я теперь иду к Штейнеру: Христос и Россия!".

[4] "Я должен заявить, что слышал лекцию Штейнера '*Христос и XX век*'. Эта лекция была точно нарочно для меня прочитана: все мои сомнения в его понимании Христа *рассеяны* этой лекцией. Его понимание не посягает на *символ веры*, ни на православное *раскрытое в разуме* учение, а углубляет, говорит о еще не раскрытом в истории < … >".

[5] "В Кельне мы прожили 3 дня, слышали три лекции. Имели получасовой разговор с Доктором. < … > Ты просто не можешь себе представить, что это за человек: его аура (свет вокруг) прямо видна глазами. Он читал лекциюо близости пришествия Христа. Такой громовой, сильной речи я не слышал никогда в жизни. У него словно разрывается лицо, из лица светит лицо и т.д. Мы были совсем п отрясены < … >".

[6] "К середине лекции голос крепнет, ладонями себя то отрезает от толпы, проводя меж собой и толпой какую-то световую линию, и после каждого проведения линии точно вырастает, то кидается на толпу—ладонями: и опять те же с Асей слышим удары по лицу. Какие-то световые клубы наполняют залу, и вот из световых клубов вижу только сквозное лицо, которое кричит нам вещи громадные—до ужаса. < … > Кончает четырехкратным криком: '*Кто понял, что такое надисторический Христос, тот не может не знать, что Иисус истории—подлинный. И Он—близится*'. На этом кончается лекция 'Христос и XX век'. Когда он кончил, я невольно вскрикнул от потрясения: 'Что ж это?!'".

[7] "< … > первый миг встречи < … > остался последним во мне: '*Штейнер говорит в сердцах тогда именно, когда все уж слова исчерпались*'".

[8] See the article of M. Odessky, M. Spivak, H. Stahl, where the history of developing the treatise and its main ideas is considered in detail, "'It must be': 'History of the Becoming of a Self-Conscious Soul' by Andrey Bely" opening the first volume (Bely 2020, vol. 1, pp. 5–84).

[9] Cf.: 1Cor. 12: 10; 1John. 4: 1–3.

[10] Cf.: Rev. 2:29; Mk. 4: 9.

[11] "Растаскиватели Евангелий на составные части, как немецкие ученые, или утверждающие, что Христа и не было, < … >—не видят того, что видит более развитой, ибо упражнявшийся в зрении глаз: они не видят стиля, который единственен в Евангелиях, на какие бы части мы ни разложили их; не слышат звука, который тоже единственен, и которому советует внимать апостол: 'Духов различайте'. Передвигающие время появления христианства на несколько столетий совершенно не имеют дара различать *времен*, на который тоже ссылаетсяапостол; '*имейте ухо, глаз, дух, ритм времени*'—вот лейтмотив, проходящий сквозь ранние памятники христианства < … >".

[12] "< … > христианство в христианстве—это то, что проницало образы христианства, данные в Евангелиях; и если Евангелия—не Евангелия, то есть Евангелие Евангелий: '*сердечное письмо*', о котором говорит Павел; и оно—стиль, дух, ритм, в котором пересекаемы Евангелия < … >".

[13] "Таково изумительное, по моему личному мнению, объяснение события сошествия Святого Духа, данное Рудольфом Штейнером как ключ к тональности темы Евангелий; ключ к ним—один: *Евангелие* от Святого Духа, как *Пятое* четырех их; ключ же к пятому Евангелию—опыт жизни во Христе тех, которые развили в себе эту жизнь, как свидетельство опыта о том, что жизнь '*Я*' во Христе—внутренняя достоверность".

[14] See chapter "'I have offered up my life according to the letter of 1913': a reply to the 'Fifth Gospel' of Rudolph Steiner" in the book: (Spivak 2006, pp. 58–68).

[15] GA—Rudolf Steiner Gesamtausgabe, Bde. 1–354. Dornach, 1955—heute. At the site of the "Library of Spiritual Sciences" the German original text and the Russian translation are provided.

[16] "Лекция первая: мы—в Импульсе; и поэтому: озирающие историю импульса в обратном порядке: от себя—до апостолов, т.е. видящие … вслед за Христом Иисусом и сердца апостолов: сердце—Круглый Стол, за которым все 12 апостолов с Христом меж ними < … >.

Лекция вторая—основа такой возможности: сошествие Св. Духа, источника Импульса; 12 апостолов в Святом Духе и 13-й апостол—Павел в Дамаске (а ведь каждый из нас теперь 'Савл', могущий стать Павлом); связь '12' с '13-м'—связь '12' в Импульсе с каждым из нас. < … > Вот—источник 4-х Евангелий: земные воспоминания сквозь призму проспанного, открытого потом,—в регионах, где и13-й, разбойник-гонитель, из Дамаска, уже видит тот же свет события; в наши дни потенциально дан в каждом 'воспоминатель', участник Голгофы; это ему сказано: 'Нынче будешь со Мною!'. < … >"

> Биография Иисуса—последние лекции, призывающие расслышать импульс Христа в себе. < ... > *К началу* лежащему до крещения, до истории, до христианства, ведет конец курса; но 'конец'—мы и XX век. < ... >
> Мы, показанные в неизбежном Пришествии—вот удар курса!"

[17] The last part of Bely's "Crisis of Consciousness" (1920) was published under the name "The Gospels as Drama" (Bely 1996).

[18] "Подчеркнем: о Христе мы знаем более всего из опыта Павла; опыт был опыт внутренний: к Павлу Христос приходил из глубины его сердца < ... >; Павел более всего понимал, что Христос Иисус—свет миру, хлеб жизни, ключ, отпирающий сердечную дверь, самая дверь, выход из нее или путь, воскресение жизни, истина и лоза; от умного света, брызнувшего в его открытое, как дверь, сердце, шел он к уразумению и личности Иисуса < ... >".

[19] "Павел здесь—эзотерик; < ... > ключ к Мудрости *'мудростей'* мира сего, с *антропизма* (язычества) и *софизма* (закона иудейского) Павлом подобран; он—подлинный *антропософ* < ... >".

[20] Rom. 12: 2.

[21] "< ... > апостол советует: 'Преобразуйтеся обновленьем ума'. Обновленье ума есть путь медитации, йога познания мысль, укрепляясь, вводится в тело сквозь сердце < ... >".

[22] "Естественно показует нам Павел, что мыогневая *'сердечная переписка'* с Христом: 'Вы ... письмо Христово, написанное ... Духом Бога Живаго ... на скрижалях сердца'. 'Скрижали'—сердечные < ... >".

[23] "< ... > обращался он в миги другие к *сердцам*; выраженье: 'от сердца к сердцу'—с какой ясной, любовной улыбкой он говорил это, когда говорил о 'младенце' Иисусе < ... >; сам он был—сердце; вернее: ум его был в месте сердца; и *умное сердце*—цвело; 'сердце', а не 'сердечный ум'".

[24] "Медитация над Именем—путь < ... >. Взывал к большему: к умению славить Имя дыханием внутренним с погашением внешнего словесного звука: к рождению—*слова в сердце*".

[25] "Он был—инспирация: не имагинация только! Ислова о *Христе*—инспирации: сердечные мысли < ... >. Доктор молчал о Христе—головой; и говорил *солнцем-сердцем*; слова его курсов о Христе—выдохи: не кислород, а лишь угольная кислота, намекающая на процесс тайны жизни. < ... > не при *этих* дверях стоял *он*—при других < ... >,—сознание мутилось. Была иная дверь—*сердце*! Он звал к *этой* двери ... < ... >".

[26] "Была иная дверь—*сердце*! Он звал к *этой* двери ... < ... >. Вне *сердечного* языка (*'вы—письмо* наше, *написанное в сердцах'*—говорит нам апостол)—молчание".

[27] "Сердце—Солнце; < ... > внутри сердца познаешь блеск солнца; оно становится Христовым сердцем. Но *Христос* пришел не для земли только, < ... > для всего Космоса: Церковь не указала на космический смысл Христа < ... >. Надо < ... > провести *звездность сквозь солнце* в земное наше сердце".

[28] Concerning Bely on the way of his initiation see: (Glukhova 2015; Oboleńska 2009; Seryogina 2015; Spivak 2006).

[29] "< ... > кто-то (кажется, Доктор) < ... > ножичком сделал крестообразный < ... > разрез на моем лбу < ... >, отчего не то капля крови со лба, не то капля елея, не то мое '*я*' капнуло в чашу, вГрааль; но эта чаша была уже не чашей, а моим сердцем, а капля была моим сознанием, канувшим в сердце: < ... > и когда капля коснулась Чаши, то Христос соединился со мной: и из меня, во мне, сквозь меня брызнули струи любви несказанной и Христова Импульса; тут я проснулся < ... >. Мне стало ясно: нет, это не сон, а подлинное посвящение".

[30] "< ... > переписка меж нами всегда, т.е. я всегда Вам пишу, всердце,—по выражению апостола Павла: 'Вы—письмо, написанное в сердцах'. (Может, цитирую не так,—на «память»); я хожу всегда как бы с письмом всердце к Вам; < ... > огромная есть потребность превратить *сердечную переписку в сердечный разговор* < ... >".

[31] See the interpretation of this quotation from the letter to Gladkov in the context of Bely's theory of the "word": (Torshilov 2015).

[32] "Пора написаний прошла; наступает пора прочтений уже в сердце написанного; нет ничего тайного, что не стало бы явным. Но кто не имеет *письмян* в сердце и откажется от понимания слов апостола ('*Вы—письмо, написанное в сердцах*'), тот меня не поймет. Мне это хорошо ведомо".

[33] I thank Oksana Chebotareva and Tom Beyer for translating the article into English.

References

Bely, Andrei (Андрей Белый). 1994. Почему я стал символистом ... [Why I Became a Symbolist ...]. In *Белый, Андрей. Символизм как миропонимание [Bely, Andrei. Symbolism as a World View]*. Publication and Notes by Larisa Sugai. Moscow: Respublika, pp. 418–95.

Bely, Andrei (Андрей Белый). 1996. *Евангелие как драма [The Gospels as Drama]*. Moscow: Russkij Dvor.

Bely, Andrei (Андрей Белый). 1997. Записки чудака [Notes of an Eccentric]. In *Собр. соч.: Котик Летаев. Крещеный китаец. Записки чудака [Collection of Works: Kotik Letaev. The Christened Chinaman. Notes of an Eccentric]*. Under the General Editorship of Vladimir Piskunov. Moscow: Respublika, pp. 280–494.

Bely, Andrei (Андрей Белый). 2000. *Собр. соч.: Рудольф Штейнер и Гете в мировоззрении современности. Воспоминания о Штейнере [Collection of Works: Rudolf Steiner and Goethe in the Contemporary World View. Reminiscences of Steiner]*. Publication and Notes by Irina Lagutina. Moscow: Respublika, pp. 256–535.

Bely, Andrei (Андрей Белый). 2014. *Начало века. Берлинская редакция (1923) [Beginning of the Century. Berlin edition (1923)]*. Edited by Aleksander Lavrov. Sankt-Peterburg: Nauka.

Bely, Andrei (Андрей Белый). 2016. *Автобиографические своды: Материал к биографии. Ракурс к дневнику. Регистрационные записи. Дневники 1930-х годов [Autobiographical Corpus: Material for Biography. A Compendium for the Diary. Registrational Notes. Diaries of 1930s]*. Литературное наследство. Т. 105 [Literary Heritage. vol. 105]. Publication and Notes by Aleksander Lavrov and John Malmstad. Moscow: Nauka.

Bely, Andrei (Андрей Белый). 2020. *История становления самосознающей души [The History of the Formation of the Self-Conscious Soul]*. Book 1, 2—Литературное наследство. Т. 112. Кн. 1, 2 [Literary Heritage. Vol. 112. In 2 books]. Publication and Notes by Mikhail Odessky, Monika Spivak and Henrieke Stahl. Moscow: IMLI RAN.

Bely, Andrei (Андрей Белый), and Alexander Blok (Александр Блок). 2001. *Переписка. 1903–1919 [Correspondence. 1903–1919]*. Publication and Notes by Aleksander Lavrov. Moscow: Progress-Pleyada.

Bely, Andrei (Андрей Белый), and Emil Metner (Эмилий Метнер). 2017. *Переписка. 1902–1915 [Correspondence. 1902–1915]*. vols. 1, 2. Publication and Notes by Aleksander Lavrov, John Malmstad and Tatjana Pavlova. Moscow: Novoe literaturnoe obozrenie.

Bely, Andrei (Андрей Белый), and Ivanov Razumnik (Иванов Разумник). 1998. *Переписка [Correspondence]*. Publication and Notes by Aleksander Lavrov and John Malmstad. Sankt-Peterburg: Atheneum; Feniks.

Bely, Bugaeva. 2013. *«Люблю тебя нежно … »: Письма Андрея Белого к матери. 1899–1922 ["I love you tenderly … ": Letters of Andrei Bely to his Mother. 1899–1922]*. Publication and Notes by Sergei Voronin. Moscow: Reka Vremen.

Bely, Florensky. 2004. Переписка <П.А. Флоренского> с Андреем Белым [Correspondence <P.A. Florensky> with Andrei Bely]. Publication and Notes by E.V. Ivanova. In *Павел Флоренский и символисты: Опыты литературные. Статьи. Переписка [Pavel Florensky and the Symbolists: Literary Experiments. Articles. Correspondence]*. Publication and Notes by Evgenia Ivanova. Moscow: Yazyki Slavyanskoj kul'tury, pp. 434–98.

Bely, Gladkov. 1988. Переписка Андрея Белого иФедора Гладкова [Correspondence of Andrei Bely and Fyodor Gladkov]. In *Андрей Белый: Проблемы творчества [Andrei Bely: Problems of Creativity]*. Publication by Svetlana Gladkova. Moscow: Sovetskij Pisatel', pp. 753–72.

Bely, Morozova. 2006. *«Ваш рыцарь». Андрей Белый. Письма к М.К. Морозовой. 1901–1928 ["Your Knight". Andrei Bely. Letters to M.K. Morozova. 1901–1928]*. Publication and Notes by Aleksander Lavrov and John Malmstad. Moscow: Progress-Pleyada.

Bogomolov, Nikolai (Николай Богомолов). 1999. *Русская литература начала XX века и оккультизм [Russian Literature of the Early 20th Century and the Occultism]*. Moscow: NLO Publisher.

Carlson, Maria. 1993. *"No Religion Higher Than Truth": A History of the Theosophical Movement in Russia, 1875–1922*. Princeton: Princeton University Press.

Glukhova, Elena (Елена Глухова). 2015. Фауст в автобиографической мифологии Андрея Белого [Faust in the Autobiographical Mythology of Andrei Bely]. In *Андрей Белый: Автобиографизм и биографические практики [Andrei Bely: Autobiographism and Autobiographical Practices]*. Edited by Klaudia Kriveller and Monika Spivak. Sankt-Peterburg: Nestor-Istoriya, pp. 230–50.

Hanegraaff, Wouter Jacobus. 2013. *Western Esotericism: A Guide for the Perplexed*. London: Bloomsbury.

Kazachkov, Sergei (Сергей Казачков). 2015. «Медитацией укрепленные мысли … »: на подступах к пониманию внутреннего развития Андрея Белого ["Thoughts Strengthened by Meditation … ": Coming to Understand the Inner Development of Andrei Bely]. In *Андрей Белый: Автобиографизм и биографические практики [Andrei Bely: Autobiographism and Autobiographical Practices]*. Edited by Klaudia Kriveller and Monika Spivak. Sankt-Peterburg: Nestor-Istoriya, pp. 27–79.

Lagutina, Irina (Ирина Лагутина). 2015. Между «тьмой» и «светом»: воспоминания о Блоке и Штейнере как автобиографический проект Андрея Белого ["Between 'Darkness' and 'Light': Reminiscences of Blok and Steiner as Blok's Autobiographical Project"]. In *Андрей Белый: Автобиографизм и биографические практики [Andrei Bely: Autobiographism and Autobiographical Practices]*. Edited by Klaudia Kriveller and Monika Spivak. Sankt-Peterburg: Nestor-Istoriya, pp. 7–26.

Lavrov, Aleksander (Александр Лавров). 1978. "Мифотворчество 'аргонавтов'" [Mythmaking of the "Argonauts"]. In *Миф—фольклор—литература [Myth—Folklore—Literature]*. Leningrad: Nauka Publisher, pp. 137–70.

Lavrov, Aleksander (Александр Лавров). 1995. *Андрей Белый в1900-е годы [Andrei Bely in 1900-ies]*. Moscow: NLO Publisher.

Mischke, Eva-Maria. 2011. «Apostle of (Self-)Consciousness»: The Figure of Saint Paul in Andrej Belyj's «Istorija Stanovlenija Samosoznajuščej Duši». *Russian Literature* I/II: 89–108. [CrossRef]

Oboleńska, Diana. 2009. *Путь к посвящению: Антропософские мотивы вроманах Андрея Белого [The Way to Initiation: Anthroposophical Motifs in the Novels of Andrei Bely]*. Gdansk: Wydawnictwo Uniwersytetu Gdańskiego.

Odessky, Mikhail (Михаил Одесский). 2011. Стратегия «символизаций» в «Истории становления самосознающей души» [Strategy of Symbolization in The History of the Formation of the Self-Conscious Soul]. *Russian Literature* I/II: 49–60.

Seryogina, Svetlana (Светлана Серегина). 2015. Штейнерианство Андрея Белого: путь к религиозной культуре [The Steinerism of Andrei Bely: The Way to Religious Culture]. In *Андрей Белый: Автобиографизм и биографические практики [Andrei Bely: Autobiographism and Autobiographical Practices]*. Edited by Klaudia Kriveller and Monika Spivak. Sankt-Peterburg: Nestor-Istoriya, pp. 102–15.

Spivak, Monika (Моника Спивак). 2006. *Андрей Белый—мистик и советский писатель [Andrei Bely—Mystic and Soviet Writer]*. Moscow: RGGU.

Spivak, Monika (Моника Спивак). 2011. Андрей Белый в работе над трактатом «История становления самосознающей души» [Andrei Bely in his Work on his Essay *The History of the Formation of the Self-Conscious Soul*]. *Russian Literature* I/II: 1–19.

Stahl, Henrieke (Хенрике Шталь). 2011. Генезис понятия самосознающей души и «История становления самосознающей души Андрея Белого» [Genesis of the Gotion of the Self-conscious Soul and *The History of the Formation of the Self-Conscious Soul* of Andrei Bely"]. *Russian Literature* I/II: 21–37.

Stahl, Henrieke (Хенрике Шталь). 2015. Медитативный опыт Андрея Белого и «История становления самосознающей души» [Meditation Experience of Andrei Bely and *The History of the Formation of a Self-Conscious Soul*]. In *Андрей Белый: Автобиографизм и биографические практики [Andrei Bely: Autobiographism and Autobiographical Practices]*. Edited by Klaudia Kriveller and Monika Spivak. Sankt-Peterburg: Nestor-Istoriya, pp. 80–102.

Steiner GA 148. 2009. Штайнер Рудольф. *Из исследования Акаши. Пятое Евангелие. 18 лекций, прочитанных в 1913–1914 гг. в разных городах [Steiner Rudolf. The Fifth Gospel: From the Akashic Records. 18 Lectures Read in 1913–1914 in Different Cities]*; GA 148. Moscow: Novalis. Available online: http://bdn-steiner.ru/modules.php?name=Ga_Book&Id=148 (accessed on 10 April 2021). (In German and Russian)

Steiner GA 307. 2018. Штайнер Рудольф. *Современная духовная жизнь и воспитание. 14 лекций, прочитанных в Илкли с 5 по 17 августа 1923 г. [Steiner Rudolf. Modern Spiritual Life and Education. 14 Lectures Read in Ilkley from 5 August to 17 August 1923]*; GA 307. Moscow: Enigma. Available online: http://bdn-steiner.ru/modules.php?name=Ga_Book&Id=307 (accessed on 10 April 2021). (In German and Russian)

Steiner GA 142. 2009. Штайнер Рудольф. *Бхагавад Гита и Послания Апостола Павла. 5 докладов, прочитанных в Кельне с 28 декабря 1912 г. по 1 января 1913 г. [Steiner Rudolf. The Bhagavad Gita and the Epistles of St. Paul. 5 Lectures Read in Cologne from 28 December 1912 to 1 January 1913]*; GA 142. Kaluga: Duhovnoe Poznanie. Available online: http://bdn-steiner.ru/modules.php?name=Ga_Book&Id=142 (accessed on 10 April 2021). (In German and Russian)

Szilard, Lena (Лена Силад). 2002. *Герметизм и герменевтика [Hermeticism and Hermeneutics]*. St. Petersburg: Ivan Limbach publisher.

Torshilov, Dmitriy (Дмитрий Торшилов). 2015. «Письмо, написанное в сердцах» в теории слова Андрея Белого и в стихах О. Мандельштама на его смерть ["An Epistle Written in Hearts" in the Theory of the Word of Andrei Bely in the Poems by O. Mandelstam on his Death]. In *Живое слово: Логос—голос—движение—жест [The Living Word: Logos—Voice—Movement—Gesture]*. Edited by Vladimir Feshchenko. Moscow: Novoe literaturnoe obozrenie, pp. 128–36.

Webb, James. 1971. *The Flight from Reason*. London: MacDonald and Co.

Zernov, Nicolas. 1963. *The Russian Religious Renaissance of the 20th Century*. London: Darton, Longman & Todd.

Article

Franny's Jesus Prayer: J.D. Salinger and Orthodox Christian Spirituality

Andrey Astvatsaturov

Department of History of Foreign Literature, Saint Petersburg State University, 190000 Saint Petersburg, Russia; a.astvatsaturov@spbu.ru

Abstract: *The Way of a Pilgrim and The Pilgrim Continues His Way*—is a Russian hesychast text that was first published in 1881 and translated into English in 1931. It has gained popularity in the English-speaking world thanks to J.D. Salinger who mentions and re-narrates it in his stories "Franny" and "Zooey". This reference has often been noted in both critical works on Salinger and studies dedicated to the book *The Way of a Pilgrim*. However, scholars have never actually attempted to fundamentally analyze the textual interconnections between Salinger's stories and the hesychast work. In this article, the text of *The Way of a Pilgrim* is read within the framework of Salinger's stories and is interpreted as being significant for his later texts. From the hesychast book Salinger borrows a number of images and presents its philosophy as a spiritual ideal. At the same time, he approaches it with a certain irony and exposes several pitfalls of incorrectly interpreting the Jesus prayer, as illustrated by Franny, one of Salinger's characters. Having brought to light the nature of Franny's mistakes and her peccant intention, Salinger reestablishes the hesychast ideal and connects it with Søren Kierkegaard's principle of theistic existentialism.

Keywords: hesychasm; Jesus prayer; Russian Orthodoxy; *The Way of a Pilgrim*; elder; Salinger; Franny; Zooey

Citation: Astvatsaturov, Andrey. 2021. Franny's Jesus Prayer: J.D. Salinger and Orthodox Christian Spirituality. *Religions* 12: 555. https://doi.org/10.3390/rel12080555

Academic Editor: Dennis Ioffe

Received: 7 June 2021
Accepted: 17 July 2021
Published: 21 July 2021

Publisher's Note: MDPI stays neutral with regard to jurisdictional claims in published maps and institutional affiliations.

The literary texts of J.D. Salinger (1919–2010) constitute a collision of various philosophical theories and religious practices among which theistic existentialism, psychoanalysis, Zen, and Vedantism occupy a special place. This collision is apparent in his *Nine Stories*. However, its revealing strengthened the feeling of the enigmatic nature of Salinger's texts and often resulted in critics justifying contradictory or at least "different interpretations" (French 1963, p. 91) of his stories. The interpretive paradox was caused, firstly, by the complete absence of any comments by the author himself, and secondly, by his primary focus not on philosophical and religious reflection, but on presenting an exclusively objective world and characters given in specific situational sensations to the reader[1]. In the Glass family stories, Salinger overcomes the enigmatic, supplementing the former concreteness with "problematic" dialogues of his characters who appear to be carriers of quite specific philosophical ideas and religious values. In the stories "Franny" and "Zooey", Salinger actually employs a direct statement, and therefore these stories, to a certain degree, appear to be expanded commentaries to his *Nine Stories*.

It is significant that in the Glass family stories another religious tradition manifests itself—one which was not obvious in *Nine Stories*—hesychasm. In "Franny", the eponymous protagonist Franny Glass tells her boyfriend Lane Coutell about the Russian hesychast book *The Way of a Pilgrim*, which is an instruction in the Jesus Prayer. She enthusiastically narrates the content of this book, and then twice in the story, when left alone, practices the Jesus prayer. In the story "Zooey", Franny's brother Zooey (born Zachary) tells his mother about the book *The Way of a Pilgrim*, but unlike his sister, he does it somewhat ironically. Hesychasm and prayer thus become the focus of Salinger's characters. It will be preliminarily noted here that hesychasm, in the case of Salinger, is controlled by the ideas of Søren Kierkegaard. As a result, in the Glass family stories one finds a combination of

Religions **2021**, *12*, 555. https://doi.org/10.3390/rel12080555

East and West, Heaven and Earth, freedom and necessity, mysticism and rationalism, as well as contemplative life and active life.

"Franny" manifests the apparent irreconcilability of these extremes. In the story, the dialogue between Franny and Lane Coutell occupies a central place. Lane, a Yale student, is a dry, philological analyst who applies classical psychoanalysis to his studies on Flaubert. In his earthiness, attachment to psychic matter, Franny opposes the Jesus prayer. She overcomes the attraction to all things human, which is the hesychastic search for ideal experiential knowledge of God.

However, the practice of the Jesus prayer, as it turns out in "Zooey", does not provide Franny relief. On the contrary, it aggravates her neurosis and results in a mental breakdown. In a conversation with Franny, Zooey offers her a slightly different path, the ideal that Kierkegaard referred to as "ethical". He talks about finding God through an earthly mission. For Franny, her relationship with Zooey becomes a way to find Christ within herself and others, a path that leads from symbolic death to spiritual transformation. In this sense, "Franny" and "Zooey" repeat the logic of *The Catcher in the Rye* that "may be read as a story of death and rebirth" (Miller 1968, p. 15).

As is known, it is thanks to Salinger that the Russian book *The Way of a Pilgrim*, the thematic center of the stories "Franny" and "Zooey", became popular in the English-speaking world. Most critics who attempted to analyze it one way or another, as well as scholars of Salinger, note the important role *The Way of a Pilgrim* plays in his stories.

The *Candid Narrative of a Pilgrim to His Spiritual Father* (*Otkrovennye Rasskazy Strannika duhovnomu svoemu otcu*) was first published anonymously. As Svetlana Ipatova notes, the book's authorship has been a matter of serious debate (Ipatova 2002, p. 301). However, the latest textological studies by Aleksei Pentkovskij (Pentkovskij 2018) convincingly prove that the author of *Candid Narrative* was Arsenii Troepolsky (1804–1870), and it was first published in 1881 in Kazan. In 1883, the second edition by Paisiy Velichkovsky was published, and the third was released in 1884. Subsequently, they were reprinted for Russian emigres by the YMCA-Press in 1930, and later editions by the same publishing house. The English translation of the book *Candid Narrative* by Reginald Michael French was published in 1931 under the title *The Way of a Pilgrim and The Pilgrim Continues His Way*. It was exactly this book that was read by Salinger's characters, members of the Glass family, first by the older brothers, Seymour and Buddy, and then by Zooey and Franny.

The *Candid Narrative*, or *The Way of a Pilgrim*, which consists of seven "stories", was provided to Russian readers by Troepolsky as an anonymous text as if it had no authorship. This was meant to emphasize religious humility and insignificance of the author before people and God. The narrator has no name and is referred to as "Pilgrim", which also accentuates his humility and generalized character of the figure that encompasses the collective experience of spiritual ascension and hesychast communion with God. Remaining an extremely earthly, ordinary layman, he at the same time acquires the common status of a pilgrim, a universal figure, who avoids a specific path.

Salinger could not help but pay attention to the intention of *Candid Narrative* since he himself tried to carefully hide his personality and biography from the public, playing down his personality and bringing it closer to anonymity in an act of humility. Salinger presented his personality indirectly, only in his works. He remained "in the public eye by withdrawing from it" (Alexander 1999, p. 26).

Unlike most hesychast texts, the "stories" of *The Way of a Pilgrim* are addressed not to ascetics, but "to the laity" (Ipatova 2002, p. 326), to a broad audience. Although the stories contain fragments of works by the authors of *Philokalia* (St. John of the Ladder, St. Symeon the New Theologian, Theoliptos, Nikiphoros the Monk, St. Gregory of Sinai, Kallistos and Ignatios the Xanthopoulos), the seventh or last story appears to be a theological dispute. In general, their author avoids theological intonation and adheres to the form of an ordinary oral conversation. In *The Way of a Pilgrim*, there are scenes that should be read allegorically, such as the fire in the pilgrim's house (an allegory for the transitory nature of the earthly), the attack of the wolf (for temptation), the willingness to pledge one's passport (willingness to sacrifice the worldly for the spiritual, the desire to put one's duty before God above civic duty), and the healing of a woman, the wife of the ruler, choking on a bone (to spiritual cleansing). These texts do not appear to be a pure allegory like John Bunyan's (1628–1688) book *The Pilgrim's Progress* (1678)[2]. This is the space of everyday events, concrete everyday circumstances, and random experiences. Events take place not in a timeless, conventional, space, but in Russia, in a specific historical epoch (the beginning of the 19th century) where the reader encounters specifically Russian types shown "in their concreteness" (Ipatova 2002, p. 324), representing distinctive social castes (landowner, priest, elder, teacher, professor, officer, merchant, fugitive soldiers, etc.). Concreteness, everyday life and realistic eventfulness are meant to create the illusion not of fiction, but of factuality, truthfulness, so that a didactic text is perceived as living proof. The practice of the Jesus prayer, conveyed in great detail, and the spiritual transformation that followed it would be seen in this light as quite attainable. Furthermore, it is addressed not to a narrow circle of monks, but for every person, ordinary laymen.

The "stories" of *The Way of a Pilgrim* are intended for ordinary people to read, for those not called upon to serve God, not chosen, and those who have not given up their earthly life. The stories retain a certain extent of complexity in regard to ascetic questions and at the same time present them as being inextricably linked to everyday situations and everyday experiences. It is likely that this is exactly what attracted Salinger. The worlds he created, peopled by characters and ordinary Americans busy seeking God or experiencing temptations or the despair of God-forsakenness (often unconscious), are also quite specific everyday spaces. Here, it is the singularity of any object that is always markedly emphasized, and experiences that are in essence religious are disguised in the shape of specific sensations, reactions or judgments. That, however, like in the *The Way of a Pilgrim*, does not exclude the allegories or allegorical gestures. The sun, sun rays from the window, and sunbeams in "Franny" and "Zooey" always symbolize the light that comes from God. The chicken sandwich that Franny orders at Sickler's, in its turn as some critics point out (Bryan 1961, p. 228; Lencina 2015, p. 118), symbolizes the Eucharist. Zooey, sitting in the sun and claiming that he always carries the sun with him is associated with Christ. The rationalist Lane Coutell "allegorically" turns his back on the rack with free Christian Science brochures by Mary Eddy.

The stories of *The Way of a Pilgrim* could have attracted Salinger due to an amusing combination of ascetic theory and practice, metaphysical reasoning, and concrete, everyday life, pieces of advice on how to transform one's theoretical understanding of prayer into spiritual action. As a matter of fact, the contradiction between "theory" and "practice" is indicated as early as in the first pages of *The Way of a Pilgrim*. The pilgrim seeks to master the ceaseless prayer and he seeks the answer to this vital question. However, every time he reads spiritual books or listens to sermons, he encounters only a general reasoning and a general understanding of prayer. All further development of the plot becomes a response to its initial search. Salinger, just like his characters, was interested specifically in those philosophical and religious theories that had practical efficacy and offered concrete practical ways to resolve spiritual problems.

However, it seems much more significant to us that in the stories of *The Way of a Pilgrim* Salinger was most likely attracted by a detailed representation and study of the nuances of experience, which was characteristic of ascetic literature. For a secular writer who meticulously constructs the psychological world of his characters, this text was an invaluable source of essential psychological patterns.

The Way of a Pilgrim is a hesychast text that asserts the possibility and necessity of contemplation of God for everyone, finding the kingdom of God within oneself, an internal connection with Christ and the ability to perceive the foundations of all things in their original design. This goal is achieved through the practice of the Jesus prayer that is constituted by the words "Lord Jesus Christ, have mercy on me". It must be performed incessantly so that it becomes an intrinsic part of one's being, and a person likens to the Holy Spirit who constantly prays. The oral prayer should transform into the prayer of the heart, and, accordingly, the mind should blend into the heart. The prayer implies humility and reveling one to God, reaching the awareness of the fact that God sees you. "Stand in prayer facing the God," notes Ignatij Bryanchaninov, "as if he sees you, stand like a criminal" (Bryanchaninov 2021, p. 275). Here it is essential to note that for the pilgrim, the name of God comprises an idiosyncratic power (Ugolnik 2016, p. 101; Pentkovskij 2018, p. 365; Bryanchaninov 2021, p. 287). The pilgrim quotes the ascetics who assert that invocation of the name of God kills not only passions but their operation as well.

At the very beginning of *The Way of a Pilgrim*, the pilgrim, inspired by the words of the Apostle Paul—"pray incessantly"—sets off on a journey in the hope of finding someone who could explain them to him. The theological books that he reads and people who he meets cannot explain the essence of prayer to him. Finally, he finds an elder who becomes his mentor, teaches him, and introduces him to *Philokalia*. Under the elder's guidance, the pilgrim begins to practice the "intellectual" prayer (*umnaya molitva*) and feels the inner changes happening within him. The elder dies but later comes to the pilgrim in his dreams and provides him with advice on how to do the right thing. Then, the pilgrim goes on a journey, visits villages, cities, meets people, including those who are skeptical about the Jesus prayer. Humbly, as it is proper for a person who has dedicated himself to God, he endures beating, persecution, and humiliation. However, more importantly, this is his inner spiritual journey that has nothing to do with his geographical logistics. The pilgrim performs a solo, independent ascension up the symbolic ladder, one that is called "a gradual transformation of the inner space into a temple" (Bagdasarov 2011, p. 91). He finds Christ, and in this discovery, he finds his true self. Prayer through practice becomes heartfelt. He stops articulating it and hears his heart resonating. The mind and heart that were separated previously, now are united and purified from passions. The changes occurring within him transform his perception of the world, which he now sees as fully divine. The pilgrim often feels the invisible light radiating from things. He reaches the level where he does not differentiate them and comprehends the wisdom of all earthly creatures.

Franny, in the eponymous story, retells the content of *The Way of a Pilgrim* to Lane, focusing on the episode of hospitality. Earlier, she goes to the ladies' room and, having locked herself in a booth, takes out *The Way of a Pilgrim* and starts reading it. The prayer itself gives her a chance to get away from the fuss and noise of life around her that haunts her ("It's getting so noisy in here I can hardly hear myself think") (Salinger 1961, p. 4) and to overcome her own passions that bind her to the world and to people ("I'm sick of just liking people") (Salinger 1961, p. 20), her false wisdom ("I'm 'just so sick of pedants and conceited little tearer-downers I could scream" (Salinger 1961, p. 17). Finally, she can now escape from her own superficial ego in which everything is concentrated: ("I'm just sick of ego, ego, ego. My own and everybody else's". (Salinger 1961, p. 29). Michael Katz notices that "It is the pilgrim's spiritual freedom that attracts the heroine to this extraordinary model" (Katz 2012, p. 539). Thus, the hesychast journey in Salinger's story is defined as an alternative to primitive conformism and as the only way to salvation of the soul.

It can be easily seen that Salinger borrows certain images, situations, and characters from *The Way of a Pilgrim* and transplants them into his stories. In the second story of *The Way of a Pilgrim*, we encounter a steward of Polish origin who speaks skeptically about *Philokalia* and the Jesus prayer: "That, if I may say so, is enough to drive you mad. Besides, it's bad for your heart (*The Way of a Pilgrim and the Pilgrim Continues His Way* 2012, pp. 74–75). Lane Coutell responds to Franny's story and her explanation of the Jesus prayer in a similar way and in almost the same terms: the prayer, in his opinion, can cause heart problems: "I mean what is the result that's supposed to follow? All this synchronization business and mumbo-jumbo? You get heart trouble? I don't know if you know it, but you could do yourself somebody can do himself a great deal of real . . . " (Salinger 1961, p. 39).

In the fourth story, when a pilgrim intends to go to Jerusalem, people insist on a deaf old man joining him as a companion. The deafness of the old man essentially symbolizes his detachment from the noise of the world. A man the pilgrim knows tells him: "He belongs to this town, he's a good old man, and what's more he is quite deaf. So much so that however much you shout, he can't hear a word. If you want to ask him anything you have to write it on a bit of paper, and then he answers" (*The Way of a Pilgrim and the Pilgrim Continues His Way* 2012, p. 87).

In the story *Raise High the Roof Beam, Carpenters*, there is also a remarkable figure of a deaf old man who takes a taxi with everyone else. Whenever asking him, Buddy writes his questions on paper. Passions run high around the old man, and yet he is completely detached, in an ascetical way, from what is happening. He is in a state of absolute peace and happiness. In the story, he becomes a companion and a living embodiment of a spiritual ideal for Buddy Glass.

By offering the reader the hesychast path as an ideal, Salinger at the same time draws the reader's attention to the fact that the Jesus prayer for some reason does not help Franny, it does not come to her with ease and leads to a nervous breakdown.

By shaping the horizon of the reader's expectation in the form of the spiritual ideal of hesychasm, from the very start, Salinger undermines this ideal. However, he does so only to reaffirm it. In "Franny" and "Zooey", the reader becomes a witness to the repeated mention of God's name (God, Christ). It parallels the manifold articulation of the fact that God's name holds miraculous power in *The Way of a Pilgrim*: "The Name of Jesus Christ invoked in prayer contains in itself self-existent and self-acting salutary power" (*The Way of a Pilgrim and the Pilgrim Continues His Way* 2012, p. 213). Franny makes it apparent: "< . . . > I mean all these really advanced and absolutely unbogus religious persons that keep telling you if you repeat the name of God incessantly something happens" (Salinger 1961, p. 39). In "Franny" and "Zooey", the divine name is mentioned on almost every page—but exclusively as an interjection or a swear word, as Buddy Glass notes, "as a familiar healthy American expletive" (Salinger 1961, p. 48). Salinger is deliberately ironic about hesychasts, implying that the repeated use of interjections cannot lead to internal change, and quantity will not transform into quality. Perhaps the repetition of the prayer will not lead to anything either, or the problem here lies not in practice but in the intention of the one who performs it.

What appears to be a factor moderating the hesychast pathos in "Franny", is the place where the heroine resorts to *The Way of a Pilgrim*. As a "secluded place" that hesychast theologians prescribe for prayer, Franny chooses the back booth in the women's restroom. Franny does exactly what the pilgrim and his teachers recommend to anyone who is praying: she finds a "secluded place", sits down, closes her eyes and steeps her thoughts into the "void-like black" (Salinger 1961, p. 22). It is worth noting that the setting chosen for the Jesus prayer can hardly be considered appropriate, it is rather ironically offensive to God, as well as the use of his name as an interjection. However, we can emphasize that what matters is not so much the prayer itself but the performer's attitude and state of mind.

The composition of "Franny" also accentuates the controversial nature of the ideal chosen by the heroine, or rather the ideal as presented in her mind. For the sake of discussion, the story can be divided into two parts, which are isomorphic, and are reflected in each other as in mirrors. In the first part, Lane shares with Franny his quite primitive psychoanalytic version of Flaubert. Franny does not approve of the very structure of his thinking; she clears her throat several times and interrupts Lane with inappropriate remarks regarding food and drink. In the second part, Franny in turn tells Lane about the pilgrim, and now he disapproves of the direction of her thoughts. He eats frog legs and constantly interrupts her with remarks about food. Warren French rightly points out that "One reason "Franny" is an unusually well-balanced satire is that Franny's absorption in the Jesus prayer appears as egotistical as Lane's in his Flaubert paper" (French 1963, p. 143). At first glance, the compositional symmetry creates binaries: psychoanalysis— hesychasm, intellect—spirit, conformism—revolt. However, what we see here is, in fact, not opposition but rather the equation of things that seem irreconcilably opposite. In all his works, Salinger treats classical psychoanalysis ironically, and it would definitely be fair to assume that he presents hesychasm as its spiritual alternative. However, in both cases, we deal with the practice exercised by ethically incompetent individuals. In Franny's case both psychoanalysis and the hesychast Jesus prayer are only designed to soothe her, to help relieve her neurotic intoxication.

For Zooey, psychoanalysis mitigates neurosis and helps one adapt to society and primitive mass culture. He tells his mother: "You just call in some analyst who's experienced in adjusting people to the joys of television, and *Life* magazine every Wednesday, and European travel, and the H-Bomb, and Presidential elections, and the front page of the *Times*, and the responsibilities of the Westport and Oyster Bay Parent-Teacher Association and God knows what else that's gloriously normal-you just *do* that . . . " (Salinger 1961, p. 108).

He is just as ironical in regard to the Jesus Prayer practiced by Franny: "And the main idea is that it's not supposed to be just for pious bastards and breast-beaters. You can be busy robbing the goddam poor box, but you're to say the prayer while you rob it" (Salinger 1961, p. 113).

As we will soon see, Zooey neglects neither psychoanalysis nor hesychasm. Dealing with Franny's case, he brings to light a characteristic common for psychoanalysis and hesychasm—the absence of a person's ethical choice and genuine religiosity. Without religious humility as an initial intention, it is impossible to set out on the practices of psychoanalysis and hesychasm. That is why, according to Zooey, Franny does not feel the purpose of the Jesus prayer while practicing it.

In hesychasm, the beginning of the spiritual journey is inseparable from religious humility. It manifests itself in finding a teacher and subjecting oneself to his will, in reading spiritual literature, in admitting that one is wrong before God and people (the latter is as essential). In the hesychast books, the necessity of submission to the teacher as an act of humility is constantly emphasized. *The Way of a Pilgrim* is also notable in this regard. It is exactly the search for teachers and the right books that the pilgrim begins his journey with It is essential that from the very beginning, humility is set as a goal.

The search turns out to be long and difficult for the pilgrim. Neither the books that he finds nor the people he meets, even those from the clergy, can guide him and explain the essence of the Jesus prayer. There are very few genuine teachers in the world, as the author of *The Way of a Pilgrim* explains to us, and most of those who claim to be teachers are incompetent. However, it does not mean that one should give up on the attempts to find God and a true mentor. In the end, having traveled for a long time, the pilgrim after all finds an elder who becomes his teacher and explains the essence of the Jesus Prayer to him. As a matter of fact, the most important purpose of *The Way of a Pilgrim*, as well as of *Philokalia*, is to affirm the importance of a spiritual teacher (Ugolnik 2016, p. 106).

In turn, Franny rejects the prayer and ends up praying without permission and religious humility. She compares Lane to the college professors who, according to her, are all inauthentic without exception. Moreover, she indirectly rejects her genuine teachers Seymour and Buddy when she says that she took *The Way of a Pilgrim* from the library—whereas in reality, she found it on the desk of the late Seymour. Zooey attracts attention to this seemingly insignificant lie, and it infuriates him:

"He said she got it where?"

"Out of the library. At college. Why?"

Zooey shook his head, and turned back to the washbowl. He put down his shaving brush and opened the medicine cabinet.

"What's the matter?" Mrs. Glass demanded. "What's the matter with that? Why such a look, young man?"

< . . . > "I asked you a question, young man. Why am I so stupid? *Didn't* she get that little book out of her college library, or what?"

"No, she didn't, Bessie", Zooey said, shaving. "That little book is called 'The Pilgrim Continues His Way', and it's a sequel to another little book, called 'The Way of a Pilgrim', which she's also dragging around with her, and she got *both* books out of Seymour and Buddy's old room, where they've been sitting on Seymour's desk for as long as I can remember. Jesus God almighty".

"Well, don't get abusive about it! Is it so *terrible* to think she might have gotten them out of her college library and simply brought them"

"Yes! It *is* terrible. It is terrible when both books have been sitting on Seymour's goddam desk for *years*. It's depressing". (Salinger 1961, p. 101)

By unmasking Franny's lie, Zooey accurately discovers her wrong intention, her initial unpreparedness for the spiritual path, her lack of proper humility. A little later, he announces to Franny that she generalizes on the teachers' account and that he used to know a professor who was a great teacher and a great scientist: "I agree with you about ninety-eight per cent of the issue. But the other two per cent scares me half to death. I had one professor when I was in college—just one, I'll grant you, but he was a big, big one—who just doesn't fit in with anything you've been talking about" (Salinger 1961, p. 161).

Like a pilgrim, he notes that there are indeed very few true teachers, but this does not mean that they do not exist and that one should not search for them. In fact, like the author of *The Way of a Pilgrim*, Zooey reaffirms the importance of the role of the teacher in a world where there are almost no true teachers anymore. Here one cannot help but agree with Elvira Osipova who observes that Zooey functions as an elder for Franny (Osipova 2018, p. 190). Furthermore, Zooey reaffirms the importance of Seymour as a teacher (guru, elder). Ihab Hassan notes "And the success of Zooey heralds both the defeat and the apotheosis of Seymour: defeat because the youngest of the Glass children has at last achieved a measure of independence from the guru of the house, and apotheosis because this is precisely what Seymour would have wished" (Hassan 1963, p. 12). However, it should be noted that Zooey, acting as both a psychoanalyst and an elder, ironically distances himself from these missions. He emphasizes that he has put on a mask, that he is playing another role as a professional actor. After all, his task is not so much to instruct Franny in the spirit of hesychasm, but rather to make her embrace her earthly vocation—a theatre actor. It is only in the light of this earthly destiny that the meaning of the Jesus prayer can be understood as a spiritual journey, not a mere practice.

In addition to renouncing her mentor, Franny makes another mistake that points to the lack of religious humility in her. Ignatij Bryanchaninov considers "the rejection of unforgiveness and condemnation of loved ones" the first step in preparation for the prayer (Bryanchaninov 2021, p. 274). He argues that humiliation, malediction, and contempt for one's neighbor entail hatred. This is a manifestation of pride. As we can see from Franny's dialogue with Lane, this is what Franny is doing. She condemns, blames, and insults those around her: Lane, college professors, students, actors with whom she has to perform. Zooey points out her mistake, which is incompatible with the essence of prayer, and while doing it he refers to the priority of the mentors (Seymour and Buddy):

"But what I don't like—and what I don't think either Seymour or Buddy would like, *either*, as a matter of fact—is the way you talk about these people. I mean you don't just despise what they represent—you despise them". (Salinger 1961, p. 162)

Bryanchaninov notes that the neighbor is the image of God and that "Christ accepts our actions in relation to the neighbor as if they were done in relation to him" (Bryanchaninov 2021, p. 269). By despising people, Franny does not see the likeness of God in them, does not discern Christ in them. As can be recalled, in response Zooey gives Franny a Fat Lady monologue, which John Antico considers an "absurd" image (Antico 1966, p. 336). The image of the Fat Lady was invented by his brother and teacher Seymour. Seymour presents it to Zooey at the moment when Zooey felt contempt for those around him and refused to clean his shoes because of this:

"The studio audience were all morons, the announcer was a moron, the sponsors were morons, and I just damn well wasn't going to shine my shoes for them, I

told Seymour. I said they couldn't see them *anyway*, where we sat. He said to
shine them anyway. He said to shine them for the Fat Lady. I didn't know what
the hell he was talking about, but he had a very Seymour look on his face, and
so I did it. He never did tell me who the Fat Lady was, but I shined my shoes
for the Fat Lady every time I ever went on the air again—all the years you and I
were on the program together, if you remember. I don't think I missed more than
just a couple of times". (Salinger 1961, p. 200)

Zooey likens a "symbol for all that is repulsive" (Gettis 1978, p. 128), an ugly Fat Lady
who is sick with cancer, on the one hand, to any weak and unattractive person, and on the
other hand, to Christ who manifests himself in this person. Discovering Christ in oneself
and in others is the real goal of the hesychast prayer.

Franny's failure with the Jesus prayer, which she performs by forcing herself and
results in her fainting and a nervous breakdown, also stems from the fact that Franny
is deprived of love for God and does not even realize it. In the second part of *Pilgrim
Continues his Way*, at a certain point, the pilgrim gets lazy, unwilling to pray, and feels
that prayer comes to him with great effort. He decides to confess his sins and turns to the
confessor handing him a paper with all his sins listed. The confessor rejects this confession
as inauthentic and tells the pilgrim that he did not write the main thing:

> "You have not acknowledged, nor written down, that you do not love God, that
> you hate your neighbour, that you do not believe in God's Word, and that you are
> filled with pride and ambition". (*The Way of a Pilgrim and the Pilgrim Continues
> His Way* 2012, p. 165)

These words perplex the pilgrim but soon he would understand their meaning.

In a conversation with Zooey, Franny confesses her sins and weaknesses, but Zooey
draws her attention to the fact that she does not really love God and Jesus Christ as his
embodiment. He recalls that as a child, Franny denied Christ after she learned that he
overturned the tables in the synagogue, threw idols around, and uttered words about the
birds of the heavens, placing them beneath a human in terms of hierarchy. As it turned
out, Jesus Christ did not align with her sentimental idea of him. She judges him and fell
out of love with him. Zooey explains that this rejection of God remained with her since
people do not change much from childhood. He encourages her to see God as he is and to
pray to the real God, not a sentimental version of him: "If you're going to say the Jesus
Prayer, at least say it to *Jesus*, and not to St. Francis and Seymour and Heidi's grandfather
all wrapped up in one. Keep *him* in mind if you say it, and him only, and him as he was
and not as you'd like him to have been" (Salinger 1961, p. 169). Zooey considers Christ to
be the least sentimental and the most genuine teacher. He sees the purpose of the Jesus
prayer in passing this very knowledge to people.

Thus, Franny makes a mistake about which the elder tells the pilgrim during their first
meeting—the mistake is her seeking to measure God by a human yardstick. She projects
her sentimental ideas about God onto Jesus Christ, thereby unconsciously manifesting
pride.

The sentimental sensualism that colors Franny's perception of Christ is also evident in
her attitude to the stories of the pilgrim. Zooey unambiguously points it out when retelling
the contents of the book to his mother:

> "He is a very simple, very sweet little guy with a withered arm. Which, of course,
> makes him a natural for Franny? With that goddam Bide-a-Wee Home heart of
> hers". (Salinger 1961, p. 110)

In "Zooey", Franny's sentimentality and sensualism are emphasized by her treatment
of Bloomberg the cat. This treatment even makes up a small separate line in the story,
which appears to be a kind of derogatory commentary to Franny's character that implicitly
presents the heroine's inner world as sentimental and aesthetic. Having woken up, Franny
immediately begins to show sentimental feelings to the cat by caressing and kissing him.
Bloomberg reacts with hostility, but Franny is convinced that the cat enjoyed it.

""Good *morning*, Bloomberg dear!" she said, and kissed him fervently between the eyes. He blinked with aversion. "Good *morning*, old fat smelly cat. Good morning, good morning, good morning!" She gave him kiss after kiss, but no reciprocal waves of affection rose from him. He made an inept and rather violent attempt to cross over to Franny's collarbone. He was a very large mottled-gray 'altered' tomcat. "Isn't he being affectionate?" Franny marvelled. "I've never *seen* him so affectionate". She looked at Zooey, possibly for corroboration, but Zooey's expression, behind his cigar, was noncommittal" (Salinger 1961, p. 129).

Her sentimental and aesthetic concept of the cat appears to be at odds with the real image of the earthly creature, just as well as her concept of Christ does not align with his real essence. Franny continues to caress Bloomberg but once she starts processing the meaning of Zooey's words, she pays less and less attention to the cat. When the moment of repentance comes, when she talks about ruining Lane's evening (not authentic repentance, but a sentimental and sensual one), she hugs Bloomberg again:

"Here there was a marked break in her voice, and she began to be very attentive to Bloomberg again. Tears, presumably, were imminent, if not already on the way". (Salinger 1961, pp. 149–50)

Zooey pays attention to this sentimental treatment of Bloomberg and notes the following:

"Say your prayer if you want to, or play with Bloomberg, but give me five minutes of uninterrupted silence". (Salinger 1961, p. 156)

The Jesus Prayer and playing with Bloomberg are thus ranked together. Zooey's remark reveals the common ground in those actions from the perspective of the driving force. It is the sentiment of a person whom Kierkegaard identified as "aesthetic". When Zooey tells Franny directly about her sentimental mistake and she understands the meaning of his words, Bloomberg is no longer there:

"Franny was now facing directly into the sound of Zooye's voice, sitting bolt upright, a was Kleenex clenched in one hand. Bloomberg was no longer in her lap". (Salinger 1961, p. 165)

Franny's playing with the cat, thus, explicates the corrupt basis of her religious impulse and attitude to the spiritual world: Franny succumbs to a dangerous temptation, against which the author warns the reader at the very beginning of *The Way of a Pilgrim*. An elder who appeared in a dream of the wanderer who had just been robbed by fugitive soldiers, told him:

"Let this be a lesson to you in detachment from earthly things for your better advance towards heaven. This has been allowed to happen to you to save you from falling into the mere enjoyment of spiritual things". (*The Way of a Pilgrim and the Pilgrim Continues His Way* 2012, p. 22)

Franny's religiosity is explained precisely by her "spiritual lust", a sensual impulse, while a Hesychast who prays should free himself from his will and remain in a state of dispassion.

It is also interesting that while explaining to Lane the essence of the Jesus prayer, Franny correlates it with Eastern practices. In *The Way of a Pilgrim*, the same is done by the steward who, essentially, is an opponent of the Jesus prayer. The steward compares the ascetic hesychasts to the magicians of India, and that results in a humble but decisive objection from the pilgrim. The magicians who distorted the meaning of God's message, in his opinion, cannot be equated with the holy fathers who fulfill God's will[3].

Therefore, Franny clearly misreads the stranger's stories, she follows the prayer literally but not its essence, and distorts its meaning with the wrong intention. It is exactly for this reason that she does not achieve the purpose of the prayer, it does not bring her peace; on the contrary, it aggravates her neurotic state. The prayer that Franny performs, as well as her state of mind before prayer, are in essence pride and denial of Jesus Christ. This is why Franny symbolically rejects the chicken sandwich and chicken broth. The reference to a chicken sandwich in several of Salinger's texts (*The inverted Forest*, *Just before the War with Eskimos*), as well as that of chicken broth in "Zooey", degrade the signs of the Eucharist to the level of everyday life. It is a transfiguration of "a mundane situation into the Holy Sacrament" (Bryan 1961, p. 228).

The failure of Franny's "project" of the Jesus Prayer does not mean a total denial of it. Silence—as an extremely important element of hesychast asceticism—increases and becomes almost tangible as the story develops. Ihab Hassan points out that silence in the Glass Stories is not only a theme "but also a principle of their form" (Hassan 1963, p. 18)[4] Pauses appear more and more often: the word 'silence' itself is repeated many times until silence captures the whole world of the story and the characters peopling it. Having calmed down and come to terms with herself, Franny lies in silence for a long time before falling asleep.

The hesychast ideal is almost achieved but only after Franny comes to the realization of the need to fulfill her earthly mission. It is here that Salinger supplements hesychasm with the religious and ethical concepts of Kierkegaard.

Salinger, as it is known, did not leave us any direct evidence of his philosophical interests. However, traces of his interests are obvious and have been noted by several researchers. One of his interests, as indicated by Thomas Brinkley (Brinkley 1976, pp. 49–55), was the philosophy of Danish philosopher Søren Kierkegaard (1813–1855). It played an important part in Salinger's worldview and its beliefs are reflected in his early texts Judging by the way Salinger describes his characters, he was attracted to Kierkegaard's philosophy of the concept of the three stages of the human "self" (aesthetic, ethic, religious). The character of Salinger's early works is precisely an "aesthetic" person who experiences despair and tragedy. The idea of the three stages of human existence is substantiated by Kierkegaard in his treatise *Either/Or* (1843). An "aesthetic" person is an artist of life, immersed in pleasure, which can be sensual, intellectual, and even spiritual. An "aesthetic" person strives to control life, but suffers defeat on this path and experiences despair, a feeling of his own powerlessness. It is precisely this which occurs with Salinger's characters. The next stage is ethical. It requires a person to choose himself, to acquire his own existence, to compile the world, which at the aesthetic stage was disintegrated into fragments. *Either/Or* concludes with a fragment that describes the premonition of the religious stage, which presupposes a leap of faith, paradox, belief in the absurd and an individual's intimate connection with God. The idea of a religious life is elaborated upon in subsequent theological works by Kierkegaard.

In *Either/Or*, Kierkegaard distinguishes between the truly religious and the mystical. A mystic, in his understanding, is a kind of an aesthetical, sensual person, and in this context, it becomes clear why Franny's spiritual impulse is sensual, and her attitude to religion is sentimental. However, there is no obvious contradiction to the hesychast problematics. Contradiction and challenge arise when we recall Kierkegaard's words that the mystic is not "concrete" to himself and God (Kierkegaard 1987, p. 223). Zooey constantly emphasizes his concreteness and embodiedness. He smokes a cigar and claims it to be the ballast that binds him to the earthly world. He constantly emphasizes the need for an earthly mission for Franny. First, he talks about a film he starred in, where a farmer who has betrayed his calling dies, and then Zooey encourages Franny to return to the theater. Here, the Protestant ideal of active life comes into play, which opposes the ideal of contemplative life, the ideal paramount for hesychasm. We are also dealing with choice, a person's transition to the stage that Kierkegaard calls "ethical". "Franny,—as Brinkley puts it—at this point seems to be "choosing" wrongly, i.e., in the direction of an airy, insubstantial possibility in the realm of mysticism" (Brinkley 1976, p. 95). The ethical stage is where a person finds himself, takes a deliberate step that is inextricable with necessity. Here an individual sees himself as a concrete embodiment of God, as a product of the external world, one with certain talents and capabilities (Kierkegaard 1987, pp. 225–26). Kierkegaard talks about a person finding himself in the concrete and the internally infinite. According to Zooey, for Franny, that thing is everyday life and acting:

"I'd like to be convinced—I'd love to be convinced—that you're not using it as a substitute for doing whatever the hell your duty is in life, or just your daily duty". (Salinger 1961, p. 169)

Having found herself in concreteness and singularity, Franny can find Christ in herself, stand up before God and achieve religious humility. It is then that the true meaning of the Jesus prayer that Zooey talks about will manifest itself. A person will be endowed with the understanding of Christ.

Thus, Salinger establishes an ethical way of personality development for the implicit reader. The end of the story, however, remains open. Franny experiences joyfulness and peacefulness, taking pleasure in the dial tone which to her sounds like a "possible substitute for the primordial silence itself" (Salinger 1961, p. 202). It appears that she is coming to terms with a world that previously seemed disorganized to her. Here, there are two possible interpretations: (1) she finds herself on the threshold of a spiritual transformation, on the threshold of the ethical stage which Kierkegaard discusses; (2) she remains within the aesthetic stage. And, if this is so, then the new pleasure she is experiencing should be understood as her previous aesthetic attitude towards the world, in which only aesthetic assessments have changed.

In this manner, Salinger does not undermine the ideal of hesychasm, as it may seem at first. Rather, he constructs a new ideal in which hesychasm is one of the possible elements. Salinger reconciles it with Kierkegaard's concept of the ethical, and by doing so he unites contemplative life and active life, the world of the spirit and the world of the body, East and West. The hesychast ideal is combined with that of Kierkegaard, the contemplative life.

Funding: This research received no external funding.

Conflicts of Interest: The author declares no conflict of interest.

Notes

[1]   The later story "Teddy" is an exception in this sense.

[2]   Regarding the opinion that Bunyan's book was a possible source for the author of *Candid Stories*, see: S. Ipatova (Ipatova 2002, pp. 312–15, 323–24), and A. Pentkovskij (Pentkovskij 2018, p. 347).

[3]   Regarding this paradoxical statement, see Z. Ugolnik (Ugolnik 2016, p. 115).

[4]   About the problem of how silence becomes a principle of poetic form, see (Ioffe 2005).

References

Alexander, Paul. 1999. *Salinger. A Biography.* New York: St. Martin Griffin, 352p.

Antico, John. 1966. The Parody of J. D. Salinger: Esme and the Fat Lady Exposed. *Modern Fiction Studies* 12: 325–40.

Bagdasarov, Roman. V. 2011. *Mistika Russkogo Pravoslaviya.* Moskva: Veche, 320p.

Brinkley, Thomas Edwin. 1976. J. D. Salinger: A Study of His Eclectism—Zooey as Existential Zen Therapist. Ph.D. thesis, The Ohio State University, Columbus, OH, USA; 298p.

Bryan, James E. 1961. J. D. Salinger: The Fat Lady and the Chicken Sandwich. *College English* 23: 226–29. [CrossRef]

Bryanchaninov, Ignatij. 2021. *Slovo o cheloveke. O chudesakh i znameniyakh.* Moskva: Eksmo, 640p.

French, Warren. 1963. *J. D. Salinger.* New Haven: Twayne Publishers, 191p.

Gettis, Alan. 1978. Psychotherapy and the Fat Lady. *Journal of Religion and Health* 17: 127–29. [CrossRef] [PubMed]

Hassan, Ihab. 1963. Almost the Voice of Silence: The Later Novelettes of J. D. Salinger. *Wisconsin Studies in Contemporary Literature* 4: 5–20. [CrossRef]

Ioffe, Denis. 2005. Liki tishiny. Novejshaya istoriya russkogo eksperimental'nogo stihoproizvodstva SUB SPECIE HESYCHIAE: tri skhematicheski vzyatye poetiki umnogo delaniya. *Russian Literature* LVII: 293–314.

Ipatova, Svetlana. A. 2002. 'Otkrovennye Rasskazy Strannika Duhovnomu Svoemu Otcu': Paradigmy Syuzheta. *Hristianstvo i Russkaya Literatura* T.4: 300–35.

Katz, Michael R. 2012. The Fiery Furnace of Doubt. *Southwest Review* 4: 536–45.

Kierkegaard, Søren. 1987. *Either/Or. Part II.* Edited and Translated with Introduction and Notes by Howard V. Hong and Edna H. Hong. Princeton: Princeton University Press, 472p.

Lencina, Eva. 2015. Cartografía de un recorrido espiritual: el símbolo de la alimentación en la obra de J.D. Salinger. *La Palabra* 27: 109–23. [CrossRef]

Miller, James E. 1968. *J. D. Salinger*. Minneapolis: University of Minnesota Press, 48p.

Osipova, Elvira. F. 2018. Selindzher, Dostoevskij i vostochno-hristianskaya tradiciya. *Literatura Dvuh Amerik* 4: 184–94. [CrossRef]

Pentkovskij, A. M. 2018. Istoriya teksta i avtor Otkrovennyh rasskazov strannika. In *Bogoslovskie Trudy*. Moskva: Izdatel'stvo Moskovskoj Patriarhii Russkoj Pravoslavnoj Tserkvi, pp. 343–448.

Salinger, Jerome D. 1961. *Franny And Zooey*. New York: A Bantam Book.

The Way of a Pilgrim and the Pilgrim Continues His Way. 2012. R. M. French, trans. Introduction by Metropolitan Anthony of Sourozh. London: Society for Promoting Christian Knowledge, 261p.

Ugolnik, Zachary. 2016. Internal Liturgy: The Transmission of the Jesus Prayer in the 'Philokalia' and 'the Way of a Pilgrim (Rasskaz Strannika). *Religion & Literature* 48: 99–13.

Article

Merezhkovsky's Neo-Christianity of the Third Testament: From Symbolist Historiosophy to Radical Politics

Vadim Polonsky

Gorky Institute of World Literature of the Russian Academy of Sciences, 121069 Moscow, Russia;
v.polonski@imli.ru

Abstract: This article places Dmitry Merezhkovsky's Chiliastic concept of Three Testaments into a unified structure. The author analyzes the writer's integral system of Christological, anthropological, and historiosophicidiomyths and meta-symbols. He studies the religious, philosophical, and aesthetic genesis of the semantic transformation of traditional theological constructions and the doctrinal compilation of Russian *fin de siècle* culture dominant elements. It is shown how religious Modernist mythmaking alters political reality in Merezhkovsky's mind and draws him towards radical ideologies of the extreme left and right.

Keywords: Merezhkovsky; Third Testament; symbolism; chiliasm; New Religious Consciousness

Citation: Polonsky, Vadim. 2021. Merezhkovsky's Neo-Christianity of the Third Testament: From Symbolist Historiosophy to Radical Politics. *Religions* 12: 456. https://doi.org/10.3390/rel12070456

Academic Editor: Dennis Ioffe

Received: 6 May 2021
Accepted: 16 June 2021
Published: 22 June 2021

Publisher's Note: MDPI stays neutral with regard to jurisdictional claims in published maps and institutional affiliations.

1. Introduction

Dmitry Merezhkovsky lived from 1865 to 1941 and was one of the founders of Russian Symbolism. His late works were framed by an integral religious and philosophical concept. Gradually forming in the second part of the 1890s, it was completed in the beginning of the new century and would survive without any notable changes until his final days.

In Russia, this type of *Weltanschauung* was connected with the so-called "New Religious Consciousness", which was simultaneously a foreshadowing of a cultural and religious Renaissance at the turn of the century. It was a partly parallel phenomenon of the "Catholic renewal" (*"le renouveaucatholique"*) in the West and was equally opposite to secular Positivism, 19th century Materialism, and to the strict dogmatic Church tradition.

According to Merezhkovsky, the base of this worldview is a universal, symbolic, and mythological construct. He perceives all world history as an opposition and antithesis of two primal sources, the "abyss of flesh" and the "abyss of spirit". The first is epitomized in Paganism, in the ancient cult of a heroic personality, which can be referred to as a "Man-God" (*chelovekobozhestvo*), who neglects the spirit. The second lies in historical Christianity, in asceticism, which dismisses the flesh. Both primal sources are imperfect. A real "spiritual revolution" is necessary, which is an amalgamation and synthesis of two abysses, two truths.

As Merezhkovsky points out, this synthesis can only be possible within a future new Church. Merezhkovsky himself and his wife, Zinaida Gippius, a loyal companion and spiritual collaborator through the years of their marriage, called this new church "Universal". Taken from the Apostle John (whose revelation about the Holy City of Jerusalem that will come down from heaven was especially significant for the God-seeking utopias of the Russian Symbolists). Alternatively referred to as "the Church of the Third Testament", whose concept emerged from the strong chiliastic tradition, from Montanus of Phrygia (II AD) to the Calabrian monk Joachim of Fiore (12th century).

This doctrine states the First Testament was the Old Testament of God the Father; the Second was the New Testament by God the Son, Jesus Christ; followed by the Third Testament, the Testament of the Holy Spirit, which is the covenant of Freedom that follows Law and Grace. That is how the sacred mystery of the Holy Trinity will be explained. The historical process will accordingly face its end and "a New Heaven and a New Earth" will come as promised in the Book of Revelation, as the thousand-year reign of Jesus Christ.

Merezhkovsky became a Prophet of the forthcoming Kingdom of the Third Testament. Since then, each and every facet of his literary and social activity was focused to that end. That is why it is close to impossible to have a grasp of either his works or the mere facts of his biography, without being fully aware of his doctrine.

The core of Merezhkovsky's position rests on the three stages of dialectics and accords with the Symbolist principles of peculiar interpretations of Baudelaire's *correspondances* (not to mention that mutual projections and "consistency" among the manifestations of the different semantic lines replace by themselves the hierarchy of values), which is: every phenomenon should be established as a conflict between thesis and antithesis, which is finalized in synthesis.

2. On the Way to the New Church and "Religious Community"

As a Symbolist Merezhkovsky had always remained an adherent of the idea of "life-creation" (*zhiznetvorchestvo*). Referring to the permeation of *noumenal* and *phenomenal*, he rejected any borderline between *eidoi* and reality and was consistently striving to bring to life his literary and philosophical narratives. In his fiction, he was also trying to impose into historical characters (such as Byzantium Emperor Julian the Apostate, Leonardo da Vinci, Peter the Great and son, Grand Duke Alexei, Napoleon, and Dante) thoughts and actions that would be more suitable for their symbolic rather than historical role, i.e., the role which is assumed to be theirs in accordance with Merezhkovsky's own religious myth of the Third Testament.

In a conceptual Thesaurus created by Merezhkovsky, the main is the antithesis "contemplation vs. action": Dialectics calls for the transformation of the first into the second. Indeed, it was not surprising, and it was also absolutely logical, that when at the very end of the 19th Century the chiliastic doctrine of the three Testaments was formed in Merezhkovsky's mind, the necessity of the practical realisation of its results had emerged as well.

In her diary "On the Past" Zinaida Gippius writes:

In October 1899 Dmitry Sergeyevich Merezhkovsky came to me and said: "No. We need a new Church".

We talked a lot about that afterwards, and that was what we learned: the Church is necessary as the face of evangelical, Christian religion, the religion of Flesh and Blood.

The existing Church in its base cannot satisfy either us or the people who are close to us in time (Gippius 1999, p. 89).

It did not take long for the transition to action. By 1901 Merezhkovsky and Gippius created something which resembled the domestic church with chiliastic dogmas and their own rituals. They did it together with their mutual friend Dmitry Filosofov. He was essentially a member of their family. The literary couple connected to him on the basis of a "spiritual affinity" but there might have been traces of a sublimated Modernist Eros (Matich 2008, pp. 170–225).

They established their own liturgical order, the succession of prayers, made their own liturgical garments, obtained a communion chalice and held at home the very first service on Maundy Thursday in 1901 (Gippius 1999, pp. 95, 97–98; Gaidenko 2000).

In this regard Olga Matich has candidly noted that "the Triple Union of Merezhkovskys exemplified the idea of a neo-Christian Holy Trinity. The mediator was Gippius, playing the role of Sophia, Blessed Virgin and the Holy Spirit." (Matich 1999, p. 114).

Using the esoteric language of the "Trinity," the chiliastic doctrine of three testaments in its entirety was called "The Main [Doctrine]" (*Glavnoye*). It could be discussed directly only with those who were part of the narrow circle of initiates. As for the wider circles, they were provided with overall more subtle and careful messages, which did not reveal the radical nature of the "Third Testament" beliefs. Instead, it was presented as no more than an "upgrade" of the Church and Christianity.

In order to expound their ideas, Merezhkovsky and Gippius launched the Religious-Philosophical Meetings in St Petersburg (1901–1903). These meetings marked the first Russian attempt since post-Petrine history in the 18th century of informal interactions and doctrinal discussions between the secular "God-seeking" *intelligentsia* and representatives of the official Russian Church. Furthermore, the Merezhkovskys were in fact those who began the whole new "genre" of church-secular intellectual and social dialogue. Continuing in the West through the trial of "catholic renewal" as happened for example, in the case of the famous "Decades of Pontigny" organized by Paul Desjardins in 1910 (Chaubet 2000).

The Russian Revolution of 1905 (the First Russian Revolution) drastically radicalized Merezhkovsky's *Weltanschauung*, and he started propounding the ideas of a "new religious community." In the beginning of 1906, the Merezhkovskys fled Russia and during the next two and a half years they preached their sermon from Paris. The anti-clerical atmosphere in the French capital generated by the controversial separation of Church and State, initiated by the 1905 law, contributed to the sharp politicization of religious and social issues. Their sermon, most vividly embodied in the collected French language edition by Merezhkovsky, Gippius, and Filosofov, *Le Tsar et la Révolution* (1907), primarily focused on "the religious justification for the Russian revolution and clearly expressed an anti-monarchy and anti-clerical perspective" (Pavlova 1999, p. 7).

Completing the international questionnaire of the magazine *Mercure de France* in 1907, "On the Religious Issue," Merezhkovsky stressed that "the future religious revolution will resemble the one that happened when Christianity was born." [*La future révolution religieuse sera pareille à celle qui s'accomplit lors de la naissance du christianisme*] (Merezhkovsky 1907, p. 68).

It is difficult to disagree with E. Goncharova when she said that:

The Russian Revolution was seen by the Merezhkovskys as a chance to release their own religious intentions. The revolution should slide from the social and political surface into the religious depths. Meanwhile, the revolutionary struggle is regarded as one of those types of sacrificial fight for world transformation (Goncharova 2009, p. 10).

Interestingly enough, here the Symbolist writer expresses a, typical for mythological mentality, interchangeability of a subject and an object; those who sacrifice others have obtained the status of the main victim. It is not a coincidence that at the same time, during the years of the first Paris residence, the Merezhkovskys drew closer to Boris Savinkov. A notorious political activist, leader of the Fighting Organisation of the Socialist Revolutionary party, he was in charge of assassinations (or attempted assassinations) of a number of prominent high-ranking officials of the Russian Empire (Goncharova 2009, pp. 5–88). In Merezhkovsky's essays of that time, the way of revolutionary terror is considered as the Via *Dolorosa* to the Calvary of full self-sacrifice. For Merezhkovsky, the hidden holiness of notorious atheists, i.e., revolutionists and terrorists whose activities he compared with Jacob wrestling with God (Jacob forced his Celestial Adversary to bless him) in above questionnaire, is unconditional: "That is why, whilst it seems strange, modern atheists, warriors with Christ, are closer to Him than modern Christians; not seeing His Image, not knowing His Name, they clasp Him in the fight, sense Him, unite with Him." [*Et voilà pourquoi, bien que ceci semble étrange, les athées contemporains, les lutteurs avec Christ, sontplus proches du Christ que les chrétiens actuels: sans voir Son visage, sans connaître Son nom, ils le serrent dans la lutte, le touchent et s'unissent à lui*] (Merezhkovsky 1907, p. 71).

Until the revolutionary events of 1917, Merezhkovsky held faith to his ideas and made few changes. He actively nourished his liaisons with socialist revolutionists and other radically opposed politicians of the last years of the Empire. However, come the Bolshevik October Uprising and his first experience of Soviet power, he revised his political position. Nonetheless Merezhkovsky's religious and missionary activity was cemented even stronger to its core during the last two decades of his life following emigration in 1920.

3. Works Abroad: Chiliastic Symbolism and Post-Apocalyptic Historiosophy

In the late, emigrant, period of his life, Merezhkovsky's concealed potential was revealed. As a result of that, the expansion of two genre constructions appeared—first, his meta-historical prose, which included his essays and philosophical pieces, such as *The Mystery of* the Three: Egypt and Babylon (1925), *Western Mystery of Atlantis—Europe* (1930), and also *Jesus the Unknown* (1932–1934), and second, his biographies (taken in a broad sense), such as *Napoleon* (1929), *Dante* (1939), *Little Theresa* (1941), and collections *Faces of Saints: from Jesus to Nowadays* (1936–1938) and *The Reformers* and *Spanish Mystics* (1940–1941).

Relying upon the New Testament code, Merezhkovsky distinctively separates the concepts of history, myth, and mystery, endowing them with a whole new thematic meaning. History is an empirical reality, which is doomed to fall into the apocalyptic abyss of universal destruction if humankind does not manage to ascend to grasp the mystery. The universal action, the eternal archetype of dynamic existence based on a construction of celestial redemption is framed according to the ideology of the three Testaments by Joachim of Fiore. Which predetermines the essence of everything happening in Time. The path from the fog of history to the ontological clarity of the revelation of the final truths in the mystery goes through its iconic sign, or symbol, which is a myth in human culture. The mystery is hidden under the cover of a myth. "The truth of a myth lies in a mystery" (Merezhkovsky 1999, p. 14). "Prophets know that the myth is not a lie, but prototype of a divine truth, implemented in a mystery" (ibid., p. 174).

In emigration, Merezhkovsky's dialectical chiliasm gained eschatological elements and became a source of a line of subordinate ideological implications (such as non-religious Christocentrism, Ecumenical church, bisexuality, femininity-maternity-spirit etc.), defining an informative side of practically in his late works.

Karl Löwith wrote:

It is well known how deeply Russian thinkers of the nineteenth century have been influenced by Hegel and Schelling. It is therefore not surprising to find many parallels among them, for instance, in Krasinski's Third Realm of the Holy Spirit and in Merezhkovsky's Third Testament Christianity (Löwith 1949, p. 210).

Of course, there were other sources out there for Merezhkovsky to adopt the ideas of Joachim of Fiore: Henrik Ibsen was undoubtedly one of them. Ibsen was the author of a chiliastic dilogy, *Emperor and Galilean* (1874), which Merezhkovsky was certainly familiar with while working on his novel on Julian the Apostate in the early 1890s (Fridlender 1995). Another is the eschatology of Vladimir Solovyov (Gaidenko 2001, pp. 327–55; Mozgovaya 1996) and all of Solovyov's "landscape" of Russian intellectual *fin de siècle* culture, starting from the scandalously famous Anna Schmidt who created her ludicrous opus, *The Third Testament* (Kozyrev 1996). This line—of a double succession—was quite evident for Merezhkovsky's contemporaries, and Andrei Bely confirmed it: Merezhkovsky "twisted Ibsen and Vladimir Solovyov into a confession: the Third Testament—The Testament of a Spirit" (Bely 1997, p. 31).

Yet unlike Junior Symbolists, Merezhkovsky did not truly adopt the main myth making ideas of Solovyov's Sophiology. His creations do not need that immanent necessity in Sophia as a mystical medium. On the contrary, Solovyov cannot fathom a metaphysical drama of world duality as well as his doctrine of "positive all-unity" without it. Solovyov was absolutely unfamiliar with verbal formalism, which was one of the characteristics of Merezhkovsky's thinking, making it very often so scholastic and lifeless. Merezhkovsky was one of a few *fin de siècle* thinkers who perceived Sophia with a certain distraction from its Gnostic/mystical roots. His version of Eternal Femininity is more rational and its origins can be found in the Goethe–Nietzschean image of Aphrodite–Venus from the trilogy *Christ and Antichrist* (1895–1905). It has to be understood allegorically, as a one-dimensional incarnation of a synthesis of heaven and earth in the Church of the Second Coming. According to Merezhkovsky, this image splits into a series of doubling and flowing into each other derivatives, from chthonic "Mother Goddess" to the Virgin:

Mighty Mother Goddess, which is the mother of not only our second generation of humankind, but the first, too—Crete-Aegean Britomartis, Hellenic Aphrodite the Divine, Urania, Babylonian Ishtar, Canaanite Astarte, Iranian Anahita—the eternal Mother Virgin, with the Child (Merezhkovsky 1996, p. 160).

Synthesising them all, Merezhkovsky created a symbol of the Holy Spirit.

Due to that peculiar eschatologically stressed view of history, Merezhkovsky found himself perceiving the modern era as the time of an inevitable finale. The apocalyptic destruction of "the aeon of this age," which will be followed by "The Final Judgment" and an installation of the "Kingdom of the Third Testament". A community of Saints, a redemption of heaven and earth by the creative power of the Heavenly Spirit. The numerological symbolism of the number three and Trinitarian dialectics form the basis of a historiosophical proto-myth of Merezhkovsky. Hence, they shape the content of practically all of his late prose. In his book *The Mystery of the Three: Egypt and Babylon* Merezhkovsky presents his concept of "Three-in-One":

The Orphics called the Kabirs by three names or one of three: Axier, Axiokersa, Axiokers: Zeus the Heavenly Father, Demeter the Mother Earth, and the Son of Heaven and Earth Dionysus. The same Three in the Eleusinian Mysteries, just in a different combination: Father Dionysus, Mother Demeter and Son Iacchus. (Merezhkovsky 1999, p. 20).

In Merezhkovsky's perception, the poem by Heraclitus is also an embodiment of "the Mystery of the Three". In his interpretation, the idea of God the Shepherd spread from Babylon to Israel through the trihypostatic deity, Tammuz–Adonis–Adonai, and then reached modern civilization in Christ's image. C. H. Bedford describes the essence of the Trinitarian system of Merezhkovsky as following:

God the Mother is the term, which provides the wholesomeness for Christianity, because God the Father could not give birth to a Son without an involvement of divine femininity. Merezhkovsky approved of the significance of the terrestrial Mother of Christ, but saw in Her a symbol only, which referred to the Divine Mother, who can complete the Trinity; the Trinity in God begins and ends with Mother the Spirit. (Bedford 1975, p. 161).

The researcher elaborates that the apocryphal New Testament according to the Jews determines the connection between Merezhkovsky's concepts of the Holy Spirit and the doctrine of eternal motherhood. There were words in there about "My Mother, the Holy Spirit," that belonged to Christ. There is no argument against it, especially taking into account the aforementioned source, which was quoted by the author regularly and in different contexts. This apocryphal prayer will undergo a symbolic re-enactment and will become one of the leitmotivs of his historiosophical writings. Merezhkovsky establishes the Trinitarian principle as the main law ruling the world and mankind, including the past, the present, and the future. Space, chemical reactions/organic life, and the universal evolution, where the antipodal processes of integration and differentiation are merged together, are the Triune.

For Merezhkovsky, mysteries, i.e., a sacred meta-narrative of Divine self-emptying, His kenosis, His sacrifice for the sake of salvation of His creation in order to its next reunion with the Creator, establish a proto-myth of history, which is the code to a Symbolist description of the inner mechanism of unfolding the path of the world in time. Incarnation, or the birth of Jesus Christ, which signifies the middle way of the Trinitarian historic and mysterial narrative, is the core of this proto-myth. As a Christian thinker, Merezhkovsky is more than a traditionalist here. This scheme, however, gets more non-standard when he deconstructs the actual historical chronotope with his sacred symbolism. The Mediterranean, which connects three parts of the world, is, according to Merezhkovsky, the heart of the Earth, and two lines drown from Memphis to Constantinople and from Babylon to Rome have created the cross, a proto-image of the Calvary Cross, upon which the path of world history is unleashing itself.

The cult history of a pre-Christian man is a shadow, a Platonic *anamnesis*, a recollection of the future mystery of Incarnation and Redemption. At the very beginning of the first part of the trilogy, Merezhkovsky reveals his views on the significance of pre-Christian history as a myth/proto-image in regard to the coming of the Savior of the world—and he does it with formulaic brilliance:

The whole matter of the world mystery-myth about the suffering God is an event, which has not happened once but always does, is happening again and again in the life of the whole world and mankind. "Now these things never happened, but always are.' (Sallust. De diis et mundo, IV)[1]. "The world history is an aeon whose essence is eternal, and the beginning and the end, the reason and the purpose is Christ." (Schelling). Christ is concealed in Paganism: He opens up in Christianity. Christ is Revelation, Apocalypse of Paganism. Those who blind cannot see the Sun but can feel its warmth. The Pagan's Christ is the Son of the blind. World history is a geometrical space where the Body of Christ is built (Merezhkovsky 1999, p. 18).

A similar view of the mutual correlation between Paganism and Christianity is, in fact, a leitmotif of the primary religious and philosophical intuitions of the Russian Silver Age, which predefines a pathos of the whole Dionysian theme in the essays and academic works of Vyacheslav Ivanov, for example. This view was clearly shown by Merezhkovsky in his *Julian the Apostate* (1895). It is originated from the immense tradition of the early Christian apologetics, from Justin the Martyr to Clement of Alexandria (the treatise *The Stromata*), who apparently was the most frequently quoted by Merezhkovsky. Those ideas were adopted by the early Christian tradition due to the necessity to explain the visual similarity between the mysteries of the bloodless sacrifice of the Savior and the liturgical Eucharist, on one side, and the orgiastic[2], on the other. Merezhkovsky, a master of quotation, constantly feels the need to refer to the authorities in order to validate his own place in the line of succession in "Christian gnosis":

Once learning from the Judaic prophets about the forthcoming arrival of the Lord, demons decided to invent a fable about Dionysus the god for distribution among Hellenes and other peoples, especially in places where, as they knew, everyone would believe their prophecies," thought Julian the Apostate, who lived in the 2nd century AD. < ... >"Demons" were preaching about that "Son of God," Dionysus the dismembered instead of Jesus the Crucified, further and earlier that Julian assumed, from the very early days of mankind until the Appearance of Christ, from Egypt and Babylon to Peru and Mexico. St. Clement of Alexandria gently argues, "Both Hellenistic and barbarian wisdom can see eternal truth in certain dismembering and crucifixion, but not in that which was described in Dionysian fables but in that which teaches the theology of eternal Logos (ibid., pp. 506–7).

Explaining the consequence of the writing order of his historiosophical trilogy, Merezhkovsky points out that "the Mystery of the East" (the Mystery of the Beginning) has led him to "the Mystery of the End" (the Mystery of the West), and since Jesus is "the Alpha and the Omega, the beginning and the end," the mystery of the East and the West is reduced to the mystery of Christ, whose solution will light up the meaning of history.

Merezhkovsky declares that historical Christianity put the true meaning of the mysteries into oblivion. That is why the main task of the new, eschatological, revelation is to show the history of Jesus the Unknown, to unveil the core of "the Testament from the Saviour," to bestow upon the world the gospel of the "mystery of the three".

The first part of *The Mystery of the Three* is an attempt made by Merezhkovsky, who was always thinking by antithesis, to find the origins of one of the basic dichotomies—belief and knowledge, mysticism, and rationality—in the religions of ancient Egypt and Babylon.

In Merezhkovsky's thinking, Egyptian cults bring up the semantics of a solar insight, harmonious cosmos, and Love. It is worth noticing, however, that those exact implications are quite unusual for the mainstream perception of Egypt in Russian literature, despite a variety of already existing "variants".[3] Harmonious radiance could be hardly ranked as one of the main attributes of the "Russian Egypt." But Merezhkovsky himself does indeed understand how ambiguous all the eschatological categories could be in regard to the civilization of the Nile. He emphasizes that the religion of Egypt did not have the experience of being infiltrated by the metaphysics of evil and, therefore, did not have that dialectical inclination towards eschatology.

Babylon is in opposition to Egypt. Its religion is the religion of the night, and Gilgamesh in Merezhkovsky's portrayal looks like the Faustian character that has comprehended that knowledge cannot be accomplished either in happiness or in eternity. Although the Babylonian symbolic character discovered that the way to Salvation lies through suffering, he was unable to give the world the true faith. Therefore, the Old Testament Israel became the synthesis of the antithesis "Egypt vs. Babylon", whose spiritual experience included the truth of faith and reason, day and night, Love and Suffering. That is how the dialectical circle of meaning-generation was completed and the Triad of mystical revelation of the pre-Christian era was formed.

4. Theological Accompaniment: The Theme of Eros and Sex

The principle of triplicity is layered onto the concept of sex, which is sublimated and connected to the entire religious doctrine in a sense of the Modernist paneroticism of the *fin de siècle* era. C. H. Bedford and T. Pachmuss rightly pointed out that for Merezhkovsky (and Zinaida Gippius), the category of sex took on a special meaning when they were in exile. At that time Merezhkovsky's speculations around the metaphysics of Eros got closer to the phallic/neo-religious ideas of his frenemy, Vasily Rozanov (Bedford 1975, p. 104; Rosenthal 1975, p. 92). However, the phallic inclination of Merezhkovsky's "philosophising" can be traced back to the 1890s, when his conceptual apparatus obtained the idea of the "sacred flesh".

However, before that, the influence of Rozanov's physiology was preceded by the philosophy of Eros by Vladimir Solovyov. As with Sophiology, Solovyov's impact on Merezhkovsky—to be precise, with his famous article *The Meaning of Love* (1892–1894)—was rather formal. Merezhkovsky adopted the basic strategy on "sex spiritualisation" without digging deeply into Solovyov's mysticism of a personality's determination via its dissolution in love to the other. In the beginning of the new age Merezhkovsky's sacred flesh found its place in the terminological vocabulary of the "God-seeking" intelligentsia and was used as a generic term in the discussions of nearly all the main topics at the St Petersburg's Religious and Philosophical meetings in 1901–1903. The very name of the writer was connected in the collective mind of his contemporaries, the Junior Symbolists, with a tendency to erotic expansion into religious spheres. Simply by virtue of the fact that each and every key notion in Merezhkovsky's system unfolds into its opposite according to the logic of the inner antithesis, the "holiness of sex" bears the "abyss of sex," and behind it, there is another one, the "abyss of Sodom".

In his letter to Alexander Blok on 1 July 1902, young Sergei Solovyov expresses himself quite vividly on the matter:

Dmitry Sergeyevich <Merezhkovsky> is greatly talented, but he forgets why exactly dogs are not allowed in church and cats are. He doesn't mind to let in notorious Isis, bless her heart, totally forgetting about her objectionable behaviour, and also Fyodor Pavlovich Karamazov as an underside of Christ etc. In general, I am feeling some concern for future historians of Russian literature, who would mark its phases in chronological order as Romanticism, Naturalism, Phallism etc. (Literaturnoy enasledstvo 1980, p. 330).

In his emigrant period, while canonizing Rozanov's image and heritage within a framework of his own, chiliastic, concept, Merezhkovsky reconciles and takes into account a number of Rozanov's main ideas that he has rejected before. For example, Rozanov's doctrine of sex, interpreted through a religious and cosmological perspective, became a leitmotif of Merezhkovsky's historiosophy. The mystery of sex "dissected by Rozanov," transformed by Merezhkovsky into "The Mystery of the Two," which is preparing on a sacred level the coming of "The Mystery of the Three," the era of the Holy Spirit. Reflecting upon the category of sex, Merezhkovsky strips it off from Rozanov's inevitable mundane physiology and, according to the laws of his own, "anti-natal," perception, transfers it into an abstraction, which does not have any blood pumping or flesh trembling, but the game of interpretations of the various literary sources and ancient texts.

Like Rozanov, Merezhkovsky in the last years of his life proclaimed the "sanctity" of pre-Christian phallic cults. However, unlike Rozanov, who offered to Christianity "inoculation by flesh," which had never been the part of it in the first place, Merezhkovsky interpreted the whole issue according to the triadic principle with which he was so familiar. Initially Judeo-Christianity knew the whole "mystery of sex," but then, in the era of the historical progression of the Church, that knowledge was lost and absorbed by ascetic modus of spiritual practice. Finally, "in the Holy Realm on earth and Heaven" that integrity will be found again as a consequence of the Third Testament.

In the neo-Christian ideology of Merezhkovsky, sex connects to a mystical concept of Love in a way that semantically conjoins to all the Greek words that exhibit all the variants of the meaning of the word in Russian. These are traditionally distinguished by Biblical exegetical scholars and theologians, including the words such as *Eros* (romantic/sensual attraction), *philia* (soulful affection, fondness), and *agape* (creative, all-forgiving, sacrificial, at best—Divine Love). Essentially, Merezhkovsky's theory directly refers to the idea that sex in its metaphysical manifestation is immanent to the eternal state of God the Trinity. Persona establishes the Mystery of the One; sex portrays the Mystery of the Two; community embodies the Mystery of the Three (*tresfaciunt collegium*).

"Sex" is also the basis of epistemological constructions in *The Mystery of the Three*. The holiness of sex is a "transcendental mystery," where a man "discovers communion with the heavenly world": "sexual longing is longing for knowledge, curiosity towards the transcendental" (Merezhkovsky 1999, p. 108), "sex is only possible for a human being, blood-and-flesh 'touching other worlds,' proximity to the transcendental entities. There is birth in sexual love, but there is also death, because everything that was born once, dies; death and birth are two paths to one place, or one path to there and here" (ibid., p. 29). Sex was perceived by Merezhkovsky as an epiphany of the Trinity in a human body, a primal sensual experience of the introduction to the ontological, non-separable and non-merged, Unity of the Three in One.

All these statements became the basis for the characteristics of the whole Pagan period of history with its orgiastic physiology as the era of the "Testament of the Father." There is no need to quote numerous examples made by Merezhkovsky about pre-Christian manifestations of sanctity of sex, from the "divine" (temple) prostitution of priestesses of Ishtar to the Egyptian *bestialia*, in which he, like Rozanov before him, saw the real example of viewing animals as creatures who are closer to the "divinity" than a man who has lost the ability of "heavenly joy of the earth." However, most evidently, the holiness of sex in the pre-Christian era incarnates for Merezhkovsky (as well as for Rozanov) in the Old Testament circumcision:

Circumcision is courtship between man and God, direct and fleshly. A marital testament, a matrimonial union, a sexual intercourse between man and God. It is strange and it is dreadful. And how can we talk about this using our damned words, which are shamelessly wild and anatomically cadaverous? That is exactly when "the tongue truly is sticking to the larynx, and the paper under the ink is burning, smouldering, crumbling" (V. Rozanov). However, is eating the Divine Body, nourishing humans with the Flesh and Blood of God less terrible and strange? "How odd are these words! Who can hear this?" His disciples were horrified when they heard this for the first time (ibid., p. 106).

For the era of "the Testament of the Father," this terrifying sacrament became similar in terms of an insight of mysterial depths with the Eucharist in the era of "the Testament of the Son." Circumcision and the Eucharist reflect on each other as the mystical cores of the two consequential Testaments. The sacrament of the Bloodless Sacrifice, which marks the victory over the flesh of the "mundane world" infested by sin and death, has its own pre-image, i.e., the fiery point of an exalted affirmation of not only flesh, but sex.

It is clear that Merezhkovsky's doctrine, in all its unorthodoxy, does not accept Rozanov's idea from the 1900s and, especially, from the period of *The Apocalypse of Our*

Time (1917–1918), that Christ is the repudiation of the Father and that the New Testament broke with the Old. Following this logic, Merezhkovsky proclaims that Lord the Savior, circumcised according to the Law, finalized and strengthened the union between man and God with the Blood of the sacrament. Therefore, in circumcision the entirety of the affirmation of sex is merged with a succession between the two Testaments. In support of the thesis of the primary acceptance of sex by Christ, Merezhkovsky constantly refers to a symbolic image where Christ is shown as the Bridegroom and the Church as the Bride. This exact aspect is stressed in his exegesis of the New Testament. The actual example of this is his interpretation of the evangelical episode about healing the bleeding woman (Mark 5: 25–28), who "came up behind Him, because she was ashamed of her illness, hid it from people as they hide from each other their eternal 'shameful wound,' i.e., sex" (Merezhkovsky 1996, p. 228). Merezhkovsky explains this episode as follows: Christ took upon Himself the "shameful wound" of sex, which means that He accepted the "Testament of the Father".

According to Merezhkovsky (and he is close to Rozanov of the 1900s—the first part of the 1910s in here), Paganism has always revealed the deep connections between sex and the Resurrection. Hence, he adds his own definition of "the mystery of love in the Resurrection." This idea has been illustrated not only with various examples from the Pagan cults (the resurrection of Osiris by Isis, Tammuz by Ishtar, and the like), but also with the special interpretation of the Resurrection of Christ. Merezhkovsky especially stresses that the first person to discover that Jesus's tomb was empty and to meet him after the Resurrection, was a woman, Mary Magdalene. He elaborates that Mary Magdalene's love was stronger than death and the Resurrection itself became the miracle out of this love. Here the epigone scheme of Solovyov's Sophiology was getting rather obvious "the mystery of revealing the Eternal Feminine in the Eternal Masculine became clear" (ibid., pp. 324–25).

Love can be perceived as "antinomical," including also the "fleshly" feature, which, as was typical for Merezhkovsky, was expressed rather secretively, as a notorious example of the "metaphysical ambiguity" of his conceptual framework. This ambiguity and blurriness of the term "love" in Merezhkovsky's writings (he knew Greek quite well, by the way) are facilitated by him on purpose: he deliberately does not distinguish *agape* and *Eros* (and the last one is not found in the texts of the New Testament). By Merezhkovsky, *Eros* is some kind of perceptive emanation of *agape*. In his article "The Beatitudes" he writes: "*Agape*, so incomprehensible and terrifying for us, in His love that we did not experience, heavenly and earthly, became simple, easy and joyous in our love, earthly only—*Eros*". (Merezhkovsky 1933, p. 208).

In connection with this statement, P. Gaidenko has noted that:

Rozanov as a fighter against Christ in a certain sense is more consistent and "with all his fearless, almost shameless and cynical curiosity" is more honest and direct than Merezhkovsky: he rejects Christianity as a religion of the "immaculate conception" naming it a religion of death, non-existence, for the sake of the Astarte cult of "holy sex," "sacred semen." Opposing the Old Testament to the New, Judaism to Christianity, Rozanov simultaneously combines the Old Testament religion with the phallic cult of Cybele, which in reality is absolutely alien to the Testament by Moses and not only by Christ (Gaidenko 2001, p. 371).

As for the concept of the three testaments by Merezhkovsky, one of the mysteries of their "last revelation" is the connection of Cybele as a Pagan omen of the "Eternal Mysteries" not only with Moses's faith, but also with the Gospel of Christ. This ambiguity leads to the idea of the "consecration" of the demonic nature of Paganism. However, on the logical level it also helps to avoid a typical Gnostic alternative that was present as a subtext in Rozanov's writings in the 1900s and was almost openly manifested in *The Apocalypse of Our Time*. If there is a gap between the "physiological" Old Testament and the "spiritual" New Testament, then Christ rejects the world with its "mystical" phallism and inclinations to "flesh". Therefore, either the Creator (*Demiurge*) of this world is absolutely not the Merciful God, or Jesus is the only begotten Son of a completely different Father[4].

Despite being equipped with all the classical texts from antiquity, the erotically modernized Christology of Merezhkovsky, also heavily saturated by mythopoeic vapors, has another source—the widely popular *fin de siècle* exercises on the "mysticism of sex." Merezhkovsky is trying to link the "images of ideas" of Nietzsche, Vladimir Solovyov, Vasily Rozanov, and Otto Weininger in order to unite these in the ideological monolith of theological historiosophy, based on an inevitable semantic correction (in fact, quite radical) of the ecclesiastical dogmas and theological speculations. The synergistic union of man and God through divine kenosis, "depletion," which has appeared in the Incarnation of "the Holy Spirit and the Virgin Mary," is nearly replaced by circumcision and the sacrament of redemption become mundane. The Lord does not resurrect with divine power, which therefore executes the eternal will of the Trinity, but rather with the help of "spiritualised" earthly love of a common woman. This ought to be noted that first and foremost, as due to the doubling of "Sophiological" thought.

Various and exhaustive reflections by Merezhkovsky on soteriological themes are summarized by B. G. Rosenthal as following: "Sex is absolutely necessary for salvation < ... > sex must reveal the secret of creation itself" (Rosenthal 1975, pp. 107, 109).

The study of androgyne[5] created in the manner of Plato's *Symposium* and the subsequent Gnostic tradition became a peculiar synthesis of the philosophemes "feminine-masculine origins of God" and "sex in love." It is also true that Merezhkovsky's immediate background at the time of adopting the actual subject was redefined by contemporaneous sources—i.e., the concepts of bipolarity and bisexuality by Otto Weininger. Overall, androgyny is quite a logical derivative of the total antithetical image of the worldview and a man produced by a mythos-creator and historyosophist.

Merezhkovsky postulates the presence of metaphysical features of the opposite sex in every living creature. In his mind, the ethos of both sexes is kept in Platonic/vertical proportions "Every man harbours a secret woman. The ethereal charm of a man is femininity and of a woman is masculinity." (Merezhkovsky 1999, p. 112). He takes this category out of the frameworks of the universal principle of the created world. According to Merezhkovsky, being androgynous is a sign of a deity. He finds its manifestations not only in Pagan cults[6] but also in the God of Israel from the Old Testament. That the Biblical Hebrew word ELOHIM is used in the Pentateuch in the plural Merezhkovsky interprets in the aforementioned sense, as an expression of the masculine/feminine duality of the Divine nature[7]. The same concept is based in the interpretation of the act of creation of a man in the Image of God with His immanent androgyny:

"And God said, Let us make man in our image, after our likeness." The Three Gods—Two in One—create a man in Their Image—two in one. "So God created man in his own image, in the image of God created he him; male and female created he them." (Gen. 1: 26–28). At first—"him," Man-woman, then "them", a man and a woman: two sexes in one creation was the meaning of the Image of God in a man. It seems it is impossible to convey the dogma about divine androgyny clearer: an Androgyne makes an Androgyne (ibid., p. 460).

Hence, according to the eternal plan of the Creator towards His creation, Adam was a "divine creature," an androgyne, and androgyny is a necessary and inseparable attribute of a *persona*. Analyzing the mythological body of Pagan cults, schematically simplifying the process of mythogenesis and not discussing such an important part of archetypical duality as the Twin duality (mostly same-sex), Merezhkovsky is trying to stress that a primitive man was androgynous in almost all etiological systems of ancient times. He split into two opposite sexes as a result of the Fall into sin and breaking up with the Divine Communion.

Masculine/feminine was "revealed" not only in Pagan deities and Jehovah, but also certainly in the Incarnated Logos, Lord Jesus Christ. With his typical casualness in choices and interpretations of the sources, Merezhkovsky, when talking about this side of Christ's persona, without further ado rejects the canonical gospel texts due to their "asexuality" and goes for apocryphal texts instead. He finds evidence about the "divine beauty" of Christ. Its origins, according to Merezhkovsky, lie exactly in the fact that this beauty does not specifically belong either to a man or a woman, but to the combination of masculine and feminine in the most glorious harmony" (Merezhkovsky 1996, p. 231), following the principle of Heraclitus, "the opposite is the unified".

Father and Mother in Son is the simplest geometrical figure of Androgyny and the Trinity in divine Eros-Logos as if it has been seen by the eyes of Numen of Sex. Because in our earthly geometry of the three dimensions sex is split by space into feminine and masculine, and in the fourth dimension, which is eternity, sex was before its birth, and shall be, after death, reunited and rebuilt in the primary oneness (Merezhkovsky 1999, p. 377).

As we can see, the thesis about the masculine/feminine in Christ was necessary for Merezhkovsky in order to complete his study about resurrection as the return to a primal state of personal wholeness—androgyny—in a logical way. In order to confirm his own thoughts and reflections, the writer gives a quote-leitmotif from a visibly Gnostic apocryphal Gospel of Thomas[8] where it has been said that the Heavenly Kingdom will become reality "when two shall be one < . . . > masculine will be feminine, and there will be neither masculine nor feminine." The shadow of this "third," "wholesome," sex in modernity Merezhkovsky sees in hermaphroditism and homosexual love, in Rozanov's complex of *People of the Moon light*:

Men love men, women love women, because feminine shines through masculine for the first ones and, vice versa, masculine shines through feminine for the second; as if the gold dust of primary androgyny, a whole man, an Androgyne, who is bigger than a modern man, split-in-two, to a man and to a woman, sparkles in the dark ore of two separate sexes. All this, which is possibly larger, primary-wholesome, perfect, enchants the "people of the moonlight" in homosexual love (Merezhkovsky 2007, p. 199).

Needless to say, the intellectually pedantic, dry, entirely bourgeois, and righteous Merezhkovsky was never an apologist for "sodomite frolics." The mystery of "people of the moonlight" as well as other frighteningly "shocking depths," like temple prostitution or anthropophagy, became the subject of his sublime poetic aphorisms and meditations only because, according to his doctrine, they were chosen as mythological signs of the genuine Mysteries. In the world of Merezhkovsky's Proteus-like antitheses, they could become a cloudy window to eternity but also could serve as satanic perversions of authentic meanings. And even the words quoted above—about "homoeroticism"—have undergone intellectual sublimation. The writer zealously denies every physical expression of homoerotic love, which he describes as an opposition to spiritual androgyny. Merezhkovsky's reflection, yet again, is inclining towards the antagonism of "noumenon vs phenomenon": he sees the Biblical incarnation of homosexual "love" in sacred but also secular history (Sodom) as a symbol of the distortion of the Divine truth in a diabolical mirror. Merezhkovsky states: "Europe is Sodom" (Merezhkovsky 2007, p. 36), because the homosexual principle prevails there. According to Merezhkovsky, in the East of Europe, in the USSR, Communism with its gregarious pseudo-*sobornost* (pseudo-spiritual community) of "Hams who came there"[9] imposes exclusive femininity, which is faceless and fruitless in its sole proprietorship. On the contrary, a similar impersonality and infertility are embodied in the West, in its self-sufficient masculinity, which is militarism. If the two of them collide, the result might be apocalyptic; the world would end up with the Second World War and the ultimate death of humankind. That is when Merezhkovsky's mythopoeic started exploring new semantic spheres and proceeded from the domain of "mysticism of sex" to pure historiosophy (where similar mythologeme works anyway) by using the same codes but slightly rearranging them.

5. The Heavenly Kingdom as the Looking Glass Land

In *Jesus the Unknown*, Merezhkovsky develops in depth the concept of the Heavenly Kingdom. He sees the main discrepancy between Christ and Christianity in its understanding. The decline of eschatology in Christianity, the disease of *kázhenie* (i.e., perception of God and devil as mythological creations) led Christianity to the loss of every trace of experiencing an inseparable union of God and the world, to a separation of mundane and heavenly, fleshly, and spiritual. For a further, more vivid, illustration of his thought, Merezhkovsky introduces the ethnic symbols of "Judean" (syncretic) and "Aryan" (taken in Plato's discourse) imaginings of the Heavenly Kingdom. "The Third Testament" will be executed in the sensual-spiritual Heavenly Realm where it will transform into the "Ecumenical Church of Godmanhood". The Law transcended by Love will be finalized by Freedom, and the "Unknown Eucharist of Jesus the Unknown" will be accomplished, and every man will really become part of the Divine Body in eternity. It is curious to note, however, that in the chapter "The Heavenly Kingdom," the most controversial in the whole text of *Jesus the Unknown*, Merezhkovsky comes to fairly Orthodox conclusions, similar to those made at the Council of Chalcedon, about the Heavenly Kingdom, which, pre-existing in eternity, is establishing itself in real history in real time.

The actual present and the historical time in general look like an "upside-down" quote of the future of the Third Testament. This present exists in the field of a paradox, as a mirror opposition to the Heavenly Kingdom:

... come out of this world, out of three dimensions and come into that other world, into the fourth dimension, where bottom becomes top and top bottom, when right becomes left and left right: where everything is vice versa. Only "getting upside down," "falling down," only "head first," to everyone else's horror, who pretends to stand steadily, who is "reasonable," you can enter, fly in, fall in, from this world into that one, from the kingdom of men to the Heavenly Kingdom, from earthly sorrow to heavenly bliss (Merezhkovsky 1996, p. 262).

The pathos of the present is exactly here. On its providential tops (Merezhkovsky's historicism does not know anything else) the present is experiencing the tragic conflict between the "pre-knowledge" of truth, which is rising within, and the impending doom of the mundane "here and now".

History for Merezhkovsky, as for the other Russian Modernists, is drama and the theatre of symbols in which a man is either a player or a marionette. It does not matter for him if his concept does not get along with the "historic accuracy" as well as with the logic of mythological concepts, because everything "historical" makes sense only in the correlation with "mysterial". Mysterial and historical were merged altogether only in the ministration of Jesus Christ. In all other cases they are tragically divided. Talking about the unbreakable connections between history and mystery in one of the canonical Gospels, Merezhkovsky has noted:

These two metals are merged by Matthew in fire that is the hottest, into the strongest amalgam, therefore, they are not two anymore but one; that one that happened once in time, in history, exists and will continue in eternity, in mystery; the first day of the Lord is the Heavenly Kingdom that has already come (ibid., p. 259).

An analysis of the existence of the New Testament theme in Merezhkovsky's discourse becomes an additional basis for the conclusions that by the 1920s became the compilation in regard to practically all the texts of the Russian *fin de siècle* culture, such as the philosophy of the All-Unity, "Russian Cosmism," actual Symbolism in its different hypostases, Rozanov's "physiology," "Christianised" Nietzscheism, transformed Sophiology, and sexual theories by Weininger as well as many other systems of views, which provided lots of material forthe doctrinal mosaics of Merezhkovsky.

Merezhkovsky becomes almost the most consistent spokesman of *couleur locale* of the Symbolist historicism, which was described by one of the modern researchers as follows:

The aestheticization of life, from one side, and the transformation of aesthetics into the daily program, from the other, has been shown as the most vivid in the historical worldview. The history of the world as a whole is an aesthetical fact, an artefact, and the "text." The correlation between one event and another is perceived as quotation. Symbolic "quoting" of one event via another is the action of the transformation of a denotative event to an event of denotation (Isupov 1992, p. 89).

Originating from Romanticism, the Symbolist mythologeme of the author-Demiurge had been consistently shifted by Merezhkovsky into the field of religion. Yet still, the deep-rooted role of the writer first and foremost was established around the metaphysical distillation of the global Symbolist aestheticism in order to give it global recognition. Generally speaking, Georgy Adamovich had every reason to note that:

Without Merezhkovsky, Russian Modernism could become truly Decadent, and it was he who from the very beginning made it strict, serious, and clear. < ... > There was the return to the greatest themes of Russian literature, to the large-scale themes overall (Adamovich 1996, p. 27).

6. Poetics and Politics on the Metaphysical Background: The Symbolist-Chiliast between Dante and Mussolini

By 1930 Merezhkovsky was almost the only intellectually significant Russian figure in exile remaining following the tradition of the Modernist religious mythology and seeking to integrate the neo-Christian ideology.

The question about the results of an interaction between Art and the "God-seeing" fields during the late period of Merezhkovsky's life in Paris gets even more interesting considering his efforts to integrate the neo-Christian ideology, which grew up in the soil of the Russian Silver Age—in its sterile and preserved version—into the drastically different cultural context of Europe between the two world wars. We will focus only on a few significant aspects of this broad theme, connected to Merezhkovsky's book, Dante (1938), but beforehand let us make a general remark.

In Merezhkovsky's version, chiliastic mythology shows typical ambivalence. On the one hand, it inclines to a certain "historical Monophysitism," to full rejection of the providential essence of historical statehood. On the other hand, however, the ideological constructions of Merezhkovsky have always had that obvious Positivist nuance, the aspiration to make metaphysical categories more ordinary and, especially, the category of the "future age," the aeon of the Millennial Kingdom. In the author's mind, the "rose" of religious and Symbolist metaphysics acquired a scent only if propagated by the "wild branch" of the political volatility of the era.

Along with that, chiliastic mythology, as well as the other *fin de siècle* Modernist constructions, reveals a tendency to expansive deforming impact first on the actual aesthetical sphere, and then, as a reflectionon life, per se, in nearly all its manifestations.

As a way of highlighting the seriousness of the wild branch to Merezhkovsky is the story while he worked on his Dante, is his apparent admiration of the character and work of Benito Mussolini, then Italian Prime Minister. Mussolini had invited the Merezhkovskys to Italy to make certain investigations funded through a scholarship by the Fascist government[10]. In May 1936, during a time of intense work on the first part of the book, Merezhkovsky wrote a letter of acceptance and thanks to Mussolini. The letter is written in French and is of considerable historical interest and was first published in a monograph by T. Pachmuss (Pachmuss 1990, pp. 9–14). A few phrases and expressions from this letter were transferred afterwards into the "Introduction" to the Italian translation of *Dante*, which was released with the inscription, "To Benito Mussolini. <This Book> about Dante the Prophet to the executor of the prophecy" [*A Benito Mussolini. Realizzatoredellaprofeciaquestosu Dante profero*] (Merezkovski 1938, p. 3).

In this very diplomatic epistle, accepting the historical determination and justification of the concordat between the Fascist regime and the Vatican, Merezhkovsky unfolds the dialectics of his own views on the relationship between the power of Caesar and that of the high priest. Criticising Dante's dual model of the necessity of an unmerged cooperation of the Emperor and the Pontiff, he makes the following conclusion:

< ... > the final resolution of the religious problem in the relationship between the Church and the State, the Cross and the Eagle, does not consist in forming an empirical or abstract alliance, not in a Concordat and, especially, not in their remoteness from each other, but in the perfect union, corresponding with the main Christian dogma, which is the union of two eternal powers, human and divine, earthly and heavenly in Christ's Persona, the Priest of priests and the King of kings. (Pachmuss 1990, p. 14)

Reflecting upon the phenomenon of a poet in the meta-historical context, Merezhkovsky foresees the ontological and providential "affinity" between Mussolini and the spirit of the great Florentine poet. The concordat between Rome and the Vatican, Mussolini's intransigence against the "Bolsheviks' barbarism" and the Duce's "affection" during the first audience with the writer that facilitated and resulted in the scholarship funding.

From the outset the letter is remarkable:

I am very well aware that I need to thank you for the pleasure I have been granted through the invitation to come to Italy not with words but with deeds, with my work, a book about Dante. I came in order to collect documents about him, but the best ever evidence of all about Dante is you. To understand Dante, one should live by him, but it is possible only with you, in you . . .

"I have been living with Dante all my life," you told me once. In your actions, you have implemented this life with Dante . . .

There is pre-established harmony between him and yourself. Your souls are initially and indefinitely related; they are destined to each other for very eternity. Mussolini in meditation is Dante. Dante in action is Mussolini (Pachmuss 1990, p. 9).

Further on, these flatteries change into three insistent demands he makes on the dictator. First, in the political sphere, to commence war against Russian Bolshevism ("satanic Communism"). In the social sphere, to implement in reality the all-human hopes and aspirations for universal peace via Dante's messianic ideas, avoiding the last and the most disastrous war. Finally, in the religious sphere, to attain the union of the historical churches under one Ecumenical Church, and execute the prophecy of his great predecessor and prepare the grounds for the coming of the new "Revelation of the Holy Spirit."

The author tenaciously imposes on the addressee the messianic and apocalyptic spirit of the Russian "God-seeking" Symbolists, doing so from the chiliastic perspective, which the Europe of the 1930s was not familiar with. It was Mussolini, who according to Merezhkovsky, should become the providential power, the one and only able to protect the world from the tragic "fate of Atlantis" in the pessimistic, Spenglerian, sense. It is worth noting that the most important epithets that Merezhkovsky uses towards Mussolini are these French lexemes *Pacificateur* and *Consolateur*, which literally mean "peacemaker" and (sic!) "comforter." Hence, for a person brought up in the Russian Orthodox Christian tradition, there is a transparent allusion to the verse 14:26 of the Gospel of John, where the Savior gives to His disciples the promise of the coming of the Holy Spirit, which is named by the polysemantic Greek word, *parakletos* ("the intercessor, who is called for help and is standing nearby . . . ") and is translated in Old Slavonic (but not necessarily in other languages) exactly as "Comforter." It is obvious that Mussolini could not understand this wordplay with a sacred intertext and with a mild allusion to the Slavic/Russian subtext. However, the author of the prophetic epistle to the dictator remained permanently monological and very often assumed monolingualism of any of his addressees for no good reason. Apparently, as a Symbolist, in this fragment Merezhkovsky gave a reference to the Orthodox prayer, the invocation of the Third Hypostasis of the Trinity "O Heavenly King": addressed to Mussolini, it should be regarded in the perspective of the upcoming era of the "Third Testament," "The Realm of the Heavenly Spirit," where "Duce" is the predecessor, according to Symbolist logic.

In Russian literature there is a tradition of the cultural communication between those in power and the artist. A writer addresses words of flattery and exaltation to the head of State in order to gently impose not only random political advice, but to some extent his own intellectual patronage. In one way or another, this narrative was used by Nikolay Karamzin and Tsar Alexander I, by Alexander Pushkin and Tsar Nicholas I, and even by Boris Pasternak and Stalin. Merezhkovsky's letter can be seen as a similar attempt to apply this role to him.

Yet at the same time Merezhkovsky goes beyond the limits of the strictly Russian cultural situation. His narrative looks fairly similar, but also holds another, symbolic, component, which refers to the Italian mediaeval and Renaissance tradition. The artist is seeking the duke or tyrant's patronage by inflating his intellectual ago. This model in Merezhkovsky's case obtains a certain Dante-esque meaning. The letter to Mussolini in the core of its "poetics" directly corresponds to two famous epistles written by Dante in 1311, *The Letter to the Princes and Peoples of Italy* (Dante wrote it on the occasion of Henry VII's campaign in Italy in order to be crowned as the Holy Roman Emperor in the Church of San Marco in Rome) and *The Epistle to Henry VII the Emperor*.

Merezhkovsky quotes the first letter in his book about Dante (Merezkovski 1938, p. 112); as for the second one, whilst not naming it literally, he undoubtedly implies it, simply because its subject is described in detail in the chronologically relevant part of the biography (ibid., pp. 113–20). Both epistles are examples of the scholastic epistolary rhetoric of the 13th–14th centuries, whose main claim was to project real events onto examples from the Sacred History by using the typical quotes and reversed allusions from the divine texts.

The rhetoric of scholastic Mannerism in the Dante-esque style became very congenial to Merezhkovsky, taking into account the peculiarity of the cultural *mise en scène* he is trying to recreate—and quite sincerely—in his letter to Mussolini. As for the second epistle by Dante, the letter to Henry VII, its content can already be the direct analogue to Merezhkovsky's letter to "Duce." In this epistle, after comparing the Emperor to Biblical and ancient heroes, Dante addresses to him the concerns of an exile from treacherous Florence. He reminds Caesar of the providential meaning of the Empire in building the Divine Salvation and requests the Emperor to leave Milan for the sake of completing the peace process and to turn his spears to Florence, the nest of the sin of Judas, the place of malice and the resistance to the unity of the Caesar ecumene, instead.

Merezhkovsky as an exile recreates the same situation, trying on Mussolini the role of the "secular Messiah," the collector of lands, and calling him to the crusade against Russian Bolshevism as the main threat on the way to global union. That is how the writer extrapolates a historical narrative that already became a fact of literature into the current political reality. Functionally, the roles remain the same, only their conceptual substance somewhat changed: it has to correspond with the "ready-made" historiosophical concept about Dante who declared the prophecy about an execution of "The Mystery of the Three" in the future, craving for the All-Unity in the real life and hanging his hopes on the "Imperial" spirit and the "Imperial" actions. He, however, did not perceive the empire itself in the eschatological way, and in this case the poet's aspirations were lower towards the shining truth of the "Third Testament." Only Merezhkovsky managed to "decipher" all Dante's prophecies that were hidden from Dante himself, and to relate a proclamation about the religious and historiosophical mission of "Messiah the Emperor" to the "Titan figure" of Mussolini. Hence, the model of the cultural communication "Merezhkovsky/Mussolini" is deliberately oriented on the metaphor of the "Magi who came from the East" (i.e., Russia) to worship the "born Messiah" (Mussolini) and to witness the execution of the Old Testament's soteriological prophecies.

That is how the purely personal preferences of Merezhkovsky the thinker and writer in the social and political European discourse of the 1930s yet again revealed the tendency, which was quite common in chiliastic Symbolism overall, of the replacement of "life" by literature and its dubious conversion that led not only to unpleasant historical misunderstandings, but also to real tragedies.

The entire religious symbolization of the reality by Merezhkovsky ended up adopting the sacred status in nearly all of his personal activities, including social and political life. The actions of the "military commanders on the ground" can only be interpreted with the help of the baroque and symbolical "sacred code," which could be imposed on these commanders, who did not always have a particular desire to accept it. That is how most of those peculiar conflicts of cultural communication between Merezhkovsky and his political addressees started.

Of course, the "romance" between Merezhkovsky and Mussolini, where the Dante-esque subtext was prevailing, thanks to the Russian Symbolist, was doomed from the very beginning due to the very essence of the cultural and historical mythology of Merezhkovsky. It became the core subject in reflections on the matter by Yuri Terapiano:

Mussolini replied to Merezhkovsky's call to start a campaign against Bolshevism, "the utmost devilish evil of the world": "Italy, with its limited military forces, cannot start a fight with the mighty Soviet Army." And Merezhkovskythen said, "he was right in the sense of the bad infinity, but in the metaphysical sense he turned out as a misbeliever." That is what Merezhkovsky was telling us on one of the "Sundays" at his house, when he came back from his second, unsuccessful, trip to Italy. "At first, I thought that Mussolini was the incarnation of the spirit of the earth and a providential personality, and guess what? He turned out to be the usual materialistic politician—and a vulgar one!" Merezhkovsky shouted. Before that, similarly vulgar were other politicians, such as Piłsudski, and before him Hitler. They all thought about "history," i.e., the "bad infinity," and were unable to see "the end of history." And Merezhkovsky, who lived by metaphysics, could not fathom that other people were absolutely unfamiliar with that, and tried to see that metaphysical alikeness in every politician (Terapiano 1953, p. 31).

In the 1930s, Merezhkovsky definitely saw his mission not in literary activities but in prophetical service in the social and political arena. The Terapiano has noted that Merezhkovsky wanted to be judged not by the literary laws, but by his religious and philosophical experiences, which directly originated from the chiliastic concept of the Third Testament; in his last years, he bitterly complained about the critics who did not understand him, trying to use strict literary criteria towards his work:

He felt like he was the forerunner and also the main ideologist of the coming Kingdom of the Spirit. Dictators, like, for example, Jeanne d'Arc, must execute their mission, and he, Merezhkovsky, must give away the guidelines. Was it naïve? For sure, it was naïve in the historical sense, but in the metaphysical level, where Merezhkovsky lived, "naïve" became wise and "absurdist" turned into the most important. That is what Merezhkovskybelieved (ibid.).

Funding: This research received no external funding.

Institutional Review Board Statement: Not applicable.

Informed Consent Statement: Informed consent was obtained from all subjects involved in the study.

Data Availability Statement: Not applicable.

Conflicts of Interest: The author declares no conflict of interest.

Notes

[1] Sallust. *Concerning the Gods and the Universe.*

[2] First of all, Osiris, Dionysian and like mysteries that suggest a sacrificial dismembering.

[3] For a detailed overview of the Egyptian theme in Russian literature and for an anthology of selected texts see (Panova 2006).

[4] Merezhkovsky had already pointed out at the logical inevitability of this alternative in Rozanov's framework in the 1900s: (Merezhkovsky 1908, p. 95).

[5] About the cultural role of the androgyne mythologeme in neo-Romanticism and Symbolism, see (Crouzet 1984). For a brief consideration of the same question (also an abstract) on the material of motivic reminiscences and allusions, see (Kozlov 1994).

[6] Isis and Osiris, Ishtar and Tammuz, the image of one entity with two faces, Hera and Zeus, on ancient coins.

[7] Traditional Christian exegetics interprets this grammatical form as a secret notion of the Three Hypostases of One God (the verb in this grammatical construction is used in the singular). It is worth noting that Merezhkovsky, inspired by theology of "The Three in One," being very well aware of the literature on the subject and certainly knowing about this interpretation, did not find it necessary to make even a quick reference to it.

[8] One can find the connections to it in the works of Clement of Alexandria, whom Merezhkovsky highly appreciated.

[9] A reference to his article "The Coming of Ham" from 1906.

[10] For more details about the history of the book and the "Dante-esque" context of the relationship between Merezhkovsky and Mussolini see (Caprioglio and Spengel 1989; Dodero Costa 1999; Polonsky 2006, 2011, pp. 251–65).

References

Adamovich, Georgy. 1996. *Odinochestvo i svoboda*. Moscow: Respublika, ISBN 5-250-02570-6.

Bedford, Harold C. 1975. *The Seeker: D.S. Merezhkovsky*. Kansas: University of Kansas Press, ISBN 9780700601318.

Bely, Andrey. 1997. *Andrey Bely o Bloke: Vospominaniya. Statyi. Dnevniki. Rechi*. Moscow: Avtograf, ISBN 5-89612-001-X.

Caprioglio, Nadia, and G. Spengel. 1989. Dmitrij Merezkovskij e Dante Alighieri. In *Dantismo russo e cornice europea: Atti dei convegni di Alghero-Gressoney (1987)*. Firenze: Olschky, pp. 341–51. ISSN 0066-6807.

Chaubet, François. 2000. *Paul Desjardins et les Décades de Pontigny*. Paris: Presses Universitaires du Septentrion, ISBN 2-85939-606-3.

Crouzet, Michel. 1984. Monstres et merveilles: Poétique de l'Androgyne, à propos de Fragoletta. *Romantisme* 14: 25–42. [CrossRef]

Dodero Costa, Maria-Luisa. 1999. O knige Merezhkovskogo "Dante". In *D.S. Merezhkovsky. Mysl' i slovo*. Moscow: Nasledie, pp. 82–88. ISBN 5-201-13314-2.

Fridlender, Georgy M. 1995. D. S. Merezhkovsky I Genrik Ibsen (U istokov religiozno-filosofskikh idey Merezhkovskogo). In *Fridlender G. M. Pushkin. Dostoyevsky. "Serebriany vek"*. Saint-Petersburg: Nauka, pp. 411–34.

Gaidenko, Piama. 2000. D. S. Merezhkovsky: Apokalipsis "vsesokrushayuschey religioznoy revolutsii". In *Voprosy literatury*. n° 5. Moscow: FSRL "Literaturnaya kritika", pp. 98–126. ISSN 0042-8795.

Gaidenko, Piama. 2001. *Vladimir Solovyov I Filosofiya Serebrianogo veka*. Moscow: Progress-Traditsia, ISBN 5-89826-076-5.

Gippius, Zinaida N. 1999. *Dnevniki: V 2 kn. Kn.1*. Moscow: Intelvak, ISBN 5-93264-006-5.

Goncharova, Elena I. 2009. *"Revolutsionnoe khristovstvo": Pis'ma Merezhkovskogo k Borisu Savinkovu. Vstup. st., sost., podgot. Tekstov i comment. E.I. Goncharovoy*. Saint-Petersburg: Pushkinsky Dom, ISBN 978-5-91476-009-7.

Isupov, Konstantin G. 1992. *Russkaya estetika istorii*. Saint-Petersburg: Izd-voVyschihk guman. kursov.

Kozlov, Sergey L. 1994. Lyubov k androginu. Blok-Akhmatova-Gumilyov. In *Tynianovsky sbornik. Piatye tynianovskie chteniya*. Riga: Zinatne, pp. 155–72. ISBN 5-7966-1033-3.

Kozyrev, Alexey. 1996. Vladimir Solovyov i Anna Schmidt v chaianii "Tret'yego Zaveta". In *Rossiya i gnosis: materialy konferentsii*. Moscow: Rudomino, pp. 23–42.

Literaturnoy enasledstvo. 1980. *Literaturnoy enasledstvo, t. 92: Aleksandr Blok: Novye materially i issledovaniya, kn. 1*. Moscow: Nauka.

Löwith, Karl. 1949. *Meaning in History, The Theological Implications of the Philosophy of History*. Chicago: University of Chicago Press.

Matich, Olga. 1999. O. Khristianstvo Tret'yego Zaveta i traditsia russkogo utopizma. In *D.S. Merezhkovsky. Mysl' islovo*. Moscow: Nasledie, pp. 106–18. ISBN 5-201-13314-2.

Matich, Olga. 2008. *Eroticheskaya utopia: novoe religioznoye soznaniei fin de siècle v Rossii*. Moscow: NLO, ISBN 978-5-86793-642-6.

Merezhkovsky, Dmitry S. 1907. La Question religieuse. In *Mercure de France*. n° 1. Paris: Mercure de France, V, pp. 40–71.

Merezhkovsky, Dmitry S. 1908. *Ne mir, no mech: K budushchey kritike khristianstva*. Saint-Petersburg: M.V. Pirozhkov.

Merezhkovsky, Dmitry S. 1933. Blazhenstva. In *Chisla*. Paris: Cahiers de l'étoile, pp. 195–209.

Merezhkovsky, Dmitry S. 1996. *Iisus Neizvestny*. Moscow: Respublika, ISBN 5-250-02601-.

Merezhkovsky, Dmitry S. 1999. *Taina Tryokh*. Moscow: Respublika, ISBN 5-250-02718-0.

Merezhkovsky, Dmitry S. 2007. *Taina Zapada: Atlantida—Europa*. Moscow: Exmo, ISBN 978-5-699-21997-1.

Merezkovski, Dmitri. 1938. *Dante. Trad. di Rinaldo Küifferle*. Bologna: Zanichelli.

Mozgovaya, Ya E. 1996. V. S. Solovyov i D. S. Merezhkovsky. In *Otechestvennaya filosofiya: Opyt, Problemy, Orientiry issledovaniya, vyp.19*. Moscow: AON, pp. 112–25.

Pachmuss, Temira. 1990. *D.S. Merezhkovzky in Exile. The Master of the Genre of Biographie Romancée*. New York: P. Lang, ISBN 0-8204-1254-6.

Panova, Lada. G. 2006. *Russky Egipet. Aleksandriyskaya poetika Mikhaila Kuzmina: V 2 kn*. Moscow: Vodoley Publishers, ISBN 5-902312-80-9.

Pavlova, Margarita. 1999. Mucheniki velikogo religioznogo protsessa. In *Tsar i revolutsiya. Pod. red. M. A. Kolerova*. Edited by Dmitry Merezhkovsky, Zinaida Gippius and Dmitry Filosofov. Moscow: OGI, pp. 7–54. ISBN 5-900241-42-4.

Polonsky, Vadim V. 2006. Mezhdu metafizikoy, istoriey i politikoy: religioznaya mifologiya v pozdnem tvorchestve D. S. Merezhkovskogo. In *Voprosy literatury*. Moscow: FSRL "Literaturnaya kritika", pp. 186–99. ISSN 0042-8795.

Polonsky, Vadim V. 2011. *Mezhdu traditsiey i modernizmom. Russkaya literature rubezha XIX–XX vekov: istoriya, poetika, kontekst*. Moscow: IMLI RAN, ISBN 978-5-9208-0389-4.

Rosenthal, Bernice G. 1975. *Merezhkovsky and the Silver Age: The Development of a Revolutionary Mentality*. The Hague: Springer, ISBN 978-9401183536.

Terapiano, Yury. 1953. *Vstrechi*. New York: Izd-vo im. Chekhova.

Article

Synthesizing Religions: Vasily Rozanov's "Phallic Christianity"

Henrietta Mondry

Department of Global, Cultural and Language Studies, University of Canterbury,
8041 Christchurch, New Zealand; henrietta.mondry@canterbury.ac.nz

Abstract: Vasily Rozanov was one of the first Russian writers of the fin de siècle to create a nexus between the study of the history of world religions and the history of sexuality. He viewed Christianity's asceticism as a source of the disintegration of the contemporary family. This article examines Rozanov's strategy to synthesize religions and to use pre-Christian religions of the Middle East as proof of common physical and metaphysical essence in celestial, human, animal, and mythological human/animal/divine bodies. I argue that while his rehabilitation of the physical life by endowing it with religious value was socially positive, his self-proclaimed "mission of sexuality", when politically motivated, was manipulative and incorporated the notion of the atavistic 'survivals'. In conclusion, I explain that Rozanov's monistic search for the divine in the physical body as well as his strategy to synthesize religions were additionally driven by his personal doubts in the preeminence of Christian eschatology.

Keywords: Vasily Rozanov; Christianity; Judaism; sexuality and religions; human–animal monism; proto-posthumanism

check for
updates

Citation: Mondry, Henrietta. 2021.
Synthesizing Religions: Vasily
Rozanov's "Phallic Christianity".
Religions 12: 430. https://doi.org/
10.3390/rel12060430

Academic Editor: Dennis Ioffe

Received: 3 May 2021
Accepted: 6 June 2021
Published: 9 June 2021

Publisher's Note: MDPI stays neutral
with regard to jurisdictional claims in
published maps and institutional affiliations.

1. Introduction

Vasily Rozanov (1856–1919) was one of the first writers of the Russian fin de siècle to view religious beliefs in relation to the history of sexuality. Rozanov stated that his generic "new philosophy" or "philosophy of life" is based on the phenomenon of life itself, which he observed with "curiosity and surprise" (*Opavshie list'ia* (*Fallen Leaves*), Rozanov 1970a, p. 144) (1912/1914). Rozanov observed life through the life of the body, which, in his terminology, was both a "phenomenon" and a transcendental "noumenon" ("Iz zagadok chelovecheskoi prirody" ("From the mysteries of human nature"), p. 29) (1901). In spite of the often contradictory nature of his parataxic narratives, his writing shows a stable preoccupation with drawing examples from diverse religions to substantiate his point of view, no matter how fluctuating the evaluative side of this position might be.[1] Rozanov's search for examples of alternatives to Christian attitudes to the physical body followed the line of argument which became methodologically acceptable by the end of the nineteenth century among ethnographers and cultural evolutionists. It became conventional to use both synchronic and diachronic approaches to explain differences between and within modern societies and to "mine non-Western cultures for 'survivals' from the past, whether as primitive practices to regulate, or as a 'window onto the past' study" (Yoshiko Reed 2014, p. 120). Rozanov turned his attention to the conceptualizations of the human and animal body in relation to the links with the divine across religions, and often based his opinions on readings of Scriptures, the Talmud, as well as on the study of ancient artefacts and pictorial representations taken from pre-Christian sources.[2] His professed missionary goal was to rehabilitate the body in all its biological activities by breaking boundaries between the physical and metaphysical essence. In doing so, he challenged Cartesian dualist thinking. Moreover, he specifically used Spinozian religious monist ideas to support his project to endow physical corporeality with spirituality.

Current body theorizing, including posthumanist thinking, is often grounded in a revival of the Spinozian monist model of the body.[3] The monist approach recognizes the existence of the same substance in all objects of nature, and does not create dualist categories

271

out of physical and metaphysical, flesh and spirit, or mind and body. For Spinoza, there is one infinite substance—God. This substance is expressed in extension and in thought, and is as corporeal and physical as it is mental. Rozanov regarded Spinoza as the most original of all the European philosophers. Additionally, Spinoza's Jewishness was of special importance for Rozanov, as it provided him with the means of comparing Christian and Judaic attitudes towards the body. For Rozanov, the Cartesian denigration of the body was synonymous with European Christian asceticism, while in contrast Spinozian mystical pantheism was congenial to his vitalism (Volzhskii 1995).[4] Grosz summarizes dualist philosophy as an understanding that assumes that there are "two mutually exclusive types of things, physical and mental, body and mind, that compose the universe in general and subjectivity in particular" (Grosz 1994, p. 7). By polarizing mind/soul and the body, Descartes denied the metaphysical aspect of the body—he regarded nature as soul-less. Importantly, Rozanov singled out Descartes's views as hindering a holistic view of material life and creating discriminatory hierarchies between spirit and matter. Notably, for Descartes, animals were devoid of soul and epitomized matter devoid of spirit, while Rozanov specifically chose (mythic) animals to promote the idea of the monist essence in the divine and physical spheres. His interest in the ancient religions of the Middle East was linked to his search for a connection between God, or other divine figures, and human and/or animal essence. In the following paper, I first explain the key ideas of his project to reform Christian asceticism, then turn to his interpretations of the relevant themes taken from the Old Testament, Jewish religious customs, and Ancient Egyptian beliefs. Next, I examine his elucidations of the religious value of (composite) animals found in ancient artefacts and pictorial images. The last section assesses Rozanov's subjectivities in his mission to synthesize religions.[5]

2. Rozanov's Mission of "Phallic Christianity"

Rozanov's main interest in the search for manifestations of the divine in human corporeality was the result of his expressed concern about the negative effect of Christian asceticism on contemporary society. He arrived at the idea that the breakup of the contemporary Russian family was a result of the somatophobia of Christian beliefs. Moreover, he argued that these harmful views on the physical life of the body were imposed by the Russian Orthodox Church and its clergy, which failed to promote the importance of love and care between family members, including parental love and harmonious relations between spouses.[6] His quest to look for unity of the physical and metaphysical in the living body was driven by what he perceived as pressing practical necessity. The New Testament and the Church's interpretation of the Holy family were, for Rozanov, key to the decline of the contemporary family. In *Liudi lunnogo sveta* (*People of the Moonlight: Metaphysics of Christianity*) (1909/1911) he argued that the moment "semenality" was introduced to "the Image of Jesus Christ", the Image will be destroyed (Rozanov 1990, p. 59). Jesus, for Rozanov, is the prime example of the annihilation and absence of sexuality and sex, asexuality being a phenomenon that defined Christianity as a religion, and which separated it from the religion of the Old Testament.

In his early work, Rozanov had already focused on sexuality and procreation, and sensationally argued that sexuality itself was of divine origin. In view of his daring pronouncements, it is not surprising that his book, *V mire neiasnogo i nereshennogo* (*In the World of the Unclear and Undecided*) (1901), was taken out from circulation a month after its publication, and he was labeled by the high-ranking ministers, responsible for censorship, as "a terrible pornographer" (Gollerbakh 1922b, p. 49).[7] In "Brak i khristianstvo" (Rozanov 1995a) ("Marriage and Christianity") (1898), he shocked his readers by proclaiming that the very essence from which sexual organs are made was not of "phenomenal" but "supra-natural" (*sverkhestestven*), even cosmic origin:

Quite often, the thought occurred to me, and still occurs to me, that the very "clod of earth" from which that place is made has a totally different origin from the other parts of the body (this is why, during the *usual, phenomenal time and with the usual eye, we cannot even look at it*) and it is to other parts of the body the same as iron from a *meteorite* is to *ordinary* iron. ("Brak i khristianstvo" ("Marriage and Christianity"), p. 119)[8].

Italicizing important words and concepts, he argued that human sexuality was a manifestation of the (*"noumenal 'nost'"*) noumenality of the human body, the latter he explained as "the second and most important, but hidden from the rational conscience, part of things" ("Iz zagadok" ("From the mysteries of human nature") p. 29).

In his search for helpful examples, Rozanov turned to the Hebrew Bible, and even the Talmud for indications of the Judaic attitude to human sexuality, love within the family, and the role of procreation. Similar to many of his contemporaries, following Tylorian methods advocated in *Primitive Culture* (Tylor 1873), Rozanov explored contemporary cultures for 'survivals' of ancient beliefs, and his views of contemporary Jewish life were based on his amateurish interpretations of the Talmud and the Hebrew Bible.[9] For Rozanov, Judaism exemplified an attitude to the family which he would like to see emulated by the Russian family. Notably, while pursuing his project, Rozanov claimed that he discovered examples of divine–human interaction in the Hebrew Bible, which he used as an antidote to contemporary somatophobia. One revealing example is found in his essay "Kontsy i nachala" ("Ends and Beginnings" 1902) in which he alludes to the Old Testament story of Watchers, whose divine origins did not prevent them from copulating with human women. As elucidated by Howard Eilberg-Schwartz in his groundbreaking study *God's Phallus and Other Problems for Men and Monotheism* (Eilberg-Schwartz 1994), there are several myths in the Old Testament that ponder erotic relationships between humans and divine beings. He specifically refers to the myth in Genesis 6: 1–4 (J) which recounts the meeting between the sons of God and humans:

The sons of God saw how beautiful the daughters of men were and took wives from those that pleased them. The Lord said, "My breath shall not abide in man forever, since he too is flesh; let the days allow him one hundred and twenty years". It was then, and later too, that the Nephilim [Fallen Ones] appeared on earth—when the divine beings cohabited with the daughters of men, and bore them offspring. They were the heroes of old, the men of renown. (Genesis 6:1–4 (J))

Eilberg-Schwartz regards this episode as a fragment of a larger mythology that is now lost, yet its presence in the Bible is meaningful. Eilberg-Schwartz stresses that the relationship between divine males and human women results in offspring as men of renown. This aspect would appeal to Rozanov, who was arguing for a need for a strong family that would take emotional care of children. However, according to Eilberg-Schwartz, the story of Watchers in the Bible does not have positive connotations in terms of progeny, as cohabitation between angels and human women does not bring good results in the logic of the Hebrew Bible. The sons of God are later described as fallen angels, or Watchers, who are responsible for the fall of humankind. The placement of this story before the flood reflects an editorial judgment that the incident contributes to God's decision to wipe out humanity: "God brings the flood to wipe the slate clean" (Eilberg-Schwartz 1994, p. 129).

For Rozanov, the Biblical episode with the Watchers has useful and positive meaning as it also helps him to imagine the compatibility of celestial and earthly corporeality. The story, for him, is not a myth, but an example of historical past. He argues that writers with a mystical inclination, such as the Russian romantic poet Mikhail Lermontov, could express this plot of sexual contacts between the divine angels/demons and humans. According to Rozanov, the plot of Lermontov's romantic poem "Demon" (1839) is a manifestation of the poet's latent mystical knowledge. In the poem, Demon falls in love with young woman Tamara, a love which Rozanov sees as a proof of the remnants of the distant past, applying the methodology of searching for 'survivals' of the past in contemporary, albeit fictional narratives. It is symptomatic that Rozanov turns to Lermontov as a suitable candidate in his search for a forgotten past, due to his foreign ancestry and personal family history: one of his ancestors was a clairvoyant who came to Russia from Scotland, where he had practiced his arcane craft. This presupposes that he had latent knowledge or insight into the beginnings of humanity.

However, these creative interpretations, based on insights into Old Testament stories, could become politicized, as was the case of Rozanov's maintaining that Jews have an atavistic predisposition to blood rites—a classic case of quasi-scientific Tylorism which looked for 'survivals' of primitive instincts and drives in contemporary peoples.[10] In a collection of articles, written during the infamous blood libel trial, the Beilis Affair (1911–1913), Rozanov maintained that contemporary Jews could have unconscious desires, expressed by their "olfactory and tactile attitude to blood"—a notion which he used as a title for the book *Oboniatel'noe i osiazatel'noe otnoshenie evreev k krovi* (*The Olfactory and Tactile Attitude of Jews to Blood*) (1913/1914) (Rozanov 1998). It is symptomatic that it was during the Beilis Affair that Rozanov declared his search for phallic Christianity was a missionary act, meant to defeat the (imagined) advance of contemporary Jewry.

In his best-known text, *Fallen Leaves* (1913), revealing a mimetic underpinning of his views on Judaism and Jewish materiality, he argued that, for Christianity to triumph, a return to the corporeality of Judaism was essential:

Christianity expressed in itself and revealed an inner content of non-semenality, just as Judaism in the Old Testament had revealed semenality" (*Opavshie list'ia* (*Fallen Leaves*), p. 21).

If Christianity continues to deny sex, the consequence will be an increase in the number of Jewish triumphs. This is why my starting to preach sex is so timely. Christianity must become phallic, at least in part (children, divorce, i.e., putting the family in good order, increase in the number of families). (*Opavshie list'ia*, p. 132).

In his mission, Rozanov defined Judaism as the religion of a people whose covenant with God was connected with their very flesh, indeed "with the extremity of flesh", the covenant of circumcision (ibid., p. 132). Strikingly, in circumcision, he saw not only a symbolic but also a material link between the sexuality of God and that of human beings. He believed that the Jews understood the mystery of how the divine is revealed in a human being, knowing that it resided in sexuality. He conflated the imagined God's phallus with human genital organs, and linked sexuality in nature with the mystical world of divine sphere.

In terms of recent academic opinions, Rozanov's turning to circumcision in relation to his project to prove the divine aspects of human procreation is not unjustified. Eilberg-Schwartz writes that, in the Judaic priestly writing, circumcision is treated as a symbol of male fertility, of God's promise to make Abraham the father of multitudes:

It is not an accident that the symbol of the covenant is impressed on the male organ of generation, rather than the ear or nostril. By exposing the male organ, the rite of circumcision makes concrete the symbolic link between masculinity, genealogy, and reproduction (Eilberg-Schwartz 1994, p. 202).

In this line of argument, however, circumcision does not necessarily mean that God's body is sexed. Eilberg-Schwartz sees one of the paradoxes of the emphasis on human procreation in the fact that the father God is sexless. He argues that the two myths of human creation defining what it means to be human pose a paradox, as the human Adam was made in the image of God who does not procreate. This factor does not hinder Rozanov's mission, as for him it was strategically paramount to make a point about the divine nature of procreation as encoded in circumcision.

Looking for further examples in the rituals of contemporary Jewry that relate to the treatment of the body, Rozanov turns to the ritual cleansing bath—*mikvah*. According to his logic, *mikvah* is yet another practice which encodes the divine nature of sexuality:

By this means [the *mikvah*] as yet nothing is actually said to the Jews, but they are given a thread, and if they grasp it and follow it each one can reach the idea, the conclusion, the identity, that 'these things' (the organs and their functions), although they are not shown to anyone and to pronounce their names aloud is indecent: they are nevertheless sacred. (*Uedinenoe* (*Solitaria*), p. 26)

By making these quasi-anthropological observations, Rozanov hoped to reform his contemporaries' perception of sexuality. His intended message was to help people remove the sense of shame in their attitude to the sexed body. This, by implication, would lead to the development of a loving and emotionally healthy family in contemporary Russia contributing to the country's social stability.

As well as Judaism, Ancient Egypt was a source in Rozanov's search for spiritualized materiality. Continuing with his ideas of the meaning of circumcision, by studying drawings made during the Napoleonic expedition to Egypt, he concluded that ancient Egyptians practiced circumcision. In "Obrezanie u Egiptian" ("Circumcision in Ancient Egypt") (1917) he polemically noted that all contemporary European commentators of Ancient Egyptian religion deliberately silence this fact out of contempt for the ritual itself. This silencing, for Rozanov, meant the lack of admission that the cultures of the Middle East "already four thousand years ago were developing the universal and global theme and idea of Fatherhood—an idea that is of utmost comforting importance to all humans" ("Obrezanie u Egiptian", Rozanov 2002, p. 92). His conclusion implicitly reveals his search for religious beliefs demonstrating the importance of strong emotional family ties and parental love.

3. Looking for God in an Animal

Egyptian art additionally provided Rozanov with examples of non-anthropocentric religious beliefs, which erased boundaries between human and animal bodies. Symptomatically for his project, his interest in the divine essence of sexuality made him arrive to the conclusion: "Look for God in an animal" already in his work on *The Family as Religion* (1898) (Rozanov 1995e, p. 67). The blurring of the boundary between animals and humans plays in his writing the role of a "synthesizer". The result of Rozanov's deciphering various Ancient Egyptian artefacts was a creation of the tripartite equation, animal = man = god, which became a cornerstone in his mission of sexuality.

Rozanov realized that within the European Christian tradition, animals have been treated as inferior others. The Church Fathers—Augustine, Clement of Alexandria, and John Chrysostom—claimed that humans and animals had nothing in common.[11] According to their teachings, animals had no souls, could not therefore have an afterlife. With the advent of the Scientific Revolution, the line separating animals from humans grew finer, but, at the same time, the triumph of Darwinism deprived animals and humans alike of souls, and maintained the taxonomy in which non-human animals were inferior to human animals. Rozanov pointed out the absence of living, earthly creatures in Christian mythology and in Russian Orthodox art, and linked this with Christianity's deliberate neglect of the body, and its sterility:

It is not by accident that the ancient temples were full of calves, sheep, pigeons—of a healthiness that was still pre-human, whereas the new [temples] are full of the lame, the blind and the weak. It is not by accident that the Gospels are sprinkled with so many stories of healing, and that Christ began his "new" healing by chasing the animals out of the temple ("Po tikhim obiteliam" ("In Quiet Monasteries"), p. 113).

In "Irodova legenda" ("Herod's legend") (1898), Rozanov makes clear the religious and philosophical meaning of his interest in animals. He attacks Descartes for considering the animal to be merely "an improved machine" ("Irodova legenda", Rozanov 1995b, p. 41) that lacks feelings, and notes that in the European philosophical tradition the animal is equated with physiology; he posits Descartes's dualism as the "antithesis" of the beliefs of Ancient Egypt, where the "animal is god" (ibid., p. 41). He also notes that there is no concept of a "sacred animal" (ibid., p. 43) in European religious and philosophical thought and that, likewise, in the scientific discourses of Darwin and Wundt the animal is merely a physiological object, "an outgrown body" (ibid., p. 41). Only in the Eastern religions, he says, is there a "constant intertwining of the animal and the human in God"—in the East "there has always been an understanding of an animal as a religious and mystical category, not as a physiological phenomenon" (ibid., p. 45).

As the vehicle for challenging this equation of sexuality with animality—and the devaluation of both—, Rozanov chose a mythical Egyptian sphinx:

Here in Petersburg, near the Nikolaevsky Bridge, there are two sphinxes past which you cannot walk without a feeling of excitement. The way their limbs are arranged is so imperishably lifelike! The smile from four thousand years ago is a smile for the gloomy and sad people of Petersburg; it is as if the cheerful young faces of the sphinxes are seeking to burst out laughing at a slow-witted generation of people (and according to Chrysanthos there is no satisfactory theory to explain the animal worship characteristic of the Egyptians). The idea expressed by the sphinx is "Look for God in an animal", "Seek Him in life", "Seek Him as the Lifegiver". "I am the great cat" (the sun-god Ra speaking of himself); but a god—as every Petersburg citizen can see—is completed at the front as a human being and consequently the full idea expressed by the sphinx reads "God-Human Being". ("Sem'ia kak religiia" (Family as religion), p. 73)

In this narrative, Rozanov creates a tripartite syllogism of man = animal = god, destigmatizes the body, and equates the physical with the metaphysical. He goes on to argue for the divine nature of human sexuality in a passage that equates the pagan sphinxes with the mystical animals of the Biblical Apocalypse:

But here... but here with the Egyptian sphinxes who died in the past we encounter the expectations of the Apostle that are promised to the future: that Someone will come who "will bring fire down from heaven" and of whom, seemingly, people will say: "Who resembles this beast? He gave us fire from heaven". "A beast"—and here again you are to look for what is divine in an animal, and this is what the sphinxes show us: and the fire from heaven, heavenly fire, is the fire of the flame of marriage ties, as understood in their heavenly origin, by which humankind are bound together, having started a religion of births to replace a religion of dying ("Sem'ia kak religiia", pp. 74–75).

Rozanov further returns to the theme of the animal in his theological discussion in "Russkaia tserkov" ("Russian Church") (1906) (Rozanov 1994b), where he notes that Christianity, especially its Russian Orthodox form, has become "monophysitic" (p. 14) despite the fact that the church rejected monophysitism—the dogma that Christ was of *one essence only* (divine) and therefore was not truly human—as heretical many centuries earlier. Christianity, Rozanov notes, has denied everything that is "human, everyday, earthly" leaving only that which is "celestial, divine, superhuman" (p. 15). In view of his main concern about Russian society, he also states that Russian Orthodox art in particular lacks representations of animals, because animals are powerful embodiments of the earthly:

"In the pictorial art of Orthodoxy (unique and original, disseminated everywhere) one never sees the animals in the Nativity Scene: the cattle, the shepherds, the little donkeys. On the whole the animal principle has been torn away, rejected by Orthodoxy with terrible force" ("Russkaia tserkov" ("Russian church"), p. 14).

Rozanov saw this as evidence that presence of animal flesh was unacceptable to the Eastern Orthodox churches. This he juxtaposed to western Christianity, stating that churches in Catholic Italy provided him with examples of cathedrals containing representations of animals. During his travels to Italy, he was struck by the many depictions of animals in ecclesiastical buildings there. Regarding the many portrayals of lambs in Italian church paintings, he concludes that, by rejecting the flesh and the joy of life, Christianity in the East has lost the opportunity of seeing "the way things are connected" (*Ital'ianskie vpechatleniia*) (*Italian impressions*) (1901) (Rozanov 1994a, p. 37). This search for the monistic connection of matter and materiality with religious beliefs is foundational for his life philosophy, with its search for physical forms revealing a synthesis between the divine and the earthly.

In the "Post-scriptum" to *Italian Impressions*, Rozanov describes his search for an alternative kind of faith and perception of life. A religious ceremony that made a particular impression upon him was the one in which the Pope himself shears a lamb, the wool from which is used to make vestments (*pallia*) for distribution to the archbishops. Indeed, he records that he found the Papal lamb-shearing ceremony more pleasing than any other dogma in Catholicism; the mere fact of an animal being brought into a Christian church was surprising to him, but what struck him as particularly significant was that the lamb was placed on the altar table. This fact leads him to compare the Catholic cathedral with a Jewish temple, using the ceremony as an allegory for the connection between animals, humans, and God and as symbolic of a pan-religious synthesis: "This is almost as delightful as the animals in Solomon's temple" (*Ital'ianskie vpechatleniia*, p. 125).

His synthesizing methodology can be found in his description of the statue of the black Diana in the museum in Rome, where he blurs animal/human/divine categories:

The Diana of Ephesus is not a flight of the spirit, but a philosophy... The black goddess is dressed in white clothing which comes down to the steps... but as you look closely at this clothing you see that it is made up of tiny animal figures: bulls, sheep, birds and even enormous flies and bees. For scholars I will observe that it would not be out of place to compare a statue made by the Greeks with the way Baal is depicted by the Assyrians and Phoenicians of whom the prophet Ezekiel tells us that behind this statue were compartments or small boxes in one of which were kept live doves, in another sheep, in another—something else, and in the last one, the seventh, human beings (the statue was of enormous dimensions) (*Ital'ianskie vpechatleniia* (*Italian impressions*), p. 432).

Soul, according to Rozanov, is something that animals and humans have in common. By blurring the boundaries between pagan and Christian categories, he believes he has restored the mystical status that the human/animal body supposedly held in the pagan world.

Unlike some of his contemporary professional and amateur Egyptologists, Rozanov did not travel to Egypt. However, his discoveries through the study of drawings received some recognition for his interpretations by contemporaries (Gollerbakh 1922a). For example, although a famous artist Aleksandr Benua was skeptical about Rozanov's knowledge of European art, he gave high praise to his work on Egyptian artifacts: "he worked out what was mysterious in the bas-reliefs and hieroglyphics which provided evidence of the beliefs of the Egyptians and in so doing revealed his gift for penetrating the deepest of their secrets" (Benua 1995, p. 142).

Pursuing his quest to synthesize religious beliefs about the body, Rozanov looks for similarities between Judaism and Ancient Egyptian beliefs, finding in the presence of the animal an equation between sexual matter and spirit. His thinking in one example develops as follows: the human "face" (litso) is, in Judaism, a path into the Cosmos, while in Egyptian drawings the face, or a person's head, is freely interchangeable with that of various animals. But the human being has more than one "face"—human genitalia, the second "face" of the body, is unique to every individual too. Both cultures recognize the importance of this "sexual face" through the ritual of circumcision ("Iz sedoi stariny" ("From grey-haired antiquity"), Rozanov 1995c, p. 361). Furthermore, the Egyptians' "usual metamorphosis of the human body" through the addition of an animal head, speaks of the organic unity of animals, humans, and deities. His book, *Iz vostochnykh motivov* (*From the Eastern Motifs*) (1916) abounds with illustrations painstakingly copied from various historical sources, and each one is provided with commentary. Significantly, a drawing of a woman with the arms and legs of an animal is catalogued as "The future" (Rozanov 1995d, p. 407, Figure 42), while another, featuring stylized vertical figures with plant, animal, and human parts, is inscribed as "A sample of physiological-religious painting" (ibid., p. 416, Figure 48). He formulates this thematic grouping as "a human, botanical, theological being" (p. 417). For Rozanov, these instances proved that, in Ancient Egypt, there was an understanding of the single essence of the physical and cosmic worlds:

And they united with the world without writing dissertations '*De rerum essentia*' but having depicted everywhere that the '*essentia*' of the whole world is one, that there is no 'multiplicity of essence', but singularity of essence, mono-essence. And it is precisely this that the universality of Egypt represents. (*Iz vostochnykh motivov*, p. 421).

4. Finding a Place for the Self through Connecting with Human/Animal/Divine Bodies

In the last year of Rozanov's life, his desire to connect with the divine through linkages with real and mythological animals became most explicit. The weaker his own body grew, the more he desired to link it with animal bodies, to create a oneness with the animal world. In a letter to art historian Erik Gollerbakh, penned in a cold room where Rozanov experienced illness and hunger in 1918, while thinking of food, he imagined an idyll with close relationships between humans and animals:

I would like to be Polyphemus and to look after goats and sheep, and I would suck milk from them with MY OWN MOUTH. By the way, for a long time now I have been tempted to drink milk from the cow with my own mouth, to suck from her udder. It is so beautiful. Truly—beautiful. And I am quite sure, very HEALTHY. This thought has attracted me for twelve years. In future, better times, people will give their aged parents a live goat or a cow to suck from. ("Pis'mo 27" ("Letter 27"), p. 545.)

Of special interest is Gollerbakh's comment on this passage, in which he states that it reveals Rozanov's characteristic "gustatory", "olfactory", and "tactile" "attitude" to everything "living" ("Pis'mo 27", p. 545). This commentary clearly alludes to the title of Rozanov's book *The Olfactory and Tactile Attitude of the Jews to Blood*, obliquely uncovering mimetic underpinnings of Rozanov's perception of Judaism and Jewry. While the book was written at the time of the blood libel, the Beilis Affair, and was perceived by contemporaries as a political and opportunistic slander, it revealed Rozanov's phylogenetic views on contemporary Jews as an ancient people. In his comment Gollerbakh exposes an important mythological aspect of Rozanov's perception of the physical world, and the human and animal body. The fact that Rozanov imitates Polyphemus, a character from the Ancient Greek myths, shows yet again that his vision was underpinned by his search in past cultures for alternative views on living matter. Fittingly, in this passage he recalls that Polyphemus was dubbed a "Divine shepherd", thus bringing the religious aspect into the theme of human-animal relations.

Through this fantasized connection between human lips and animal udder, Rozanov breaks antithetical categories. He forms assemblages with the animal at the same time envisioning a proleptic posthumanist futurity based on his understanding of the premodern cultural beliefs about inter-species communications. Notably, while thinking about the family of the future with its caring relations between children and old parents, he includes domestic animals as part of the family. While he evokes Ancient Greek mythology in the image of Polyphemus, this utopian family idyll also implicitly alludes to the attitude to domestic animals in the Hebrew Torah. In Rozanov's visualization of the idyll, domestic animals are treated as part of the household, which is in line with the ethos of the Torah which grants animals rest on the sabbath.[12]

After the Bolshevik Revolution, Rozanov continues to conceptualize theogony and cosmogony by religious synthesis underpinned by his search for the phallic beginnings of the universe. The divine/animal/human correlation is part of his synthesis. Tellingly, in his last letters to Gollerbakh, he asked for a drawing of an antique statuette which embodied a much-desired synthesis of religions, of which mystical animals and pagan deities are part:

I am in urgent need of a statuette (small). Horus in the shape of an eagle with a phallus that has a lion's head at the end. Please copy it on paper and send it to me (Rozanov, "Pis'mo 16", p. 526)

In another, later letter to Gollerbakh ("Pis'mo 30" 6. 10. 1918) (Rozanov 1970b), musing about the link between mysteries of sexuality and religion, he recalls the episode when he met the (in)famous Russian mystic, monk, and political charlatan Grigory Rasputin. According to Rozanov, Rasputin allegedly told him that he was struck by fear when he first met Rozanov. Rozanov interprets this fear as evidence of yet another of Rasputin's "revelations", another "miracle", in that he recognized something mystical and "noumenal" in Rozanov ("Pis'mo 30", p. 551). In turn, Rozanov labels Rasputin's essence as "noumenal" and links it to the Egyptian deity Apis. Rozanov's interest in Apis is related to his interest in phallic cults, and it emerged when he discovered illustrations of this phallic animal/god in an Egyptian drawing from the Napoleonic expedition to Egypt. Characteristically, Rozanov describes his first reaction as a desire to kiss the "balls" of this sacred bull in an act of mystical-religious elation ("Pis'mo 30", p. 551). Since he sees in Rasputin an embodiment of the phallic Apis, in order to experience the phallic power of religious beliefs, he takes part in Rasputin's notorious parties. Rasputin's performances of quasi-orgiastic dances[13] were, for Rozanov, a manifestation of the "Theogony of Rasputin-Apis-Dionysus-Adonai" ("Pis'mo 30", p. 551)—a striking synthesis of the Siberian mystic sectarian practices and Ancient Egyptian, Greek, and Judaic religions. Noteworthy is the fact that Jesus Christ is markedly absent in this synthesis. In order to secure a place in this almost all-encompassing theogony, Rozanov develops a formula "I am Grishka, Grishka is Apis" (ibid., p. 551). In this way, he fantasizes crossing the boundaries between his own body and the divine/animal body. By doing this he puts the tripartite formula animal = man = god into operation in relation to his own body.

Rozanov's professed mission of sexuality aimed at saving the Christian Russian family became redundant after the Bolshevik Revolution, which separated the church from the state. His late writing made clear that his search for god in the human/animal body was to a large degree motivated by eschatological concerns about the immortality of the body. This concern was well understood by a Russian-Jewish historian and literary critic Mikhail Gershenzon, who knew Rozanov well. Gershenzon characterized Rozanov's religious position by using the expression "maybe", explaining that Rozanov presumed that any religious belief could be true.[14] Of relevance to Gershenzon's point is the fact that Rozanov wrote about his intention to convert to Judaism in his last work, *The Apocalypse of Our Times*, a few months before his death. He also at the time ordered that his anti-Jewish writing be burned and made a written apology to the Jewish community. While his mission of sex was socially driven by his worries about the disintegration of the Russian family as the supporting pillar of Russian society, his interest in the physical body had eschatological underpinning. His notion of phallic Christianity was connected with his uncertainty over the divine origins of Jesus as a messiah.

Rozanov died from malnutrition at the Sergiev Posad Monastery where he was given refuge after the October Revolution. In his last, already dictated thoughts, he compared his visceral experience of disintegrating body as being inside "the black waters of Styx" (Ivanova 1990, p. 84). Concomitantly he also described his suffering as Hell, thus synthesizing two diverse religious mythologies. After his death his family repeatedly maintained that he died peacefully as a Christian. While it was important for the family to maintain respectability, Rozanov's dying as a Christian does not contradict his views on Christianity which, for him, was a religion of the cult of the death. At the same time there was a telling rumor circulating among his friends that on his deathbed he asked for a figurine of a phallus, the Talmud, and an icon. While the story was apocryphal and invented, it nevertheless encapsulates Rozanov's unorthodox quest for synthetic religious beliefs.

Rozanov embarked on his mission of sexuality during a time characterized by a proliferation of scientific discourse on sexuality and the institutionalization of body politics. He developed a comparative approach to the history of religions and sexuality by using syncretistic multicultural sources. By defying the human/animal dualism, and instead creating an alternative trinity of animal/human/god(s), he challenged the hierarchies of religious, philosophical, and scientific categories as they were expressed in the body politics of his time, and promoted a monistic cohesion of organic matter and metaphysical essence. In his elucidations, he moves in the posthumanist direction, thus contributing to contemporary thought.

Funding: This research received no external funding.

Conflicts of Interest: The author declares no conflict of interest.

Notes

1. For an overview of Rozanov's stance on Judaism and Jewry see Glouberman (1976), on Jewry and sexuality see Engelstein (1994) and Mondry (2010, 2021), on Jewish religion and blood libel see Kurganov and Mondri (2000); Murav (2017).
2. On Rozanov's interest in the representation of animals in art see Mondry (1999).
3. On Spinozian monism as opposed to Cartesian dualisms in application to body theorizing see Grosz (1994).
4. Murav notes that Rozanov viewed Jews' vitality in opposition to tiredness of the Russian people. Murav (2017, pp. 151–71).

Religions **2021**, *12*, 430

5   Ure (2011) asserts that Rozanov rejected Christian eschatology.

6   On Rozanov's views on marriage see Putnam (1971) and Matich (2005).

7   Ober-Procurator of the Holy Synod K. Pobedonostsev himself was involved in stopping the book's publication. See Gollerbakh (1922b, p. 49).

8   All translations from Rozanov's texts are by me.

9   For a discussion see Mondry (2010, pp. 15–19).

10   See a discussion on Tylorism and Rozanov's contribution to the discourse of atavistic survivals during Russian ritual murder trials in Mondry (2021, pp. 73–90).

11   For the history of the reception of animals in Christianity see Salisbury (1994).

12   Rozanov used the Torah as an example in his animal advocacy. See Mondry (2021, pp. 73–78).

13   On Rasputin's dancing, intoxication and orgies see Radzinsky (2000, pp. 383–84).

14   See Prishvin (1995, p. 110).

References

Benua, Aleksandr S. 1995. Religiozno-filosofskoe obshchestvo, Kruzhok Merezhkovskikh. In *V.V. Rozanov: Pro et Contra*. Edited by D. K. Burlaka. St. Petersburg: Russkii Khristianskii Gumanitarnyi Institut, vol. 1, pp. 132–43. ISBN 5-88812-003-0.

Eilberg-Schwartz, Howard. 1994. *God's Phallus and Other Problems for Men and Monotheism*. Boston: Beacon Press, ISBN 0-8070-1225-4.

Engelstein, Laura. 1994. *The Keys to Happiness: Sex and the Search for Modernity in Fin-de-Siecle Russia*. New York: Cornell UP, ISBN 0801499585.

Glouberman, Emanuel. 1976. Vasilii Rozanov: The Antisemitism of a Russian Judophile. *Jewish Social Studies* 38: 117–44.

Gollerbakh, Erik. 1922a. V. V. Rozanov kak istorik iskusstva i kollektsioner. *Sredi Kollektsionerov* 2: 36–39.

Gollerbakh, Erik. 1922b. *V.V. Rozanov. Zhizn' i Tvorchestvo*. St. Petersburg: Poliarnaia Zvezda.

Grosz, Elizabeth. 1994. *Volatile Bodies: Toward a Corporeal Feminism*. Bloomington: Indiana University Press, ISBN 9780253208620.

Ivanova, Evgenia. 1990. O poslednikh dniakh i konchine V.V. Rozanova. *Literaturbaia Ucheba* 1: 70–88.

Kurganov, Efim, and Genrietta Mondri. 2000. *Rozanov i Evrei*. St. Petersburg: Akademicheskii Proekt, ISBN 5-7331-0225-x.

Matich, Olga. 2005. *Erotic Utopia: The Decadent Imagination in Russia's Fin de Siecle*. Madison: The University of Wisconsin Press, ISBN 0-299-20880-x.

Mondry, Henrietta. 1999. Beyond the Boundary: Vasily Rozanov and the Animal Body. *Slavic and East European Journal* 43: 651–73. [CrossRef]

Mondry, Henrietta. 2010. *Rozanov and the Body of Russian Literature*. Bloomington: Slavica, ISBN 978-0-89357-370-6.

Mondry, Henrietta. 2021. *Embodied Differences: The Jew's Body and Materiality*. Boston: Academic Studies Press, ISBN 9781644694855.

Murav, Harriet. 2017. The Predatory Jew and Russian Vitalism: Dostoevsky, Rozanov, and Babel. In *Ritual Murder in Russia, Eastern Europe, and Beyond*. Edited by Eugene Avrutin, Jonathan Dekel-Chen and Robert Weinberg. Bloomington: Indiana UP, pp. 151–71. ISBN 9780253026408.

Prishvin, Mikhail M. 1995. O V.V. Rozanove. In *V.V. Rozanov: Pro et Contra*. Edited by Dmitrii K. Burlaka. St. Petersburg: Russkii Khristianskii Gumanitarnyi Institut, vol. 1, pp. 102–32. ISBN 5-88812-003-0.

Putnam, George F. 1971. Vasilii V. Rozanov: Sex, Marriage and Christianity. *Canadian Slavic Studies* 5: 301–25. [CrossRef]

Radzinsky, Edavard. 2000. *Rasputin: The Last Word*. Crows Nest: Unwin and Alan, ISBN 1-86508-193-0.

Rozanov, Vasilii V. 1970a. Opavshie list'ia. In *Izbrannoe*. München: A. Neimanis, pp. 81–426.

Rozanov, Vasilii V. 1970b. Pis'ma k E.F. Gollerbakhu. In *Izbrannoe*. Munich: A. Neimanis, pp. 515–64.

Rozanov, Vasilii V. 1990. *Liudi Lunnogo Sveta. Metafizika Khristianstva*. Moscow: Druzhba narodov, ISBN 5-285-00016-5.

Rozanov, Vasilii V. 1994a. Ital'ianskie vpechatleniia. In *Sredi Khudozhnikov*. Edited by Aleksandr Nikoliukin. Moscow: Respublika, pp. 19–159. ISBN 5-250-02461-0.

Rozanov, Vasilii V. 1994b. Russkaia tserkov'. In *V Temnykh Religioznykh Luchakh*. Edited by Aleksandr Nikoliukin. Moscow: Respublika, pp. 8–30. ISBN 5-250-02465-3.

Rozanov, Vasilii V. 1995a. Brak i khristianstvo. In *V Mire Neiasnogo i Nereshennogo*. Edited by Aleksandr Nikoliukin. Moscow: Respublika, pp. 117–39. ISBN 5-250-02507-2.

Rozanov, Vasilii V. 1995b. Irodova legenda. In *V Mire Neiasnogo i Nereshennogo*. Edited by Aleksandr Nikoliukin. Moscow: Respublika, pp. 40–53. ISBN 5-250-02507-2.

Rozanov, V. V. 1995c. Iz sedoi drevnosti. In *V Mire Neiasnogo i Nereshennogo*. Edited by Aleksandr Nikoliukin. Moscow: Respublika, pp. 356–402. ISBN 5-250-02507-2.

Rozanov, Vasilii V. 1995d. Iz vostochnykh motivov. In *V Mire Neiasnogo i Nereshennogo*. Edited by Aleksandr Nikoliukin. Moscow: Respublika, pp. 339–424. ISBN 5-250-02507-2.

Rozanov, Vasilii V. 1995e. Sem'ia kak religiia. In *V Mire Neiasnogo i Nereshennogo*. Edited by Aleksandr Nikoliukin. Moscow: Respublika, pp. 67–82. ISBN 5-250-02507-2.

Rozanov, Vasilii V. 1998. Oboniatel'noe I osiazatel'noe otnoshenie evreev k krovi. In *Sakharna*. Edited by Aleksandr Nikoliukin. Moscow: Respublika, pp. 276–410. ISBN 5-250-02674-5.

Rozanov, Vasilii V. 2002. Obrezanie u Egiptian. In *Vozrozhdaiushchiisia Egipet*. Edited by Aleksandr Nikoliukin. Moscow: Respublika, pp. 90–91. ISBN 5-250-01840-8.

Salisbury, Joyce. 1994. *The Beast Within: Animals in the Middle Ages*. New York: Routledge.

Tylor, Edward B. 1873. *Primitive Culture*. London: Murray.

Ure, Adam. 2011. *Vasily Rozanov and the Creation*. London: Bloomsbury.

Volzhskii, Aleksandr S. 1995. Mysticheskii panteizm Rozanova. In *V.V. Rozanov: Pro et Contra*. Edited by Dmitrii K. Burlaka. St. Petersburg: Russkii Khristianskii Gumanitarnyi Institut, vol. 1, pp. 418–55. ISBN 5-88812-003-0.

Yoshiko Reed, Annette. 2014. From Sacrifice to the Slaughterhouse: Ancient and Modern Approaches to Meat, Animals, and Civilization. *Method and Theory in the Study of Religion* 26: 111–59. [CrossRef]

Article

Kazimir Malevich's Negative Theology and Mystical Suprematism

Irina Sakhno [1,2]

1. Department of Theory and History of Culture, Peoples' Friendship University of Russia (RUDN University), 117198 Moscow, Russia; sakhno_im@pfur.ru or isakhno@hse.ru
2. HSE Art and Design School, Higher School of Economics, National Research University, 115054 Moscow, Russia

Abstract: This article examines Kazimir Malevich's Suprematist art in the context of negative (apophatic) theology, as a crucial tool in analyzing both the artist's theoretical conclusions and his new visual optics. Our analysis rests on the point that the artist intuitively moved towards recognizing the ineffability of the multidimensional universe and perceiving God as the Spiritual Absolute. In his attempt to see the invisible in the formulas of Emptiness and Nothingness, Malevich turned to the primary forms of geometric abstraction—the square, circle and cross—which he endows with symbolic concepts and meanings. Malevich treats his Suprematism as a method of perceiving the ineffability of the Absolute. With the *Black Square* seen as a face of God, the patterns of negative theology rise to become the philosophical formula of primary importance. Malevich's *Mystical Suprematism* series (1920–1922) confirms the presence of complex metaphysical reflection and apophatic thought in his art. Not only does the series contain icon paraphrases and the Christian symbolism of the cross and mandorla, but it also advances the formulas of the apophatic faith of the modern times, since Suprematism presents primary forms as the universals of "the face of the future" and the energy of the non-objective art.

Keywords: Kazimir Malevich; apophatism; suprematism; *The Black Square*; icon; mandorla; sacred mysteries

Citation: Sakhno, Irina. 2021. Kazimir Malevich's Negative Theology and Mystical Suprematism. *Religions* 12: 542. https://doi.org/10.3390/rel12070542

Academic Editor: Dennis Ioffe

Received: 4 June 2021
Accepted: 30 June 2021
Published: 16 July 2021

Publisher's Note: MDPI stays neutral with regard to jurisdictional claims in published maps and institutional affiliations.

1. Introduction

Knowledge and the religious experience of God as the basic matrices of Christian self-cognition are the philosophical and theological codes which have always shaped the nature of religion and the foundations of centuries-old liturgical traditions. The feeling of divine presence and the belief in God's utter transcendentality are crucial for comprehending the world via divine revelation:

God as the Transcendent power is infinite, absolutely remote and aloof from the world; no regular and methodical pathways lead us to Him, which is precisely why He can become so endlessly close to us in His mercy; He is the closest to us, the most intimate, the most internal and the most immanent, He is closer to us than we are to ourselves. The God outside of us and the God within us, the absolutely transcendent becoming the absolutely immanent. By saying this, the only thing we deny is that seeing God is a mandatory and natural thing for those seeking Him. But the search, together with preparing oneself and finding the divine within, are attained by a human effort—the effort that God expects from us (Bulgakov 1994, pp. 24–25).

This passage by the Russian philosopher and theologian Fr. Sergii Bulgakov refers us to the tradition of negative theology, first[1] developed in the books ascribed to Dionysius the Areopagite, a Christian saint and the first bishop of Athens who lived in the 1st century CE.[2] Dionysius held that God cannot be imagined through concrete knowledge and specific ideas, since divine likeness is impossible to depict and the "primordially perfect wisdom"

(Dionysius 1987, p. 156) greatly exceeds the images created in words and intellect. The divine nature is above comprehension, and thus, God is unknowable and incomprehensible:

How can we speak of the divine names if the Transcendent surpasses all discourse and all knowledge, if it abides beyond the reach of mind and of being, if it encompasses and circumscribes, embraces and anticipates all things while itself eluding their grasp and escaping from any perception, imagination, opinion, name, discourse, apprehension, or understanding? The Godhead is superior to being and is unspeakable and unnameable. (Dionysius 1987, p. 53).

This sharply metaphoric text indicates that God is immeasurable and transcendentally unknowable. The author sees the hidden divinity in its boundlessness and in lying beyond comprehension, as divinity is beyond both everything substantive and insubstantive. By denying the particularly nominative (since God cannot be comprehended and signified in a specific way), negative theology attempts to move beyond the bounds of the understandable and the customary and reject the existing knowledge of the divine. God's trans-substantivity cannot be an object of reason, nor can it be comprehended by means of signs and symbols of this world. As Apostle Paul puts it, "Who only hath immortality, dwelling in the light which no man can approach unto; whom no man hath seen, nor can see: to whom be honour and power everlasting. Amen" (Holy 2010, 1 Tim. 6, 16). What negative theology offers is a different path of theosis. Unachievable through individual effort, this grace is "from above, and cometh down from the Father of lights, with whom is no variableness, neither shadow of turning" (Holy 2010, James 1, 17). In his treatises, St. Gregory Palamas further expands many ideas of the Areopagite's negative theology:

The divine nature is incomprehensible, as He is above all existence and any particular image, and inexpressible in words. In his substantive nature, God is far removed from everything, absolutely above any word or reason, any unity and any belonging, irrespective of anything, incomprehensible, impartial, invisible, incognoscible, nameless and absolutely inexpressible. (St. Gregory Palamas 2009, p. 27).

Examining the Corpus Areopagiticum, Palamas meditates on God's nature and energy and comes up with a concept of "uncreated divine light" as one these "energies" (see Kern 1996, p. 36). On the one hand, the incomprehensible divine indicates that the understanding God's presence in the world lies utterly beyond human comprehension. On the other, the incomprehensible nature of God through divine energy is revealed to the created world, and these revelations make the world copresent with the immanent and the eternal. Beyond this ambivalence lies the question of how to grasp the ungraspable. St. John of Damascus' response to it was as follows:

The Divinity, then, is limitless and incomprehensible, and this His limitlessness and incomprehensibility is all that can be understood about Him. All that we state affirmatively about God does not show His nature, but only what relates to His nature. And, if you should ever speak of good, or justice, or wisdom, or something else of the sort, you will not be describing the nature of God, but only things relating to His nature. There are, moreover, things that are stated affirmatively of God, but which have the force of extreme negation. For example, when we speak of *darkness* in God we do not really mean darkness. What we mean is that He is not light, because He transcends light. In the same way, when we speak of *light* we mean that it is not darkness. (St. John of Damascus 2010, p. 172)

The symbolic insights and mystical intuition, comprehending the Light in the Darkness—such is the way of negative and mystical theology. V.N. Lossky in his magisterial *The Vision of God* observed that every theology is "mystical" (Lossky 2006, p. 11), since it presents the divine mystery through revelations. In this respect, the apophatic or negative theology is the perfect path since the "complete ignorance" allows "the unknowable nature of God" to be reached (Lossky 2006, p. 126). Since God is transcendental, He cannot be grasped by reason or be an object of cognition: "it is by ignorance that we know the One who is above all that can be an object of knowledge" (Lossky 2006, p, 126). Negative theology as the "way of accessing the first cause of all things" and the method of "negating the negation and retrenchment of the subtraction" (Hadot 2005, p. 216), according to

Pierre Hadot, is linked to intellectual intuition and mystical vision. The "unutterable" and "inexpressible" mark the mystical order and the experience of the transcendental (Hadot 2005, p. 233). Nicholas of Cusa saw God as the infinite Absolute, who is "the form of forms, the being of being, the foundation and essence of all things; in this mode of being all things in God are the absolute necessity" (Nicholas of Cusa 1979, p. 117). God is the absolute maximum which surpasses all understanding, and His being can be contemplated only at the height of the *docta ignorantia*. Thus, the notion of God is moving closer to Nothingness. A question rightfully asked at the moment is how negative theology is related to the Russian avant-garde art, with its radical aesthetics and total formal experimentation. Was Russian avant-garde an act of theomachy[3], or was it trying to construct a new religion "out of nothing"[4]? Is it possible to see a metaphysics of its own, or an act of sacralization of the everyday, in the works of avant-garde reformers of art? All of these questions can hardly be addressed within the space of one article, but we may try to focus on the most significant challenges by reflecting upon the numerous theoretical works by Kazimir Malevich. The impact religious conscience and Old Russian icon painting had upon many artists of the Russian avant-garde is a commonly known fact, expanded upon by many Russian and Western scholars.[5] The discourse of such an impact is still attractive to many researchers, since it allows us to rethink the influence Christianity had on the rise of new visual symbolism and crucial philosophical and religious dichotomies, such as God/revolution, Christianity/Communism, Icon/Nothing. Nikolai Berdyaev considered Communism a type of religious faith (Berdyaev 1955, p. 129), and it seems to us far more than a paradox. Both the fervent energy of a new-born faith and a creative ecstasy are discovered within the Russian avant-garde, not only on the level of social and aesthetic utopia, but also within the body of sacred symbols which open the door into the transcendental and the mystical. Avant-garde art displays new forms of the religious feeling, and this concentrated energy and the experience of the perfect and the unreal leads to "setting oneself free from the whole created world, and beyond it, to creating out of Nothing" (Berdyaev 1990, p. 9). Malevich sees Suprematism as a new religion and the new artistic matrix of the world, both of which call for social change and the "Revolution of the Spirit".

El Lissitzky was perfectly right when he compared Suprematism and Christianity in his article titled *Suprematizm Mirostroitel'stva* (Suprematism of the World-building):

And if the Communism of today, which recognizes Work as its master, and Suprematism which puts forward the square of creative work, go hand in hand, then subsequently, Communism will fall behind in its movement. Suprematism, having covered all of life, will lead people out of the authority of work, of the authority of heartbeat, will liberate all of their creativity and will take the world further to the perfection of pure action. This is what we are expecting from Kazimir Malevich (Lissitzky 2003).

2. "The World as an Unencompassable Whole": The Apophatism of Kazimir Malevich

> I am looking for God for myself in myself
> God all-seing all-knowing all-powerful
> the future perfection of the intuition of the universal
> global supermind (Malevich 2004a, p. 440)

Kazimir Malevich is a universally recognized genius. A prophet, a thinker, an artist—the creator of Suprematism saw in himself a man of "the world of new transformations" (Malevich 1995b, p. 109). He interpreted Suprematism as a "source for creation of the world", with its Utopian model as a laboratory for the new religion of pure sentiments where "every form is a world" (Malevich 1995c, p. 53). The "concept of self-negation" (Epshtein 2013, p. 208), which is characteristic of the Russian culture in general, metamorphosed into an apophatic search for the Absolute in the Russian avant-garde when the conventional vision of the world was no longer possible at the turn of the 20th century. Not only do the visual codes change, but a whole new experimental artistic practice rises, aimed at the revolution of the Spirit and apophatic optics. This new eschatology demanded new, radically different symbols, which marked both the customary objects and notions, and the transformations, which reveal the true meaning of the apophatic—"naming the unnamable, pointing towards the invisible, knowing the unknowable" (Mikhailova 2000, p. 167). The Russian avant-garde features the apophatic discourse in both its philosophical/theological sense and in its poetics of the sacred, when denying the classical canons and reaching beyond Time and Space marked the attempt to "see the unvisible" in the formulas of Emptiness and Nothingness, of sacred geometry and symbolism of mystical theology[6]. Transcendental, otherworldly space as a new form of immersion and silent communication with God was the new reflexive method of rhetoric, which, in Russian avant-garde, was introduced by the notion of silence. In his 1923 article *Suprematicheskoe zerkalo* (A Suprematist mirror), Malevich declared, "There is no being in me, nor outside me, nothing can be changed by anything else, since there us nothing that could change, or be changeable. The essence of difference. The world as objectlessness" (Malevich 1995d, p. 273).

Malevich presents the Universe as the world of human difference, borderless, the one where God, the soul, religion, science and art equal nothing. This level zero of the crucial meanings of life aimed to illustrate another of Malevich's iconic assumptions—"there is nothing knowable, and the eternal Nothingness does exist" (Malevich 1995e, p. 242):

To study reality is to study what does not exist, what is unintelligible; and whatever is unintelligible for a human, is non-existent; consequently, it is the non-existent that must be studies. Studies will prove that things do not exist, and yet the infinity of them does, a "nothing" at the same time is a "thing" (Malevich 1995e, pp. 241–42).

The mystery of the Universe and "God as an Absolute of nature's perfection" (Malevich 1995e, p. 243) are both things incognoscible. The man who strives towards God as an ideal is on a conscious quest for harmony and cosmic calm. He is struggling to understand the meanings which are impossible to find, as they lie beyond human conscience. In fact, Malevich as an original thinker advances his own interpretation of divine ontology—not of a Christian God, but of a transcendental personality, perfection incarnate and incognoscible:

God cannot have a human meaning; since reaching him as the final meaning does not allow to reach God, since God is the final limit, or, more correctly, the limit of all meanings lies before God, and in God there is no meaning already. Thus, in the end, all human meanings pointing towards God the meaning, will be crowned with thoughtlessness, hence God is the non-meaning rather than meaning. It is his thoughtlessness that must be seen as objectlessness in the Absolute, in the final limit. Reaching finality is achieving objectlessness. It is indeed unnecessary to strive for God somewhere in the spaces of the celestial, since He is present in every human meaning, as every meaning of ours is, at the same time, a non-meaning (Malevich 1995e, p. 243).

Malevich is thinking out of the box, not as an orthodox Roman Catholic who has a good command of the Gospel and the canonical liturgical tradition. The Eucharist as a form of communion with God and realizing His presence transformed into the new covenant of the Great liturgy as "God's dance" (Malevich 1995f, p. 147), wherein the Master (a poet or an artist) partakes in the perfect nature. His notion of God forms a new metaphysics and an apophatic methodology based on the practice of negation. However, the Master does not reject God as the incognoscible perfection, but rather, the Church and the iconography of Christ. He aims to purge the particular from the religious thought and clear the Church of the unnecessary accessories: "The icon as it is appears as a low-cultured barbarity, and thoughtless worship of them belittles and obscures the spiritual in the Master who through the painted face has made himself part of the future highest being of his own spirit" (Malevich 2004b, p. 92).

We must note here that Malevich's own religious views underwent an evolution, as the status of the apophatic was changing within his philosophic doctrine.[7] He always interpreted the Church canons as a visionary, as a prophet of a new reality and of the transcendental future of the new order. Religion, in Malevich's idea, strives to turn a human into a "non-being" and reveal the true way to the immortality of the soul as a part of death-defying God. However, nothing can be proved or comprehended. Apparitions can be destroyed, but God is indestructible, and it is thus impossible to topple Him. In another crucial text of his, *Suprematizm. Mir kak bespredmetnost', ili vechny pokoi* (Suprematism. The World as Objectlessness, or the Eternal Rest (1922), Malevich further expands on the link between God, non-objective art and Suprematism: "No icon depicts the zero. The nature of God is zero good, and that is good in itself. Zero [is] a circle of transformation of everything objective into non-objective" (Malevich 2000, p. 84). This is how he comes to formulate an important thesis on Suprematism as a "zero limit" and Nothingness liberated. The artistic world, according to Malevich, has become "new, objectless, pure" (Malevich 1995c, p. 53) in the same way as a Deity obtains a new form of existence. Rest and meditative contemplation, the infinity of space and the esoteric emptiness—the essence of total perfection—all of these are similar to the eternal "Nothingness": "God is rest, rest is perfection: everything has been achieved, the worlds have been built, motion has been set in eternity" (Malevich 1995e, p. 257). The dichotomy of the objective and the objectless appears here as part of the dialectic of the spiritual and material: "The action of the object-focused consciousness is an empty action. There is no reason for viewing this action as a kind of a highest cause and thus privilege it—no more, at least, than granting this right to spiritual and even religious worldview. In general, both of these worldviews face the same objectlessness" (Malevich 1995e, p. 255).

Several questions can be brought up at this point, such as how this objectlessness is linked to the divine creation of the imperfect world and to God himself. Why does Malevich grant a special ontological status to it? Is mysticism of the artist at all compatible with Christianity? Some answers to these questions can be found both in Malevich's theoretical reflections and his Suprematist practice. He sees objectlessness as a feature of divine rest and of the existential Universe:

Nothingness, i.e., the resting God, and it happened to be so that Nothingness was God, and passing through perfection, it returned to the nothing it had been, because that is what it still was. "Nothing" cannot be researched or studies, since it is "Nothing", but in that "Nothing", man, "the thing" revealed itself. Since "the thing" cannot comprehend anything, it becomes "nothing"; now what follows from this—that man exists or that God exists as "nothing", as the objectless?" (Malevich 1995e, p. 264)

This passage clearly demonstrates that the apophatic intention of God's unknowability is transformed into a separate category of "nothingness", understood as "non-being". Correspondingly, the Creator is not an object of our perception, but the unknowable primary origin of the world. "Nothingness" is not only the infinity of the Absolute and a symbol of the complete unknowing, but also the perfect transcendentality. Malevich sees Suprematism as a method of comprehending the ineffability of the Absolute. In this context, many researchers see a link between his philosophy and Buddhist–Hindu mysticism. Cornelia Ichin argues that Malevich's notion of emptiness hails back to the Taoist ideas of inaction, setting the self free of passions and returning to the original substance. By removing the focus on objects in art, Malevich in his *Black Square* comprehends the mystery of emptiness and nothingness (Ichin 2011, pp. 48–56). By trying to see through space and discern the invisible, the artist-thinker transcends the bounds of the conventional and focuses on the sign of the absolute in total emptiness. God appears in the shape of the black square. As the artist recollected later, by staring at the mystery of the black space, he "knifed down art, put its body in the coffin and sealed it with the black square" (Malevich on himself (Malevich o sebe) 2004, p. 77).

3. Sacred Mysteries of Suprematist Primary Forms: The Square, Circle and Cross

For more than a hundred years, the mysteries of Malevich's *Black Square* have been attracting the interest of both viewers and scholars of art. Hundreds of studies have been published in many languages. Scholars have long been trying to decipher this work of art as a provider of a number of meanings, as well as to suggest which tools of theoretical analysis will be most useful in this act of deciphering. Heated debates started in the wake of the X-ray fluorescence and infrared tests carried out in November 2015 at the State Tretyakov Gallery. The tests proved that underneath the *Black Square* are two layers of paint: a Cubofuturist image and a Proto-Suprematist one. An inscription was found, which said, "Africans fighting at night". According to art historians, this is a reference to the painting by Alphonse Allais (1880s), which showed a black rectangle titled "The battle between Africans in a cave in the dead of night".[8] Was this the poet's joke or an appropriation of the experience of liberating art from its figurative and object-focused manacles? It is hard to guess why Malevich overpainted these palimpsest levels: Was it that he had no other canvas to use? Or was he trying to redo the two previous versions? Or maybe it was a gesture which symbolized the victory of Suprematism as the most perfect form of art?

It is well known that the *Black Square* was first exhibited at the "Last Futurist Exhibition: Oh—Ten (0,10)" in St. Petersburg. It was displayed in what is known as "the red corner" of the room, where traditionally, an icon would be placed[9] (see Vakar 2015; Shatskikh 2000; Levkova-Lamm 2004). The provocative character of the new plastic form brought about fierce debate, involving many members of the artistic intelligentsia. In a January issue of the *Rech* newspaper, the doyen of the Mir Iskusstva (World of Art) group, painter and art historian Alexander Benois gave his impression of the painting as follows:

Without a catalogue number, high under the ceiling, upon the place of an icon, hangs Mr. Malevich's "work"—a black square framed in white. It is, without doubt, the very "icon" that Messrs. Futurists plan to replace the Madonnas and shameless Venuses with. This is the "power over the forms of nature" that is the logical consequence of more than just Futurist art alone. A black square in a white frame is not only a mere joke, a simple challenge, a small accidental event in the house on Field of Mars. It is an act of self-assertion by a power named "abomination of desolation", a power that takes pride in the fact that its haughtiness, its hubris and its trampling down on everything tender and loving will make us all perish" (Krusanov 1996, p. 262).

What Benois saw in this "farce of contemporary culture" (Krusanov 1996, p. 262) was nothing more than an abomination and the decay of culture. He saw the Black Square as an anti-Christian act in defiance of Orthodoxy. Malevich's response was concise: " I have nothing else but a naked, frameless—like a pocket—icon of my time, and struggling is hard. But the happiness of being different from you gives me energy to move on and on, deeper into the emptiness of deserts. For there is transfiguration waiting for me there" (Malevich on himself (Malevich o sebe) 2004, p. 85). This minimalist form of abstract geometrical shape has since become an emblem of new art, where "the mystical optics of black shape" (Levkova-Lamm 2004, p. 17) was represented as both part of a personal story and a non-objective apophatic symbol. The Black Square is a symbol of a hidden perfection. Natalia Rostova sees its religious mystery in the fact that "Malevich appears here as a negative theologian, but not within a Christian tradition dominated by the idea of the transcendent God. He comes as a new kind of negative theologian, who himself acts the impossible way" (Rostova 2021, p. 27).

Even at the very foundations of theoretical reflections on the figure of black square, the problem of negation makes itself quite conspicuous. The logic of witnessing how artistic matter turns into non-objective art is such that the simplest of geometrical forms attains the status of the philosophical Universe. The artist's formula, "The living square, the royal infant" (Malevich 1995c, p. 53), contains two important assumptions. "The living square" symbolizes the flexible and changeable form, the process of dispersing and destroying the wholeness of an object: "Here Deity commands the crystals to transfer into a new form of existence" (Malevich 1995c, p. 47). The metaphor of the "royal infant" has many associations of its own, including references to Russian icon painting traditions. Malevich offers his own interpretation of classical icon painting by appealing to his semiotic practice. It is not accidental that he mentioned "new signmaking" in Suprematism so often (Malevich 1995g, p. 162).

On the one hand, the *Black Square* can be interpreted as a paraphrase of an icon and the "new icon of the transcendental" (Tarasov 2017, p. 125). Inessa Levkova-Lamm sees a similarity between the Square and the Orthodox icon of the Holy Mandylion (c. 1657), now in the collections of the Yaroslavl Museum-Reserve. On the icon, the face of Jesus the Saviour appears in the center of a black and light ochreish squares (Levkova-Lamm 2004, p. 114). Many other references to Christian liturgy and Orthodox practice have been found; they are especially frequent in Malevich's early frescoes and pre-Suprematist paintings (Lozovaia 2011; Mudrak 2016; Marcadet 2000; Bowlt 1991; Kurbanovsky 2007; Spira 2008; Tarasov 2002, 2011). The artist was quite vocal when talking about the influence of the Novgorod icon style and icon painting in general. He reflected upon the emotional impact of peasant art and of the new meanings rising in his painting in the wake of his interest in Russian icons. "I saw in it the whole spirituality of the peasant times", he wrote, "I understood peasants through the icon, I saw their faces as those of common people, rather than saints. [I understood] the colour and the attitude of an icon painter" (Malevich on himself (Malevich o sebe) 2004, pp. 28–29). "Icon archetypes" (Sarabyanov 1993, p. 168) have also been found in the *Black Square*—not only in its visual imagery and iconographic narrativity, but also in the absence of the proto-image and the apophatic interpretation of the painting. The presence of the divine is discovered in the very rhetoric of the black square, and the face of God is recognized through empty space. An entry into the world of the collapsed things and signs through the annulment of the object introduces a new dimension of the Suprematist objectlessness, which, in turn, takes the viewer a step closer to comprehending the transcendental spiritual Absolute. Having seen the face of God in the square, the artist relies on patterns of negative theology in constructing his primary philosophical formula: " What we call reality is in fact infinity which has neither weight nor measure, neither space nor time, nothing absolute and nothing relative, nor anything shaped into a form. It can be neither imagined nor known. Nothing is comprehensible, and yet the eternal "nothingness" exists" (Malevich 1995e, p. 242).

On the other hand, Malevich assumes that the existence of non-existence is ambivalent. The non-objective nature of Suprematist primary forms is interpreted as the being of non-being. Malevich saw Suprematism as the religion of pure action. In 1920, he wrote to Mikhail Gershenzon, "I am not sure what you will think of my opinions, but I interpret the three squares and the cross as the foundations of art, and moreover, of everything, and also as a new thing—religion. I also see [in it] the New Temple (Malevich on himself (Malevich o sebe) 2004, pp. 127–28). The square sets the pace for the new time as it is the "embryo of all opportunities" (Kovtun 1990, p. 105). It is the square that helps Malevich preach the new Suprematist religion and proclaim the power of the symbol of emptiness, bowing before the authority of the Zero. *The Black Square* is the limit beyond which there is no rationality or logic of verisimilitude, and the world appears as Infinity.[10] While setting the transcendental supraterrestrial space and the evasive motion, Malevich stares into the abyss of metaphysical emptiness and discovers that the zero point is polymorphous and inexhaustible: "I have been transformed in the zero form and move beyond the zero point to art, i.e., to Suprematism, to the new artistic realism of non-objective art" (Malevich 1995c, p. 53). *The Black Square* is the zero matrix reflecting the foundations of the Universe, and the spiritual Absolute as the liberated Nothingness symbolizes the apophatic non-being of God: "If religion has comprehended God, it has also understood the zero" (Malevich 1995d, p. 273). We see the metaphysics of zero as the formula of apophatism and of the mystical phenomenon which describes the process of negation, understood as the endless production of secret meanings. As Malevich wrote to Gershenzon, "The square is now a living one, creating a new world of perfection, and I see it in a different light now, as something other than art. I've had an idea that if mankind has always been portraying God in their own image, then the Black square is the image of God as the essence of his perfection on the new path of today's beginnings" (Malevich on himself (Malevich o sebe) 2004, p. 125). In his attempt to decipher the secret of the black square,[11] the artist often mentioned revelations and a mystical experience while creating the eponymous painting: "I have invented nothing. I felt the night within me and saw the new thing I called Suprematism. It came as a black plane, shaped as a square" (Malevich 1998a, p. 30). The formula of plastic minimalism and the conciseness of the Suprematist figure lead us to a conclusion that Malevich manifested the black square as an apophatic symbol of otherworldliness, as a form of the ineffable and inexpressible which cannot be reduced to a clear definition. At the same time, the transcendentally signified (God) appears as a negation of all previous forms of the objective world. Having accumulated all of his theoretical breakthroughs in the Suprematist model, Malevich has turned to primary forms, which would be relevant to his new system of art.

The Black Square is open to many interpretations, which can be converted into a discourse of morphological functions of Suprematist geometry. Metaphorical constructs telling of the world's first elements (interpreted as the quintessential signs of human existence) contain figures made of squares, which the artist explains in an interesting way in his book *Suprematizm: 34 risunka* (Suprematism in 34 drawings):

In its historical development, Suprematism passed through three stages of black, colored and white. The three Suprematist squares are establishing certain worldviews and world sentiments. The white square, besides the purely economic movement of the shape of the whole new white worldbuilding is also an impetus towards justifying worldbuilding as a "pure action", as self-knowledge in the purely utilitarian perfection of the "all-man". In communal life it has more meanings: black as the sign of economy, red as the signal of revolution and white as pure action. The white square I painted gave me an opportunity to study it, which resulted in a brochure on "pure action". The black square defined the economy which I introduced as art's fifth dimension. Isn't it strange that the three squares show the way, while the white square brings the white world (worldbuilding), thus establishing the sign of purity of artistic life (Malevich 1995h, pp. 187–88).

The empiric outcomes of this classification allow us to go beyond seeing it only as "the methodological foundation of the concept if developing the Suprematist plane" (Goryacheva 2020, p. 13). We can also see it as a description of mechanisms to record the potential infinity of the metaphysical. Discovering all of the capabilities the squares have and tracing the multitude of their links constitute the new meanings of the sacred symbols. A reminiscence of the black square can be found in "pure colour art" (Malevich 1995h, p. 150)—in an abundance of modifications of the white and red squares. Within the space of a Suprematist model, the multidimensional and multivalent character of primary geometric forms appears on various levels of stratification and complexity. The minimal structure is the white square—*Beloye na Belom* (White on White) or *Belyi Kvadrat na Belom Fone* (A White Circle with White Background)—painted in 1918 and first displayed at Malevich's personal 10th State Exhibition "Non-objective art and Suprematism" in Moscow. The new optics of pure sensations in the liberated space was an illustration of Buddhist emptiness, transcendental abyss and apophatic non-existence. Malevich was far from the purely religious interpretation of the white color as a symbol of divine light and formal symbolization. The Via Negativa as a method of negation of the knowability of God (first appearing as early as in Neo-Platonism), and negation of clear definition of God in the language of human notions—both of these produced the new optics of pure sensations and were the paradigm of actualizing various levels of transcendentality. Researchers have long been emphasizing a typological similarity between Malevich's Suprematism and the works of the great mystics of the East the Russian artist cannot have read, such as Lao Tze, Zhuang Zhou or Huineng, and also the negative path of knowing God and the mystical teachings of Meister Eckhardt, etc.[12]

For us, of high importance is the very fact that the artist was capable of complex methodological reflection and apophatic thinking. In our view, the white circle was a message, which the artist himself saw as a productive one. The mythology of whiteness as a symbol of multidimensional space and of the presence of God articulates the balance between the sacred and profane, the real and the mystical. The white square is drastically different from the black one, since the white abyss is a leap to infinity, to objectless nature and a new form of conscience: "I tore through the blue lampshade of colored limitations and stepped into the white. Float on! The free white abyss, the infinity lie in front of you". (Malevich 1995a, p. 151).

A rotating square formed a circle,[13] and the transformed white and black squares, a cross. Suprematist figures of the square, cross and circle are secret mystical signs, which lead us to more questions than answers. At this point, we would venture to suggest a number of hypotheses. The circle can be seen as a black hole, a symbol of cosmic transition to another transcendental space and time continuum. Or as a black sun, which, as a symbol, contains the autoreflexive message on the transfer from the real to the unreal. The circle both displays the new plastic ideas of Suprematism and presents, in accordance with Slavic mythology, a face of Chernobog (the Black god), the twin of Belobog (the White god of light and creation). Both gods were made by the Creator God to perform their own functions, but both strove to bring the world to a state of balance. A number of sacred contexts can be discerned here as well. The black circle symbolized non-existence, the beginning and the end of the macrocosm, an ontological infinity. What we see here is an apophatic formula which implements the notion of the black circle as a point of rest, and the idea of non-manifested divine existence. Darkness is born together with the Light, since Light rises out of Darkness, out of mystery, out of the innermost, of absolute calm. Malevich made a most eloquent point of it in his letter to Mikhail Gershenzon on 13 October 1924:

Light and darkness from the point of view of objectlessness is the same substance, different on no more than two counts. The sun compared to reason is a dark spot, but there is something which will make reason itself be the same. Darkness is neither good nor evil; it rather is rest, and woe to him who comes to vex darkness with light, for it is the light that creates representations that men assault with their bayonets and save themselves. The world lies in darkness; I understand darkness which has neither will nor representation—and it was Schopenhauer who wrote a book titled *The World as Will and Representation*. Of course, I haven't read it—but I read the title in the window of a bookshop. I haven't given it a proper thought, but I trust the world can exist only where there is neither will nor representation. And where the latter two stand, there can be no world, only a struggle between representations. (Malevich 2000, p. 354)

Malevich's negative theology is closely linked to the notion of "religion of pure action" he often talked about. Contrary to the Taoist notion of "pure experiences", the artist sees the Suprematist primary forms as new visual codes of objectlessness. The religious world, according to Malevich, is the world of the unknowable Absolute where union with God is only possible by means of negating matter and objectness. Fully realizing that knowing God and theosis are both impossible, Malevich codifies this impossibility into a set of statements where the religion and artistic text are both converted into the zero point of meaning—"the infinite nonsense". In a letter to Mikhail Gershenzon on 11 April 1920, he wrote:

For many years I have been focused on my own progress in painting, leaving the religion of the Spirit aside, and now I am entering or returning to the world of religion, and I do not why it is happening now. I go to churches, look at [the images of] saints and the whole living spiritual world—and I see within myself—or maybe in the whole world—that the moment of change has come for religions. I saw it this way: just as art was progressing to the purest form of performance, the world of religion was approaching the religion of the same; all the saints and prophets were motivated by the same but could not bring it to life. Reason prevented them by focusing on the ends and point of everything—and the whole work of the world of religion crashed against these two parts of the wall of reason. The work became finite and could no longer reach infinity. The entrance of religion into pure action is for me a mandatory requirement; the endlessness of the work of the spirit of religion is a universal substance. Then its power will be no longer contained within itself, for prayer will not be limited with its ends and meaning—and will turn into the action of endless nonsense (Malevich 2000, p. 340).

Malevich's semiotics of the cross is linked to the many meanings a symbol has within the author's epistemology. These semantic articulations found their best expression in Malevich's series "Mystical Suprematism" (1920–1922) [Figures 1 and 2].

Figure 1. *Mystic Suprematism (Black Cross on Red Oval)*, 1920–1922. The collection of the Stedelijk Museum of Amsterdam. © Stedelijk Museum.

Figure 2. *Mystic Suprematism (Red Cross on Black Circle)*, 1920–1922. The collection of the Stedelijk Museum of Amsterdam. © Stedelijk Museum.

In the focus of the paintings is the four-pointed Orthodox cross with a slanted crossbeam. This is a reference to the Orthodox visual version of the Crucifixion, where each of Christ's feet was nailed to the cross with a separate nail—unlike the crossed feet and the single nail for both feet on Catholic crucifixes. This is the first riddle of Malevich's crosses—the artist was a Roman Catholic, and we could have expected he would use Catholic imagery. The second riddle is in Malevich's use of another important symbol of orthodox iconography. In his *Cherny krest na krasnom ovale* (Black Cross on Red Oval), Malevich places the cross within a red oval resembling the mandorla[14]—the light around the body of the Savior. In Christian iconography, it symbolized the divine glory and holiness, the victory of spirit and light coming from the Savior. Its color of red is reminiscent of the fiery nature of the divine: "For the Lord thy God is a consuming fire, even a jealous God" (Deut 4:24). This symbol of Christ transfigured and His divine majesty are also found in Andrei Rublev's *The Savior Enthroned in Glory* (1408, made for the Assumption Cathedral in Vladimir). The blue-green mandorla and the red field around the Savior in this icon clearly gravitate towards the simplest of geometric figures—the square and cross. The symbolism of the icon has given rise to ambivalent interpretations: Oleg Tarasov, for instance, talked of "the ambiguous nature of signs and symbols in the Christian worldview" and the complex structure of the text of the Orthodox icon. (Tarasov 2011, p. 120). In the Russian icon painting, we see a multitude of symbols of sacred space, which mark ambiguous discourses of theology. The important thing for us is the cultural meaning of the two geometric figures, since the cube or rectangle symbolized the earth, and the circle, the sky. Art historians currently interpret *The Savior Enthroned in Glory* in the context of early Christian iconography: Jesus is seated against the background of the Universe (the sky and angels), and the earth is represented by a large red square, with its four corners standing for North, South, East and West. It is in this combination of the earthly/profane and the heavenly/sacred that the mandorla captured this duality and symbolized the Transfiguration of Christ.

Malevich as a Catholic must have known the Christian iconography quite well. However, it would have been a mistake to only look for the religious context in his mystical paintings. The mandorla for the artist was more of a mystical space, and the cross, rather an ideologeme of Suprematism, which postulated the new sign system of the non-objective art. By liberating the Suprematist art from objectness, Malevich followed the notion of "dissimilar likeness" advanced by Pseudo-Dionysius, since God is unknowable and impossible to signify. The square, circle and cross were formulas of the apophatic faith of Modern times, since "Suprematism is also a prism, through which, however, no 'thing' could be seen" (Malevich 1998b, p. 41). Primary forms were the universals of the "new face" of the future, signs of the multidimensional Universe, which was, in turn, Malevich's shorthand for a special world structure of the transcendental. The descriptions of Suprematist narratives lay in the transformation of matter's previous state and its forms and in the justification of the new energy of objectlessness: "Having traced the Suprematist line, and energy as the major line of life, I saw they were identical in their dynamism, and not in their ideology. I then made a graph of the movement of color, and three points became clear in Suprematism: the rainbow, the black and the white. The white square is some kind of border of the rising motion" (Malevich 1998b, p. 50).

4. Conclusions

Malevich's negative theology still remains a moot point. A researcher who aims to comprehend the art of the non-objective and interpret the deeper layers of the texts in the context of theological exegesis and the apophatic thought will have to grapple with both the clear and evident surface meaning, and with deeper, palimpsest-like layers, which demand a more in-depth scholarly approach. We believe that Malevich has intuitively reached the ideas of negative theology by relying primarily on the outcomes of the artistic system he had discovered, and by rejecting the objectness of art. Having transcended the traditional boundaries of art, he felt the presence of the invisible world of the spiritual Absolute and tried to add the new sacred and symbolic aspect to the geometrical abstraction. Malevich imagined Suprematism as the artistic discovery of the mystical space and as a reflection of the invisible world, a "pure form" system which could transform reality. The artist saw his God through the prism of his own subjective experience and apophatic thought, relying on the negation of "object-focused" consciousness and on the mystical experience of the Absolute. The latter's utter ineffability found its reflection in the abyss of *The Black Square*. The impossibility to express the sacred or divine in the language of Suprematism pointed towards the transcendental nature of the spiritual, which could be made explicit on the level of metaphysical experience. The mystical theology formula of the *Unio mystica* as a direct union and communication with God, which mystics described in their works, is understood by Malevich as the "supranaturalist" effect (in his original words). In the supranaturalist space/time continuum, all forms are reduced to zero: "I was transformed into zero shape and went beyond the 0–1" (Malevich 1995c, p. 34).

To go beyond the limits of the visible world, open new horizons and be transferred to other cosmic dimensions—all of these refer to a mystical and religious experience of contacting something infinitely larger and unnamable. Then, the artist would have to symbolically express these feelings in zero-point texts. Having intuitively felt the boundaries and limits of the knowable, the worldly and divine, Malevich, in his abstract geometry of the black square, cross and circle, took these primary forms up to the level of apophatism, both in the artistic and symbolic aspects. Malevich repeatedly postulated that *The Black Square* is not a painting: "There can be no talk of painting in Suprematism— it has long become obsolete, and the artist himself is a living prejudice from the past." (Malevich 1995h, p. 189)

What we deal with here is a philosophical text which creates new meanings and accumulates a certain kind of metaphysical energy. Malevich wanted to use the language of the new art to express the apophatic infinity, and to render the unknowable Universe in the formulas of Suprematist minimalism. His own search for God shines through the semantic structure of the white square, which crushes the conventional ideas of polymorphism in art. The manifested phenomenon of apophatic emptiness and cosmic silence reflect the experience of sacred space and otherness as an invisible and objectless form: "The infinite white of Suprematism allows the ray of sight to proceed further without any limit" (Malevich 1995h, p. 187).

Funding: This paper has been supported by the RUDN University (Peoples' Friendship University of Russia) Strategic Academic Leadership Program.

Institutional Review Board Statement: Not applicable.

Informed Consent Statement: Not applicable.

Data Availability Statement: Not applicable.

Conflicts of Interest: The author declares no conflict of interest.

Notes

[1]   It must be added that elements of apophatic thought can be found in Plato's Socratic dialogue "Parmenides". The dialogue reproduces a talk between 65-year-old Parmenides, 40-year-old Zeno, Socrates who was 20 at the time, and Aristotle, then just a youth at the Great Panathenaea in 450 BCE. In the first hypothesis, the philosopher talks of the abstract and universal Unity which is "unlimited, if it has neither beginning nor end" (Plato 1997, p. 137d). In the dialogue between Parmenides and Aristotle we can easily discern the main tenet of negative theology: the via negativa as a means of postulating the Unity, "[P]: So neither name nor account belongs to it, nor is there any knowledge or perception or opinion of it. [A]: It appears not. [P]: So it is neither named nor spoken of, nor will it be an object of opinion or knowledge, nor does anything among things which are perceive it". (Plato 1997, p. 142a) For more details, see (Dodds 1928; Rist 1962).

[2]   Theologians and scholars of religion are still debating the origin of the corpus of texts published under the name of Dionysius the Areopagite (including the *Divine Names, Mystical Theology, Celestial Hierarchy, Ecclesiastical Hierarchy* and *Ten Epistles)* and its dating. Some think they are 5th century forgeries, while others guess their putative real authors, such as Severus of Antioch, Dionysius the Great, Ammonius Saccas, Peter the Iberian, or John Philoponus. For more on this, see (Koch 1900; Stiglmayer 1928; Devreesse 1930; Puech 1930; Nutsubudze 1942; Honigmann 1952; Golitzin and Bucur 2013; Kharlamov 2016).

[3]   One of the first to address this issue was Galina Belaya in a small article titled "Avangard kak Bogoborchestvo" [Avant-garde as creating God] (Belaya 1992).

[4]   I. Klyun made an interesting point in his brochure "Tainye poroki akademikov" [Secret sins of academic artists]: "A huge challenge arose in its full might—to create form out of *Nothing*". See (Kruchenykh et al. 1916, p. 29).

5 (Ampilova 2009; Bychkov 1998; Goryacheva 1993; Sarabyanov 1993, 1998; Tarasov 1992, 2011; Marcadet 2000; Sarabyanov and Shatskikh 1993; Strigalev 1989; Mudrak 2016; Schreier 1966; Bowlt 1991; Spira 2008).

6 It must be noted that, although Malevich has been the subject of numerous monographs and articles, his apophatic thought mostly stays under the radar in Russian scholarship. The reason for this goes beyond the difficulty of comprehending his vast philosophical and theoretical heritage: the very problem is quite provocative. For a long time, the art of the avant-garde has been studied in Russia solely in the context of theomachy and social utopianism. The issues of negative theology and Suprematism have been treated, albeit cursorily, in (Bychkov 1998; Mikhailova 2000; Shatskikh 2000; Levkova-Lamm 2004; Kurbanovsky 2007; Lozovaia 2011; Rostova 2021).

7 Miroslava Mudrak has noted a significant influence of Byzantine liturgy and Christian tradition of icon painting on Malevich's early Symbolist frescoes. She also described a synthesis of Oriental and European iconographical motives in his art. See Mudrak, Miroslava. 2016. *Kazimir Malevich i vizantiiskaia liturgicheskaia traditsiia [Kazimir Malevich and the Byzantine liturgical tradition]*. Iskusstvo, № 2 (597). pp. 50–67.

8 See (Voronina and Rustamova 2015).

9 On the dating of Malevich's Black Square, see (Goryacheva 2020).

10 This might be a reference to Robert Fludd's "Utriusque cosmi maioris scilicet et minoris Metaphysica, physica atque technica Historia" (1617). Fludd, a mystic and astrologer, saw the black square as a symbol of the darkness of the Universe—a macrocosm where eternal Darkness reigns supreme.

11 It must be noted that the articles and treatises by Malevich devoted directly to the *Black Circle* have not yet been discovered. Tatyana Goryacheva has suggested that in his Vitebsk years, Malevich wrote an article titled *Solntse i Cherny Kvadrat* [The Sun and the Black Square], or, according to a different source, *Beloe Solntse i Cherny Kvadrat* [The White Sun and the Black Square], but no manuscript of it is currently known. See (Goryacheva 2020, p. 27).

12 See (Andreeva 2019; Bychkov 1998; Vakar 2015; Ichin 2011; Gill 2016; Lozovaia 2011; Levina 2015; Shatskikh 2000; Kurbanovsky 2007; Nakov 2010; Bowlt 1991; Spira 2008; Tarasov 2002, 2011; Milner 1996; Bering 1986; Krieger 1998).

13 The *Black Circle* (1915) is now part of the collections of the State Russian Museum in St. Petersburg. It went on display together with the *Black Cross* and *Black Square* at the 0.10 exhibition.

14 Mandorla (Italian for "almond"), or "vesica piscis" (Latin for "swim bladder")—a symbolic depiction of oval-shaped shining, or an almond-shaped halo around the body of the Savior, which appears on the icons of Transfiguration and Ascension.

References

Ampilova, Anna. 2009. *Suprematizm i ikonopis': k probleme formy i soderzhaniia v izobrazitel'nom iskusstve nachala XX veka [Suprematism and Icon Painting: On form and Content in Pictorial Art of Early 20th Centiry]*. № 4. Moscow: Prepodavatel' XX vek, MPGU, pp. 370–74.

Andreeva, Ekaterina. 2019. *Kazimir Malevich. «Chernyi kvadrat» [Kazimir Malevich: The Black Square]*. St. Petersburg: Arka.

Belaya, Galina. 1992. *Avangard kak bogoborchestvo [The Avant-Garde as Theomachy]*. № 3. Moscow: Voprosy literatury, pp. 115–24.

Berdyaev, Nikolai. 1955. Kommunizm i khristianstvo [Communism and Christianity]. In *Istoki i smysl russkogo kommunizma [The Origins of Russian Communism]*. Paris: YMCA-PRESS.

Berdyaev, Nikolai. 1990. *Krizis iskusstva [The Crisis of Art: A Reprint Editon]*. Moscow: SP Interprint.

Bering, Kunibert. 1986. *Suprematismus und Orthodoxie. Einflüsse der Ikonen auf das Werk Kazimir Malevics*. No. 2/3. Minneapolis: Ostkristliche Kunst, Fortress Press, pp. 143–55.

Bowlt, John. 1991. *Othodoxy and the Avant-Garde. Sacred Images in the Work of Goncharova, Malevich and Thier Contemporaries*. Christianity and the Arts in Russia. Edited by M. Velimirovic and W. Brumfield. New York: Cambridge University Press.

Bulgakov, Sergii. 1994. *Svet nevechernii: Sozertsaniia i umozreniia [Unfading Light: Contemplations and Speculations]*. Mysliteli XX veka [20th Century Thinkers]. Moscow: Respublika.

Bychkov, Victor. 1998. Ikona i russkii avangard nachala XX veka [The icon and the Russian avant-garde in early 20th century]. In *KorneviShche OB. Kniga Neklassicheskoi Estetiki*. Moscow: IF RAN, pp. 58–75.

Devreesse, Robert. 1930. *Denys l'Aréopagite et Severe d'Antioche*. Archives d'histoire doctrinale et littéraire du moyen âge. Paris: Librairie Philosophique J. Vrin.

Dionysius, Pseudo. 1987. *The Complete Works*. New York: Paulist Press.

Dodds, Eric. 1928. *The Parmenides of Plato and the Origin of the Neoplatonic 'One'*. Classical Quarterly, No. 22. Cambeidge: Cambridge University Press, pp. 129–42.

Epshtein, Mikhail. 2013. Russkaia kul'tura na rasput'e. Sekuliarizatsiia i perekhod ot dvoichnoi modeli k troichnoi [Russian culture at the crossroads. Secularisation and the tranfer from the dual to the ternary model]. In *Religiia Posle Ateizma. Novye Vozmozhnosti Teologii [Religion after Atheism: New Opportunities for Theology]*. Moscow: AST Press, pp. 159–222.

Gill, Charlotte. 2016. A "Direct Perception of Life". How the Russian Avant-Garde Utilized the Icon Tradition. *IKON* 9: 323–34. [CrossRef]

Golitzin, Alexander, and Bogdan G. Bucur. 2013. *Mystagogy: A Monastic Reading of Dionysius Areopagita*. Collegeville: Liturgical Press.

Goryacheva, Tat'iana. 1993. Malevich i Renessans [Malevich and the Renaissance]. *Voprosy Iskusstvoznaniia* 2–3: 107–18.

Goryacheva, Tat'iana. 2020. Pochti vse o «Chernom kvadrate» [Almost everything about the Black Square]. In *Teoriia i Praktika Russkogo Avangarda: Kazimir Malevich i ego Shkola [Theory and Practice of the Russian Avant-Garde: Kazimir Malevich and his School]*. Edited by T. V. Goryacheva. Moscow: AST.

Hadot, Pierre. 2005. Apofatizm, ili Negativnaia teologiia [Apophatism, or Negative Theology]. In *Dukhovnye uprazhnenia i antichnaia filosofia [Spiritual/Per. s frants. pri uchastii V.A. Vorob'eva. Moscow]*. St. Petersburg: Stepnoi Veter, pp. 215–26.

Holy, Bible. 2010. *Paul´s First Letter to Timothy, The Epistle of James*. King James Bible: 400 th Anniversary Edition. Oxford: Oxford University Press.

Honigmann, Ernest. 1952. *Pierre l'Iberian et les écrits du Pseudo-Denys l'Aréopagite*. (Classe des leres et des scien-ces morales et politiques. Mémoires, 47, fasc. 3). Bruxelles: Con Académie royale de Belgique.

Ichin, Cornelia. 2011. *Suprematicheskie Razmyshleniia Malevicha o Predmetnom Mire [Malevich's Suprematist Reflections on the World of Objects]*. № 10. Moscow: Voprosy filosofii, Izdatel'stvo Nauka, pp. 48–56.

Kern, Archimadrid Kiprian Konstantin. 1996. *Antropologiia sv. Grigoriia Palamy. [The Anthropology of St. Gregory Palamas]*. Introduction Article by A. I. Sidorov. Moscow: Palomnik.

Kharlamov, Vladimir. 2016. Pseudo-Dionysius the Areopagite and Eastern Orthodoxy: Acceptance of the Corpus Dionysiacum and Integration of Neoplatonism into Christian Theology. Theological Reflections. № 16. *Euro-Asian Journal of Theology* 2016: 138–54. [CrossRef]

Koch, Hugo. 1900. *Pseudo-Dionisius Areopagita in seinen Beziehungen zum Neoplatonismus und Misterienwesen*. Forschungen zur christlichen Litteratur und Dogmengeschichte. 1 Bd., t. 86, 2-3 Heft. Mainz: Verl. von Franz Kirchheim.

Kovtun, Yevgeniy Fedorovich. 1990. *Nachalo suprematizma [The Rise of Suprematism]*. Malevich khudozhnik i teoretik. Moscow: Sovetskii khudozhnik.

Krieger, Verena. 1998. *Von Der Ikone Zur Utopie: Kunstkonzepte der Russischen Avantgarde*. Köln: Böhlau.

Kruchenykh, Alexey, Ivan Klyun, and Kazimir Malevich. 1916. *Tainye poroki akademikov [Secret Sins of Academic Artists]*. igns: From Iconoclasm to New Theolo. Moscow: Budetlian.

Krusanov, Andrei Vasilievich. 1996. Russkii avangard: 1907–1932. (Istoricheskii obzor) [The Russian avant-garde: 1907–1932: A historical survey]. In *Boevoe desiatiletie [The Decade of Battles]*. St. Petersburg: Novoe Literaturnoe Obozrenie, 3 vol., vol. I.

Kurbanovsky, Aleksey. 2007. A. *Malevich's Mystic Signs: From Iconoclasm to New Theology*. Sacred Stories: Religion and Spirituality in Modern Russia. Edited by Mark D. Steinberg and Heather J. Coleman. Bloomington: Indiana University Press.

Levina, Tatiana. 2015. Ontologicheskii argument Malevicha: Bog kak sovershenstvo, ili vechnyi pokoi [Malevich's ontological argument: God as perfection, or the eternal rest]. *Artikul't* 19: 18–28.

Levkova-Lamm, Inessa. 2004. *Litso kvadrata: Misterii Kazimira Malevicha. [The Face of the Square: The Mysteries of Kazimir Malevich]*. Moscow: Pinakoteka.

Lissitzky, El. 2003. Suprematizm mirostroitel'stva [The Suprematism of World-Building]. In *Al'manakh Unovis. [Almanach UNOVIS]*. A Facsimile Edition. Preparation, Introduction and Comments by T. Goryacheva. Moscow: SkanRus.

Lossky, Vladimir Nikolayevich. 2006. *Bogovidenie [The Vision of God]*. Translated by V. A. Reshchikova. Comp. and Introd. by A. S. Filonenko. Moscow: AST.

Lozovaia, Lidia. 2011. *Bogoslovskie aspekty russkogo avangarda: Kazimir Malevich i Vladimir Sterligov [Theological Aspects of the Russian Avant-Garde: Kazimir Malevich and Vladimir Sterligov]*. No 950. Karazina: Koinoniia/Vestnik KhNU im. V. N, pp. 333–49.

Malevich, Kazimir. 1995a. Suprematizm. "Iz kataloga desiatoi gosudarstvennoi vystavki bespredmetnoe tvorchestvo i suprematizm" [Suprematism. "From the catalogue of the 10th state exhibition "Non-objective art and Suprematism]. In *Stat'i, manifesty, teoreticheskie sochineniia i drugie raboty. 1913–1929 [Articles, Manifestos, Theoretical Works and Other Texts, 1913–1929]*. Moscow: Gileia.

Malevich, Kazimir. 1995b. Ya prishel [I have come]. In *Stat'i, manifesty, teoreticheskie sochineniia i drugie raboty. 1913–1929 [Articles, Manifestos, Theoretical Works and Other Texts, 1913–1929]*. Moscow: Gileia.

Malevich, Kazimir. 1995c. Ot kubizma i futurizma k suprematizmu [From Cubism and Futurism to Suprematis]. In *Stat'i, Manifesty, Teoreticheskie Sochineniia i Drugie Raboty. 1913–1929 [Articles, Manifestos, Theoretical Works and Other Texts, 1913–1929]*. Moscow: Gileia.

Malevich, Kazimir. 1995d. Suprematicheskoe zerkalo [The Suprematist mirror]. In *Stat'i, Manifesty, Teoreticheskie Sochineniia i Drugie Raboty. 1913–1929 [Articles, Manifestos, Theoretical Works and Other Texts, 1913–1929]*. Moscow: Gileia.

Malevich, Kazimir. 1995e. Bog ne skinut. Iskusstvo, tserkov', fabrika [God has not been toppled: Art, Church and Factory]. In *Stat'i, Manifesty, Teoreticheskie Sochineniia i Drugie Raboty. 1913–1929 [Articles, Manifestos, Theoretical Works and Other Texts, 1913–1929]*. Moscow: Gileia.

Malevich, Kazimir. 1995f. O poezii [On poetry]. In *Stat'i, Manifesty, Teoreticheskie Sochineniia i Drugie Raboty. 1913–1929 [Articles, Manifestos, Theoretical Works and Other Texts, 1913–1929]*. Moscow: Gileia.

Malevich, Kazimir. 1995g. O novykh sistemakh v iskusstve [On new artistic systems]. In *Stat'i, Manifesty, Teoreticheskie Sochineniia i Drugie Raboty. 1913–1929 [Articles, Manifestos, Theoretical Works and Other Texts, 1913–1929]*. Moscow: Gileia.

Malevich, Kazimir. 1995h. Suprematizm. 34 risunka [Suprematism in 34 drawings]. In *Stat'i, Manifesty, Teoreticheskie Sochineniia i Drugie Raboty. 1913–1929 [Articles, Manifestos, Theoretical Works and Other Texts, 1913–1929]*. Moscow: Gileia.

Malevich, Kazimir. 1998a. «Nashe vremia iavliaetsia epokhoi analiza» ["Our time is the era of analysis"]. In *Stat'i i Teoreticheskie Sochineniia, Opublikovannye v Germanii, Pol'she i na Ukraine. 1924–1930 [Articles and Theoretical Works Published in Germany, Poland and Ukraine]*. Moscow: Gileia.

Malevich, Kazimir. 1998b. Mir kak bespredmetnost' (Fragmenty) [The World as Objectlessness (Fragments)]. In *Stat'i i Teoreticheskie Sochineniia, Opublikovannye v Germanii, Pol'she i na Ukraine. 1924–1930 [Articles and Theoretical Works Published in Germany, Poland and Ukraine]*. Moscow: Gileia.

Malevich, Kazimir. 2000. *Suprematizm. Mir kak bespredmetnost', ili Vechnyi pokoi: s prilozheniem pisem K. S. Malevicha k M. O. Gershenzonu (1918–1924) [Suprematism. The World as Objectlessness, or the Eternal Rest; with an Appendix of Letters from K.S. Malevich to M.O. Gershenzon (1918–1924)]*. Moscow: Gileia.

Malevich, Kazimir. 2004a. «Ia nachalo vsego» ["I am the beginning of everything"]. In *Proizvedeniia Raznykh Let: Stat'i. Traktaty. Manifesty i Deklaratsii. Proekty. Lektsii. Zapisi i zametki. Poeziia [Works of Various Years: Articles, Treatises, Manifestos, Declarations, Projects, Lectures, Notes and Records. Poetry]*. Moscow: Gileia.

Malevich, Kazimir. 2004b. Zametka o tserkvi [A note on the Church]. In *Proizvedeniia Raznykh let: Stat'i. Traktaty. Manifesty i Deklaratsii. Proekty. Lektsii. Zapisi i Zametki. Poeziia [Works of Various Years: Articles, Treatises, Manifestos, Declarations, Projects, Lectures, Notes and Records. Poetry]*. Moscow: Gileia.

Malevich on himself (Malevich o sebe). 2004. *Sovremenniki o Maleviche: pis'ma, dokumenty, vospominaniia, kritika [Contemporaries on Malevich: Letters, Documents, Memoirs, Criticism]*. Moscow: RA (OAO Tip. Novosti).

Marcadet, Jean Claude. 2000. *Malevich i Pravoslavnaia Ikonografia [Malevich and Orthodox Iconography]*. Moscow: Yazyki knizhnoi kul'tury, pp. 167–174.

Mikhailova, Maria. 2000. *Apofatika v Postmodernizme [The Apophatic Thought in Post-Modernism]*. St. Petersburg: Eidos, pp. 166–79.

Milner, John. 1996. *Kazimir Malevich and the Art of Geometry*. New Haven: Yale University Press.

Mudrak, Miroslava. 2016. *Kazimir Malevich i Vizantiiskaia Liturgicheskaia Traditsiia [Kazimir Malevich and the Byzantine Liturgical Tradition]*. Iskusstvo, № 2 (597). Moscow: Alya Tesis, pp. 50–67.

Nakov, Andrei. 2010. *Malevich: Painting the Absolute*. Burlington: Lund Humphries.

Nicholas of Cusa. 1979. Ob uchenom neznanii (De docta ignorantia) [On Learned Ignorance (De docta ignorantia)]. In *Filosofskoye Nasledie [The Philosophical Heritage Series]*. Moscow: Mysl'.

Nutsubudze, Shalva. 1942. *Taina Psevdo-Dionisiya Areopagita [The Mystery of Pseudo-Dionysius the Areopagite]*. Tbilisi. Tbilisi: Academy of Sciences of the Georgian SSR.

Plato. 1997. *The Dialogues of Plato*. Parmenides: Yale University Press, vol. 4.

Puech, Henri-Charles. 1930. *Liberatus de Carthage et la Date de L'apparition des Écrits Dionysiens. Annales de l'École des Hautes Études, section des sciences religieuses*. Paris: Gand, L'Ecole.

Rist, John. 1962. *The Neoplatonic «One» and Plato's Parmenides//Transactions and Proceedings of the American Philological Association*. Baltimore: Johns Hopkins University Press, vol. 93, pp. 389–401.

Rostova, Natalia. 2021. Religioznaia taina «Chernogo kvadrata» [The religious mystery of The Black Square]. In *Filosofia Russkogo Avangarda: Kollektivnaia Monografia [The Philosophy of the Russian Avant-Garde: A Collective Monograph]*. Moscow: RG-Press.

Sarabyanov, Dimitri Vladimirovich. 1993. *Russkii Avangard Pered Litsom Religiozno-Filosofskoi Mysli [Russian Avant-Garde and the Religious-Phil Dimitri Vladimirovich Osophic Thought]*. Iskusstvo Avangarda: Materialy Mezhdunarodnoi Konferentsii. Aleksei Stukalov: Ufa, pp. 3–23.

Sarabyanov, Dimitri Vladimirovich. 1998. Kandinskii i russkaia ikona [Kandinski and the Russian icon]. In *Mnogogrannyi mir Kandinskogo [Kandinski's Multifaceted World]*. Moscow: Nauka, pp. 42–49.

Sarabyanov, Dimitri, and Alexandra Shatskikh. 1993. *Kazimir Malevich: Zhivopis'. Teoriia. [Kazimir Malevich: Art and Theory]*. Moscow: Iskusstvo.

Schreier, Lotar. 1966. Die Ikone Wassily Kandinsky. In *Errinerungen an Sturm und Bauhaus*. Edited by L. Schreier. München: Langen Müller, pp. 72–85.

Shatskikh, Alexandra. 2000. Malevich posle zhivopisi [Malevich after painting]. In *Suprematizm. Mir kak Bespredmetnost', ili Vechnyi Pokoi: s Prilozheniem Pisem K. S. Malevicha k M. O. Gershenzonu (1918–1924) [Suprematism. The World as Objectlessness, or the Eternal Rest; with an Appendix of Letters from K.S. Malevich to M.O. Gershenzon (1918–1924)]*. Moscow: Gileia.

Spira, Andrew. 2008. *Avant-Garde Icon. Russian Avant-Garde Art and the Icon Painting Tradition*. Aldershot and Burlington: Lund Humphries.

St. Gregory Palamas. 2009. *Polemika s Akindinom [Gregorii Acindyni refu Tationes Duae Operis Gregorii Palamae]*. Series: Smaragdos Philocalias; Holy Mt. Athos: The New Thebais Monastery.

St. John of Damascus. 2010. *Writings*. The Fathers of the Church Series; Washington: CUA Press, Cambridge: Cambridge University Press, vol. 37.

Stiglmayer, Joseph. 1928. *Der Sogenannte Dionysius Areopagita und Severus von Antiochien*. Moscow: Scholastic, III, Freiburg, pp. 1–27, 161–89.

Strigalev, Anatoliy. 1989. «Krest'ianskoe», «Gorodskoe» i «Vselenskoe» u Malevicha [The «Peasant», «City» and «Universal» in Malevich]. № 4. Moscow: Tvorchestvo, Moscow Uchpedgiz, pp. 26–30.

Tarasov, Oleg. 1992. *Ikona v Russkom Avangarde 1910–1920-kh Godov [The Icon in the Russian Avant-Garde of the 1910s and 1920s]*. № 1. Moscow: Iskusstvo, Alya Tesis, pp. 49–53.

Tarasov, Oleg. 2002. *Icon and Devotion. Sacred Spaces in Imperial Russia*. Translated and Edited by Robin Milner-Gulland. London: Reaktion Books.

Tarasov, Oleg. 2011. *Framing Russian Art. from Early Icons to Malevich*. Translated by R. Milner-Gulland, and A. Wood. London: Reaktion Books.

Tarasov, Oleg. 2017. *Spirituality and the Semiotics of Russian Culture: From the Icon to Avant-Garde Art*. Modernism and the Spiritual in Russian Art. New Perspectives. Edited by Louse Hardiman and Nicola Kozicharow. Cambridge: Open Book Publishers, pp. 115–29.

Vakar, Irina. 2015. *Kazimir Malevich. "Chernyi kvadrat"*. [*Kazimir Malevich: The Black Square*]. Moscow: Gos. Tret'iakovskaia galereia.

Voronina, Elena, and Iina Rustamova. 2015. Rezul'taty tekhnologicheskogo issledovaniia «Chernogo suprematicheskogo kvadrata» K.S Malevicha (prilozhenie) [The outcomes of a technological study of the Black Suprematist Square by K.S. Malevich: An appendix] In *Vakar I. Kazimir Malevich. Chernyi kvadrat [Kazimir Malevich: The Black Square]*. Moscow: Gos. Tret'iakovskaia galereia.

Article

The October Revolution as the Passion of Christ: Boris Pasternak's Easter Narrative in *Doctor Zhivago* and Its Cultural Contexts

Svetlana Efimova

Slavic Philology Department, Faculty of Languages and Literatures, Ludwig Maximilian University of Munich, 80539 Munich, Germany; Svetlana.Efimova@slavistik.uni-muenchen.de

Abstract: This article offers a new interpretation of Boris Pasternak's novel *Doctor Zhivago* in the cultural and historical context of the first half of the 20th century, with an emphasis on the interrelationship between religion and philosophy of history in the text. *Doctor Zhivago* is analysed as a condensed representation of a religious conception of Russian history between 1901 and 1953 and as a cyclical repetition of the Easter narrative. This bipartite narrative consists of the Passion and Resurrection of Christ as symbols of violence and renewal (liberation). The novel cycles through this narrative several times, symbolically connecting the 'Easter' revolution (March 1917) and the Thaw (the spring of 1953). The sources of Pasternak's Easter narrative include the Gospels, Leo Tolstoy's philosophy of history and pre-Christian mythology. The model of cyclical time in the novel brings together the sacred, natural and historical cycles. This concept of a cyclical renewal of life differs from the linear temporality of the Apocalypse as an expectation of the end of history.

Keywords: Orthodox Christianity; philosophy of history; myth; Passion of Christ; Easter; cyclical time; Boris Pasternak; *Doctor Zhivago*; Leo Tolstoy

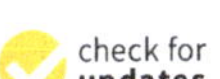
check for
updates

Citation: Efimova, Svetlana. 2021. The October Revolution as the Passion of Christ: Boris Pasternak's Easter Narrative in *Doctor Zhivago* and Its Cultural Contexts. *Religions* 12: 461. https://doi.org/10.3390/rel12070461

Academic Editor: Dennis Ioffe

Received: 11 May 2021
Accepted: 21 June 2021
Published: 24 June 2021

Publisher's Note: MDPI stays neutral with regard to jurisdictional claims in published maps and institutional affiliations.

1. Introduction

In his contribution to the third volume of *Christianity and the Eastern Slavs* that saw light with UCLA Press in 1995, Lazar Fleishman says: "The question of Boris Pasternak's Christian faith—and first of all in his novel *Doctor Zhivago*—is one of the most painful, complicated and contentious ones in literary scholarship" (Fleishman 1995, p. 288). A quarter of a century later, researchers exploring religious motifs in Pasternak's novel still grapple with paradoxes: "Pasternak thus challenges, complicates and renews the reader's image of Christ, even as he largely affirms the Christ of faith from the four evangelists' accounts" (Givens 2018, p. 204). In my article, I suggest a reading of *Doctor Zhivago* from a new perspective, with emphasis on the interrelationship between religion and philosophy of history throughout the novel. This interrelationship has never been the subject of a separate investigation. My interpretation also places Pasternak's novel in the broader context of historical and religious consciousness in Russian 20th century culture.

On 13 October 1946, Boris Pasternak mentioned in a letter to his cousin Olga Freidenberg that he had started work on a novel. He named his two key aims: "to paint a historical image of Russia over the last forty-five years, and at the same time [...] this work will be an expression of my views on art, on the Gospels, on one's life in history and on much more" (Mossman 1982, pp. 254–55). In the author's characterisation of his own project, the historical and theological topics are inseparable: the history of Russia in the first half of the 20th century was to become material for an interpretation of the Gospels.

Completed in 1955, the novel *Doctor Zhivago* does, indeed, cover the historical period from 1901 to 1948 or 1953,[1] from the first Russian Revolution of 1905 to the Stalinist purges, the Second World War and the first post-war decade. However, most of the events in the novel take place between 1917 and 1922: the central historical points of reference are the

February and October Revolutions of 1917. Already in the first, prose part of the novel, the reader's attention is drawn to numerous theological allusions. The Christian references are especially pronounced in the poetic cycle "The Poems of Yurii Zhivago" with which the novel concludes. The motifs of crucifixion and resurrection are central to the whole cycle; the subject of the last five poems is the Passion of Christ.[2]

Although the Christian references play a central role in the novel, they elude a simple interpretation. Over time, scholars have reached a degree of agreement in reading the life story of the main character, Yurii Zhivago, as a Gospel narrative, with Zhivago himself appearing in *imitatio Christi*.[3] However, as early as in 1974, Thomas F. Rodgers noted an "association of Christ with a mass of incidental persons" in Pasternak's novel, which to him suggested a "mystical association of the revolution with Christ's passion" (Rogers 1974, pp. 384, 387). This observation has not been developed further in subsequent research; however, as this article will show, it is central to my interpretation of the novel.

The question concerning a possible connection between the Russian Revolution and religious consciousness was raised in Russian philosophy in the first half of the 20th century; for example, in the political theology of Nikolai Berdiaev, Sergei Bulgakov and Semen Frank (Poole 2013). Research on religious representations of the revolution in the Russian literature of the first half of the 20th century focuses primarily on the Apocalypse and the history of apocalyptic sects, on the idea of establishing God's Kingdom on earth (Rosenthal 1980; Bethea 1989; Etkind 1998). In *Doctor Zhivago*, the main biblical source is the Gospels,[4] on the basis of which Pasternak creates an Easter narrative of crucifixion and resurrection and develops his particular interpretation of the Book of Revelation.

The last poem of the novel ("Garden of Gethsemane") directly connects the fate of Christ with the course of historical time. The last two stanzas, the ones that conclude the whole novel, are based on a comparison between history and parable: "the passing of the ages is like a parable," and it is in the name of its "majesty" that the sacrifice of Christ is made, who will rise on the third day after his death (Pasternak 1958, p. 558).[5] Thus, the conclusion of the novel emphasises the meaningfulness of history as a "parable" that can be interpreted from a religious perspective. This is the logic that guides my interpretation of the Revolution of 1917 in *Doctor Zhivago* and the general conception of history in the novel.

I open with a short survey of the symbolic meaning of Easter in the cultural and historical context of 1917 in Russia (2). I then analyse Pasternak's interpretation of the Easter narrative as a particular model of cyclical temporality that includes the sacred, the historical and the natural (biological) cycles (3).[6] The novelist's main sources here include Leo Tolstoy's philosophy of history, the Gospels, and the pre-Christian mythology which his cousin Olga Freidenberg researched in her work on mythopoetics (4). The Easter narrative in *Doctor Zhivago* is not limited to the revolution and the Civil War; it comes to a climax in the epilogue to the novel that depicts the Second World War and the years of the Thaw (5). I will then touch upon the political significance of the text (6). In conclusion, I will present a short summary of the religious model of history in the novel that is constructed as a cyclical repetition of the Easter narrative (7). The cyclicity of time in Pasternak's text is different from the linear model that the expectations of the Second Coming follow. Pasternak's concept of time leads to the understanding of a repeated renewal of life as a cyclical process.

2. Easter in the Cultural and Historical Context of 1917

There existed a Marxist tradition that cast Jesus Christ in the role of a proletarian, with early Christianity foreshadowing the socialist revolution. Friedrich Engels, Rosa Luxemburg and Karl Kautsky were among the proponents of this interpretation. Anatolii Lunacharskii, who, in October 1917, became Head of the People's Commissariat for Education, expressed the same views in many of his publications and speeches (Bergman 1990). In the first quarter of the 20th century, the great majority of Russians were practicing Orthodox Christians, which made them especially likely to interpret the revolutionary changes in a religious light. The festival of Easter, considered to be "the heart of Orthodox Christianity" (Bulgakov 1989, p. 285) and the main holiday in the Russian Orthodox calendar (Esaulov 2006, p. 65), acquired a special political significance in the revolutionary era.

Following the Revolution of 1905 in Russia, the State Duma (the nation's first elected parliament) was established, and the freedom of the press was expanded. In 1906, Easter was celebrated on 2 April (15 April according to the modern Gregorian calendar), thus overlapping with the elections to the first State Duma that took place in March–April 1906. Against this background, the liberal press metaphorically connected the festival of Easter with a political resurrection of free Russia (Baran 1993, pp. 295, 333). By a strange coincidence, in 1917 Easter again fell on 2 April (15 April according to the modern calendar), just one month after two momentous events occurred as a consequence of the February Revolution. These were the abdication of Tsar Nikolai II, and the formation of the Provisional Government on 2 March (15 March according to the modern calendar).

The celebration of "the Red Easter of the Revolution" in 1917 became a testimony to "the merging of religious and political consciousness" in Russian society (Kolonitskii 2012, pp. 57, 81). The February Revolution was perceived as a sacral event, as a resurrection of Russia that was sanctified by the Resurrection of Christ. This is evidenced by press publications of the time, archival documents, private correspondence and memoirs, private and official Easter blessings that were sent out, statements by some religious figures, and even Easter cards that in 1917 were decorated with revolutionary symbols (Kolonitskii 2012, pp. 57–86).[7]

It is in this cultural and historical context that we can read Boris Pasternak's poem "The Russian Revolution" (written in 1918).[8] It opens with an enthusiastic welcome addressed at the revolution that came in the spring: "How good it was to breathe you in March." This revolution is associated not only with the sun, the sound of merry springs and the smell of melting snow, but also with the "catacombs" of early Christians, as well as with "the sacred night" separating "the last days of fasting" from each other. The link between the political events and "the last days of fasting" introduces the motif of Easter, thus emphasising the Christian significance of the February Revolution: "And the Socialism of Christ breathed deep and free" (Pasternak 2004a, pp. 224–25). The second part of the poem refers to the revolutionary events of the final months of 1917, and the atmosphere changes abruptly: now, it is images of violence, blood and death that are in the foreground.

After the October Revolution, and with the start of the Civil War, religious metaphors in the Russian society changed. A typical example of this metaphoric shift is an article by the literary critic and philosopher Razumnik Ivanov-Razumnik, "The Two Russias," dated November 1917. It was published in 1918 in the second issue of *The Scythians*, a collection of literary and critical texts, two issues of which saw light in Petrograd in 1917–1918. The article opens with the author reminiscing about the events of the spring of 1917: "The Russian February Revolution was delivered painlessly and greeted with joy and enthusiasm by the whole nation" (Ivanov-Razumnik 1918, p. 201). However, the topic of the article is the turn that the history of Russia took in October 1917. It is described with the help of a metaphor which is repeated multiple times: the way of the cross as the way of the Russian Revolution and society. Ivanov-Razumnik quotes the Gospel According to St. Luke and concludes: "And everything that is said in the eternal book about the way of the cross—all this can be repeated, word for word, about the Nation's way of the cross leading into the great and terrifying days of the Russian revolution" (Ivanov-Razumnik 1918, p. 204). However, the philosopher concludes by reminding his readers that Calvary is a promise of the resurrection to come, and compares the belief in the victory of the revolution with the Christian faith in the resurrection of Christ.

The metaphor of the Passion of Christ was also widespread among the poorly educated worker–writers of the revolutionary era. In many of their poems, worker–poets described the fate of the proletariat as martyrdom, crucifixion, and the way of the cross in revolutionary struggle (Steinberg 1994, pp. 221–23). The fact that these images were so popular with such diverse social groups points to a key particularity of Orthodox Christianity: "The Orthodox path of the Christian is *the path of the cross and of struggle*. In other words, it is the path of patience, of the bearing of sorrows, persecutions for the name of Christ" (Pomazansky 1984, p. 327; italics in the original).

The religious imagery that was commonly used for descriptions of both the February and the October Revolutions constitutes a bipartite Easter narrative: the Passion of Christ and the Resurrection. The religious and historical metaphor of the way of the cross implies hope for a new Easter. Thus, in April 1918 Andrei Belyi wrote his poem "Christ Has Risen," the title of which refers to the traditional Easter greeting in Russia.[9] The poem first talks about the Passion of Christ, and then about the victims of the revolution who are likened to the crucified Son of God. The whole text can be read as a symbolic message announcing the resurrection of Christ, to which the speaker responds in the third stanza with the following tripartite pronouncement: "It is. It was. It will be" (Belyi 2006, p. 9). This unity of the three tenses reflects the cyclicity of the Orthodox calendar. A repetition of the festival of Easter reminds the reader about the repetitive nature of the revolutionary events in early 20th century Russia with its three revolutions in 1905 and in 1917.

Pasternak was a contemporary of these events, and the responses to them that we have quoted above; his contribution to the cultural discourse of the time was the poem "The Russian Revolution". An understanding of this cultural and historical background is essential for an analysis of the Easter narrative and the cyclical nature of time in *Doctor Zhivago*.

3. The Easter Narrative and the Cyclical Nature of Time in *Doctor Zhivago*: Religion, Nature, History

The cycle "The Poems of Yurii Zhivago", which concludes the novel and emphasises its religious dimension, is made up of twenty-five poems. The subject of seven of them is the Gospel story of the Passion of Christ. These poems provide a symbolic key to an interpretation of the events in the prose part of the novel; therefore, we should consider them in more detail.

In the first poem of the cycle ("Hamlet"), the actor playing Hamlet at the same time identifies with Christ, repeating his prayer from the Gospels: "If it may be, I pray Thee, Abba, Father/Grant it: let this chalice from me pass" (Pasternak 2004d). Jesus, foreseeing his own crucifixion, speaks these words in the Garden of Gethsemane when praying to God the Father. Following the spirit of the Gospels, the last stanza of the poem declares that "there is no turning from the road." The topic of the way of the cross is continued in the third poem, "In Holy Week" (Pasternak 2004d), which describes the Orthodox church service in the days preceding Easter. The poet recreates the Good Friday evening service: the "procession," carrying forth "the Shroud," as well as the ritualistic burial of Christ ("A God is being buried") (Pasternak 2004d). The poem concludes with the message that resurrection will overcome death, i.e., with an announcement of the approaching festival of Easter.

The Easter narrative of the Passion and of the Resurrection, recreated in the first and third poems, is repeated at the end of the poetic cycle (Polivanov 2015, pp. 219, 223). Yurii Zhivago's last five poems take us back to the time of the Gospels. We become witnesses to Jesus's journey from Bethany ("Miracle") and his entry into Jerusalem ("Evil Days"); we hear Mary of Magdalene address him ("Magdalene"). There is a direct intertextual connection between the first and the last poems of the cycle. The last poem ("Garden of Gethsemane") is a return to a scene from the first poem, "Hamlet". Jesus has a premonition about the suffering that awaits him and appeals to his Father with a prayer: "And sweating drops of blood, He prayed to the Father/That from this deathly cup He be exempted" (Pasternak 2004d). The poem concludes with Jesus's words, who is ready to sacrifice himself in the name of the "parable" of history and who announces his own coming Resurrection. Thus, in the cycle "The Poems of Yurii Zhivago", the Easter Narrative is reproduced twice, and this repetition is emphasised through an intertextual reference to the Agony in the Garden. The poetic part of the novel reproduces the cyclical Orthodox calendar, thus symbolising the recurrent nature of the Passion and the Resurrection.

The religious cycle, in turn, is inscribed into the natural cycle of seasons: both Easter narratives are accompanied by poems about spring ("2. March"; "21. The Earth"). They do not share a narrative line, but the natural imagery suggests the passing of a full year ("7. Summer in Town"; "10. Indian Summer"; "12. Autumn"; "14. August"; "15. Winter Night"; "18. The Star of the Nativity") (Sukhikh 2013, p. 569). Nature not only provides a background to the Easter narrative, but becomes its agent in the poem "In Holy Week." Trees in the woods and urban gardens are described as active participants of the sacred passion play. Trees in the forest are "Like solemn worshippers at prayer," trees in the city are straining to "peer through churchyard railings," and gardens "leave their boundary walls" while "a God is being buried" (Pasternak 2004d; Pasternak 1958, p. 525; this translation combines both published versions—S.E.). The end of the poem is astonishing in its inner logic: the last stanza implies a causal relationship between the coming of spring and the Resurrection. Overcoming death will become possible once snow starts to melt:

At midnight man and beast fall dumb

On hearing springtime's revelation:

Once weather clears, then just as soon

Can death itself be overcome

By the power of Resurrection. (Pasternak 2004d)

Spring becomes an integral part of the Easter narrative in Pasternak's reading of it. In the poem "The Earth", the life stories of numerous people enter this narrative, together with the spirit of the spring that fills the homes of Muscovites, so that one can hear "April's casual discussions/With dripping waters of the thaw" (Pasternak 2004d). The subject of these "casual discussions" of April is human suffering, and these lines make us think of the Passion Week: "For April knows thousands of stories/Of human sorrow" (Pasternak 2004d; Pasternak 1958, p. 552; this translation combines both published versions—S.E.). Thus, the story of suffering and hope for renewal, for rebirth, are transferred from Jesus Christ and his lyrical double (for example, in the poem "Hamlet") onto thousands of human stories, all of them different and at the same time united in sorrow.

"The Poems of Yurii Zhivago" do not describe historical events, even though "the war" and "devastation" are briefly mentioned in one of them ("Dawn") as references to the 20th century history (Pasternak 1958, p. 549). Nevertheless, as I have already pointed out above, in the last poem of the cycle ("Garden of Gethsemane") the Easter narrative is directly linked to the "parable" of history, inviting the reader to project religious motifs from the poetic cycle onto the prosaic part of the novel. "The Poems of Yurii Zhivago" are based on three interconnected temporal cycles: the cycle of the death and rebirth of nature, the Orthodox cycle of the Passion and the Resurrection, and the cycle of suffering and joyful renewal in human life. These three cycles are the compositional basis of the prose part of the novel, too, where the course of history is another major theme.

The events of February 1917 find Yurii Zhivago in a military hospital where he is recovering from an injury he received while serving as a military doctor at the front. Spring came early that year: "It was a warm day at the end of February" (Pasternak 1958, p. 125). It is as if nature itself (the warm weather and the change from winter to spring) had some deep connection with the anticipation of political changes: "For several days the weather was variable, uncertain, with a warm, constantly murmuring wind in the night, smelling of damp earth. During those days there came strange reports from G.H.Q. [...]. Telegraphic communications with Petersburg were cut off time and again. Everywhere, at every corner, people were talking politics" (Pasternak 1958, p. 126). For Zhivago, those days brought with them not only recovery, but also a meeting with Lara, who was destined to become the love of his life; this is also when the news of the publication of his first book in Moscow reached him.

At the end of February, nurse Lara was remembering the revolutionary events of 1905 in Moscow that she had witnessed. And just as she was thinking to herself: "And now they were shooting again, but how much more frightening it was now!", patients burst into the room with the latest news from Petersburg: "The revolution!" (p. 128). Thus, in the novel, the February Revolution is symbolically connected with spring, with momentous events in the heroes' lives, with recovery, renewal, and at the same time, with a repetition of revolutionary events (1905 and 1917).

These motifs come to a climax in a conversation between the main characters in the summer of 1917, when Zhivago explains to Lara his understanding of the revolution that took place a few months previously:

> Freedom! Real freedom [...], freedom, dropped out of the sky, freedom beyond our expectations [...]. And it isn't as if only people were talking. Stars and trees meet and converse, flowers talk philosophy at night, stone houses hold meetings. It makes you think of the Gospel, doesn't it? [...] Everyone was revived, reborn, changed, transformed. You might say that everyone has been through two revolutions—his own personal revolution as well as the general one. It seems to me that socialism is the sea, and all these separate streams, these private, individual revolutions, are flowing into it—the sea of life, the sea of spontaneity. (pp. 146–47)

Here, we see, yet again, how the temporalities of natural cycles, human life, and a religious experience merge. A rebirth of nature (the world of trees and flowers) is inseparable from the rebirth and transformation of individuals. Nature is no longer just a passive background; it acquires a voice and becomes an active participant of the events, as in the poem "In Holy Week." An awakening of nature, a rebirth of the individual, and a reference to the Gospels—these are the motifs which in the novel are connected with the imagery of Easter, even though the festival is not mentioned directly.

Now the history of the February Revolution is added to this tripartite temporality (natural, individual, and religious). The political event is perceived as a harmonious component in the life of nature and each individual person; it becomes a metaphor of internal transformations and a re-awakening of nature. The philosophy of history that the novel conveys suggests that the revolution is experienced by each individual in a different way. This comes to the fore at two key moments in the text. Early in the novel, Nikolai Vedeniapin explains that, in his opinion, the modern understanding of the concept of history begins with Christ, and that the Gospels provided a philosophical grounding for it. Then, in the spring of 1917, when serving at the front a few months before the revolution, Yurii Zhivago and his friend Mikhail Gordon discuss Christianity, having abolished the idea of "a nation" and replaced it with "the mystery of the individual" (p. 122). The protagonist's perception of the February Revolution reflects this understanding of the nature of things: a 'collective' political event brings together a multitude of individuals rather than a nation. All these individuals are seized by the spirit of revival that has come with the spring, but for each of them it is a different experience. Yurii Zhivago's reflections lead us to the Christian understanding of socialism that was expressed by Pasternak in his 1918 poem "The Russian Revolution": "And the Socialism of Christ breathed deep and free" (Pasternak 2004a, p. 224). There are a multitude of individually experienced revolutions, just as there are a multitude of forms of individual suffering in Yurii Zhivago's poem "The Earth" about the days preceding Easter: "For April knows thousands of stories/Of human sorrow" (Pasternak 2004d; Pasternak 1958, p. 552; this translation combines both published versions—S.E.). This logic of unity connects the experience of suffering and rebirth that corresponds to the Passion and the Resurrection in the Easter narrative.

We should remember that in Pasternak's poem "The Russian Revolution" (written in 1918), the February Revolution, which ended on 2 March 1917, is symbolically connected with the last days of Lent and with the month of March. Zhivago's poem "In Holy Week" is set at the same point in the annual cycle, and the poem immediately preceding it is called simply "March." The connection between Easter and March which is emphasised in Yurii Zhivago's poetic cycle is not obvious from the Russian Orthodox perspective: for Easter to be celebrated so early, it would need to be according to the Julian calendar style. The earliest date that Easter could fall on according to the Gregorian (Russian Orthodox) calendar is 4 April (Polivanov 2015, p. 219). I see this special role of the month of March in Yurii Zhivago's poems as a thread that links Pasternak's poem "The Russian Revolution," the actual dates of the February Revolution as a historical event, and its depiction in the prose part of the novel.

The February Revolution in the novel is symbolically connected with spring, with a rebirth of life and a blooming of nature, whereas the October Revolution is a manifestation of the power of winter. When Yurii Zhivago buys the special issue of a newspaper announcing the establishment of the dictatorship of the proletariat, Moscow is caught in a mighty snowstorm: "the snow thickened and the wind turned into a blizzard" (p. 192). However, again, the narrator stresses the unity of nature, human beings and history: "There was something in common between the disturbances in the moral and in the physical world [...]. Here and there resounded the last salvoes of islands of resistance" (p. 192). The crushing force of the events that followed the revolution and brought with them a collapse of the habitual order of life is the elementary force of the early winter that followed the late October days: it was "dark, hungry, and cold" (p. 195). More than that, in the narrator's memory, three winters merged into an image of post-revolutionary Russia, into one terrible winter: "There were three of them, one after the other, three such terrible winters, and not all that now seems to have happened in 1917 and 1918 really happened then—some of it may have been later. These three successive winters have merged into one and it is difficult to tell them apart" (p. 195).

Winter and snowstorms are associated in the novel not only with the October Revolution, but also with death. The novel opens with the funeral of the 10-year-old Yurii Zhivago's mother on the eve of the festival of the Protection of the Holy Virgin, i.e., on 13 October. The boy spends the night after the funeral with his uncle in a monastery. He is woken by the sound of a raging snowstorm: "Outside there was [...] nothing but the blizzard, the air smoking with snow" (p. 4). There is a suggestion of a white (funeral) shroud that turns snow into a symbol of death: "Turning over and over in the sky, length after length of whiteness unwound over the earth and shrouded it" (p. 4). And the only counterbalance to the unruly elements is the uncle talking about Christ to console the crying boy. Pasternak was free to pick any day for the funeral of his protagonist's mother, and it is significant that both key events of the novel (the death of the mother and the revolution) take place in October. This autumn month is not immediately associated with snow, but in both cases the author introduces a snowstorm which becomes all the more significant for happening in autumn rather than in winter.

Snow is likened to a shroud on at least one more occasion in the novel. It happens in the twelfth section ("The Rowan Tree"). By this time, Yurii Zhivago is a prisoner of war, serving as a doctor in a unit of red partisans. At the centre of this part of the novel is the inhuman violence which, as the narrator emphasises again and again, is perpetrated by all participants of the conflict—red revolutionaries and white anti-revolutionaries alike. Here, too, an early winter welcomes the protagonist to the kingdom of snow falling "with a convulsive, insane haste," covering "in a moment the broad expanse of the earth [...] with a white shroud" (p. 359–60; translation modified—S.E.).

The twelfth section of the novel is where the Gospel motifs of the Passion of Christ are concentrated: eleven conspirators are discovered and shot in a partisan unit. This event, which has a key symbolic importance in the novel, also takes place in late autumn, when the trees in the woods have already lost all their leaves. The clearing where the men are shot is a slightly elevated piece of land with "prehistoric" boulders; it reminds Yurii Zhivago of a place specially designated for sacrificial rituals and invites associations with Calvary (p. 353). Before the execution, one of the condemned conspirators throws an accusation at the partisan Sivobliui, who was both an instigator of the conspiracy and a traitor: "Judas! Christ-killer!" (p. 355). Following the same logic of the Gospels, the anarchist Vdovichenko's last words are a proud statement of faith: "We die as martyrs for our ideals at the dawn of the world revolution. [...] Long live world anarchy!" (p. 355).[10]

More than that: one of the 'executed' men actually 'rises from the dead'—the young Terentii Galuzin, left for dead, eventually came to and managed to escape.

The partisan Vdovichenko was not a conspirator, but his influence in the unit was a threat to the authority of the commander, and he, too, ends up being shot. The blinding snowstorm erupts just as another character of the novel, the partisan Svirid, is reflecting on the fate of the executed men. Snow and blizzard in the novel are not only symbols of suffering and death; they also symbolise the October Revolution. According to the Gospels, there are but three days between the Crucifixion and the Resurrection of Christ. However, set against the calendar cycle of nature, the two parts of the Easter narrative are associated with two seasons: the season of snow in autumn and winter, and the season of rebirth in spring and summer. The Christian worldview becomes one with the myth of nature dying and coming back to life.

This image of the revolution as a snowstorm, the execution of the condemned men, and the blizzard in the *twelfth* section of the novel ("The Rowan Tree") are intertextually connected with Aleksandr Blok's poem "The Twelve" (written and published in 1918). It is well known that in this poem, revolutionary violence is presented as part of the devastation waged by a snowstorm, and the text concludes with the image of Jesus Christ appearing from the blizzard. Out of numerous interpretations of this mystical image, I would like to single out the one offered by the artist Vasilii Masiutin in his introduction to an edition of the poem which he illustrated for a publication by the Russian émigré publishing house "Neva" in Berlin in 1922. This interpretation did not win broad recognition, but it is an example of a reading by a contemporary; it also bears an astonishing similarity to the conception of time in *Doctor Zhivago*.

Masiutin believed that contemporary readers arguing about whether or not Blok's poem expressed his sympathy with the October Revolution were wrong. According to him, the poem is not an evaluation of the revolution; rather, it is concerned with the power of "the inevitable" and "the predetermined" (Blok 1922, p. 1). He sees in the title not only a reference to the twelve apostles, but also a recognition of the ineluctable flow of time: twelve months in a year, twelve hours in a day, twelve hours in a night. The passing of time is a sequence of light and darkness, winter and spring, hopelessness and salvation. According to Masiutin, the end of the poem is a double procession to Calvary—not only of Christ, but also of the red guards who are shooting at him, and who will become victims of the revolution in their turn. However, the logic of the Gospels and of the number twelve suggests that the way of the cross is also a promise of hope: "A spring morning of the Resurrection will come after the dark snowy night" (Blok 1922, p. 2). We can see here, just as in the novel *Doctor Zhivago*, a correlation of the Orthodox calendar and the natural time. For Blok, as for Pasternak, the motif of the Passion of Christ from the Easter narrative corresponds to a snowy, cold winter, i.e., the death of nature.

Masiutin's interpretation shows that in the cultural and historical context of the revolution, a contemporary of Blok could respond to the poem "The Twelve" in the way that the novel *Doctor Zhivago* with its natural and religious symbolism suggests. Jean-Luc Moreau expressed a similar opinion in one of his essays: "Where Blok ends, Pasternak begins"; Christ "completes in the poems of *Doctor Zhivago* the cycle of his Passion begun in 'The Twelve'" (Moreau 1970, p. 238). This argument also assumes that the religious and mythological symbolism of the Russian Revolution associates the beginning of the Passion of Christ with a snowstorm.[11]

The observations offered above are supported by the fact that throughout the novel, a juxtaposition of winter and spring is presented as the cycle of suffering and rebirth. In the revolutionary winter of 1917, Yurii Zhivago falls ill with typhoid fever and spends two weeks between life and death. Delirious, he sees himself writing a poem about the three days that passed between the Deposition in the Tomb and the Resurrection. This is a history of struggle, on one side of which is "hell," "dissolution" and "death," and on the other is "the spring and Mary Magdalene and life" (p. 207). The poet's recovery starts with his telling himself, in his delirium: "Time to wake up and to get up. Time to arise, time for the resurrection" (p. 207). This is how the spring of the resurrection is symbolically juxtaposed to the illness of the revolutionary winter. The spring of 1918 brings with it change and a sense of happiness for Yurii Zhivago. After his recovery, he and his family decide to leave Moscow for Ural. At this point, the motif of Easter appears in the novel: the packing for the journey passes for "a spring cleaning for Easter" (p. 209). However, again, a cyclical change of seasons brings together the historical, the natural, the psychological and the religious dimensions.

This cyclical conception of time in the novel manifests itself also in the fact that an exhausting illness with delirium returns one more time in Yurii Zhivago's life, in the thirteenth section of the novel. After his arrival in the Ural town of Iuriatin in early spring, Zhivago's recovery becomes his rebirth. This rebirth symbolically corresponds to the transition from winter to spring, and at the same time—from bloodshed to peace. The text emphasises the connection between the change of seasons and the course of the Civil War, the coming of peace: "The winter had just gone [...]. The Whites had recently gone, left the town, surrendering it to the Reds. The bombardment, bloodshed, and wartime anxieties had ceased. This too was disturbing, and put one on one's guard, like the going of the winter and the lengthening of the spring days" (p. 376; translation slightly modified—S.E.). Thus, we have established that the cyclical model of time in the novel includes the closely interrelated cycles of death and rebirth of nature, the Passion and the Resurrection of Christ, suffering and rebirth in the lives of individuals, violence and peace, destruction and resurrection in the course of history. The question we can ask now concerns the narrative devices used in the novel in order to justify this view of the cycle of historical events and their connection with the Easter narrative.

In the beginning of *Doctor Zhivago*, when Nikolai Vedeniapin reflects on how history in its modern sense started with Christ, the narrator remarks that all that happens in the world simultaneously occurs in what "some called the Kingdom of God, others history" (p. 13). Here, the meaningfulness of history is explained through its connection with the sacred. Towards the end of the book, when Yurii Zhivago contemplates the nature of history, he explains it through the logic of the world of plants: "He reflected again that he conceived of history, of what is called the course of history, not in the accepted way but by analogy with the vegetable kingdom" (p. 453). In this natural life, the dying in a snowy winter is juxtaposed with the rebirth that spring will bring with it: "In winter, under the snow, the leafless branches of a wood are thin and poor [...]. But in only a few days in spring the forest is transformed, it reaches the clouds" (p. 453). What is special about trees in a forest is that they grow, but a human being, unable to observe the process of growth, always perceives it as static at any given moment. This mysterious growth becomes a symbol of history in the novel: "And such also is the immobility to our eyes of the eternally growing, ceaselessly changing history, the life of society moving invisibly in its incessant transformations" (p. 453). Here, the forest becomes a symbol of history; in Yurii Zhivago's poem "In Holy Week", it is a participant of a sacred passion play, with trees "like solemn worshippers at prayer" (Pasternak 2004d).[12]

Zhivago's reflections on the similarity between history and the plant kingdom are not just abstract deliberations; they are connected with the history of the Russian Revolution. Thus, he remembers the summer of 1917 as a time when nature, history and God were in a harmonious agreement: "the revolution had been a god come down to earth from heaven, the god of that summer" (p. 454; translation slightly modified—S.E.). In the novel, the natural, religious and historical images of the world merge into one. During the Civil War, when people lose their human face and reality itself feels unreal, the protagonist makes a paradoxical observation: "only nature had remained true to history" (pp. 378–79).[13] Nature retains a sense of habitual, meaningful existence; its calendar cycle promises the arrival of spring, and with it a historical rebirth.[14] This coming together of nature and history is a manifestation of the belief in a "shared unity of life" [*vseedinstvo zhizni*] which is evocative of the "philosophy of life", as developed by Arthur Schopenhauer and Henri Bergson, among others (Briukhanova 2010, p. 214).[15]

In his reflections on history as if it were the plant kingdom, Yurii Zhivago follows Leo Tolstoy's philosophy of history as it is expressed in *War and Peace*: "Tolstoy thought of it in just this way, but he did not spell it out so clearly. He denied that history was set in motion by Napoleon or any other ruler or general, but he did not develop his idea to its logical conclusion. No single man makes history. History cannot be seen, just as one cannot see grass growing. Wars and revolutions, kings and Robespierres, are history's organic agents, its yeast" (p. 454). In order to explain the significance of this direct reference to the Russian classic, I will allow myself a slight digression to talk about Tolstoy's philosophy of history.

4. The Sources of *Doctor Zhivago*: *War and Peace* and Myth

Here, I am interested in the influence of *War and Peace* not so much on the genre and the system of characters in *Doctor Zhivago* (see, e.g., Polivanov 2015) as on the religious philosophy of history in it. Taking my lead from the research on Tolstoy written by Gary Saul Morson, Donna Tussing Orwin, and Christian Münch, I will focus on four key aspects of the image of the world in *Doctor Zhivago* that can be traced back to *War and Peace*.

(1) The first aspect is the unity of nature, history, religion and human life that I analysed above. The same idea of unity is at the base of the philosophical conception of *War and Peace*. On the one hand, Tolstoy depicts "the unity of the self and nature"; on the other hand, political history in the novel is "a part of God's will", and simultaneously "a mysterious force of nature" (Tussing Orwin 1993, p. 100). The history of humanity in *War and Peace* is connected with the images of river and water (Tussing Orwin 1993, pp. 101–2). Water is also one of the central symbols in the novel *Doctor Zhivago*: socialism is likened to a sea which is made up of the multiple streams; the imagery of water, melting snow and creeks play an important role in the poems "March" and "In Holy Week."

(2) The second aspect is a paradoxical reconciliation of chance and determinism in a unified conception of history. Scholars of Pasternak's work have noticed on more than one occasion that the lives of the characters in *Doctor Zhivago* are determined by coincidences and accidents, although at the same time the novel is driven by the belief that everything is pre-determined (Shcheglov 1991; Lavrov 1993; Sukhikh 2013). In the beginning of the third volume of *War and Peace*, Tolstoy writes: "Man lives consciously for himself, but is an unconscious instrument in the attainment of the historic, universal, aims of humanity" (Tolstoy 1952, p. 343). According to Tolstoy, history is subject to divine logic, but "the principles governing human events are incomprehensible to the human mind" and are only open to God (Morson 1987, p. 92). Individuals perceive themselves as free agents and cannot foresee the future; their reasoning sees the course of events as accidental (ibid.). Thus, Tolstoy combines a belief in a higher logic of the course of history with a non-teleological vision of it: one cannot hope to describe history by assuming a linear progression towards a goal that is known to humans.[16]

Following this logic, Tolstoy recognises a higher meaning of such catastrophic events as war—a meaning that is hidden from humans. Thus, for example, his narrator makes the following remark concerning the bloody Battle of Borodino: "and that terrible work which was not done by the will of a man but at the will of Him who governs men and worlds continued" (Tolstoy 1952, p. 467). In the life of Tolstoy's heroes, joy follows sorrow, and Pierre Bezukhov, when taken prisoner of war by the French troops, has an epiphany: "Life is God. Everything changes and moves and that movement is God. [...] To love life is to love God. Harder and more blessed than all else is to love this life in one's sufferings, in innocent sufferings." (Tolstoy 1952, p. 608).

This love of life, despite historical catastrophes, defines the spirit of the novel *Doctor Zhivago* and its heroes, caught unawares by the Russian Revolution. All of the philosophical, symbolic content of the novel leads the reader to the conclusion that the October Revolution, despite all its tragic consequences, was inevitable, just as the Passion of Christ was inevitable in the Gospels (Rogers 1974, p. 389). A version of the epigraph to the novel has been preserved in Pasternak's archive—a line from Paul Verlaine: "Aime tes croix et tes plaies/Il est sain que tu les aies" ("Love your crosses and your wounds/For he is holy who has them") (Pasternak 2004c, p. 645).[17]

The inevitability of revolution is determined not by the will of humans to transform life, but by a higher logic of life itself, a logic which is beyond human comprehension. This is what Yurii Zhivago tries to explain to the commander of red partisans Liberius Mikulitsyn: "Reshaping life! People who can say that have never understood a thing about life—they have never felt its breath, its heartbeat—however much they have seen or done. [...] But life is never a material, a substance to be molded. If you want to know, life is the principle of self-renewal, it is constantly renewing and remaking and changing and transfiguring itself, it is infinitely beyond your or my obtuse theories about it" (p. 338). This idea of predetermination is also central to the poem "Hamlet," where Christ obeys the will of the Father, and the actor must play his part in a play that had been written without his involvement: "But alas, there is no turning from the road/The order of the action has been settled" (Pasternak 2004d). Considered together, the different episodes of the novel embody the same understanding of existence that we find in *War and Peace*: "Everything changes and moves and that movement is God" (Tolstoy 1952, p. 608).

(3) The third aspect is the cyclical nature of life and history that I analysed above with reference to *Doctor Zhivago*. The course of events in *War and Peace* is also based on the concept of "an unending circular motion": "The equilibrium and family happiness established at the end of *War and Peace* is no more the goal of life than the state of motion" (Tussing Orwin 1993, p. 107). For example, the life of Pierre Bezukhov is cyclical in that he experiences recurrent periods of spiritual crisis. Searching for a way out of them, Bezukhov becomes fascinated with freemasonry, social reforms, a plan to assassinate Napoleon and, finally, he experiences a religious revelation when taken prisoner by the French troops. In the epilogue of the novel, we understand that his involvement with a secret political society is yet another such infatuation, yet another cycle in Pierre's life, and the young Nikolenka who admires him so much will follow in the footsteps of his father, Andrei Bolkonsky, whose dream had been to become a hero in the war against Napoleon.

The cycles in the life of Tolstoy's heroes are mirrored in the cycles of nature, with spring playing a key role, just as in Pasternak's text, bringing with it associations of a revival and a spiritual rebirth of characters. A tree that dies in the autumn and comes back to life in the spring is a symbol of history for Pasternak and a symbol of life for Tolstoy, when Andrei Bolkonsky compares himself to an old oak that is born again with the coming of spring.

(4) A combination of the vitality of nature and sacrality of religion in Tolstoy's image of the world leads us to the fourth important aspect—Tolstoy's striving for a "holistic awareness" that would help comprehend the unity of existence by bringing together seemingly contradictory principles, such as faith and reason (Münch 2014, p. 324). This synthesis of different interpretations of the world is at the basis of the novel *Doctor Zhivago*: nature and faith, the physician Zhivago's interests in the natural sciences and in religion, Orthodox Christianity and mythology.[18]

The Christian component in Pasternak's novel is connected with the tradition of the Russian Orthodox church; citations from Orthodox liturgy abound in the text (Raevsky-Hughes 1995, p. 318).[19] At the same time, the novel has many elements of folklore and fairy-tales (Lavrov 1993), and its images of nature evoke the myths of plants dying and coming back to life. Pasternak brings together religion and myth in a unified image of the world, a single Easter narrative.

We can find a possible origin of this synthesis of opposites in the writings of Pasternak's cousin Olga Freidenberg, who worked on mythopoetics. Many comments made by Pasternak as he was working on *Doctor Zhivago* are known to us from his extensive correspondence with Olga Freidenberg. In February 1947, Pasternak responded to an outline of Freidenberg's talk "The Origins of Greek Lyrics" that she had sent to him following her presentation at the University of Leningrad. In her talk, Freidenberg discussed the transformation of myth into poetic metaphor. Pasternak noted that Freidenerg's analysis was very similar to what he was looking to convey in his novel. He was excited: "Three pages of your summary—it's a thing of bottomless depth and a real breakthrough, akin to the Communist Manifesto or the Apostolic Letters" (Mossman 1982, p. 267). What is remarkable here is not just the mention of the Bible in one sentence with the Manifesto by Marx and Engels, but also the obvious congruence of thought of Pasternak and Freidenberg. They were engaged in an intellectual and creative dialogue: Freidenberg was sending her scholarly texts to her cousin, and "many thoughts and plot moves" were suggested to Pasternak in the course of their discussions (Bykov 2007, p. 50).

Freidenberg's works on mythopoetics include some writings on the Gospels. In 1930, she completed her article "The Entry into Jerusalem on a Donkey," where she suggested a connection between the image of the donkey in the Gospels and the Sumerian god of fertility. In the same year, she published in the Soviet journal *The Atheist* (!) an article entitled "The Gospels as a Version of the Greek Novel," particularly interesting when read in conjunction with *Doctor Zhivago*. Importantly, the article is based on Freidenberg's doctoral dissertation *The Origins of the Greek Novel* that she defended in 1924 (Braginskaia 2018, p. 96) and which is mentioned more than once in her correspondence with Pasternak. In the autumn of 1924, Pasternak asked his cousin for a copy of *The Golden Bough* by Sir James George Frazer (Mossman 1982, p. 72)—a famous comparative study of mythology and religion.

In her 1930 article, Freidenberg, just like Frazer, reads the story of the crucifixion and resurrection of Christ with reference to a myth about a god of trees and plants that dies and is reborn. She claims the erotic elements of the ancient Greek novel share mythological roots with the evangelical motif of resurrection, because in agrarian mythology the act of fertilisation and the emergence of seedlings from the soil are interpreted as a resurrection.[20] Freidenberg illustrates her thesis with observations on links between spring and resurrection in the Christian tradition. Without identifying her source, she makes a reference to "one of the church fathers" and quotes *Octavius*, an early writing in defence of Christianity by Marcus Minucius Felix (Braginskaia 2018, p. 96): "See, therefore, how for our consolation all nature suggests a future resurrection. [...] We must wait also for the spring-time of the body" (Freidenberg 1930, p. 147). Thus, in Olga Freidenberg's publications, we find an analysis of the connection between Christianity and myth that was creatively realised in the Easter narrative of *Doctor Zhivago*. Pasternak also transferred this religious and mythological narrative onto his understanding of history as the plant kingdom that dies in the autumn and comes back to life in the spring, similarly to Christ going through death and resurrection.

In 1930–1931, Pasternak wrote the second and third parts of his short novel *Safe Conduct*. It is in the second part of the text that his famous comparison of the Bible with "the notebook of humankind" can be found (Pasternak 2004b, p. 207). Pasternak explains his understanding of the similarity by saying that "each new generation interprets and describes its own reality with the help of the Bible" (Bodin 1976, p. 3). In addition to the obvious parallel with *Doctor Zhivago*, we can note here that the key concept which in Pasternak's opinion best describes the connection between the Bible and various eras in the history of humankind is "a legend." He says: "I understood that the history of culture is a chain of equations in images, pairwise connecting the next unknown with something that is known, and this known element, the one thing that remains consistent throughout the sequence, is a legend that is the foundation of a tradition, while the unknown element, the one that is always new, is the present moment of the current culture" (Pasternak 2004b, p. 207). The concept of a legend is close to the idea of myth turned metaphor, i.e., to mythopoetics—Olga Freidenberg's subject of research. If we read Freidenberg's article on the Gospels (published in 1930) together with the passage I have just quoted from Pasternak's *Safe Conduct* (published in 1931), we can derive a tripartite model: the myth–the Bible–the legend. The Gospels absorb elements of pre-Christian mythology and, in turn, become a source of legends, such legends functioning as interpretative models across historical eras. The same legend being actualised in different historical periods is yet another aspect of Pasternak's concept of cyclical time.

5. The Easter Narrative of History from 1917 to the Thaw

There is a famous episode in the eleventh part of *Doctor Zhivago*: Yurii Zhivago discovers a red partisan that has been killed, and a wounded White Guard, who both wore a charm with the text of Psalm 90 around their necks. The psalm was believed to have the power to protect against bullets, and the episode stresses the fact that all the participants of the Civil War were victims seeking the protection of the same God. More than that, this motif of shared suffering connects the Civil War and the repressions of the 1930s: "decades later prisoners were to sew [the text of the psalm] into their clothes and mutter its words in jail when they were summoned at night for interrogation" (p. 335). One of the novel's central motifs is the suffering that the post-revolutionary era brought with it. Zhivago's nemesis, the revolutionary fighter Strelnikov, used to passing ruthless verdicts at the court-martial, talks to Zhivago while expecting to be arrested and executed by his own comrades in arms. He calls himself a "martyr," explaining that the revolutionaries had risen against the world of "dirt," "misery" and "degradation of human beings," but ended up becoming tragic victims of history, in their turn (p. 459; translation modified—S.E.).

Historical events in the novel are organised cyclically: the pre-revolutionary era of servitude, poverty and humiliation, against which the revolutionaries rebelled, ends with the jubilation of the February Revolution and the welcoming of Easter. The October Revolution brings with it the Civil War, as a new turn in the spiral of national suffering. This is followed by the relief of the 1920s, and the beginning of the NEP (New Economic Policy) is associated in the novel with the coming of spring, the sun, and the domes of the Cathedral of Christ the Saviour. History, nature and religion merge again in a symbolic unity: "The doctor and Vasia arrived in Moscow in the spring of 1922 at the beginning of the NEP. The weather was fine and warm. Sunshine glancing off the golden domes of the Church of the Saviour played on the square below" (p. 473). The Easter narrative of "The Poems of Yurii Zhivago" presents the reader with a certain interpretative key that can help decipher various episodes of the novel connected with the theme of spring. Thus, the images of spring and cathedral domes in Moscow in the early NEP era are associated with Easter, without the festival being named directly.

The repressions of the 1930s represent the next stage in the cyclical history of the Passion. Lara is arrested after Zhivago's death; his two friends, Mikhail Gordon and Innokentii Dudorov, spend time in labour camps. As with the October Revolution, the northern labour camp appears in the novel as a symbol opposed to that of spring. It is a kingdom of white snow: "A wilderness of snow. [...] An open snow field with a post in the middle and a notice on it saying: 'GULAG 92 Y.N. 90'—that's all there was" (p. 506). Unlike the camps, the Second World War in the novel is symbolically connected with the summer and with the approaching victory: the one war episode is set in "the summer of 1943, after the breakthrough on the Kursk bulge and the liberation of Orel" (p. 504). Major Dudorov explains to his comrade-in-arms Gordon how he understands the historical significance of the war. According to Dudorov, after the repressions of the 1930s, "the war came as a breath of fresh air, a purifying storm, a breath of deliverance" (p. 507). The people who had been tempered by the adversities, the people who became "the moral elite of the generation," threw themselves into a meaningful battle "with abandon," pursuing a higher, real goal (p. 508; translation modified—S.E.).

Here, history is perceived again as a force of nature (a storm, fresh air); at the same time, the war makes the Christian symbolism more prominent. Dudorov shares with Gordon the story of the "martyrdom" of his bride Christina Orletsova (p. 508). Christina, who was with the partisans, "got inside the German lines and blew it [a fortified building— S.E.] up, and was taken alive and hanged" (p. 505). On the one hand, this fate is evocative of the famous story of the Soviet partisan Zoia Kosmodemianskaia, executed by the Nazis (Mukhina 2019). On the other hand, the young woman's name literally means 'a follower of

Christ': she was the daughter of a priest who had fallen victim to the Stalinist repressions. Moreover, in his conversation with Dudorov, Gordon adds: "They say the Church has canonized her" (p. 505). Thus, the partisan Christina embodies the concept of martyrdom as *imitatio Christi*, which is one of the foundations of the Orthodox culture.[21]

This interpretation of the Great Patriotic War as a spiritual rebirth in *Doctor Zhivago* should be considered in the cultural and historical context of the Soviet Union of the 1940s. During the war, the Soviet government stopped persecution of the Orthodox church, whereby the turning point was the year 1943, when the 'war episode' in the novel is set. In September 1943, Stalin had a meeting with three Metropolitan Archbishops of the Russian Orthodox church; four days later, a new Patriarch of Moscow and all Rus' was elected—for the first time since 1925.[22] By placing the only war episode in the novel in 1943, with a reference to the Orthodox church and the story of the partisan–martyr Christina, Pasternak made the war into an element of his religious interpretation of Russian history.

The execution of Christina, similarly to the crucifixion of Christ, represents the first part of the Easter narrative. In the logic of this narrative, the Passion must be followed by a resurrection. At first glance, the motif of Easter seems to be absent from the war episode of the novel; nevertheless, *Doctor Zhivago* is a text in which the symbolic meaning of many events can be decoded with reference to the Easter narrative. Dudorov says that, despite Christina's death, witnessing the heroic self-sacrifice of the war generation fills him "with happiness" (p. 508). The feeling of happiness in the novel is frequently accompanied by the symbolism of spring and Easter. Dudorov's reflections are also an anticipation of victory, the approach of which is signalled by the mention of the breakthrough on the Kursk bulge (1943) in the opening passages of the war episode. I think that this episode contains a coded reference to a fact of which Pasternak's contemporaries were mostly certainly conscious: the Orthodox Easter of 1945 fell on 6 May. It almost coincided with the day on which the victory of the Soviet Union in the Second World War was celebrated (9 May), thus imparting upon the victory the status of a sacral event. The Soviet government allowed Easter services, and to a certain extent the situation of 1917 was repeated, when the February Revolution and Easter merged into one in the consciousness of the contemporaries.[23]

The action of the last chapter of the "Epilogue" is set at a moment in time which is imprecise in the extreme: "Five or ten years later" (p. 518). In other words, it is the summer of 1948 or 1953. This indeterminacy is surprising, considering the significant difference between "the year when Stalinist postwar repressions were coming into high gear, and the year of Stalin's death" (Polivanov 2015, p. 142). I would argue that the content of the chapter partly resolves the problem of the precise dating. The chapter in which we meet Gordon and Dudorov one autumn evening reading Yurii Zhivago's poems consists of four paragraphs. The third paragraph is just one sentence describing the atmosphere of the post-war years. The starting point is the Easter feeling of the victory of 1945 that is implied in the previous chapter: "Although victory had not brought the relief and freedom that were expected at the end of the war, nevertheless the portents of freedom filled the air throughout the postwar period, and they alone defined its historical significance" (p. 519). It is worth noticing that here "the postwar period" is referred to retrospectively, as something that is already in the past, and the focus is on their "historical significance," i.e., a macro-perspective on history as a change of eras. The concluding paragraph recreates a completely different atmosphere: "To the two old friends, as they sat by the window, it seemed that this freedom of the soul was already there, as if that very evening the future had tangibly moved into the streets below them, that they themselves had entered it and were now part of it" (p. 519). The transition from the anticipation of freedom to the experience of freedom is perceived as a transition from one historical era to another, to "the future." Unlike the "postwar period," this new era is not over yet; it is just beginning.

Thus, based on the content of this last chapter, it seems appropriate to assume it is set in 1953. These two passages describe a transition from the "postwar period" to the Thaw with its feeling of a new freedom; it is the closest the characters of the novel were to the actual time in which the author lived. Therefore, why did Pasternak introduce this indeterminacy into the chronology; why did he speak of "five or ten years"? The writer hoped to have *Doctor Zhivago* published in the USSR and offered the novel to the magazines *Znamia* and *Novyi mir*. With this in mind, it made sense not to make the link between the epilogue and the year of Stalin's death too obvious.

The epilogue sets the reader a riddle while at the same time hinting at a solution. The novel concludes with the "music of happiness" of the now much older heroes: they feel a "sentimental tenderness and peace" when "[t]hinking of this holy city [Moscow] and of the entire earth" (p. 519; translation modified—S.E.). The happiness that the heroes feel can be a sign of the new era of freedom following the death of Stalin in 1953. However, the motif of a "holy city" that continues the religious symbolism of the war episode signals the presence of yet another element in the epilogue. The epilogue ends with a phrase that links the prose part of the novel with "The Poems of Yurii Zhivago" the heroes read: "And the book they held seemed to know all that, to confirm and encourage their feeling" (p. 519; translation modified—S.E.). This phrase suggests that the clue to understanding the characters' feelings is to be found in Zhivago's poems. More than that—the poems written by Zhivago, who died in 1929, acquire a great symbolic significance: this book "seems to know" everything about the atmosphere of 1953.

It is worth remembering that the first three poems recreate the Easter narrative of the Passion and the Resurrection, with the poems "March" and "In Holy Week" connecting Easter with March.[24] In my analysis so far, I have looked into the special role of March within the context of the February Revolution that ended on 2 March 1917. However, if Zhivago's poems give us a clue to understanding Gordon's and Dudorov's feelings at the end of the epilogue, could it be that March is somehow connected with 1953? Stalin died on 5 March 1953, and this event was perceived as a symbolic starting point of the political reforms in the country.

On 7 March 1953 Pasternak wrote a letter to Varlam Shalamov, or rather—to Shalamov's wife, Galina Gudz', with the request to pass it on to her husband. In this letter, he unexpectedly prefaces his response to the death of Stalin with reminiscences about the February Revolution: "When the February Revolution broke out, I was in a god-forsaken provincial town, on the Kama river, at some plant. [...] This last tragic event also found me outside of Moscow, in the winter woods, and the state of my health will not allow me to travel to the city one of these days to say good-bye. [...] All the words are full to the brim with significance, with truth. And it is quiet in the woods" (Pasternak 2005, p. 721). The context does not make it clear what these "words" that became "full to the brim with significance" are; the phrase remains mysterious. It is possible that Pasternak meant the "words" of his novel; at the time, he was working on *Doctor Zhivago*, which he also mentions in his letter.[25]

Even though Pasternak speaks of "winter woods" in the letter (apparently a reference to the actual weather conditions at the time), it is worth noting that the political turning point coincided with the arrival of spring—just as in 1917. The event of 1953 is attributed a great significance, which is, however, not explained, and the mention of this significance is structurally close to the image of the woods. In *Doctor Zhivago*, a forest appears as a symbol of history and at the same time as a participant of the Easter narrative in the poem "In Holy Week". The fact that in the same poem March is linked to the festival of Easter reminds us that, in 1953, the Russian Orthodox Lent started on 16 February, and that the same year Easter fell on 5 April. In 1917, exactly one month passed between the end of the February Revolution and Easter (2 March–2 April); in 1953, exactly one month separated the death of Stalin and Easter (5 March–5 April). I think Pasternak was aware of all these parallels when, in his letter, he linked Stalin's death with his memories of the February Revolution.

In the summer of 1917, Zhivago recognises the significance of the February Revolution that had ended just a few months previously, in March: "Freedom! Real freedom, not just talk about it, freedom, dropped out of the sky, freedom beyond our expectations" (p. 146). Gordon and Dudorov have a similar feeling at the end of the "Epilogue" in the summer of 1953: "To the two old friends, as they sat by the window, it seemed that this freedom of the soul was already there" (p. 519). The expression "freedom of the soul" at the end of the novel evokes the connection between the February Revolution and individual experience: "Everyone was revived, reborn, changed, transformed" (p. 146) Moreover, Yurii Zhivago emphasises the accidental character of the February Revolution, and in the novel the accidental is predetermined: it is "freedom, dropped out of the sky, freedom beyond our expectations, freedom by accident, through a misunderstanding" (p. 146).[26] The key events in Pasternak's novel, of which Stalin's death is one (without being mentioned directly), obey not the plans of human beings, but the higher natural and religious logic of life itself. Thus, in the context of the novel, the February Revolution and Stalin's death are connected with each other through the triple semantics of the beginning of spring, a political renewal, and the approaching festival of Easter.[27] The death of Stalin as a harbinger of the resurrection of Russia[28] acquires a sacral significance in Pasternak's religious concept of history.

On 14 March 1953, Pasternak wrote a letter to the Chairman of the Soviet Writers' Union, Aleksandr Fadeev. The letter was evasive, with some ambiguous formulations (Bykov 2007, pp. 531–32), but I am particularly interested in two sentences that refer to Moscow saying good-bye to Stalin: "And this second city, a city within a city, a city of wreaths that has been erected on the square! As if the whole plant kingdom came to pay its respects, as if all of it showed up for the funeral" (Pasternak 2005, p. 723). Just like in the novel *Doctor Zhivago*, nature here is not a passive background, but an active participant of events: it has come to attend the funeral. More than that: the expression "plant kingdom" (*rastitel'noye tsarstvo*) connects Pasternak's letter with the novel.[29] I have already suggested that in *Doctor Zhivago* the plant kingdom is a metaphor for history: Zhivago "reflected again that he conceived of history, of what is called the course of history, not in the accepted way but by analogy with the vegetable kingdom" (p. 453). Pasternak's words from his letter to Fadeev can be understood with reference to this metaphor: not only Stalin himself, but the whole era, is being seen off by the plant kingdom of Russian history.

Let us come back to Yurii Zhivago's poem "March," which is to be read in connection with the end of the February Revolution and the death of Stalin. One word that is most applicable to the state of nature described in the poem (even though it is not used in the text) is 'thaw.' The Russian word for Thaw (*ottepel'*) refers to the warm weather (*teplaia pogoda*) in the winter or else in the early spring, when snow and ice start melting. March is the first month of spring, and the poem's imagery conveys a sense of snow melting, of ice in the sun: "the snow is wasting," there is "the thrumming of melting icicles," "cathectic icicles hanging on to gables," "the chattering of rills that never sleep" (Pasternak 1958, pp. 523–24; translation modified—S.E.). We are, yet again, dealing here with a situation when a key symbol is not named directly, but is instead encoded in images of the novel.

The word 'thaw' became a political symbol following the publication of Il'ia Erenburg's eponymous short novel in the spring of 1954, before Pasternak completed his work on the novel. In the writings of Pasternak, the images of spring and thaw acquired a political significance as early as in 1918, in the poem "The Russian Revolution." There, the speaker feels the "ice-breaking breath" of the revolution, hears "the dripping of melting snow," sees the sun reflected in brooks, and speaks to the Russian Revolution directly: "How good it was to breathe you in March" (Pasternak 2004a, p. 224). After Erenburg's opportune choice of the title for his short novel made the word 'thaw' a metaphor for the liberalisation of the society, everything fell into place for Pasternak. To quote his own letter to Shalamov, "all the words filled to the brim with significance."

Interestingly, the joyful arrival of spring at the end of Erenburg's *The Thaw* is accompanied by the image of willow: "First of all—the winter is over. [...] Second—willow trees are burgeoning"; "a girl is carrying a willow branch" (Erenburg 1954, pp. 138, 141). In Orthodox Christianity, the willow, which blooms before other trees in the spring, is the symbol of Palm Sunday, the festival celebrating Christ's entry into Jerusalem the week before Easter. Even though there are no religious motifs in Erenburg's short novel, the image of a willow tree is associated not only with the spring, but also with the approaching Orthodox Easter.[30] Yurii Zhivago's poem "The Earth," which is also connected with the Easter narrative, features a willow tree.[31] Thus, *The Thaw* by Erenburg and *Doctor Zhivago* echo each other not because there is a causal relationship between the two texts, but because they share a cultural and historical context.

6. *Doctor Zhivago* and Politics

Before we move on to a general conclusion, a few words on the political dimension of the novel are in order. Immediately upon publication, *Doctor Zhivago* was rejected not only by the official Soviet critical establishment, but also by the right-Russian émigré groups "who felt nostalgic for the pre-revolutionary Russia" (Fleishman 2009, p. 289). Dmitrii Pronin's review in the Russian-language American newspaper *Russian Life* (7–8 January 1959) is representative of this reaction. Pronin suspected Pasternak of collusion with the Soviet government. In the reviewer's opinion, this would explain the negative portrayal of Tsarist Russia and the White Guard's crimes in the Civil War, whose cruelty rivalled that of the Red Army (Fleishman 2009, pp. 291–96). This response is not surprising, considering Pasternak's complex political views that could not be reduced to any particular ideology.

Rejecting monarchy, the novel's protagonist enthusiastically welcomed the February Revolution. He also saw the October Revolution as historically inevitable, even though with time he grew disillusioned with its results (Polivanov 2015, pp. 142–48).[32] The novel's key ideals—freedom and the value of individual life—echo the declared principles of the February Revolution with its assertion of the primacy of the democratic world order.

Writing about Pasternak's life, Lazar Fleishman speaks of the writer's hopes for "a transformation of the Soviet regime that would direct it to a more humanist and democratic path of development" (Fleishman 2006, pp. 685, 704). The novel's treatment of the 'resurrection' of Russia in the spring of 1917 as a metaphor of liberation was common in the cultural context of the February Revolution. The changes that the spring of 1953 brought with it could be seen as a 'resurrection' in the sense of a return to the ideals of the February Revolution, suggesting that it might be possible for Russia to choose democracy.

It is worth noting that Pasternak's vision of the future of Russia was different from that of another Soviet Nobel Prize winner, Aleksandr Solzhenitsyn,[33] as is evidenced by the latter's article "Reflections on the February Revolution" (written in 1980–1983). The same historical event that Pasternak sees as sacral acquires demonic features in Solzhenitsyn's interpretation: it is a "fatal night" that destroyed Russian history (Solzhenitsyn 1995, p. 482). According to Solzhenitsyn, the February Revolution prevented the monarchy from going into battle against what was to become the October Revolution (ibid., p. 484). The exiled writer was also convinced that the spring of 1917 brought with it a catastrophic "loss of the national consciousness" (ibid., p. 503).

Pasternak's protagonist sees the same events completely differently. For him, the individual comes before the nation. He and his friend Mikhail Gordon believe that Christianity has replaced the idea of "a nation" with "the mystery of the individual" (p. 122). I think Yurii Zhivago (and Boris Pasternak) would not be sympathetic to the political programme proposed by Solzhenitsyn in his 1990 article "Rebuilding Russia." The country, writes Solzhenitsyn, should become democratic, but the model best suited to its needs is to be found in the pre-revolutionary institutions of local self-government such as *mir* (peasant communities) and *zemstvo* (elective assemblies) (Solzhenitsyn 1995, p. 585). Solzhenitsyn's "extremely conservative programme for national regeneration" and his "concern with preserving Russian values" (Rowley 1997, p. 328) would not have appealed to Pasternak's heroes, convinced as they were that there is no room for nations in the Christian world.

7. Conclusions

We can now present a short summary. "The Poems of Yurii Zhivago" play the role of a symbolic clue that helps decipher the events of the prose part of the novel. *Doctor Zhivago* retells the history of Russia as a religious parable modelled on the Gospel story of the life of Christ. This parable is founded on the idea of a cyclical movement of history, akin to the principle at the basis of both the natural calendar and the Orthodox religious practices. In the Russian history of the years 1901–1955, periods of suffering, martyrdom, and political and military violence were followed by years of renovation, liberation and rebirth. In the novel *Doctor Zhivago*, this bipartite Easter narrative passes through several cycles: (1) pre-revolutionary Russia and the February Revolution; (2) the October Revolution with the Civil War, followed by the relief brought by the NEP period; (3) the repressions of the 1930s and the Second World War (the way it is interpreted in the novel); and (4) the post-war years and the Thaw.[34] In the second part of each cycle, the reader can discern echoes of the hopes for political progress that Boris Pasternak nourished at the time. For example, the writer was inspired by his meetings with Anatolii Lunacharskii and Lev Kamenev, both of whom occupied senior positions in the Communist Party and rejected revolutionary terror (Fleishman 2006, p. 642). For some time in 1933–1934, Pasternak felt sympathetic to the Soviet power, when he thought he detected signs of the government moving towards liberal reforms (ibid., p. 685).

At the same time, *Doctor Zhivago* depicts several annual micro-cycles (the temporary relief of suffering that spring brings with it) and the grand historical macro-cycle from the Passion of the October Revolution to the Resurrection of 1953. The end of the novel marks a symbolic return of the Russian history to the March of 1917, to the Easter days of the February Revolution. It seems reasonable to conclude that the dominant motifs of the two springs in the novel (the spring of 1917 and that of 1953) correspond to the two Easters in "Yurii Zhivago's Poems."

According to the overarching symbolic logic of the novel and the philosophy of history it expresses, the happy conclusion of *Doctor Zhivago* in the epilogue is not the final point of a linear progression, but just another loop in the cyclical structure of time. It is not by chance that the narrator chooses to use the verb "seem": "To the two old friends, as they sat by the window, it seemed that this freedom of the soul was already there" (p. 519).

The revolutionary Strelnikov is one character in the novel who believes in the linear model of time—like a "straight line into a better future" (Bethea 1989, p. 251). He interprets the events of the Civil War as the Last Judgement, i.e., with reference to the apocalyptic narrative of the Second Coming. Fighting for a better future goes hand in hand with an apocalyptic sense of the end of history, as when Strelnikov tells Zhivago: "These are apocalyptic times, my dear sir, this is the Last Judgement. This is a time for angels with flaming swords and winged beasts from the abyss" (p. 252).

At the same time, it is obvious that the Book of Revelation is the gravitational centre of Yurii Zhivago's worldview, of his reflections on death and ways to overcome it. Unlike Strelnikov, Zhivago suggests an alternative reading of the Apocalypse, where the linearity of time can be overcome. In a conversation with the gravely ill Anna Ivanovna, Zhivago questions a literal understanding of the resurrection that the Book of Revelation promises to the dead: "Resurrection. In the crude form in which it is preached to console the weak, it is alien to me. I have always understood Christ's words about the living and the dead in a different sense. Where could you find room for all these hordes of people accumulated over thousands of years? The universe isn't big enough for them; God, the good, and meaningful purpose would be crowded out" (p. 67). According to Zhivago, the triumph of resurrection that St John the Divine promises to the followers of the true religion is not a dream of the future, but the essence of life itself in its cycle of dying and being born anew: "But all the time, life, one, immense, identical throughout its innumerable combinations and transformations, fills the universe and is continually reborn. You are anxious about whether you will rise from the dead or not, but you rose from the dead when you were born and you didn't notice it" (pp. 67–68).

Thus, Strelnikov and Zhivago embody not only two different understandings of time—linear and cyclical—but also two corresponding interpretations of the Book of Revelation. In the novel's parable of history, the emotional collapse and suicide of revolutionary

Strelnikov is juxtaposed with Gordon's and Dudorov's repeated returns to Yurii Zhivago's poems that explain to them the meaning of the new era.

In my introduction, I pointed out that a religious interpretation of the revolution in Russian literature often includes references to the Apocalypse and the apocalyptic (millenarian) sect (Etkind 1998, p. 20). Understood thus, the Apocalypse is linked with the linear teleological model, the final point of which is the general salvation at the end of history, the Last Judgement. In *Doctor Zhivago*, the cyclical repetition of the Easter narrative creates an alternative conception of history that is not linear but cyclical, and whose narrative model is not the Second Coming, but the Gospel story of the Passion and the Resurrection of Christ. Thus, when talking about connections between a philosophy of history and religion in Russian culture, I suggest distinguishing between the Easter and apocalyptic narratives. These two narrative models can be in a complementary relationship with each other, but they have different foundations.

Funding: This research received no external funding.

Acknowledgments: The author is grateful to Natalia Jonsson-Skradol for her careful translation and copy-editing of the paper. Thanks are also due to the anonymous reviewers as well as to Dennis Ioffe for their valuable comments on an earlier version of this article.

Conflicts of Interest: The author declares no conflict of interest.

Notes

These start and end dates of the novel's action are based on precise references in the text and on the age of the characters (Sukhikh 2013, pp. 536–37, 564). The year in which the last scene is set cannot be determined with absolute certainty, because the narrator places the episode "five or ten years" after the action of the previous section (Pasternak 1958, p. 518). In this article, I argue that the final scene takes place in 1953.

"Miracle," "Evil Days," "Magdalene I" and "Magdalene II", "Garden of Gethsemane".

This observation is made, either in passing or as part of a more detailed argumentation, in most works on *Doctor Zhivago*. I will mention here by way of example just John Givens's monograph on the image of Christ in Russian literature, in which one chapter is dedicated to Pasternak's novel and the figure of Yurii Zhivago (Givens 2018, pp. 177–204).

Not by chance do some scholars call the novel "the Gospel of Boris Pasternak" (Leiderman 2010, p. 801; Sukhikh 2013, p. 534).

Throughout the text, references to the translation of Pasternak's novel are given by page numbers only.

Ivan Esaulov points out the central role of Easter in *Doctor Zhivago*, the structure of which is based on "the Orthodox yearly cycle" (Esaulov 2006, p. 73). However, Esaulov does not make a connection between this temporal cycle and depictions of history and nature in the novel. In his book on the image of Christ in Russian literature, John Givens twice makes a passing reference to Pasternak's magnum opus as an "Easter novel" (Givens 2018, pp. 178, 180).

Boris Kolonitskii cites several sources from the spring of 1917, the authors of which spoke of the "resurrection" (*voskresenie/voskreshenie*) of Russia or of the Russian people. Without implying a particular political system, this emotionally charged metaphor was associated with the idea of liberation—a rejection of the chains of oppression. The idea of a "resurrection" postulated not only a return to a previous era, but also a transformation. Historical evidence from the time includes, for example, an Easter card printed in 1917 that bore the following inscription: "Christ has risen. Long live the republic!" (Kolonitskii 2012, pp. 79, 82, 83).

An authoritative Russian source dates this poem to 1918; it was first published in 1989 (Pasternak 2004a, pp. 225, 455).

Despite the presence of some apocalyptic motifs in the poem, its overall religious imagery is contained by the story of Christ's crucifixion and resurrection (Etkind 1998, p. 397).

As Thomas F. Rogers noted, the eleventh section of the novel also features the Gospel motifs of Judas's betrayal and the Last Supper (Rogers 1974, pp. 385–86). There is one more similar reference in a speech held before new recruits, who are told they are about to make their first steps on the way of the cross that "stretches out before [them]" for the salvation of the motherland (p. 322).

I should point out that Christmas does feature in the novel, but associations with it are not always positive. For example, Anna Ivanovna, in whose family Yurii Zhivago grew up, dies at Christmas. Thus, in the text, the Christmas festival is secondary to the Easter narrative.

In this connection, it is also important to remember that, as a child, Yurii Zhivago often imagined the world around him as a forest, with God as a "keeper of the forest" (p. 87).

Early in the novel, Nikolai Vedeniapin makes the frequently quoted comment that "man does not live in nature but in history" (p. 10; translation modified—S.E.). In the general context of the novel, this statement can be interpreted as a reference to people's living not *just* in nature, but in nature which is at the same time history, i.e., in history.

[14] David Bethea pointed out that the transformation of "brute history into Christian History" in the text is related to "the seasonal cycle, the cycle that [...] is epitomized by the natural and religious connotations of Easter" (Bethea 1989, p. 251).

[15] The 20th century philosophy of life was partly grounded in the teachings of Friedrich Wilhelm Joseph Schelling (Briukhanova 2010, p. 13). Pasternak studied philosophy at the universities of Moscow and Marburg, and his writings show the influence of Schelling (Gasparov 1992, p. 101; Smirnov 1996, pp. 60–61).

[16] Hence, the ambivalent attitude of Tolstoy (and Pasternak) to Hegel's teleological philosophy of history (for a discussion of Tolstoy and Hegel see Klimova 2017). Scholars are split on how this is reflected in *Doctor Zhivago*. Igor Smirnov reads the novel as a refutation of Hegel's philosophy (Smirnov 1996, p. 193), whereas Barry Scherr finds evidence in the text of its author's having been influenced by Hegel's *Early Theological Writings* (*Hegels Theologische Jugendschriften*) that first saw light in Germany in 1907 (Scherr 1991).

[17] In the authoritative Russian source used here, the original French text by Verlaine seems to have been reprinted with a typo ("saint").

[18] The links between Christian and mythological motifs in the imagery of the forest and the garden in *Doctor Zhivago* are analysed in Anna Skoropadskaia (2006) doctoral dissertation.

[19] In the late 1940s, Pasternak often attended Orthodox church services in Moscow and in Peredelkino (a suburb of Moscow, where Pasternak had a country house), and studied the Bible and liturgical literature in Church Slavonic (Fleishman 1995).

[20] In my opinion, Freidenberg's analysis of the similarity between the ancient erotic novel and the Gospels may well be the origin of the paradox that John Givens formulated thus when discussing Zhivago: "After all, how can a man who loves and abandons three women and the five children he sired with them be anything like Christ?" (Givens 2018, p. 201).

[21] This tradition originated with the first Russian saints ("passion-bearers," as they are referred to in Orthodox sources), the princely martyrs Boris and Gleb, who were murdered in 1015. For an exploration of the theological and cultural aspects of this tradition, see Dirk Uffelmann (2010) monograph.

[22] Lazar Fleishman suggested that the growth of Pasternak's interest in Christianity could be related to the rehabilitation of the religion that inspired hope for changes in the country (Fleishman 1995).

[23] The famous Soviet Victory Parade on the Red Square in Moscow took place on 24 June 1945, on the day of the Orthodox festival of Holy Trinity (the fiftieth day after Easter).

[24] There are no obvious religious motifs in the poem "March," but the ones immediately preceding and following it are "Hamlet" and "In Holy Week", the second of which also mentions March.

[25] Yurii Zhivago's poem "March" was written in 1946. In April 1954, it was among the ten poems printed in the magazine *Znamia* under the title "Poems from the Prose Novel *Doctor Zhivago*". Il'ia Erenburg's short novel *The Thaw* appeared in the next issue of the same journal.

[26] The question of the extent to which this understanding of the February Revolution reflects the actual historical circumstances of the event is beyond the scope of this article.

[27] Pasternak was particularly productive in the summer and autumn of 1953; he felt inspired and worked on the novel with enthusiasm (Bykov 2007, p. 712). In September 1953, he wrote in a letter to the Zhuravlevs that in the preceding summer months he had experienced that "fruitful bliss" of creativity that he had known only once before, when working on the poetry collection "My Sister, Life. The Summer of 1917" (Pasternak 2005, p. 746; Bykov 2007, p. 712). Most of the poems that were included in that collection had been written following the February Revolution, in the summer of 1917.

[28] In the novel, this notion of a 'resurrection' is programmatically connected with the more general idea of a liberation, just as it was in the texts and historical artifacts from the spring of 1917 that were analysed by Boris Kolonitskii (2012, pp. 79, 82).

[29] Ivanova (2001) and Bykov (2007) both noted references to some motifs from *Doctor Zhivago* in the letter.

[30] *The Thaw* was printed in the fifth (May) issue of the magazine *Znamia* in 1954, shortly after Easter that was celebrated relatively late that year—on 25 April.

[31] In 1946, one of the working titles of the novel was *Boys and Girls*—a slightly modified quotation from Aleksandr Blok's poem "Willow Branches," from which Pasternak also intended to borrow an epigraph to his novel (Pasternak 2004c, p. 645). Blok's poem describes children carrying home candles and willow branches on the eve of Palm Sunday.

[32] The protagonist's ambivalent attitude to the October Revolution reflects Pasternak's own ambiguous feelings towards Vladimir Lenin. His "condemnation" of Lenin immediately following the October Revolution gave way to an understanding of the "tragically irresolvable contradictions inherent in the revolution" (Fleishman 2006, p. 641).

[33] Pasternak was awarded the Nobel Prize in Literature in 1958; Solzhenitsyn received his in 1970. The two writers are often discussed in the same context due to the important role they played in the formation of the social and political consciousness of the Russian intelligentsia (Sergeeva-Kliatis 2018).

[34] Naturally, these cycles are not identical, although in their core they repeat the same symbolic model of the Easter narrative. *Doctor Zhivago* was not Pasternak's first work that featured Easter motifs. The protagonist of the epic poem *Lieutenant Schmidt* (published in a serialised form in 1926–1927) stands at the head of the sailors' uprising on the cruiser *Ochakov*. Sentenced to death, Schmidt welcomes his fate in the spirit of Christ, on the threshold of a new historic era (Bodin 1976, pp. 2–3).

References

Baran, Khenrik. 1993. *Poetika Russkoi Literatury Nachala XX Veka*. Moscow: Progress–Univers.

Belyi, Andrei. 2006. *Stikhotvoreniia i Poemy*. Saint-Petersburg and Moscow: Akademicheskii Proekt, Progress-Pleiada, vol. 2.

Bergman, Jay. 1990. The Image of Jesus in the Russian Revolutionary Movement: The Case of Russian Marxism. *International Review of Social History* 35: 220–48. [CrossRef]

Bethea, David M. 1989. *The Shape of Apocalypse in Modern Russian Fiction*. Princeton: Princeton University Press.

Blok, Aleksandr. 1922. *Dvenadtsat'. Illustrated and Introduced by Vasilii Masiutin*. Berlin: Neva.

Bodin, Per-Arne. 1976. *Nine Poems from Doktor Živago: A Study of Christian Motifs in Boris Pasternak's Poetry*. Stockholm: Almqvist & Wiksell.

Braginskaia, Nina. 2018. O sviazi osnovnykh idei O. M. Freidenberg. *Vestnik RGGU. Seriia: Istoriia. Filologiia. Kul'turologiia. Vostokovedenie* 3: 71–97.

Briukhanova, Iuliia. 2010. *Tvorchestvo Borisa Pasternaka kak Khudozhestvennaia Versiia Filosofii Zhizni*. Irkutsk: Izdatesl'stvo Irkutskogo Gosudarstvennogo Universiteta.

Bulgakov, Sergii. 1989. *Pravoslavie. Ocherki Ucheniia Pravoslavnoi Tserkvi*, 3rd ed. Paris: YMKA Press.

Bykov, Dmitrii. 2007. *Boris Pasternak*. Moscow: Molodaia Gvardiia.

Erenburg, Il'ia. 1954. *Ottepel'*. Moscow: Sovetskii Pisatel'.

Esaulov, Ivan. 2006. The Paschal Archetype of Russian Literature and the Structure of Boris Pasternak's Novel Doctor Zhivago. *Literature and Theology* 20: 63–78. [CrossRef]

Etkind, Aleksandr. 1998. *Khlyst. Sekty, Literatura i Revoliutsiia*. Moscow: Novoe Literaturnoe Obozrenie.

Fleishman, Lazar. 1995. Boris Pasternak i khristianstvo. In *Christianity and the Eastern Slavs*. California Slavic Studies Vol. 18. Edited by Boris Gasparov, Robert P. Hughes, Irina Paperno and Olga Raevsky-Hughes. Berkeley and Los Angeles: University of California Press, vol. 3, pp. 288–301.

Fleishman, Lazar. 2006. *Ot Pushkina k Pasternaku. Izbrannye Raboty po Poetike i istorii Russkoi Literatury*. Moscow: Novoe Literaturnoe Obozrenie.

Fleishman, Lazar. 2009. *Vstrecha Russkoi Emigratsii s "Doktorom Zhivago": Boris Pasternak i "Kholodnaia Voina"*. Stanford: Department of Slavic Languages and Literatures, Stanford University.

Freidenberg, Olga. 1930. Evangelie—Odin iz vidov grecheskogo romana. *Ateist* 59: 129–47.

Gasparov, Boris. 1992. Gradus ad Parnassum (samosovershenstvovanie kak kategoriia tvorcheskogo mira Pasternaka). *Wiener Slawistischer Almanach* 29: 75–105.

Givens, John. 2018. *The Image of Christ in Russian Literature. Dostoevsky, Tolstoy, Bulgakov, Pasternak*. DeKalb: Northern Illinois University Press.

Ivanova, Natal'ia. 2001. "Sobesednik roshch" i vozhd'. K voprosu ob odnoi rifme. *Znamia* 10: 186–200.

Ivanov-Razumnik, Razumnik. 1918. Dve Rossii. In *Skify. Sbornik Vtoroi*. Edited by Andrei Belyi, Razumnik Ivanov-Razumnik and Sergei Mstislavskii. Petrograd: Skify, pp. 201–31.

Klimova, Svetlana. 2017. *N. N. Strakhov i L. N. Tolstoi—russkoe preodolenie nemetskoi filosofii*. *Chelovek* 6: 92–103.

Kolonitskii, Boris. 2012. *Simvoly vlasti i bor'ba za vlast': K Izucheniiu Politicheskoi kul'tury Rossiiskoi Revoliutsii 1917 g*. Saint-Petersburg: Liki Rossii.

Lavrov, Aleksandr. 1993. "Sud'by skreshchen'ia": Tesnota kommunikativnogo riada v "Doktore Zhivago". *Novoe Literaturnoe Obozrenie* 2: 241–55.

Leiderman, Naum. 2010. *Teoriia Zhanra. Issledovaniia i Razbory*. Ekaterinburg: UrGPU.

Moreau, Jean Luc. ; Translated by Constance Wagner. 1970. The Passion According to Zhivago. *Books Abroad* 44: 237–42. [CrossRef]

Morson, Gary Saul. 1987. *Hidden in Plain View. Narrative and Creative Potentials in "War and Peace"*. Stanford: Stanford University Press.

Mossman, Elliott, ed. 1982. *The Correspondence of Boris Pasternak and Olga Freidenberg 1910–1954*. Elliott Mossmann, and Margaret Wettlin, transs. New York: Harcourt Brace Jovanovich.

Mukhina, Anna. 2019. Zoia Kosmodemianskaia—Grunia Friedrikh—Khristina Orletsova: Transformatsiia sovetskogo mifa v drame i romane Borisa Pasternaka. *Stephanos* 5: 130–37. [CrossRef]

Münch, Christian. 2014. Glaube und Vernunft. In *Tolstoj als Theologischer Denker und Kirchenkritiker*. Edited by Martin George, Jens Herlth, Christian Münch and Ulrich Schmid. Göttingen: Vandenhoek & Ruprecht, pp. 323–38.

Pasternak, Boris. 1958. *Doctor Zhivago*. Translated from the Russian by Max Hayward and Manya Harari, "The Poems of Yurii Zhivago" by Bernard Guilbert Guerney. New York: Pantheon Books.

Pasternak, Boris. 2004a. *Polnoe Sobranie Sochinenii s Prilozheniiami. Vol. 2. Spektorskii. Stikhotvoreniia 1930–1959*. Moscow: Slovo.

Pasternak, Boris. 2004b. *Polnoe Sobranie Sochinenii s Prilozheniiami. Vol. 3. Proza*. Moscow: Slovo.

Pasternak, Boris. 2004c. *Polnoe Sobranie Sochinenii s Prilozheniiami. Vol. 4. Doktor Zhivago*. Moscow: Slovo.

Pasternak, Boris. 2004d. The Poems of Doctor Zhivago. In *Toronto Slavic Quarterly*. Translated by Christopher Barnes. Available online: http://sites.utoronto.ca/tsq/10/barnes10.shtml (accessed on 28 March 2021).

Pasternak, Boris. 2005. *Polnoe Sobranie Sochinenii s Prilozheniiami. Vol. 9. Pis'ma 1935–1953*. Moscow: Slovo.

Polivanov, Konstantin. 2015. *"Doktor Zhivago" kak Istoricheskii Roman*. Tartu: University of Tartu Press.

Pomazansky, Michael. 1984. *Orthodox Dogmatic Theology: A Concise Exposition*. Translated by Hieromonk Seraphim Rose. Platina: Saint Herman of Alaska Brotherhood.

Poole, Randall A. 2013. Russian Political Theology in an Age of Revolution. In *Landmarks Revisited: The Vekhi Symposium 100 Years On*. Edited by Robin Aizlewood and Ruth Coates. Boston: Academic Studies Press, pp. 146–69.

Raevsky-Hughes, Olga. 1995. Liturgicheskoe vremia i evkharistiia v romane Pasternaka Doktor Zhivago. In *Christianity and the Eastern Slavs*. California Slavic Studies Vol. 18. Edited by Boris Gasparov, Robert P. Hughes, Irina Paperno and Olga Raevsky-Hughes. Berkeley and Los Angeles: University of California Press, vol. 3, pp. 302–25.

Rogers, Thomas F. 1974. The Implications of Christ's Passion in Doktor Živago. *Slavic and East European Journal* 18: 384–91. [CrossRef]

Rosenthal, Bernice Glatzer. 1980. Eschatology and the Appeal of Revolution: Merezhkovsky, Bely, Blok. *California Slavic Studies* 11: 105–39.

Rowley, David G. 1997. Aleksandr Solzhenitsyn and Russian Nationalism. *Journal of Contemporary History* 32: 321–37. [CrossRef]

Scherr, Barry. 1991. Pasternak, Hegel, and Christianity: Religion in "Doctor Zhivago". In *Boris Pasternak. 1890–1990*. Edited by Lev Loseff. Northfield: The Russian School of Norwich University, pp. 26–39.

2018. Pasternak—Solzhenitsyn. Special Issue, Russian Literature, vol. 100–2. Available online: https://www.sciencedirect.com/journal/russian-literature/vol/100/suppl/C (accessed on 28 March 2021).

Shcheglov, Iurii. 1991. O nekotorykh spornykh chertakh poetiki pozdnego Pasternaka. (Avantiurno-melodramaticheskaia tekhnika v "Doktore Zhivago"). In *Boris Pasternak. 1890–1990*. Edited by Lev Loseff. Northfield: The Russian School of Norwich University, pp. 190–216.

Skoropadskaia, Anna. 2006. Obrazy Lesa i Sada v Poetike Romana B. Pasternaka "Doktor Zhivago". Ph.D. Dissertation, Petrozavodskii Gosudarstvennyi Universitet, Petrozavodsk, Russia.

Smirnov, Igor. 1996. *Roman Tain Doktor Zhivago*. Moscow: Novoe Literaturnoe Obozrenie.

Solzhenitsyn, Aleksandr. 1995. *Publitsistika. Vol. 1. Stat'i i Rechi*. Iaroslavl: Verkhne-Volzhskoe Knizhnoe Izdatel'stvo.

Steinberg, Mark D. 1994. Workers on the Cross: Religious Imagination in the Writings of Russian Workers, 1910–1924. *The Russian Review* 53: 213–39. [CrossRef]

Sukhikh, Igor'. 2013. *Russkii Kanon. Knigi XX Veka*. Moscow: Vremia.

Tolstoy, Leo. 1952. *War and Peace*. Translated by Louise and Aylmer Maude. Chicago, London, Toronto and Geneva: Encyclopædia Britannica.

Tussing Orwin, Donna. 1993. *Tolstoy's Art and Thought, 1847–1880*. Princeton: Princeton University Press.

Uffelmann, Dirk. 2010. *Der Erniedrigte Christus—Metaphern und Metonymien in der Russischen Kultur und Literatur*. Köln: Böhlau.

 *religions*

Article

'Only God Can Be": Aleksandr Vvedensky, Kant, God, and Time

Evgeny Pavlov

Department of Global, Cultural and Language Studies, University of Cantebury, Christchurch 8140, New Zealand; evgeny.pavlov@canterbury.ac.nz

Abstract: This article discusses the place of God in the poetic system of Aleksandr Vvedensky. Vvedensky's famous pronouncement on his "poetic critique" is more throughgoing than Kant's critical enterprise, and invites a comparison between the movement of Kant's thought in the Critique of Judgment, and what Vvedensky's recourse to senselessness aims to achieve. Time in Vvedensky poetics may be seen as a radical extension of Kant's philosophical system where it ultimately resides in an equally inaccessible realm on which its entire edifice is founded.

Keywords: Alexansder Vedensky; Russian avant-garde; time; God

Citation: Pavlov, Evgeny. 2021. 'Only God Can Be": Aleksandr Vvedensky, Kant, God, and Time. *Religions* 12: 658. https://doi.org/10.3390/rel12080658

Academic Editor: Dennis Ioffe

Received: 9 July 2021
Accepted: 14 August 2021
Published: 18 August 2021

Publisher's Note: MDPI stays neutral with regard to jurisdictional claims in published maps and institutional affiliations.

> The poet is a cocoon that unwinds itself in our reading, and this cocoon unwinds into an endless thread that doesn't lead anyone anywhere, whether it flashes or disappears into darkness,–not because it doesn't know the beginning, but because as we try to "walk along, or follow" this thread, we will never have enough time which in its tireless becoming an enveloping "cobweb" draws its own experience from nothing and only then from "before".
>
> "Mesh". (Dragomoshchenko 2011)

Aleksandr Vvedensky, arguably the most outstanding poet of the late pre-war Soviet avant-garde and a key member of the Chinar-OBERIU group of poets and philosophers, has been the focus of notable critical attention in the last two decades, most recently after the publication of the definitive complete edition of his surviving oeuvre.[1] He is a very difficult poet to tackle critically. As Keti Chukhrov points out, "his work not only surpasses all interpretations and analytical observations, but has the capacity to cancel their significance and explanatory pathos. This happens because Vvedensky's writings already contain within themselves those meta-positions with which one could approach them, including philosophical, theological, and strangely enough, political ones" (Chukhrov 2011, p. 145). This could be part of the reason why to date there has been little effort to identify the place of God in his works. According Vvedensky's most often cited statement, the poet was interested only in three things–time, death, and God (Vvedensky 1993, vol. 2, p. 167). Equally famous is his overarching desire to *overstep* reason, undertaking what he called "a poetic critique of reason—a more substantial one than that other, abstract critique",[2] the latter being of course Kant's First, *Critique of Pure Reason*.

Kant, a one-time Russian subject and member of St Petersburg Academy of Science, was most certainly a visible influence on the development of the Russian philosophical and religious-philosophical thought. It was not until the late 19th century that an explosion of interest in his work occurred in the Russian academia—owing to the influence of European neo-Kantianism (with one of the most important Russian philosophers of this school being, curiously, a different Aleksandr Ivanovich Vvedensky (1856–1925), the poet's complete namesake who as professor of St Petersburg University taught, among others, influential figures as Nikolai Lossky, Petr Struve, and Mikhail Bakhtin). Overall, much as the Russian philosophers were interested in Kant, they were also critical of him. According to Semyon

Religions **2021**, *12*, 658. https://doi.org/10.3390/rel12080658

Frank, "critique of Kant's philosophy and struggle against Kantianism are ... a constant theme in the Russian philosophical thought" (Zenkovsky 1999, vol. 1, p. 347). Most Russian philosophers incorporated Eastern Orthodox beliefs into their systems which were often more than tinged with irrationality. Like Kant, they recognized the limitations of human reason, but they also allowed for mystical experiences and intuitions that could give access to what in Kant is inaccessible, i.e., the realm of the noumena. It is the privileging of ontology that stopped Russian philosophy from developing its own branch of Kantian thought. According to another prominent Russian philosopher Aleksei Losev, the cognition of the hidden realm can only be achieved in a symbol, an image, through the power of imagination and "inner living agility" (Losev 1991, p. 213).

This postulate was key to the work of Russian symbolists, especially Andrei Bely who was the only one among them to have studied Kant. In their endeavours, they actively worked against what they perceived to be purely rational spirit of Kantian philosophy. "True symbolism", writes Belyi in an early article "Krititsizm i simvolizm" ("Criticism and Symbolism") (written in 1904 on the centennial of Kant's death), "begins only beyond the gates of criticism. Symbolism born of criticism, unlike the latter, becomes a living method that equally differs from dogmatic empiricism and abstract criticism by overcoming them both" (Bely 1910, p. 29). He argues that the difference Symbolism makes lies in its ability to overcome the "purely scientific" character of knowledge in Kant and ultimately bridge the schism between phenomena and noumena—this "Scylla and Charibdis of the Kantian philosophy" (ibid., p. 25)—by means of creative, intuitive cognition: "The cognition of ideas reveals in temporal phenomena their timelessly eternal meaning. This cognition joins together understanding and feeling into something different from them both, something that covers them both" (ibid., p. 29). Bely's younger colleagues who, as said, hardly read any Kant at all, saw him as a Prussian prisoner, a sinister spider whose web of rigid logic and concepts unfairly limits the freedom of poetic genius, barring it any access to the noumenal world. To paraphrase the poet Marina Tsvetaeva, "бить Канта наголову" ("rout Kant, beat Kant hollow") becomes a mission for many figures of Russian modernism who in their quest for a new poetic language think of Kant at best as a powerful enemy (and at worst, a beating boy).

The Oberiu poets were no exception in this respect. As Iakov Druskin, Vvedensky's devoted friend, colleague and astute interpreter testifies, neither Vvedensky, nor his close associate Daniil Kharms whose work has garnered a much greater critical response were particularly well-steeped in philosophy. And yet Kharms famously scribbled "against Kant" underneath his "Blue Notebook no. 10", prompting future scholars to speculate on the anti-Kantian drive of the Chinar-Oberiu ideas.[3] Druskin (he along with Leonid Lipavsky were the only real philosophers in the group) once remarked that even though Daniil Ivanovich may have dropped a quote from Kant, he never really read the philosopher's works.[4] Meanwhile, Druskin himself held Kant in the highest esteem and in his diaries listed the philosopher alongside Bach and Vvedensky among the greatest geniuses of humankind (Druskin 1999, p. 433–34). Moreover, to philosopher Druskin who considered Vvedensky greatest Russian poet of all times, Kant and Vvedensky share not only greatness but also a certain universality. Kant's task was to "abolish knowledge to make room for faith" (Kant 1993, p. 21); Vvedensky's "poetic critique of reason" in search of answers to the questions of time, death, and God pursued the very same objective. Yet, neither was Kant a religious philosopher, nor Vvedensky a religious poet. Both their projects were thoroughly critical, even if Vvedensky's was more thoroughgoing than Kant's in that it turned its critique to language.

In what follows, I would like to comment on several crucial passages from Vvedensky's *Grey Notebook* where the poet is at his most revealing on the subject of time, death, and God (1932–1933). I hope to demonstrate that instead of "anti-Kantian", Vvedensky's poetics should be described as "ultra-Kantian". Although he collapses the edifice of Kantian rationality by blowing up its structural supports, he does so in order to show us that rationality and a rationality share the same bottomless foundation that cannot be removed. According to Kant, "unconditioned necessity, which as the ultimate support of all existing things is an indispensable requirement, is an abyss on the verge of which human reason trembles in dismay" (Kant 1987, p. 418). Vvedensky's celebrated line is an eloquent affirmation of this thesis: "горит бессмыслицы звезда, она одна без дна" ("the star if senselessness is shining, it alone has no bottom").

This is how Druskin describes the poetic *bessmyslitsa* of Kharms and Vvedensky in his essay "Chinari":

"Works of Vvedensky and Kharms are linked by "the star of senselessness":

> The star of senselessness is shining,
> It alone has no bottom"

writes Vvedensky in the epilogue to his large [...] dramatic poem "God is Perhaps All Around". I distinguish semantic senselessness which distorts rules of so-called "normal" speech from situational senselessness which follows from a logical nature of human relationships and situations. Vvedensky has not only situational senselessness, but also semantic, while Kharms uses mostly that of the situational kind. (Druskin 2000, vol. 1, p. 60).

In another essay on *chinari*, "Stages of Understanding", Druskin says the following with regard to senselessness in the work of Vvedensky:

One has to understand Vvedensky's senselessness, the logic of alogicality. By itself, this word combination is senseless, for alogical is that which is not logical. Senslessness is that which has no sense, is incomprehensible. Fichte once said: we need to understand the incomprehensible as incomprehensible. Vvedensky would have said: we need not to understand the incomprehensible as incomprehensible. This is what he did say: to truly understand is not to understand. Still, alogicality has its own logic, alogical logic. But this logic would always be alogical to our reason fallen in Adam—not relatively, but absolutely alogical, *docta ignorantia* (Nicholas of Cusa), madness for reason. (Druskin 2000, vol. 1, p. 420f)

That Druskin brings up Nicholas of Cusa here is supremely significant. The German philosopher's doctrine of learned ignorance states precisely this: "Since the unqualifiedly and absolutely Maximum (than which there cannot be greater) is greater than we can comprehend (because it is Infinite Truth), we attain unto it in no other way than incomprehensibly. For since it is not of the nature of those things which can be comparatively greater and lesser, it is beyond all that we can conceive" (Nicholas of Cusa 1981, p. 8). When Druskin brings up Cusanus' concept of Divine madness, it is in the context of an absolute break between human logic and the alogical Logos. According to Druskin, as we have no logical means of passing from human wisdom to Divine madness, we must conduct this *passage* ("perekhod") "in leaps" ("skachkami"): "Each of us makes it daily, without realising that it is alogical. The poet, the philosopher make it consciously" (ibid., p. 421). From this he concludes that most often people make mistakes when they follow the logic of correct reasoning, and vice versa: "erroneous, alogical reasoning is correct" (ibid.).

To Vvedensky, in the relationship between poetry and life what matters is the "correctness of the verse line", "правильность стиха". As he explicitly states, "it is incorrect to discuss art in terms of beautiful/not beautiful. Art should be discussed in terms of correct/incorrect" (as per Druskin 1993, vol. 2, p. 167). In Kant's third Critique the feeling of the beautiful is seen as arising out of the play of the understanding and the imagination, and it only concerns the finite forms of phenomena: It concentrates on the capacity of transcendental imagination to present a form that accords with its free play. At issue in the Analytic of the Beautiful is the existence of an accord between the sensible manifold and a certain pre-conceptual unity of the supersensible (Kant 1987, p. 15). The kind of art that Vvedensky proposes is not about the imagination and its play, but rather about what Kant discusses in the Analytic of the Sublime, which turns to the realm of infinite Ideas, and this is where we are indeed faced with the notion of passage. The sublime enters at a crucial point in the Third Critique where the philosopher is in search of a passage, an *Übergang*, between the theoretical and the practical realms and to do violence to imagination as sensorily determined by clashing it with reason's supersensory demands. The sublime is a powerful reminder of the fact that the *a priori* principle, which grounds reflective judgment cannot cover up the abyss separating the worlds of nature (phenomena) and freedom (noumena). Although, the reconciliation of the two is promised in the sphere of the beautiful. In its violence, the sublime also involves a presentation, albeit a negative one, of imagination's inability to present ideas of reason. As imagination strives to progress toward infinity, "reason demands absolute totality as a real idea, and so [imagination]... is inadequate to that idea" (Kant 1987, p. 106). Yet, with the spontaneous arousal of the feeling "that we have within us a supersensible power", reason forces the mind to an invariably doomed effort to make a presentation of the senses adequate to the totality.

While the beautiful "concerns the form of the object", i.e., limitation, the sublime strives for the unlimited as it seizes us in the presence of "a formless object insofar as we present unboundedness" (Kant 1987, p. 98). The beautiful "concerns the form of the object, that is its limitation", whereas the feeling of the sublime seizes us in the presence "of a formless object to the extent that the unlimited here represents itself" (Kant 1987, p. 99). The feeling of a lack of limits that the sublime brings with itself is about disorder and a return to the chaos, which the transcendental imagination orders by imposing form. It is in this sense that Vvedensky's art directly speaks to the ideas of reason which it of course cannot represent, but at which it continuously gestures through a critique of language whose "poverty" it exhibits as a means to discredit completely our rational knowledge, the ego-centric self and its spatio-temporal reality as expressed in language.[5] Let us once again ponder Vvedensky's most important poetic pronouncement in Lipavsky's *Conversations* where the poet draws the often-cited comparison between Kant's and his own critiques:

> Можно ли на это [проблему времени] ответить искусством? Увы, оно субъективно. Поэзия производит только словесное чудо, а не настоящее. Да и как реконструировать мир, неизвестно. Я посягнул на понятия, на исходныеобобщения, что до меня никто не делал. Этим я провел как бы поэтическую критику разума—болееосновательную, чем та, отвлеченная. Я усумнился, что, например, дом, дача и башня связываются иобъединяются понятием здание. Может быть, плечо надо связывать с четыре. Я делал это на практике, в поэзии, и тем доказывал. И я убедился в ложности прежних связей, но не могу сказать, какие должны быть новые. Я даже не знаю, должна ли бытьодна система связей или их много. И у меняосновноеощущение бессвязности мира и раздробленности времени. А так как это противоречит разуму, то значит разум не понимает мира. (Vvedensky 2010, p. 593)

Could one respond to this [the problem of time] with art? Alas, art is subjective. Poetry produces only a verbal miracle, not a real one. Besides, we don't know how to reconstruct the world. I infringed upon concepts, primary generalizations, which no one has done before me. By doing so I conducted a kind of a poetic critique of reason—a more substantial one than that other, abstract critique. For example, I put in doubt that "house", "dacha", and "tower" must be connected and joined together by the concept "building". Maybe "shoulder" must be connected to "four". I did it in practice, in poetry, and thus proved it. And I saw for myself the falseness of previous connections, but I can't tell you what new ones should be. I don't even know whether there should be one system of connections or whether there are many of them. And I've got a general sense that the world is disjointed and time is fragmented. And since this contradicts reason, then reason doesn't understand the world.

This passage sums up Vvedensky's poetics as a critique of reason which certainly appears to go against the grain of Kant's First Critique where the philosopher aims to explain how synthetic *a priori* knowledge makes the phenomenal world cohere via concepts of understanding. However, the First Critique contains within itself those fundamental theoretical postulates that are at the core of Vvedensky's poetic practice. In the *Critique of Pure Reason*, synthetic *a priori* judgments (and hence our knowledge) hinge upon the idea of time as an *a priori* non-figurable form of all forms in which "alone all reality of appearances is possible" (Kant 1993, p. 54). However, according to Kant, it only has empirical reality insofar as it is a form of our internal intuition: "If we take away from it the special condition of our sensibility, the concept of time also vanishes; and it inheres not in the objects themselves, but solely in the subject which intuits them" (Kant 1993, p. 58). Time also determines the reality of our very selves: just as time without us is nothing, we are nothing without time. This is precisely what Vvedensky states in *Grey Notebook* where we read, "Время единственное что вне нас не существует. Оно поглощает все существующее вне нас. Тут наступает ночь ума. Время восходит над нами как звезда" ("Time is the only thing that doesn't exist outside us. It consumes all that exists outside us. Here the night of reason sets in. Time rises above us like a star"). (Vvedensky 1993, vol. 2, p. 78). Here we are once again confronted with the need to make a leap, to cross over from what Druskin designates "this" and "that" ("*eto*" and "*to*"–see Druskin 2000). What Vvedensky terms "correct" art is the kind that operates within this "night of reason". In Kant's first Critique, understanding, the faculty Vvedensky wants to shut down, is "comprehension of plurality in unity" in the sense of a unifying intention of imagination that reproduces past moments, in order to open the horizon of the present, keeping present what passes, creating, as it were, an illusion of the temporal flux. Vvedensky famously proposes an experiment that would demonstrate that in what we perceive as linear continuity is, in fact, a discontinuous succession of apprehension.

> Если с часов стереть цифры, если забыть ложные названия, то уже может быть время захочет показать нам свое тихое туловище, себя во весь рост. Пускай бегает мышь по камню. Считай только каждый ее шаг. Забудь только слово каждый, забудь только слово шаг. Тогда каждый ее шаг покажется новым движеньем. Потом, так как у тебя справедливо исчезло восприятие ряда движений как чего-то целого, что ты называлошибочно шагом(ты путал движенье и время с пространством, ты неверно накладывал их друг на друга), то движение у тебя начнет дробиться, оно придет почти к нулю. Начнется мерцание. Мышь начнет мерцать. Оглянись: мир мерцает(как мышь). (Vvedensky 1993, vol. 2, p. 81)

If we were to erase the numbers from a clock, if we were to forget its false names, maybe then time would want to show its quiet torso, to appear to us in its full glory. Let the mouse run over the stone. Count only its every step. Only forget the word every, only forget the word step. Then each step will seem a new movement. Then, since your ability to perceive a series of movements as something whole has rightfully disappeared, that which you wrongly called a step (you had confused movement and time with space, you falsely transposed one over the other), that movement will begin to break apart, it will approach zero. The shimmering will begin. The mouse will start to shimmer. Look around you: The world is shimmering (like a mouse) (Vvedensky 2002, p. 11).

Such non-understanding of time would cancel not only most basic logical connections but also memory. In order to make the world shimmer, one has to forget every movement of the mouse before it makes a new one. Some sixty years earlier Friedrich Nietzsche suggested in his *Untimely Meditations* that a very similar mode of perception would characterize an animal: "[Man] wonders about himself, that he is not able to learn to forget and that he always hangs onto past things. No matter how far or how fast he runs, this chain runs with him [. . .] Man says, "I remember", and envies the beast, which immediately forgets and sees each moment really perish, sink back in cloud and night, and vanish forever" (Nietzsche 1983, p. 61). Nietzsche thinks the beast is happy for existing purely in the present even though it cannot communicate this happiness because it immediately forgets what it wants to say. Vvedensky refrains from such pronouncements although he too envies the beast (cf. "Мне жалко что я не зверь . . . ". "Я с завистью гляжу на зверя...". ("I feel sorry I'm not a beast . . . ", "I look at the beast with envy . . . "). In the *Grey Notebook* he is far more ambivalent about the happiness of animals for he can only observe them from within his "glass jar of time": "Букашка думаето счастье. Водяной жук тоскует. Звери не употребляюталкоголя. Звери скучают без наркотических веществ. Они предаются животному разврату. Звери время сидит над вами. Время думаето вас и Бог. [...] Но мыоставим в покое лес, мы ничего не поймем в лесу". (Vvedensky 1993, vol. 2, p. 82) "The little bug is thinking about happiness. Water beetle is sad. Animals don't take alcohol. Animals are bored without narcotic substances. They give themselves to animal lechery. Animals time sits above you. Time is thinking about you and god. But let us leave the forest in peace, we won't understand anything in the forest" (Vvedensky 2002, p. 15). What we can do instead is try some of those substances that animals do not take and see whether we could approximate their condition and cast off the chain of time. The character named Svidersky's relates the following story:

> Однажды я шел по дорогеотравленный ядом,
> и время со мною шагало рядом [...]
> Я думало том, почему лишь глаголы
> подвержены часу, минуте и году,
> а дом, лес и небо, как будто монголы,
> от времени вдруг получили свободу.
> Я думал и понял. Мы все это знаем,
> что действие стало бессонным китаем,
> что умерли действия, лежат мертвецами,
> и мы их теперь украшаем венками.
> Подвижность их ложь, их плотностьобман,
> и их неживой поглощает туман [...]
> Яостановился. Я подумал тут,
> я не могохватить умом нашествие всех новых бедствий.
> И я увидел дом ныряющий как зима,
> и я увидел ласточкуобозначающую сад
> где тени деревьев как ветви шумят,
> где ветви деревьев как тени ума.
> Я услышал музыкиоднообразную походку,
> я пытался поймать словесную лодку.
> Я испытывал слово на огне и стуже,
> но часы затягивались все туже и туже,
> И царствовавший во мне яд
> властвовал как пустой сон.
> Однажды. (Vvedensky 1993, vol. 2, p. 77)

Once upon a time I walked poisoned down a road

> And time walked in step by my side. [...]
> I thought about why only verbs are
> subjugated to the hour, minute, and year,
> while house, forest and sky, like the Mongols
> have suddenly been released from time.
> I thought about it and I understood. We all know it,
> that action became an insomniac China,
> that actions are dead, they stretch out like dead men,
> and now we decorate them with garlands.
> Their mobility is a lie, their density a swindle,
> and a dead fog devours them. [...]
> I stopped. Here I thought,
> my mind could not grasp the onslaught of new tribulations.
> And I saw a house, like winter, diving.
> And I saw a swallow signifying a garden
> where the shadows of trees like branches make sound,

> where the branches of trees are like shadows of the mind.
> I heard music's monotonous gait,
> I tried to catch the verbal boat.
> I tested the word in cold and fire,
> but the hours drew in tighter and tighter.
> And the poison reigning inside me
> wielded power like an empty dream
> Once upon a time. (Vvedensky 2002, p. 9)

This poem presents a certain narrative, which Vvedensky discusses in a series of comments in the remainder of the *Grey Notebook*. Here, we have a sequential series of actions, unfolding, strictly speaking, one after another, i.e., as a temporal progression: "Шел, шагало, думал, понял(walked, thought, understood)—all these are verbs that are points in a simple sequence of events. Yet, it is the synthesis of unfolding actions into a single story that Vvedensky sees as the source of our habitual delusion with regard to time. According to the *First Critique*, "if I always let the preceding representations escape from my thought . . . and if I did not reproduce them as I arrive at the following representations, no complete representations . . . not even the fundamental representations, not even the most pure and completely primary ones of space and time could be produced" (Kant 1993, pp. 114–15, 133). This is indeed the death of actions, and a slipping of the world into a senseless chaos—precisely the kind of thing that we witness in Svidersky's monologue. The corpses of verbs unleash an attack of "new tribulations", that are no longer connected with one another, or with anything else for that matter, and that proliferate ad infinitum once the synthesis has stopped. The mind can no longer grasp what is occurring, and the scenario is very much that which Kant describes in the Analytic of the Sublime where imagination falters, leaving us with a negative presentation of the effort to represent infinite ideas of reason, in this case, time itself.

The moment of failure is marked not only in the actual narrative but also in the rhythmical pattern of the poem: 'Яостановился. Я подумал тут,/я не могохватить умом нашествие всех новых бедствий". This is the *locus classicus* of Hölderlin's caesura which the German poet famously defines in his Annotations to Oedipus as "the pure word, the counter-rhythmic rupture" necessary, "in order to meet the onrushing change of representations at its highest point, in such a manner that not the change of representations but the representation itself very soon appears". In relation to the Greek tragedies that Hölderlin comments on, a silent moment of truth is presented so we can glimpse the unbridgeable rift between the hero and gods. It is not easy to articulate what Hölderlin's "representation itself" would be without falling back into the associative chain of the succession of representations. However, one could say that the interruption of the sequence of representations transforms it into a *presentation* of representation, which is no longer the associative chain of imagination but the presentation of its construction [presentation in the sense of "putting on display"]. As Jacob Rogozinski persuasively argues in an illuminating analysis of the temporality of the sublime, originary time is then not reducible to transcendental imagination and the latter's violent maintenence of it in the form of a homogeneous, monotonous progression, "for if originary temporality were identical to imagination, nothing other would be possible, nothing sublime could happen" (Rogozinsky 1993).

In the commentary that follows the poem, Vvedensky says the following: "Глаголы на наших глазах доживают свой век. В искусстве сюжет и действие исчезают. Те действия, которые есть в моих стихах, нелогичны и бесполезны, их нельзя уже назвать действиями. [. . .] События не совпадают со временем. Время съело события. От них неосталось косточек". (Vvedensky 1993, vol. 2, p. 81) ("Verbs live out their last days in front of our very eyes. In art, plot and action are disappearing. Those actions that exist in my poetry are alogical and useless you can't call them actions any more. Events don't coincide with time. Time has eaten up events. No bones are left of them".) (Vvedensky 2002, p. 12). Unable to comprehend the bad infinity of illogical and useless actions, his imagination now proliferates, and he is left with mere verbal building blocks that time deprives of any referential meaning. The word "однажды" at the end of Svidersky's monologue is more than just a repetition of the "однажды" at the beginning. It is perhaps the very word he has been testing "in cold and fire" only to see it snap under the pressure of "tightening hours". As a singular occurrence torn from the thread of violent synthesis through which imagination operates, it is exhibited as an empty shell that means everything and nothing. It can be interpreted as the point in the temporal series when time momentarily halts; in conjunction with the imperfective "властвовал", it also indicates the open stretch of time in which the story ends. "Однажды" is indeed the empty dream with which we end up: Nothing which is pure time that we can never access. Events have all been eaten up, with no bones left of them, but this "once upon a time" is the allegorical last bone of the sublime event when time momentarily "showed its quiet torso". To paraphrase Walter Benjamin, it is the scull in whose language "total expressionlessness—the black of the eye sockets—is coupled with the wildest expression—the grinning row teeth" (Benjamin 1973, IV-1:112).

Chinari had another word to describe this "scull": hieroglyph. The hieroglyph, in its simplest sense, is a sign that contains several meanings, some of them mutually contradictory. By definition, it is illogical. It seeks to make the individual commit to a dynamic, richly ambiguous symbol always in the process of being transformed. For Vvedensky and for other Chinari, such a sign is valuable because of its closer proximity to the fragmented truth of our existence than does the logical world of reason.

In a footnote to the Analytic of the Sublime, Kant famously illustrates the in ability to grasp pure time by citing the sublime image of veiled Isis: "Perhaps nothing more sublime has ever been said, or a thought ever been expressed more sublimely, than in that inscription above the temple of Isis (mother Nature): "I am all that is, that was, and that will be, and no mortal has lifted my veil" (Kant 1987, p. 187). Vvedensky's commentary on the poem that contains Svidersky's monologue insists on the same:

Все, что я здесь пытаюсь написатьо времени, является, строго говоря, невер-
ным. Причин этому две. (1) Всякий человек, который хоть сколько-нибудь
не понял время, а только не понявший хотя бы немного понял его, должен
перестать понимать и все существующее. (2) Наша человеческая логика и
наш язык не соответствуют времени ни в каком, ни в элементарном, ни в
сложном его понимании. Наша логика и наш язык скользят по поверхности
времени. (Vvedensky 1993, vol. 2, p. 79)

All that I am trying to write here about time is, strictly speaking, untrue. There are two reasons for this. (1) Any person who has not understood time at least a little bit—and only one who has not understood it has understood it at all—must cease to understand everything that exists. (2) Our human logic and our language do not in any way correspond to time, neither in its elementary nor in its complex understanding. Our logic and our language slide along the surface of time.

Yet, perhaps one can try and write something, if not about time—nor on the non-understanding of time—then at the very least to try to fix those few positions of our superficial experience of time, and, on the basis of these, the way into death and general non-understanding becomes clear (Vvedensky 2002, p. 9; translation modified).

If our logic and language slide along the surface of time, it is because the irreducible veil of the phenomenal world is woven of the thread of temporality. The violence of the imagination veils itself under an illusory transparence that Vvedensky calls into question. It is only at the moment of death that its texture can be broken. To Vvedensky, this is the moment—the only moment that deserves to be called "moment"—when a *real* miracle can happen: "Чудо возможно в момент смерти. Оно возможно потому что смерть естьостановка времени" ("A miracle can happen at the moment of death because death is the stop of time".) All other miracles are merely verbal, and yet, "If we experience wild non-understanding we will know that no one will be able to counter it with clarity. Woe to us who ponder time. But then with the growth of this non-understanding it will become clear to you and me that there is no woe, neither to us, nor to pondering, nor to time". ("Если мы почувствуем дикое непонимание, то мы будем знать, что этому непониманию никто не сможет противопоставить ничего ясного. Горе нам, задумавшимсяо времени. Но потом, при разрастании этого непонимания тебе и мне станет ясно, что нету ни горя, на нам, ни задумавшимся, ни времени") (Vvedensky 1993, vol. 2, p. 79). There is a very thin line between the sublime and the monstrous, which Vvedensky's poetry never crosses even though there is a degree of madness in the sublime and definitely in his art. Being a critical project, his poetics slips neither into visionary insanity nor into what Kant calls metaphysical *Schwärmerei*, or empty flights of fancy. "Уважай бедность языка", "respect the poverty of language", insists Vvedensky in "A Certain Quantity of Conversations", as he pushes language to its limit. However, it is entirely consistent with his overall poetic position, which is based on the most radical incomprehension that schematises time, as we do not know the cost of disfiguration and fragmentation. The proliferation of impossibly arational actions that so overwhelms Svidersky is in fact the totality, the *vse* (everything), which appears to be the very last word of Vvedensky's very last surviving piece "Gde, kogda" ("Where, when"). When in *Krugom vozmozhno bog* (*God is Perhaps All Around*) Vvedensky says, "Only God Can Be" ("Быть может только Бог"), he points to the totality of all that is for which God is one name. Vvedensky's "Gost' na konie" ("Guest on Horseback"), poses the question

> Бог Ты может бытьотсутствуешь?
> Несчастье.
> God could You be absent?
> Woe. (Vvedensky 2010, p. 183)

That Vvedensky posits God in the same inaccessible realm where time and death reside is very telling. Nicholas of Cusa proposed the idea of *coincidentia oppositorum*. God, in his view, cannot be part of his own creation, and thus, his presence can only be appreciated if one acknowledges His absence (see Nicholas of Cusa 1981, p. 79).

This is immediately countered with a vision of totality:

> Нет я все увидел сразу,
> поднял дня немую вазу,
> а сказал смешную фразу,
> чудо любит пятки греть [...]
> Я забыл существованье
> я созерцал
> вновь
> расстоянье.
> Now I saw everything at once
> lifted the mute vase of the day
> I said a funny phrase,
> miracle likes to warm you heels [...]
> I forgot existence,
> I contemplated
> again
> the distance. (Vvedensky 2010, p. 183)

This is precisely the sort of uncomprehending that Vvedensky is after. Far from denuding the goddess, his irrational project, just as the rational project of Kant, reveals nothing except the veil itself, the non-figurable weave of time that is the world of phenomena and that is our human language.

In a pursuit of incomprehension, Vvedensky's "poetic critique of reason" unfolds according to the very same principles that Kant postulates for his rational philosophy, and the thoroughness of the poetic critique very thoroughly undertakes to prove Kant right. As I have demonstrated, the passage that Kant proposes in his *Analytic of the Sublime* does not lead into the world of noumena, but rather, demonstrates the texture of the veil out of which phenomena are woven: Vvedensky's ultra-Kantian poetic project shows us "the surface of time". In the epigraph for the article that I chose from a short piece on Vvedensky by the late poet Arkady Dragomoshchenko, the poet is said to unwind himself through his language, like a cocoon. The thread of temporality shows neither a beginning, nor an end, but rather that it comes from nothing, the nothing that is, in fact, everything—given in presence, in absence as death and as God—incomprehensibly, by a radical disfiguration which language alone can make visible.

Funding: This research received no external funding.

Institutional Review Board Statement: Not applicable.

Informed Consent Statement: Not applicable.

Conflicts of Interest: The author declares no conflict of interest

Notes

[1] See (Vvedensky 2010).

[2] "Я посягнул на понятия, на исходныеобобщения, что до меня никто не делал. Я провел как бы поэтическую критику разума–болееосновательную, чем та, отвлеченная". In Leonid Lipavsky, "Razgovory" (Vvedensky 2010, pp. 592–93).

[3] Very few scholars writing on the philosophy of *Chinari* and/or Oberiu failed to describe the group's views as "essentially anti-Kantian". See, for example, (Roberts 1997, p. 126). In her study of Bergson's influence on Kharms, Hilary Fink argues that Kharms discarded Kant's analytical tools in favour of "more intuitive ways of apprehending reality" which in turn led to the birth of the absurd in his works (Fink 1998). For a more involved discussion of the anti-Kantian tenor in the philosophy of *chinari* see Protopopova. See also V. Sazhin's assessment in *Sborishche druzei*, (Sazhin 2000, p. 770). Notable exceptions to the above include Skidan 2011 and Rezvykh 2014. Evgeny Ostashevsky's article offers an extremely interesting analysis of Vvedensky's poetic critique in the context of Wittgenstein's philosophy of language (Ostashevsky 2011). See also Protopopova 2007.

[4] See (Jaccard 1995, p. 369): "Канта Даниил Иванович не читал" ("Daniil Ivanovich never read Kant). Jaccard uses this testimony to suggest that given Kharms's lack of philosophical sophistication, it would not be particularly productive to read his works through the Kantian lens.

[5] "Уважай бедность языка. Уважай нищие мысли". ("Respect the poverty of language. Respect squalid thoughts". (See "Nekotoroe kolichestvo razgovorov", Vvedensky 1993, vol. 2, p. 196)

References

Bely, Andrei. 1910. *Simvolizm*. Moscow: Musaget.

Benjamin, Walter. 1973. *Gesammelte Schriften*. Edited by Hrsgb Hermann Schweppenhäuser and Rolf Tiedemann. Frankfurt: Suhrkamp.

Chukhrov, Keti. 2011. Nekotorye pozitsii poetiki Aleksandra Vvedenskogo. *Novoe Literaturnoe Obozrenie* 108: 249–57.

Dragomoshchenko, Arkady. 2011. Mesh. *Novoe Literaturnoe Obozrenie* 108: 225–28.

Druskin, Iakov. 1993. Materialy k poetike Vvedenskogo. In *Polnoe Sobranie Proizvedenii v Dvukh Tomakh*. Edited by Mikhail Meilakh and Vladimir Erl. Moscow: Gileia, vol. 2, pp. 164–84.

Druskin, Iakov. 1999. *Dnevniki*. St. Petersburg: Akademicheskii Proekt.

Druskin, Iakov. 2000. Chinari. In *Sborishche Druzei, Ostavlennykh SUD'BOIU …": Chinari v Tekstakh, Dokumentakh i Issledovaniiakh*. Edited by Sazhin Valerii. Moscow: Ladomir.

Fink, Hilary. 1998. The Kharmsian Absurd and the Bergsonian Comic: Against Kant and Causality. *The Russian Review* 57: 526–38. [CrossRef]

Jaccard, Jean-Philippe. 1995. *Daniil Kharms i Konets Russkogo Avangarda*. Translated by F. A. Perovskaia. St. Petersburg: Akademicheskii Proekt.

Kant, Immanuel. 1987. *Critique of Judgment*. Translated by Werner Pluhar. Indianapolis: Hackett.

Kant, Immanuel. 1993. *Critique of Pure Reason*. Translated by J. M. D. Meiklejohn. London: Everyman.

Losev, Aleksei. 1991. Russkaia filosofiia. In *Filosofiia, mifologiia, kul'tura*. Moscow: Politizdat.

Nicholas of Cusa. 1981. *On Learned Ignorance (De Docta Ignorantia)*. Minneapolis: Arthur Banning.

Nietzsche, Friedrich. 1983. *Untimely Meditations*. Translated by Reginald John Hollingdale. Cambridge: Cambridge University Press.

Ostashevsky, Evgeny. 2011. Logiko-filosofksii potets: Beglye zamenchanie k teme 'Vvedenskii i Vittgenshtein'. *Novoe Literaturnoe Obozrenie* 108: 228–33.

Protopopova, Irina. 2007. Chinari i filosofiia (chast' 1). In *Russkaia Antropologicheskaia Shkola, Trudy*. Moscow: RGGU, vol. 4, pp. 46–99.

Rezvykh, Tatiana. 2014. Antinomii vremeni u Aleksandra Vvedenskogo. *Logos* 3: 67–94, 99.

Roberts, Graham. 1997. *The Last Soviet Avant-Garde*. Cambridge: Cambridge University Press.

Rogozinsky, Jacob. 1993. The Gift of the World. In *Of the Sublime: Presence in Question*. Translated by Jeffrey Librett. Albany: SUNY Press.

Sazhin, Valerii, ed. 2000. "… *Sborishche Druzei, Ostavlennykh SUD'BOIU …": Chinari v Tekstakh, Dokumentakh i Issledovaniiakh*. Moscow: Ladomir.

Skidan, Aleksandr. 2011. Kritika poeticheskogo razuma. *Novoe Literaturnoe Obozrenie* 108: 257–63.

Vvedensky, Aleksandr. 1993. *Polnoe Sobranie Proizvedenii v Dvukh Tomakh*. Moscow: Gileia.
Vvedensky, Aleksandr. 2002. *The Grey Notebook*. Translated by Matvey Yankelevich. New York: Ugly Duckling Presse.
Vvedensky, Aleksandr, ed. 2010. *Anna Gerasimova*. Moscow: OGI.
Zenkovsky, V. V. 1999. *Istoriia Russkoi Filosofii*. Rostov-on-Don: Feniks, 2 vols.

MDPI
St. Alban-Anlage 66
4052 Basel
Switzerland
Tel. +41 61 683 77 34
Fax +41 61 302 89 18
www.mdpi.com

Religions Editorial Office
E-mail: religions@mdpi.com
www.mdpi.com/journal/religions